NUCLEAR ARMS TECHNOLOGIES IN THE 1990s

AIP CONFERENCE PROCEEDINGS 178

RITA G. LERNER
SERIES EDITOR

NUCLEAR ARMS TECHNOLOGIES IN THE 1990s

WASHINGTON, DC 1988

EDITORS:

DIETRICH SCHROEER
UNIVERSITY OF NORTH CAROLINA, CHAPEL HILL

DAVID HAFEMEISTER
CALIFORNIA POLYTECHNIC STATE UNIVERSITY

AMERICAN INSTITUTE OF PHYSICS NEW YORK 1988

L.C. Catalog Card No. 88-83262
ISBN 0-88318-378-1
DOE CONF 8804202

Printed in the United States of America.

Contents

Preface

The physics community has been interested in public affairs ever since the Second World War. However, that interest has tended to oscillate among several different public policy issues, going from the control of nuclear weapons, to the promotion and control of nuclear energy, to concerns about decreasing energy resources, to the call for more support for physics (sometimes to solve various technological inferiorities of the United States). Given the relatively short time scale for such oscillations in the past, typically on the order of five years, it is somewhat unusual that the current interest among physicists in arms race matters has lasted so long. The character of the interest has in fact been changing over the last eight years or so, going from appeals for nuclear arms control and disarmament to a much more professional effort to carry out technically competent analyses that would have a real impact on the process of arms buildups and control. In the light of this sustained interest among physicists in the arms race, it appeared appropriate to review the current state of the technical literature on nuclear technologies in order to improve our ability to project into the future.

The decade of the 1990s has seen several surprising events, both in technical areas and in the political arena. In the technical area, the events of the 1980s have been more predictable, as systems planned in the 1970s have begun to come to fruition in the 1980s. Strategic modernization has included: larger mobile missiles such as the MX and SS-24; smaller and more mobile missiles such as the SS-25 and Midgetman; improved SLBMs on Trident and Typhoon submarines; enhanced navigational accuracy with the NAVSTAR and GLONASS systems; improved bombers, including the B1, B2, and Blackjack; and the proliferations of ALCMs and SLCMs. Also, new technological possibilities have been explored: ASAT weapons on F-15s and missiles; enhanced verification technologies; and laser weapons at Sary Shagan and White Sands.

In comparison, the events in the political arena have been much less predictable. Who in 1980 would have foreseen the following events? The Soviets invade Afghanistan in 1979, and then retreat in 1988; in 1983 President Reagan proposes to make nuclear weapons obsolete by strategic defense; the Soviets violate the ABM Treaty by the Krasnoyarsk radar, and then suspend its construction and possibly dismantle it; charges and countercharges were filed on noncompliance with arms control treaties; President Reagan and General Secretary Gorbachev make deep-cut proposals at Reykjavik, perhaps ultimately trading Soviet heavy missiles for US SLCMs; the Soviets accept the zero–zero option for an INF Treaty; intrusive on-site and challenge inspections were carried out under the INF Treaty; a joint Soviet and American group of scientists used CORRTEX and seismic methods to monitor explosions in Nevada and Semipalatinsk during the summer of 1988.

There are of course uncertainties contained in technologies, as demonstrated by the Chernobyl and Challenger disasters. However, a comparison of these two listings of events during this decade suggests to us who are technologically oriented that it is easier to predict and analyze technological progress (or the lack thereof) in the arms race than it is to predict and understand the broader political relations between nations. From our technological homebase we personally feel more comfortable with the prediction and analysis of "our kind" of technological subject matter, since open debate in that area can at least develop reasonable numerical bounds on future technological progress. We also feel that our "customers" have at least some expertise in technology, so that a certain intellectually satisfying consensus might be reached. Of course, we may in the process have focussed on the less important issues. If so, we apologize.

Most of the chapters in this book are derived from the talks presented at the Third Short Course on the Arms Race held at the The George Washington University on April 16–17, 1988. The chapters in this book deal with the following general issues: Chapter 1 on the Effects of Nuclear

War, Chapter 2 on Conventional Forces, Chapter 3 on Nonproliferation, Chapters 4–8 on Verification Technologies, Chapters 9–12 on Strategic Defense Technologies, and Chapters 13–16 on the TRIAD Technologies. Chapters 17–19 are somewhat different, since they present two contrasting views on the overall policy of deploying nuclear weapons. Finally, Appendices A–E summarize some historical and technological facts.

This book (as did its predecessor AIP Conference Proceedings No. 104) involves the efforts of many people, whom one can never thank enough. The Short Course was sponsored by the Forum on Physics and Society of the American Physical Society, The George Washington University, and the American Association of Physics Teachers. We would particularly like to thank our local co-organizers and hosts at GWU—Professor Barry L. Berman of the Department of Physics and Professor John M. Logsdon, Director of the Graduate Program on Science, Technology and Public Policy—for their enthusiastic contributions. We greatly appreciate the financial support and hospitality of the Columbian College of Arts and Sciences, the School of International Affairs, and the School of Engineering and Applied Science at The George Washington University. In addition we would like to thank the following individuals for helping make the arrangements flow smoothly and pleasantly: Bets Crocker (Cal. Poly.), Bill Havens (APS), Don Holcomb (AAPT), Tom Johnson (US Military Academy), Lu Kleppinger (GWU), Rita Lerner (AIP Books), Kitty McCullum (UNC), Cyndee McLin (Cal. Poly.), and Jack Wilson (AAPT). Lastly, one of us (DH) would like to thank Sid Drell and his colleagues at the Center for International Studies and Arms Control at Stanford for their kind hospitality during the summer of 1988.

Dietrich Schroeer
Univ. North Carolina
Chapel Hill, NC

David Hafemeister
Calif. Poly. State Univ.
San Luis Obispo/Palo Alto

September 1988

CHAPTER 1

THE ENVIRONMENTAL EFFECTS OF NUCLEAR WAR

Michael C. MacCracken
Lawrence Livermore National Laboratory, Livermore, CA 94550

ABSTRACT

Substantial environmental disruption will significantly add to the disastrous consequences caused by the direct thermal, blast, and radiological effects brought on by a major nuclear war. Local fallout could cover several per cent of the Northern Hemisphere with potentially lethal doses. Smoke from post-nuclear fires could darken the skies and induce temperature decreases of tens of degrees in continental interiors. Stratospheric ozone could be significantly reduced due to nitric oxide injections and smoke-induced circulation changes. The environmental effects spread the consequences of a nuclear war to the world population, adding to the potentially large disruptive effects a further reason to avoid such a catastrophe.

INTRODUCTION

> "The scientific evidence is now conclusive that a major nuclear war would entail the high risk of a global environmental disruption. The sensitivity of agricultural systems and natural ecosystems to variations in temperature, precipitation and light leads to the conclusion that the widespread impact of a nuclear exchange on climate would constitute a severe threat to world food production. The socio-economic consequences in a world intimately interconnected economically, socially and environmentally would be grave."[1] United Nations, Report of the Secretary General, 1988.

Starting with the explosions over Hiroshima and Nagasaki in 1945, the prospect of the extensive destructive effects of a nuclear war have served as a primary deterrent to conflict among the major powers. Given present arsenals, even a nuclear attack focused on the opponent's nuclear and command forces would likely involve the deaths of tens of millions of people. A major countervalue exchange, the underlying premise of mutual assured destruction, would lead to deaths and injuries of a few hundred million and extensive blast and fire damage to numerous urban areas.[2] While active and functional civil defense actions might reduce deaths several fold, the destruction done to medical, water, electric, sewage, distribution, economic, and food supply networks would pose such serious problems to surviving populations that millions more would likely die during the following few months. This aspect of nuclear war impacts on the combatants seems too often to be understated.

While recognition of such disastrous consequences from the direct effects alone (i.e., the fire, blast, radiological and, less importantly, electromagnetic pulse) has and continues to be the primary basis for deterrence between the potential combatant nations, it has been the consequent global-scale environmental effects that have created the most concern among non-combatant nations. Although such environmental consequences would also affect the combatant nations, they are generally viewed as a secondary concern. Conversely, until recently[3] non-combatant nations have generally ignored the severely disruptive consequences of the destruction of economic and commercial networks, especially in our increasingly interdependent world.

To provide insight into the potential deterrent effects among non-combatants of the threat of the consequences, as well as to better understand the basis of the resulting stimulation of such nations to support arms control initiatives, it is useful to review the environmental consequences of nuclear war, particularly those extending to the global scale. These environmental effects arise from the various gaseous and particulate injections into the atmosphere (see Figure 1), where they can be dispersed over long distances. (The effects of liquid and particulate releases into water bodies and onto the land surface, including the deposited early fallout of radionuclides, are generally localized, and will not be covered in this review.) Emissions to the atmosphere will be divided based on their effects:

1. Radionuclides that cause human exposure;
2. Smoke and dust that alter climate;
3. Nitrogen oxides and other species that alter atmospheric chemistry.

The injections and consequences of each of the major emissions will be covered in successive sections. In this review of potential effects, the arbitrary assumption will be made that the nuclear war involves explosion of about 10,000 strategic nuclear weapons totalling about 5000 Mt. Such an exchange is within plausible reach given the existent arsenals and it thus is perceived as a possible exchange by non-combatant nations. The extent to which the potential environmental consequences are dependent on this assumption will be discussed further below, but it appears that (within the limits imposed by present arsenals) there is greater sensitivity of potential environmental effects to target characteristics and the assumptions concerning how weapons are allocated to and exploded over targets than to the actual number and size of weapons.

Finally, this review attempts to focus more on concepts, providing order of magnitude estimates of emissions and atmospheric consequences, calibrated based on more detailed reviews carried out by SCOPE,[3,4] Golitsyn and MacCracken,[5] and others. This is done to minimize questions arising from use of theoretical models in such calculations—the potential effects are sufficiently robust that key assumptions, critical unknowns, and the potential global-scale significance of the emissions can be illustrated schematically. Resolving the

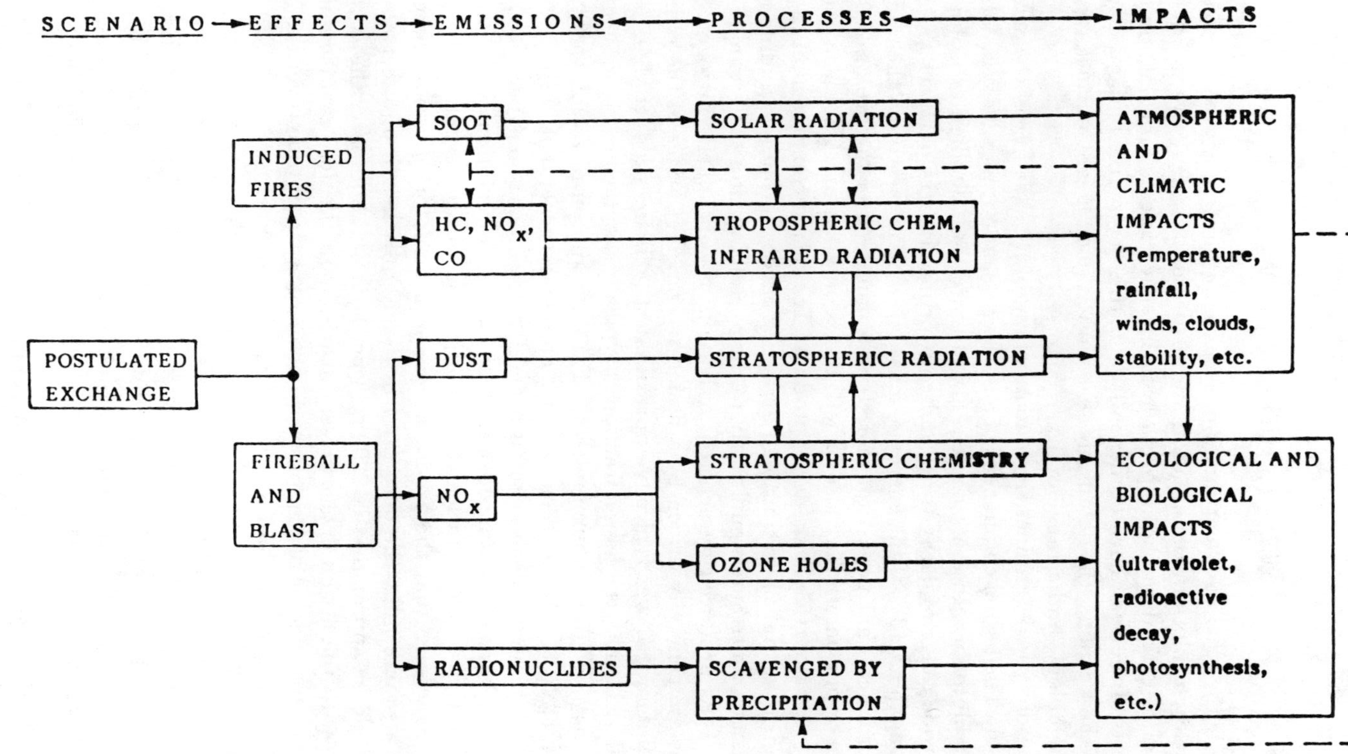

Figure 1: Effects chains for the major species emissions resulting from a nuclear war.

uncertainties and providing spatial and temporal details certainly requires theoretical approaches and models, but the most important insights are readily apparent.

RADIONUCLIDES

While the blast and heat radiation (flash) are the primary destructive agents of a nuclear explosion, the creation and dispersal of radioactive debris has the potential of extending the impacts both in space and in time. Estimation of the effects of the radionuclides depends on many factors, including the type and yield of the weapons, the details of the explosion (e.g., height, surface type, etc.), the meteorology, and a range of other factors.

Weapon type determines the amount of radioactive loading of the debris that is created. The fission component of a nuclear weapon explosion transforms uranium or plutonium into a complex mix of more than 300 lighter isotopes that have a wide range of decay times and emit primarily beta particles and gamma radiation; about 10% of the energy of a fission weapon ends up in the radioactive debris products that persist for more than a minute after the explosion.*[6] Some uranium or plutonium may also be left, but is only harmful if inhaled because it gives off alpha particles as a decay product. The fusion component of the explosion transforms light nuclei into an array of heavier isotopes, including tritium, but does not contribute significantly to the overall inventory of fission products. Carbon-14 is produced by neutron activation of atmospheric nitrogen. Fusion weapons are generally driven by a fission trigger. Estimates of the radioactive effects of a nuclear war typically approximate the fission product yield as half of the total yield (i.e., the fission fraction is typically assumed to be one-half; other choices may be scaled appropriately). For each kiloton of fission energy yield, approximately 3×10^{23} fission product atoms are generated, weighing almost 60g and at 1 minute generating of order 10^{21} disintegrations per second.[6] For reference in later estimating local fallout effects, observations indicate that if the radioactivity from a nuclear explosion were deposited evenly over a smooth infinite plane such that the products from each 1 kiloton of fission energy yield covered a square mile, the dose rate (in tissue) at a height of 3 feet above the plane would be approximately 2900 rads per hour at 1 hour after the explosion.[6] Starting with the fission debris present at one hour, the dose rate is reduced by approximately a factor of ten for every factor of seven in time (over the first six months). This is quite close to a $t^{-1.2}$ time dependence.

Several factors combine to determine how and where the radioactive debris is actually deposited and how rapidly this occurs, which in turn determines the extent of decay that has occurred prior to deposition (thereby reducing any subsequent time-integrated dose). In addition, the ground is not flat, deposition is not uniform, people are usually mobile and protected (by clothes, as well as by structures), and people and the fallout may not be collocated. As a

---

* This energy, since not part of the explosive yield, is not included in stating the yield.

result, estimating impacts can be highly dependent on assumptions and special conditions. The analysis here will discuss only important controlling factors; for a more thorough analysis, see Pittock et al.[4] and Institute of Medicine.[2]

The height of the burst (HOB) above the surface is critical in determining the size of the particles on which the radioactive debris resides, which in turn contributes to determining the rate of deposition of the particles, and the height from which the debris starts its journey towards the surface. Explosions at the surface or for which the fireball touches the surface (i.e., those for which the HOB (in feet) is less than 180 $W^{0.4}$, where W is the explosion yield in kilotons), draw large amounts of surface materials into the cloud, leading of the order of half of the hot radioactive gases to condense on relatively large particles. These relatively large particles tend to fall out of the atmosphere in the 24 hours following the explosion, creating a highly radioactive footprint often referred to as early, or local, fallout. Because it is deposited so rapidly, this fallout can contribute to the acute dose (an external gamma ray dose of approximately 450 rad received within 48 hours is often assumed lethal to 50% of a healthy population; presumably a lower dose would create similar lethality to a population simultaneously experiencing the many insults of a nuclear war).

If we assume that half of the yield of our hypothetical war is exploded at or near the surface,* that the typical yield is 500 kt, that the fission fraction is 0.5, and that half of the fission products are deposited as early or local fallout, then, using Figure 2 taken from Pittock et al.,[4] each explosion covers an area of about 350 km^2 with a minimum, unshielded dose of 450 rad within 48 hr. Assuming no overlap of fallout patterns (an assumption which, as explained in Pittock et al., is probably within a factor of a few of being correct, depending on the specific scenario), the 5000 surface burst explosions would cover an area of about 1.75 million km^2 with a 450 rad–48 hr unshielded dose, which is roughly 5% of the area of the NATO and Warsaw Pact countries. While the potential lethal area due to early fallout could be increased by a number of factors (tactical weapons effects, internal radiation, etc.), shielding and deposition of the debris over the oceans could reduce the percentage coverage. Because of the complexities of the meteorology (rainout, of the debris for example, can concentrate dosage over smaller areas) and the vagaries of the actual war scenario, significantly more accurate projections are probably not justified for the purposes of this paper. Clearly, however, local fallout would be worsened by tendencies to harden targets, thereby requiring a higher fraction of surface bursts of higher yield.

For explosions above the surface and for the half of the debris from surface bursts that ends up on small particles, the yield and HOB of the explosion determines the level of the atmosphere to which the debris cloud rises. Low-altitude (but above the surface) air bursts with yields greater than about 500 kt loft their debris into the stratosphere, where the residence time for particles in

* Surface bursts are generally viewed as preferred against hardened targets to enhance ground shock. Bursts above the surface are generally viewed as preferred against unhardened targets to maximize the area of damage.

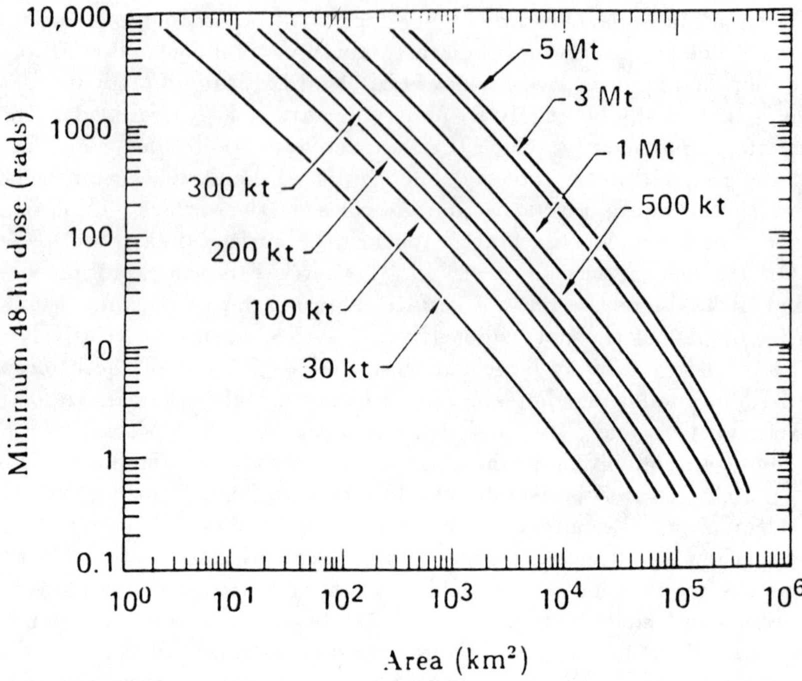

Figure 2. Fallout areas versus minimum 48-hour doses for selected yields from 30 kt to 5 Mt. The weapons were surface-burst and all-fission. These curves include an instrument shielding factor of 25%.[6] Doses within the area defined would exceed the minimum dose. (From Pittock et al.[4])

the unperturbed atmosphere is typically six months or more, a fact known from study of volcanic aerosol clouds. This part of the radioactivity spreads around the globe, mostly in the hemisphere of detonation; hence it is called global fallout. Given six months for radioactive decay, the dose rate drops by about 4 orders of magnitude.* Spreading the radioactivity from the airbursts (2500 Mt times 0.5 fission yield) and the remaining 50% from surface bursts (1250 Mt times 0.5 fission yield) over the Northern Hemisphere, depositing it all at one year and allowing for radioactive decay, gives a dose rate after one year of about 10^{-3} rad/hr, which is about one hundred times the natural background rate, and continuing to decrease. The 50-year integrated external gamma ray dose, accounting for radioactive decay and not considering protection, rough terrain, weathering, and other factors, would be of the order of a few tens of rad, which is less than the U.S. occupational safety standard of 5 rad/year. Thus, what gets into the stratosphere has relatively little practical significance, and estimates made in the 1970s, when the nuclear arsenals included many multi-megaton

* Strictly, the decay rate rule doesn't apply beyond six months.

weapons, suggested that global fallout was not a serious problem in the context of a nuclear war.[7]

Over the past decade, as the yields of nuclear weapons have decreased to average about 500 kT,[+] the importance of the radioactive debris that remains in the troposphere (roughly the lowest 10–12 km of the atmosphere, which is generally well-mixed vertically and within which precipitation systems exist) has increased. Particles in this part of the atmosphere typically have lifetimes of days to weeks, so that there is less time for both decay and for spreading. If we assume that 4000 Mt of total yield (times one half to get fission fraction) all end up in the troposphere and are deposited evenly over the approximately 10^8 km^2 between 30° and 60°N at a time of one week, the dose rate at one week is a few tenths of a rad per hour (decreased by a factor of 10 after 7 more weeks). Model calculations suggest that the total 50 year integrated dose in this latitude band (assuming no weathering or sheltering) would be of order 30–50 rad (about half during the first year), which is less than a factor of ten over the U.S. occupational standard of 5 rad/year. As shown in Figure 3, the spatial variability of this dose would be significant and could be a factor of several higher in hotspots where rainout has concentrated the radioactive deposition. Thus, although the reduction in weapons yields from megatons to hundreds of kilotons over the past decade has led to increased radionuclide doses, this intermediate fallout will likely not be life threatening in most regions, especially outside the Northern Hemisphere mid-latitudes.

These calculations have focused on external gamma ray dose. The weapons debris can also be taken up by plants and animals that are later eaten as food. Because food pathways following a nuclear war may be significantly disrupted, estimating internal dose contribution requires many uncertain assumptions. Peterson et al.[8] suggest that the internal dose due to local fallout could range from much smaller than to about equal to the external dose, depending on a wide range of factors (e.g., protection factor, diet, etc.). In general, however, it seems unlikely that the dose from contaminated food will have as severe an impact as limitations in food supply and other infrastructure problems.

In addition to radionuclides from the weapons themselves, there is the possibility that radionuclides tied up in the nuclear fuel cycle (naval reactors, power plants, reprocessing facilities, storage locations, etc.) could be released into the environment. In the vicinity of the facilities, local fallout from the assumed surface burst that would be required to overcome the containment vessel and disperse these materials,* would be the greater short-term problem, because the fuel cycle radionuclides involved are generally longer-lived than weapons debris; over the longer term, the fuel cycle radionuclides would contribute to a greater,

---

[+] The arsenal and exchange considered here are based on Tables 2.1 and 7.7 of reference 4.

* There is, of course, the potential that prolonged disruption of the electrical power network may cause a failure of nuclear power plants that could also lead to radionuclide dispersal.[9]

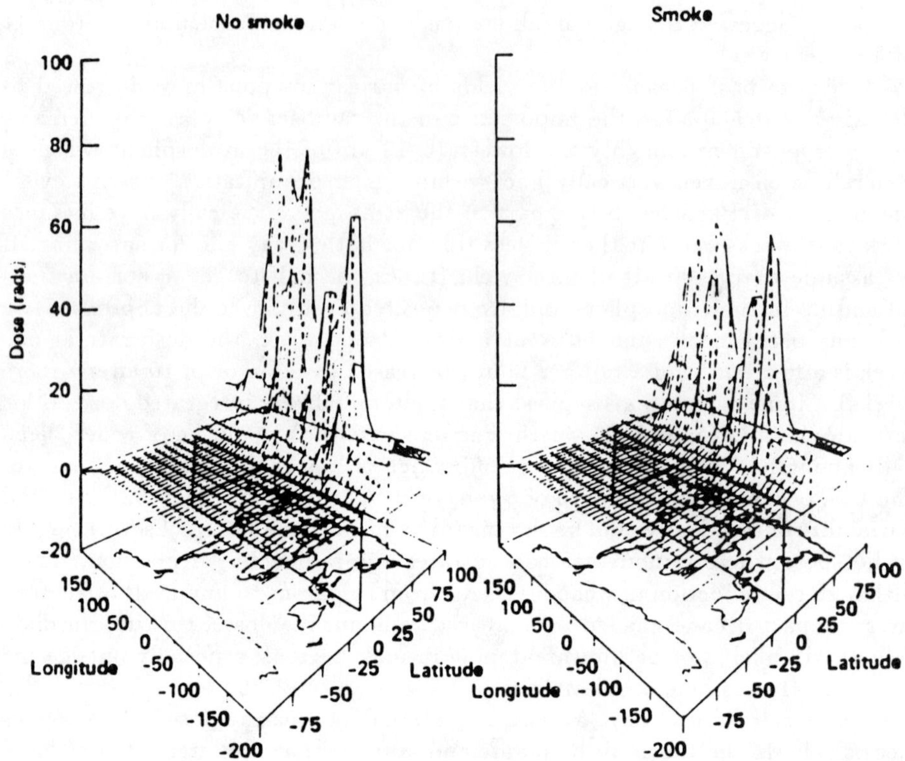

Figure 3. Comparison of radionuclide global dose distribution for cases with unperturbed and smoke-perturbed climates (tropospheric contributions only).

but prolonged, dose. For the case of intermediate and global fallout, if the nuclear explosion lofts the long-lived isotopes and mixes them with the weapons debris, rough calculations indicate that the Northern Hemisphere mid latitude 50 year external doses could be roughly tripled.[4] Quite certainly, as the Chernobyl accident demonstrated, rather large areas could have levels where, given peacetime standards, habitation would need to be prohibited. In that these areas could be far from the target area, it would seem prudent for nations to avoid direct targeting of nuclear facilities with thermonuclear weapons.

CLIMATIC PERTURBATIONS

Nuclear explosions lead to the injection of a range of substances into the atmosphere that could perturb the atmosphere, leading initially to modest changes in atmospheric conditions about the present averages (i.e., to perturbing the "weather") and, in addition, were emissions massive enough, to changes in the average state of the atmosphere over weeks, months, and even years (i.e., to perturbing the "climate"). Nitric oxide created by the nuclear fireball reacts with ozone to become nitrogen dioxide, an absorber of solar radiation. Surface materials, usually referred to as dust, are carried upward from surface and near-surface bursts by the mushroom cloud where they act like volcanic aerosol injections to whiten the sky by increasing the scattering of solar radiation. Smoke, especially black smoke created predominantly in urban fires started by nuclear explosions, is carried upward in the heat-driven fire plumes to levels where its absorption of solar radiation takes place at altitudes above most of the "greenhouse" gases that trap infrared radiation and add warmth to the planet.

The effects of nitrogen dioxide and dust injections alone on the atmosphere seem likely to involve perturbations that are not unlike the anomalous summer or winter weather experienced in the absence of nuclear explosions. That this might be the case can be seen by comparing the radiative effects of a nuclear dust loading with those caused by a major volcanic injection of sulfate-creating gases. Although the actual value would depend on specific soil conditions and other factors, observations suggest that a typical surface burst would lead to lofting of about 0.3 Tg/Mt of dust into the stabilized cloud (1 Tg = 10^{12}g = 10^6 metric tons). About 8% of this mass, or about 0.025 Tg/Mt, has a radius of less than one micrometer.[10] The dust with radius larger than 1 μm generally has a fall velocity such that it would be deposited within days to weeks, insufficient time to induce a significant disturbance to the atmospheric circulation, although like the Mt. St. Helens dust injection, local daily maximum temperatures under the dust cloud would be reduced during this period. If we assume that the 2500 Mt of surface burst explosions used above loft all of the submicron dust into the stratosphere, and then spread the dust over the Northern Hemisphere, there would be about 0.25 g/m^2 [4,11] (with more careful analyses, one would get about half of the injection calculated here). Dust acts mainly to scatter solar radiation, with an extinction optical extinction efficiency of about 2 m^2/g; when multiplied by the dust column, this gives an optical depth of about 0.5, meaning that for an average solar zenith angle of 60°, e^{-1} of the direct solar radiation would be transmitted. Were dust a perfect absorber, the absorption of this radiation in the stratosphere above the greenhouse gases could lead to substantial surface cooling (as is the case for smoke). But dust scatters of order 90% of intercepted radiation forward (downward), creating a whitening of the sky as do water clouds (which have optical depths up to ten or more). In terms of potential atmospheric effects, the optical depth of 0.5 from nuclear-injected dust can be compared to a hemispheric average optical depth of about 0.15 following the El Chichon volcanic eruption. Although unusual weather induced by an El Niño event followed, the connection is, at best, tenuous and there was virtually no effect on global average temperature. Larger eruptions, for example Krakatoa in 1883 and the much larger Tambora in 1815, also seem to have preceded unusual weather events (even unseasonal frosts), but without substantial or lasting perturbations to the underlying climate.

Because smoke particles are much darker than dust, the potential effects on the atmosphere of smoke injections are much more severe, as Crutzen and Birks,[12] Turco et al.,[10] NRC,[11] and Pittock et al.[4] all describe. These studies draw on a wide array of laboratory, field, theoretical, and numerical experiments and analyses in an attempt to improve understanding of an event filled with uncertainties and unknowns; this is one large event for which, drawing from Heisenberg, conducting the full experiment as a means of reducing uncertainties would destroy the object of the experiment. Although we can not, therefore, foresee the details, or perhaps even major aspects, the potential for large climatic effects is readily illustrated.

To estimate the potential effects, we need to estimate the amount of smoke injected and its climatic effect. If we assume 2000 500 kt nuclear explosions (or equivalent) occur over urban/suburban areas, a scenario plausible either as a direct result of declared retaliatory policy (i.e., MAD) or as a side-effect of an allout attack on military and industrial targets, the area exposed to ignition would be of order 100,000 km^2, even allowing a factor of 3 reduction to account for the limited size of such areas, open areas, and for overlap of nearby explosions. Assuming a combustible fuel loading of 30 kg/m^2 (i.e., combustion of 3000 Tg of material, which is of order 25–40% of total combustible fuel in NATO and Warsaw Pact countries[13]) and a soot emission factor of 2% (which allows for a mix of wood and petroleum-based materials such as plastics, etc. and ignores the additional smoke mass made up of scattering constituents), submicron soot emissions total 60 Tg. A major uncertainty concerns the fraction of these emissions that might be scavenged as a result of rain induced by the very rapidly rising fire plume, an effect that apparently led, for example, to the "black rain" following the Hiroshima explosion. Field experiments using forest fires and oil pool fires are being used to investigate this phenomenon. The NRC[11] report suggested that the uncertainty spanned the range 10 to 90%; uncertainties are due mainly to limited understanding and treatment of microphysical processes, including nucleation, capture, electrical processes and more. For the present exercise, assume that half of the soot will survive early time removal without being agglomerated into supermicron, easily scavenged particles.

The hot fires, behaving much like large thunderstorm clouds, would thus loft 30 Tg of sooty material into the upper troposphere (5–10 km above the surface), with some possibly reaching even into the lower stratosphere. Such soot has a specific absorption coefficient of order 10 m^2/g, leading to an absorption optical depth of about 1.2 when spread over the entire Northern Hemisphere. Penner[13] considers the uncertainties in the various factors contributing to this number and suggests a range of from about 0.1 to about 5 (her estimates assume spread over half of the Northern Hemisphere).

With no analog to such an event (except perhaps following impact of an asteroid or comet), numerical models are typically used to estimate the potential impact. Clearly, however, it will be large. Assuming an average solar zenith angle of 60°, only about 10% of the solar radiation will make it through the smoke cloud. (Spreading of the aerosol evenly over a large area and not allowing scavenging maximizes the absorption, however, so the actual effect will be less.) This factor of 10 effect can be compared to the change of about a factor of 2 decrease in hemispheric average solar absorption from summer to winter. It is not surprising, therefore, that the phrase "nuclear winter" has been adopted to describe the effect.

The importance of the altitude and high absorption coefficient of the smoke can be illustrated analytically using a highly simplified representation. Consider a model atmosphere consisting of two layers, with solar and infrared radiative fluxes represented at the top of the upper layer, between the upper and lower layers, and between the lower layer and the surface (assumed to be non-reflecting). Then, assuming radiative equilibrium of each system component (which won't actually occur due to the high thermal capacity of the ocean), the infrared radiative flux from the surface is

$$\sigma T_s^4 = S_o(1 - r_u)(1 - r_l)G$$

where

σ = Stefan-Boltzman constant
T_s = surface temperature
S_o = solar insolation at the top of the atmosphere
r_u, r_l = reflectivities of the upper and lower layers, respectively, and
G = greenhouse factor.

The Greenhouse factor is dependent on the absorptivities, a, of visible light passing through the two layers, and the infrared emissivities, e, of the two layers, such that

$$G = \frac{4 - e_u e_l - a_u[2 - e_l(1 - e_u)] - a_l(2 - e_u)(1 - a_u) + \frac{r_l}{(1-r_l)} a_u(2 - a_u)(2 - e_l)}{(2 - e_u)(2 - e_l)}$$

where multiple reflections between cloud layers have been ignored.

Plausible parameters for the present atmosphere, assuming each layer is made up of clouds and various greenhouse gases, are:

S_o = 340 W/m^2
r_u = 0.09 (30% high cloud cover, 30% reflectivity)
a_u = 0.05
e_u = 0.3
r_l = 0.24 (40% low cloud cover, 60% reflectivity)
a_l = 0.15
e_l = 0.8

These parameters give $G = 1.68$ and $T_s = 289$ K and a planetary radiating temperature of 255 K; were there no atmospheric layers and the planet completely absorbing, $G = 1.0$ and $T_s = 277$ K, illustrating the importance of greenhouse gases in warming the Earth.

If the top layer becomes totally soot-filled such that $r_u = 0$ and $a_u = 1$, which would be the extreme case for smoke injections, then T_s drops to 244 K, even colder than the planetary radiating temperature. This can occur because the only energy now available to the layers below is the infrared energy radiated downward by the top layer (which is, in this approximation, half the solar energy plus half of the absorbed infrared energy coming up from below). The surface temperature would drop even lower if the upper layer also maintains some reflectivity. Such a result was found by Turco et al.[10] who used a one-dimensional radiative-convective model to simulate a land covered planet.

Several variants provide interesting insights. Were the smoke to spread only throughout the lower atmosphere (such that $a_l = 1$ and $r_l = 0$), then T_s would decrease to only 275 K, a consequence of much of the sunlight being absorbed at altitudes below the greenhouse gases; thus, the higher the smoke is injected, the larger the temperature reduction. Were the smoke highly absorbing in the infrared such that the emissivity approached unity (which is not likely to be the case because the smoke particles tend to be submicron in size), the surface temperature reductions would be to 278 K and 286 K for soot in the upper and lower layers, respectively; thus, the low IR opacity of the smoke also exacerbates the cooling.

These calculations are for equilibrium conditions, which implies that the surface in effect has zero heat capacity during the time that the smoke is present. Because of the large ocean heat capacity, this is not the case; ocean surface temperatures would cool only very slowly as the oceans gave up their heat to warm the atmosphere, helping to buffer the cooling tendency over land. That this buffering will be substantial is evident by applying the formula for a volcanic dust injection which might increase r_u to 0.12. Such an increase, by the formula, would lead to a global average temperature reduction of about 2.4 K, whereas past volcanic aerosol injections with such a reflectivity appear to induce coolings of about one-third this amount.

When more sophisticated climate models are applied to the problem of smoke injections,[14,15,16] they all find that surface temperatures, particularly over continental interiors, are quite sensitive to smoke injections. This is especially true in summer when solar radiation is the primary determinant of surface temperature, but less so in winter when continental temperatures are maintained to a large extent by heat transport from the ocean to the atmosphere and then over land. The model calculations indicate that massive smoke injections can reduce summertime surface temperatures by 10–20 K, with episodes to near or below freezing, in the weeks following the war (see Figure 4).

Two important factors have also been identified that enhance the persistence of the perturbation, particularly in the summer. Cooling of the lower troposphere and warming of the upper troposphere by solar heating of the smoke lead to increased atmospheric stability, which suppresses the convective precipitation that would normally cleanse the smoke particles from the atmosphere. The heating also promotes lofting of the smoke to higher altitudes, even well up into the stratosphere, where it can spread globally. Once at these altitudes, the smoke lifetime increases from weeks to a year or more, introducing the potential for the smoke to induce a long-term temperature perturbation of several degrees, which would make agriculture more difficult during the recovery phase—not just in the combatant nations, but around the world.

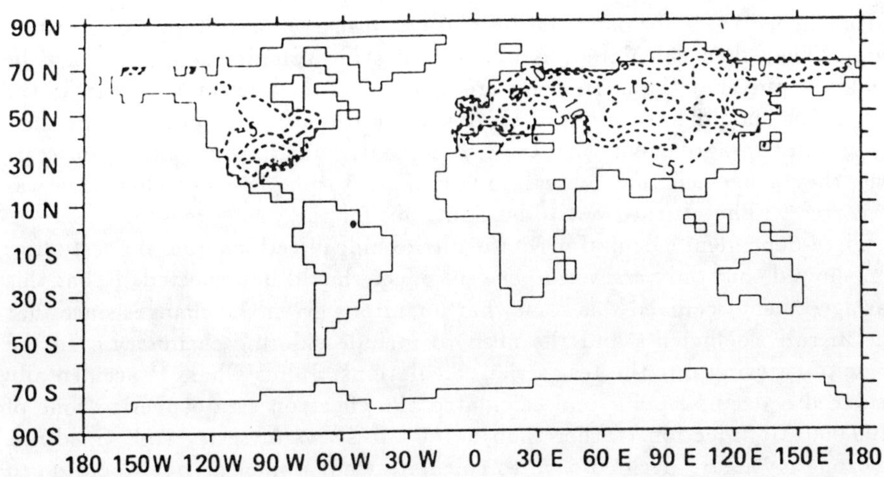

Figure 4. Distribution of surface air temperature change (smoke − control) averaged from day 1–10 following the nominal injection. Contour levels are −20, −15, −10, −5 K, with negative isotherms dashed. (From Ghan et al.[16])

The recognition of this potential for both acute and chronic climatic perturbations has led to several inquiries into the likely impacts of such changes on societal activity. Harwell and Hutchinson,[3] concurred in by a special United Nations panel,[1] suggest that such climatic effects and the societal disruption of a major nuclear war pose the threat of starvation to a majority of the world's population, making the potential for deaths from smoke effects up to several times larger than from the direct fire, blast, and radiological consequences.

ATMOSPHERIC CHEMISTRY

Nuclear explosions lead to the injection of many different gases into the atmosphere. Gases that affect the climate and chemistry of the global-scale atmosphere can be created as a result of transformations in the fireball and the ensuing fires; in the event of widespread environmental disruption, gases released by the decay of dead plants or the altering of present activities may also alter the global environment.

Initial concern about what nuclear weapons might do to the global atmosphere goes back to a calculation performed in 1943 by Konopinski and Teller to determine whether a chain reaction might accidentally be set off "that would encircle the globe in a sea of fire."[17,18]

The matter next arose in the early 1970s in a serendipitous fashion. Environmental analyses of the potential climatic effects of a fleet of supersonic aircraft led to recognition that nitrogen oxides (primarily nitric oxide and nitrogen dioxide) emissions could reduce stratospheric ozone.[19,20] As a test of this hypothesis, Foley and Ruderman[21] pointed out that of order 10^{32} molecules of nitric oxide were created per megaton explosion[22] because the rapid expansion led to cooling that quenched the reformation of N_2 and O_2, just as in an internal combustion engine. They then calculated, assuming that stratospheric chemistry would be in equilibrium, that the nitric oxide created during the nuclear test series in the early 1960s should have reduced stratospheric ozone by about 15%. Because this was not in agreement with the few percent reduction seen in the observations, they suggested that understanding of stratospheric ozone chemistry was not correct. (They turned out to be right, but for the wrong reason.)

Time-dependent calculations of the nitric oxide injections from the test series soon showed that only a several per cent effect should be expected. (That this near agreement occurred was somewhat fortuitous given the changes since that time in rate coefficients and the need to include chlorine chemistry.) In the course of carrying out the test series calculations, Julius Chang[23] accidentally misplaced a decimal point and calculated the effect on stratospheric ozone of a 100,000 Mt injection (rather than a 100 Mt series of tests) and wiped out the ozone layer. He decided then to calculate what a nuclear war would do to the ozone layer. Calculations by Chang[23] and independently by Hampson[24] led to a National Academy of Sciences review in 1975.[7] Their analysis suggested that a 10,000 Mt war involving the many multi-megaton weapons characteristic of arsenals at that time would loft of order 50 million tonne of nitric oxide to the stratosphere, which would lead to a reduction of the Northern Hemisphere stratospheric ozone loading of order 50% with a recovery time of several years. If the 10^{32} molecules/Mt injected in a 10,000 Mt war destroy 50% of the Northern Hemisphere ozone, each NO molecule has destroyed about 2×10^4 molecules of ozone (assuming no production). This high effectiveness of NO destruction of ozone results from a catalytic cycle in which

$$
\begin{aligned}
NO + O_3 &\rightarrow NO_2 + O_2 \\
NO_2 + O &\rightarrow NO + O_2 \\
\text{Net} : O + O_3 &\rightarrow 2\ O_2.
\end{aligned}
$$

These reactions proceed until the nitrogen oxides are removed from the stratosphere by transport processes or tied up chemically, both of which are relatively slow processes. Although increased incidence of skin cancer (melanoma) is the normal concern from decreasing ozone levels, that is believed correlated with the integral lifetime dose. For large perturbations, it is the acute phase that is of concern. A 50% reduction would increase the intensity of solar ultraviolet radiation reaching the surface sufficiently to lead to skin blistering in tens of minutes, an effect, it was felt, that could also significantly impact many plants and animals, including humans as they ventured out seeking food.

As nuclear arsenals changed in the 1970s, the ozone problem ameliorated. The MIRVing of strategic missiles reduced the yields of individual weapons to the few hundred kiloton range—a trend also dependent on improving accuracy—which meant that the nitric oxide would be injected mainly into the troposphere, well below the peak concentrations of stratospheric ozone at 25–30 km (Foley and Ruderman[21] estimate stabilized cloud top $C_T(\text{km}) = 22Y^{0.2}$ and cloud bottom $C_B(\text{km}) = 13Y^{0.2}$ where Y is yield in megaton). A recalculation of the effect on stratospheric ozone assuming nuclear arsenals of the mid-1980s suggested a reduction of 15–20% with recovery over a few years.*[11] Such an ozone reduction would lead to a few-fold increase in the ultraviolet flux, thereby threatening many sensitive species.[7,25] In that the annual ultraviolet erythema dose increases about 50-fold from the pole to the equator,[26] an increase of a few fold would not however, be an unsurvivable threat to humans.[27]

The recognition of the extent of fires has, however, again raised questions concerning potential perturbations to atmospheric chemistry. The fires themselves lead to nitrogen oxide and hydrocarbon emissions that, in the presence of sunlight, could create ozone and other oxidants. Calculations by Crutzen and Birks[12] suggested that ozone levels could reach 0.16 ppmv, exceeding air quality standards, but well below peak levels experienced in Los Angeles. Calculations by Penner[25] subsequently showed that, because the nitrogen oxide would be removed more rapidly than the smoke, the diminished light levels would not be sufficient to induce high tropospheric ozone levels. Of course, while removal of the nitrogen and sulfur oxides created by the fires limits the tropospheric chemical changes, it will acidify the precipitation, causing what has been referred to as the "nuclear pickle."

The smoke injections would also lead to climatic changes that would affect atmospheric chemistry. Smoke-induced absorption of solar energy, which could raise stratospheric temperatures by 50–100 K, would accelerate the Chapman photochemical cycle which destroys O_3. More importantly, the smoke-induced heating of the troposphere would cause a lofting of ozone poor tropospheric air into the stratosphere, displacing the lower stratospheric ozone reservoir upward to altitudes where it can be photochemically destroyed and carrying the fireball-created nitrogen oxides upward to levels where catalytic destruction can occur. The smoke particles themselves also serve as an ozone sink. The stratospheric displacement by tropospheric air in the Northern Hemisphere also would lead to a southward and downward displacement of the ozone rich air in the tropical and Southern Hemisphere stratosphere, possibly enhancing surface ozone values in the Southern Hemisphere.

Calculations of these many interacting fire-induced effects are just beginning, but it seems quite possible that stratospheric ozone reductions of 50% or more are possible. Interestingly, this again raises the potential for tropospheric chemistry changes, because the reduction in ozone optical depth in the ultraviolet is probably balanced by the increase in optical depth due to the smoke only over the first few weeks.[25]

There is a wide range of other possible chemical effects, many of which have not yet been explored. These include creation of pyrotoxics, build-up of

* An excursion case with 100 20 MT explosions led to a depletion of just over 40%, slowing the strong dependence of the effect on high yield weapons. It is interesting that while the trend toward lower yield weapons eased the ozone depletion effect, it exacerbated the global fallout effect by a factor of ten (although still significantly below harmful effects).

H_2S due to reduction in the oxidizing potential of the atmosphere, stratospheric chlorine and bromine injections from nuclear explosions over the oceans, carbon monoxide generation, and many more. Enhancement of the greenhouse gas potential of carbon dioxide emissions would be very small (combustion emissions are about equal to one year's fossil fuel emissions; decay of vegetation without regrowth could be a substantial contributor, but seems unlikely). At present, none of these many potential effects has been demonstrated to be significant, but that may only be due to arbitrary assumptions that have been made or aspects, particularly interactive aspects, that have been overlooked.

CONCLUDING REMARKS

Nuclear war would be a catastrophe of unprecedented proportions. The direct effects of fire and blast created by a major nuclear exchange could kill hundreds of millions, which has provided the compelling basis for avoiding nuclear war.

In addition to these direct consequences, and the very substantial disruption of the national and international social and economic fabric that would occur, a nuclear war would induce a series of environmental effects that would spread the impacts among combatant and non-combatant nations alike. Beyond regions of local fallout, global fallout would exert a relatively minor effect, unless a substantial fraction of the nuclear fuel cycle inventory were widely dispersed. Smoke and dust injections could seriously disturb the global climate, particularly the hydrologic cycle during the first several months, but depressing temperatures for of order a year. The combined fireball and fire emissions could significantly deplete stratospheric ozone, greatly impacting human activities and ecological systems for perhaps a few years.

If an additional reason to avoid war is needed, the environmental impacts will certainly be present. When "nuclear winter" was first interpreted as a freezing solid of the entire planet, there may well have been an additional and compelling moral imperative obvious to all to avoid the total annihilation of life. The present estimates do not encompass extinction, but they do indicate the potential for threatening a majority of the global population with starvation. Since the 1950s, the threats of fire, blast, destruction, and disruption have been sufficiently convincing; the potential environmental consequences substantially raise the ante for the non-combatants and, hopefully, will assure that deterrence is not simply a destructive threat between the superpowers, but is an active basis for responsible behavior among all nations.

ACKNOWLEDGEMENTS

The paper benefited from helpful comments by Charles Shapiro, Steven Ghan, and Joyce Penner. This work was performed under the auspices of the Department of Energy by the Lawrence Livermore National Laboratory under contract W-7405-ENG-48.

REFERENCES

1. United Nations, Report of the Secretary-General A/43/351, United Nations, NY (1988).
2. Institute of Medicine, The Medical Implications of Nuclear War, F. Solomon R. Q. Marston (eds.), (National Academy Press, Washington, D.C., IOM, 1986), p. 619.
3. M. A. Harwell and T. C. Hutchinson, Environmental Consequences of Nuclear War, Volume II, Ecological and Agricultural Effects, SCOPE 28 (John Wiley & Sons, Chichester, 1985), p. 517.
4. A. B. Pittock, T. P. Ackerman, P. J. Crutzen, M. C. MacCracken, C. S. Shapiro and R. P. Turco, Environmental Consequences of Nuclear War, Volume I, Physical and Atmospheric Effects, SCOPE 28 (John Wiley & Sons, Chichester, 1986), p. 360.
5. G. S. Golitsyn M. C. MacCracken, Geneva: World Meteorological Organization, WCP-142 (1987).
6. S. Glasstone and P. Dolan, U.S. Department of Defense and U.S. Energy Research and Development Administration (1977), p. 653.
7. National Academy of Sciences (NAS), Long Term World-wide Effects of Multiple Nuclear-weapon Detonations (National Academy Press, Washington, DC, 1975), p. 213.
8. K. Peterson, C. Shapiro, and T. Harvey, Paper presented at SCOPE ENUWAR Moscow Workshop, 21–25 March 1988, Lawrence Livermore National Laboratory Report UCRL-98348 (Livermore, Calif.).
9. C. S. Shapiro, Paper presented at SCOPE ENUWAR Bangkok Workshop, 9–13 February 1987, Lawrence Livermore National Report UCRL-95827 (Livermore, Calif.).
10. R.P. Turco, O. B. Toon, T. P. Ackerman, J. B. Pollack, and C. Sagan, Science 222, 1283 (1983).
11. National Research Council (NRC), The Effects on the Atmosphere of a Major Nuclear Exchange, (National Academy Press, Washington, DC, 1985) p. 193.
12. P. J. Crutzen and J. W. Birks, Ambio, 11, 114 (1982).
13. J. E. Penner, Nature 324, 222 (1986).
14. R. C. Malone, L. H. Auer, G. A. Glatzmaier, M. C. Wood, and O. B. Toon, J. Geophys. Res., 91, 1039 (1986).
15. S. L. Thompson, V. Ramaswamy and C. Covey, J. Geophys. Res. 92, 10942 (1987).
16. S. J. Ghan, M. .C. MacCracken and J. J. Walton, J. Geophys. Res., 93, 8315 (1987).
17. E. J. Konopinski and E. Teller, Los Alamos National Laboratory, Los Alamos, NM (unpublished paper) (1943).
18. W. Teller, with A. Brown, The Legacy of Hiroshima (Doubleday and Co., Garden City, NY, 1962).
19. H. S. Johnston, Science, 173, 516 (1971).
20. P. J. Crutzen, J. Geophys. Res., 76, 7311 (1971).
21. H. M. Foley and M. A. Ruderman, J. Geophys. Res., 78, 4441 (1973).

22. N. Bonner, Lawrence Livermore National Laboratory, Livermore, CA, (internal memo, 1971).
23. M. C. MacCracken and J. S. Chang (eds.), LLNL Rep. No. UCRL-51653, Lawrence Livermore National Laboratory, Livermore, CA (1975).
24. J. Hampson, Nature, 250, 189 (1974).
25. J. E. Penner, Proceedings of the International Seminar on Nuclear War 3rd Session: The Technical Basis for Peace, Erice, Italy, August 19–24, 1983 Servizio Documentazione dei Laboratori Frascati dell'INFN, July 1984, pp. 94, also available as Lawrence Livermore National Laboratory Report UCRL-89956, Livermore, CA (1983).
26. T. Mo and A. E. S. Green, Photochem. Photobiol., 20, 483 (1974).
27. H. W. Ellsaesser, Climatic Change, 1, 257 (1978).

CHAPTER 2

CONVENTIONAL FORCES AND ADVANCED TECHNOLOGIES

Seymour Deitchman
Institute for Defense Analysis

I. INTRODUCTION

There is some equivalence between tactical nuclear weapons and advanced conventional weapons. They both offer the prospect of greatly increased effectiveness in a war between armies. A hope has often been raised that with advanced conventional weapons, weapons that are an order of magnitude more effective than unguided ballistic high-explosive weapons, maybe we can do away with nuclear weapons altogether, except for their role in global strategic deterrence. There are, however, significant differences between these weapons; some of these are shown in Table 1.

Table 1. Comparison of tactical nuclear and advanced conventional weapons.

A. NUCLEAR WEAPONS

1. Mass: Little required; one weapon per concentrated target set.
2. Kill Mechanism: Blast; heat; radiation.
3. Area Effects: Everything within specific radius is affected; decreases with radius.
4. Countermeasures: Virtually none, once fired.

B. ADVANCED CONVENTIONAL WEAPONS

1. Mass: Much required--one per target within concentrated target set.
2. Kill Mechanism: Local blast or armor penetration; fragments over small radius.
3. Area Effects: Local for unitary warheads; randomly-distributed point effects for submunition warheads, same everywhere within pattern.
4. Countermeasures: Subject to technical countermeasures.

There are two main differences contained in Table 1. The first difference has to do with effectiveness. A nuclear weapon dropped into an area will destroy everything in that area. If a nuclear weapon of the appropriate size is dropped on an armored company, for example, the armored company will be completely eliminated. Advanced conventional weapons, however, have to be targetted against each individual target within the area. Consequently, if one is using conventional weapons, even good ones, against a company of tanks, one has to attack essentially every individual tank, and hit enough of them so that the company ceases to be an effective fighting unit.

The second difference between tactical nuclear weapons and advanced conventional weapons has to do with countermeasures. A nuclear weapon, once it has been fired at a target, is pretty hard to stop. The warhead can be built and fuzed in such a way that even if its guidance mechanisms and its carrier vehicle are destroyed, it can still follow a ballistic path to the target and explode there, doing a lot of damage. In contrast, advanced conventional weapons are subject to many possible countermeasures. One would like to substitute advanced conventional weapons for tactical nuclear weapons, and there is some prospect that they may both reach similar levels of very high effectiveness. But they are very different in terms of how they are used militarily, what effects they have, and what roles they might play in various deterrence postures.

II. ADVANCED CONVENTIONAL WEAPONS

Because of these differences between advanced conventional and nuclear weapons, and because of the difficulties of making the advanced conventional weapons indeed work well, the proposals to replace one by the other are fraught with controversy. It may be useful to state some of the premises I bring to the analyses of advanced conventional weapons in this paper, because one hears people give very different evaluations, which ultimately are based on other sets of premises. Hopefully the disagreements will be clarified by this paper.

One premise is that generally the needs for military capability are threat-driven (to use the jargon). They are driven by the opposition that has to be overcome, not by some abstract judgement of what weaponry might be useful to have. That means one is always looking at someone else, and is deciding how to defeat him. This concern with the enemy determines how much money one wants to spend, and a lot of other things as well. It is often said by the opponents of the trend to advanced technology in weaponry that good men will defeat machinery. Certainly machinery isn't everything; people, tactics, and training will count, and so will the chances of war. However, no one can argue that, in conventional war, well-trained, well-led troops who have good machinery can do better than well-trained, well-led troops who do not have good machinery. In the long run, the quality of the machinery is going to tell, although in any given battle it may take time to observe that effect. There are, of course, no pat solutions to the problem of how to make conventional weapons most useful. Many arguments about conventional arms have been presented in black and white, and many pat solutions have been offered. But it is a very complex topic, and one of my purposes in this paper is to convey some of that complexity.

The notion of guided weapons as a way of enhancing military effectiveness is not new. But the realization of them through modern technology began in World War II. The Germans used radio-guided glide bombs at Anzio. The United States came out of the war developing a set of radio-receiver command-guided weapons called Tarzon, Azon, and Razon, all with slightly different guidance systems. Since then there has been a sequence of development which is still in progress. These developments have fed on the improvements that technology has made in the signal-to-noise ratio. The signal, of course, is the target signature, and the noise is the background clutter. The first work was on guided missiles to shoot down aircraft with surface-to-air and air-to-air interceptions, because such interceptions take place against a fairly uncluttered background. No background is ever totally uncluttered. For example, if one is using infrared guidance, and there are clouds, then there will be reflections from the target and from the clouds, so that targetting can still be confused unless one can separate those signals. However, that particular challenge is simple compared to some other problems, as for example those faced when the next steps were taken to build weapons that could be guided to attack ships.

A major question that has consistently surfaced is whether to utilize self-contained or external command guidance to direct a missile toward the target. External command guidance is easier than self-contained guidance. But the command links are subject to interference. Therefore, one would rather have self-contained target-seeking guidance mechanisms, if at all possible. That in fact has been another part of technology development over the years; starting with command guidance and trying to make it self-contained. Therefore, the next step in missile guidance was to target stationary objects that were located on the ground. This led to fairly crude TV-guided weapons, which could seek a contrast edge or a contrast centroid, as did the Walleye glide bomb for example, and the Maverick anti-tank air-launched missile. Laser-guided bombs follow quite a different guidance scheme; they are designed to track a spot of laserlight that somebody projects onto the target. A simultaneous parallel development was to try to hit moving targets on the ground, targets that were within sight of the gunner so that he could act as the computer that distinguished the signal from the noise.

Now the development work focuses on the most difficult set of targets of all, mobile targets that are located over the horizon, and that must be discriminated from ground clutter. Only fair progress is being made on that problem. But these objects are the most important set of targets. In warfare armies still occupy territory, so the defender still has to stop armies from moving by inhibiting their machinery from moving, and by forcing the personnel to hide from attack. But, opposing armies are getting more spread out, so one cannot see them as easily as in the past. Hence one is now confronting the guidance problems for objectives which (a) will probably count the most in a military sense, yet (b) are the most difficult to accomplish.

III. A TECHNICAL REVOLUTION IN ADVANCED CONVENTIONAL WEAPONS

The technological progress described above has been very gradual, covering some 45 years of weapon development. But the cumulative impact of it has been revolutionary, and will be even more so, if all the problems can ever be successfully solved. It may be helpful to describe some of the problems that get in the way of succeeding with advanced conventional weapons. The areas of technology under discussion can be listed under two rubrics. One is electronics, where electronics includes integrated circuits, advanced computing, and detectors. Progress in electronics has gradually been coming to permit recognition of target signatures, and guidance of weapons toward those target signatures. Furthermore, the electronics of advanced computing techniques and communications are allowing the acquisition and processing of large amounts of information of various kinds from different sources. When all of this information is put together, a military force can detect, identify, and track masses of opposing forces. One is now beginning to put together the two important parts of the problem solution: knowing what the enemy is doing, and then being able to accurately target weapons at him in order to stop him from carrying out undesirable actions.

Another important area of current technology development is the improvement of materials. Improved materials underly not only much of solid-state electronics, but also allow the creation of high-performance aircraft, where high performance refers not only to speed and payload, but also to the ability to stay aloft a long time for observation purposes. Progress in materials also is involved in spacecraft, long-range missiles, and vertical take-off aircraft--in fact, in all weapons-delivery systems. Many parts of the problem are beginning to be solved through these areas of technology. But there are, of course, other technological developments that help solve problems with advanced conventional weapons, such as work on propulsion, aerodynamics, hydrodynamics and so forth.

How can one measure the payoff from all of these technological innovations? One might compare such numbers as the destructive capacity of the forces using them. It is always bit naive to talk in such simplistic terms as the destructive capacity of weapons; but one can learn something even if one talks about armies as if they were pieces on a chessboard. One might compare what a modern armed force can do, with that which the U.S. Armed Forces were able to do in Korea. If one goes through all the equations of kill probabilities and destructive capacity, then the destructive power carried by modern divisions and air wings may be on the order of ten times as high as that possessed by equivalent units deployed in Korea, only 35 years ago. Modern military forces can also be deployed much more rapidly. About three months were required to get a sizable army into Korea, now one could do that in about a month because of high-speed, high-payload aircraft. Besides the increase in destructive power and the more rapid deployment, modern forces can be utilized more effectively. Due to the advances in sensing and in command and control, one can know much better where the enemy is located, so that with the more advanced weapons one can destroy one's enemy much better; so that smaller forces can count for a lot more.

When people talk about quantity versus quality, it is these technological advantages to which they are refering. The United States now has a pretty good long-term advantage in these areas of technology. This means that the United States could be dominant in the area of conventional forces, if we can succeed in the last difficult steps of making advanced conventional weapons work in extraordinarily difficult environments. That success would then negate the predominance in land forces held by the United States' primary opposition. That opposition is, of course, also moving toward advanced conventional weapons, as will be discussed below.

Advanced conventional weapons do not come for free and it is not even obvious that they will come for certain. There are obviously technical problems that must be solved. However, if one is willing to spend the money, one can solve these technical problems at a rate comparable to the rate of progress made by advanced technologies in the civilian sector. Many examples come readily to mind describing how civilian technologies are moving ahead in the areas of technology mentioned above.

The cost of advanced technologies remains a problem, particularly since weapons developments are threat-driven, because the response is to the perceptions of the opposition. The opposition for the United States is of course not only the Soviet Union--although for the foreseeable future that might be the main opposition. However, the United States, the Western Allies, as well as the Soviet Union, are all arming much of the Third World. Given a fluid political world over the next twenty to thirty years, it is not obvious that the United States is going to have to face only Soviet weaponry if it gets into some military conflict. The weaponry facing the United States might be some of its own, and certainly in many cases will be that of its allies. Advanced-technology weapons are inevitably subject to countermeasures, and the knowledge about how to build such countermeasures will be possessed by a variety of people.

With so many uncertainties, advanced-technology weapons represent a mixed blessing. For this reason, many military people are cautious about jumping on the bandwagon. They know what the tried-and-true conventional weapons can do. For a standard ballistic projectile, if one has success in launching it (and that is most of the time) then it goes to the target with an accuracy that is known, and one can correct for the known inaccuracies by sending enough of them. Those who want to acquire advanced-technology weapons, and thereby transform the forces that might use them, must carry out a risk-benefit calculation. The problem is that once one has made the transformation to advanced-technology weapons, one cannot very easily go back to simpler technologies, because one's manufacturing capacity will have changed, and one's deployment technology will have been changed. Once a military force has converted itself to a new way of operating with the new kinds of weapons, if it finds that the new technology does not work very well, then it has a major structural problem where it will be very hard to recover the lost ground. A certain amount of conservatism about jumping on the advanced-conventional-weapons bandwagon is thus understandable. Actually, there has been less reluctance to become committed to some specific advanced technologies, for example to air-to-air missiles and to anti-ship missiles. But those involve applications that are better known, and objectives that one now knows how to achieve.

IV. COST-BENEFIT ANALYSES FOR SOME ADVANCED TECHNOLOGIES

It is possible to analyze the contribution to warfare that might be made by some specific advanced-technology weapons, and try to understand why one is thinking about this issue now. Consider for example tactical air war against ground targets. Figure 1 is a sketch of an air-defense system for a Soviet army. While this sketch is about ten years old, the pattern of coverage has remained essentially the same over the years; the only thing that has changed is that the weapons are more effective now than they used to be. One can see that at the forward edge of the battle area, the zone where the forces are facing each other, there is a lot of low-altitude defensive coverage. Then there is medium-altitude defensive coverage that extends to the territory of the opponent, and very-high-altitude defensive coverage reaching even further into the enemy territory. This defensive structure makes it very difficult for an attacker to approach the war zone at a high altitude, and then go down to attack targets. U.S. armies have similar defenses, so the Soviet Union faces similar attack problems.

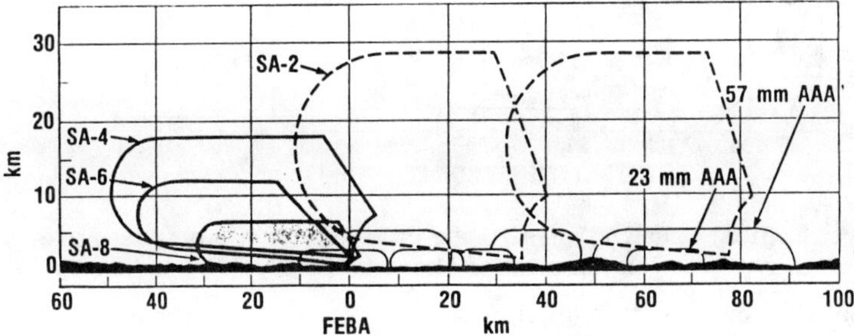

Fig. 1. Profile of the air-defense coverage of a Soviet army.

One can analyze the relative merits of various attack approaches that might be used to overcome such defenses. Consider this defensive structure from another point of view, as shown in Fig. 2. This data is also ten years old; again the only things that have changed over the years are the names and the effectiveness of the weapons performing the particular defensive functions. The numbers of weapons that cover different parts of the altitude spectrum vary considerably; there are a very few defensive weapons that are effective to very high altitudes, and very many that are effective at very low altitudes. Because of a fear of these few high-altitude defensive weapons, the U.S. Air Force, and really all air forces, have adopted the tactic of low-altitude attack. The high-altitude defenses tend to be directed by radar, and it is then desirable to attack below their radar horizon. However, that exposes the attacker to the much-larger number of low-altitude defensive weapons. These Soviet high-altitude defensive weapons are about a half to three quarters of an order of magnitude more effective than the low-altitude defensive weapons. There are however somewhere between one-and-a-half and two orders of magnitude more low-altitude defensive weapons. An expected-value calculation suggests then that, in the long run, low-level attacks are not going to pay off. Instead, one would like to destroy the high-altitude defensive weapons. Fortunately they are radar-guided; they have signatures that makes it possible to find them, and there are far fewer of them than there are of the low-altitude defensive weapons. And if one attacks from high altitudes, then one has the freedom of more airspace in which to maneuver. But the high-altitude attacks require advanced-technology weapons, because one then has the problem of finding targets at long range against ground clutter, and destroying them successfully. For example, in the Falklands war the Argentinian low-level attacks were successful because there were usually no in-depth defenses of this kind. When the British were able to put up a lot of low-altitude gunfire, then the Argentinian aircraft had to increase the standoff distance, which reduced British ship casualities.

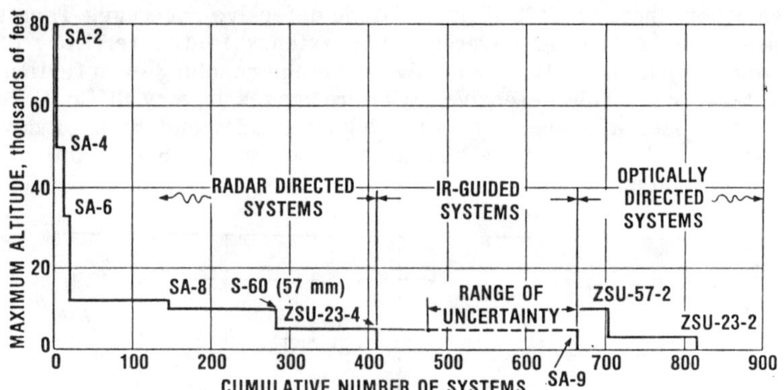

Fig. 2. Typical densities of altitude coverage of air defenses over a Soviet army, covering an area of about 50 km by 100 km (ca. 1976).

One can carry out similar analyses for ground forces such as armor. Typical calculations using simulation models suggest that, with modern weapons, armored forces can be depleted to on the order of half their starting strength in about a month. That is a very high loss rate. Instead, what one would like to do is to avoid closing with the enemy, so that one can inflict those loss rates on him while he cannot inflict them in return. And in naval warfare similar considerations apply. Air forces are not yet very good at carrying out such standoff tactics, because the required advanced technologies are not yet good enough. But such analyses say that standoff tactics are becoming more important, and one wants now to work on technologies in this area. Such over-the-horizon attacks on targets against ground clutter are very difficult, but also very important.

Another problem with advanced-technology weapons is, of course, that as the costs of the delivery vehicles, the airplanes, ships, and so on, rise rapidly, one needs greater force efficiency to make up for that. For example, an F-4 aircraft, which is part of the previous generation of aircraft, costs about twelve millions dollars in 1988 currency. An F-15, which is the present-generation equivalent aircraft, costs about 25 to 30-million 1988-dollars (depending on how it is equipped). The cost of the next-generation fighter, the Air Force's Advanced Tactical Fighter, will be in the neighborhood of 30 to 40-million 1988-dollars. It is economically foolish to send aircraft that risk great losses and that cost so much money to the attack if they have weapons that do not hit their targets. This is the source of yet another pressure to build weapons that hit their targets; it is not possible to keep re-exposing these very expensive assets to hostile fire without a significant payoff. One must learn to begin the attack more quickly, know more precisely where the target is located, and complete the attack more effectively.

V. INSTITUTIONAL RESISTANCE TO ADVANCED-TECHNOLOGY WEAPONS

However, the development and deployment of advanced-technology weapons is not moving in this direction. There is a great deal of conservatism that prevents moving rapidly towards the deployment of advanced-technology weapons. Consider how the services treat the issues of weapons and delivery

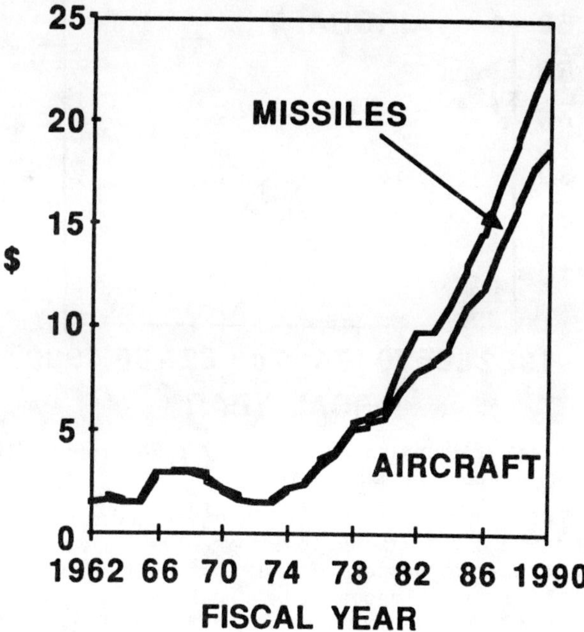

Fig. 3. Historical trend of the current-dollars acquisition budget (total obligational authority) for weapons and platforms, of the U.S. Air Force.

platforms. The analysis here will concentrate on the Air Force and the Navy, because their budgets are simpler to analyze, but the Army's situation can be shown to be roughly similar.

Figures 3 and 4 display the historical trends of the U.S. Air Force acquisition budget; both figures show how the budgets are distributed between aircraft and missiles. Figure 3 presents the actual current-dollars value of the budget, Fig. 4 presents the fractional allocations. These figures do not include unguided bombs and other standard munitions, because those do not absorb much of the budget relative to guided weapons and platforms. Over the period 1962 to 1990 the fraction of the budget allocated to weapons has been pretty constant. There was a significant increase in 1981-82. About half of that increase was due to the ground-launched cruise missile that will now be moved out of Europe under the INF Treaty. AMRAAM and the heat-seeking Maverick air-to-air missiles account for most of the remaining discontinuity. Other than that, less than 10% of the Air Force acquisition budget has been going to advanced weaponry--most of it has been going into airplanes.

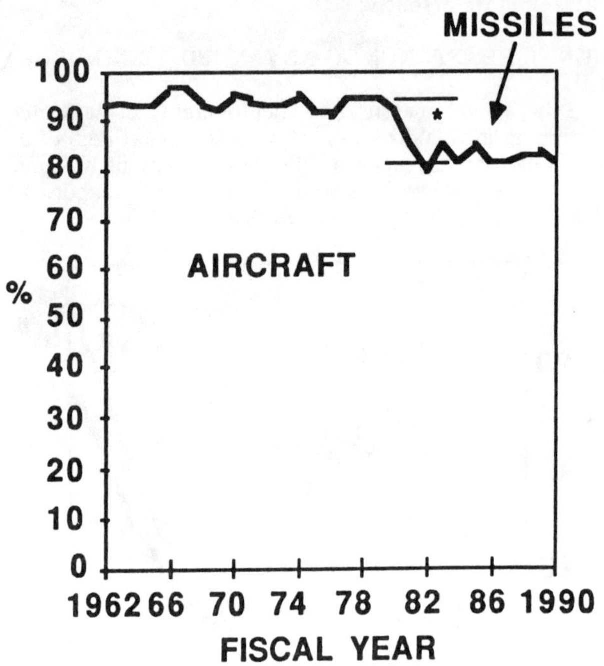

Fig. 4. Percentage distribution of the U.S. Air Force acquisition budget between weapons and platforms.

A similar picture holds for the U.S. Navy, as shown in Figs. 5 and 6--although that picture has now started to change. The increase in the slope of the percentage allocation to weapons after 1982 is due to some advanced anti-ship missiles. That trend would appear to be a healthy thing.

Overall, Figs. 3 to 6 have shown that there is a "cultural" pattern in the planning factors. Consistent fractions of the budget go to specific kinds of things. One can look at this data from a somewhat different perspective, by focusing on the expenditures on combat systems versus those for support

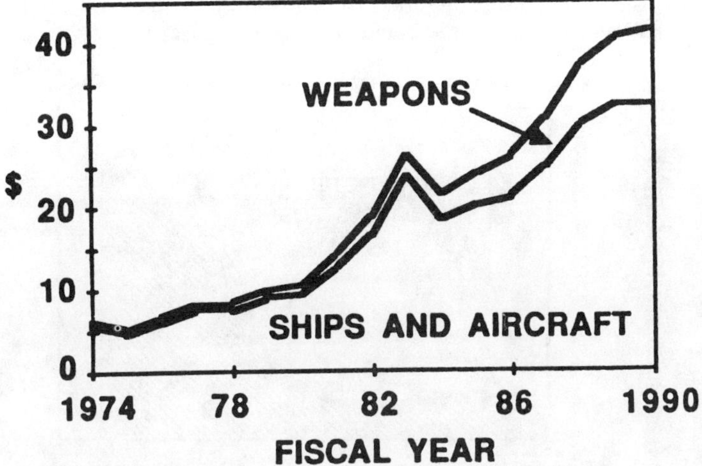

Fig. 5. Historical trend of the acquisition budget (total obligational authority), in current dollars, of the U.S. Navy for weapons and platforms.

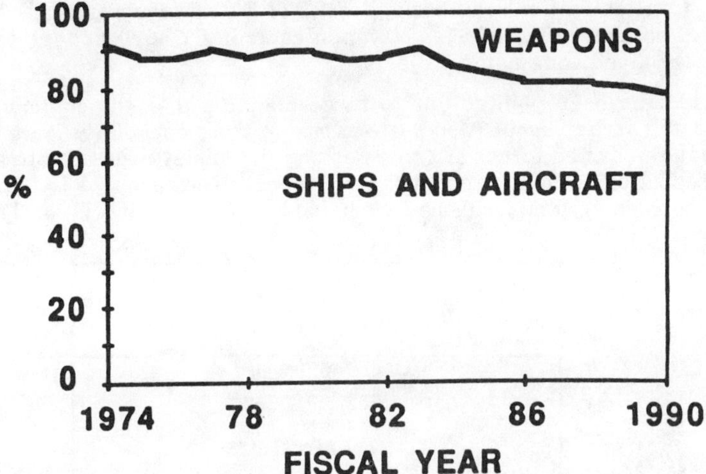

Fig. 6. Percentage distribution of the U.S. Navy acquisition budget between weapons and platforms.

systems. Consider for example the U.S. Air Force budget data shown in Figs. 7 and 8. In the acquisition budget, combat systems include weapons systems, combat aircraft, missiles, and ammunition. Support systems include surveillance, target acquisition, C^3I, EW, logistic and general support systems and dedicated platforms (e.g. AWACS). These figures show that over a 30-year period the relative expenditures have been roughly constant, although there are some fluctuations. The acquisition pattern for the Navy is essentially the same, except that the Navy devotes an even smaller part of its budget to support systems--as shown in Figs. 9 and 10.

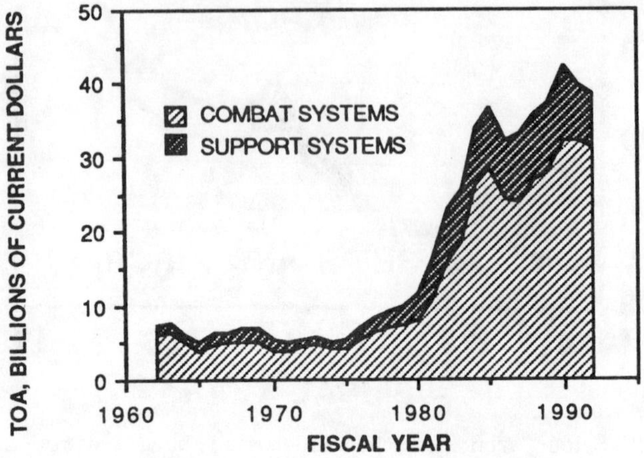

Fig. 7. Historical trends of the acquisition-budget allocations for the U.S. Air Force for combat systems and support systems. Shown is the total obligational authority for RDT&E plus procurement. There is a possible error of $1-2 billion in each of FY 1980-82, due to undetermined code-name projects.

It has often been pointed out to force planners that the command-and-control and the target-acquisition systems are getting constantly more expensive, and yet the armed forces are not spending the money necessary to acquire them in proportion to how much they are needed. The response is that these are very expensive systems, and they are being bought, as shown in in Figs. 7

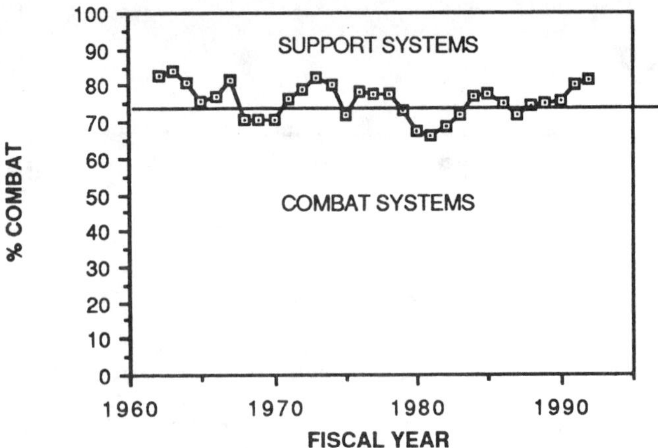

Fig. 8. Percentage distribution of the U.S. Air Force acquisition budget (RDT&E plus procurement) between combat systems and support systems. The mean percentage over this period was 73.6%.

and 9, where the absolute allocations for support systems is indeed rising. But when these rising support expenditures are analyzed as a proportion of the total budget, as in Figs. 8 and 10, then the expenditures on combat systems, on ships, airplanes and weapons, are overwhelmingly larger. For example, those blips in Fig. 9 in the early 1980s are due to the two carriers that were bought at the beginning of the Reagan administration, i.e. due to the acquisition of combat systems. The ratio of expenditures on combat and support systems is a sociological phenomenon, not a necessary and militarily-driven division of resources.

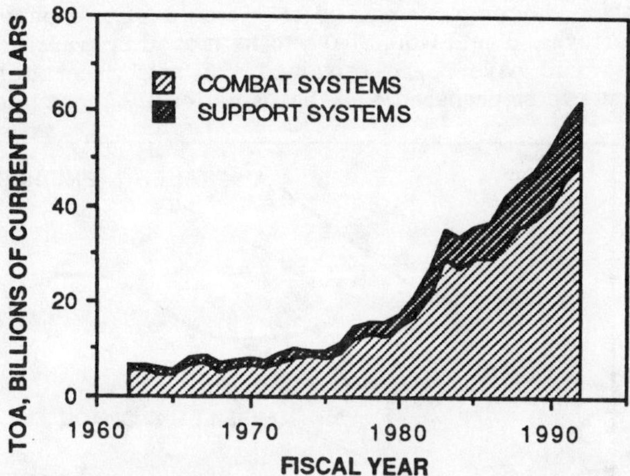

Fig. 9. Historical trend of the acquisition budget allocations of the U.S. Navy for combat systems and support systems. Shown is the total obligational authority for RDT&E plus procurements. There is a possible error of $1-2 billion in each of FY 1980-82, due to undetermined code-named projects.

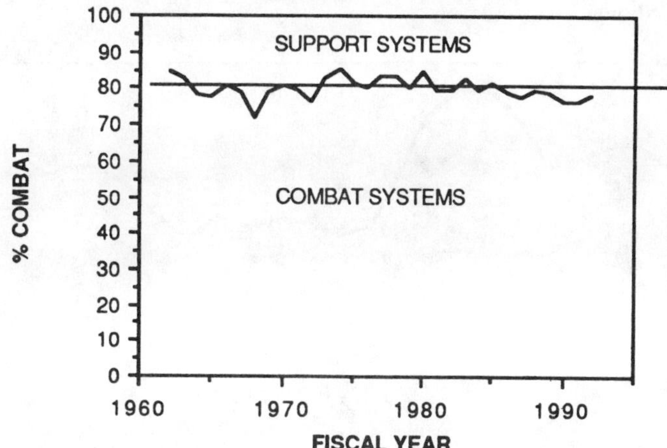

Fig. 10. Percentage distribution of the U.S. Navy acquisition budget (RDT&E plus procurement) between combat systems and support systems. The mean percentage over this period was 80.1%.

So, there is clearly a lot of conservatism to be overcome before advanced technology weapons will be bought in appropriate numbers. An additional problem is that such acquisition changes would not only involve changes in the military system, they would also result in changes in the civilian sector. This can be illustrated by means of Fig. 11, which displays employment in the aerospace industries. The employment devoted to missiles and space systems is only about 20% of that involved in producing airframes and engines. If the acquisition pattern for advanced technology weapons were ever to change on a large scale, then the employment pattern would also change. To change employment patterns, plants would have to be moved or transferred, engineers and physicists would have to be retrained. A new generation of technical people would have to be prepared, and that is not easily accomplished. The

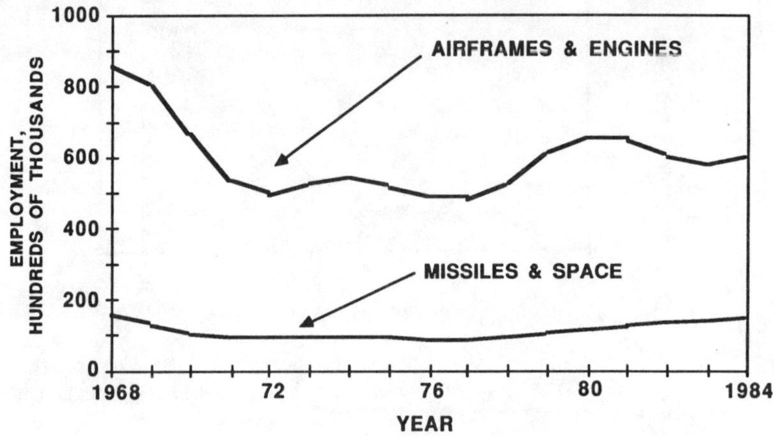

Fig. 11. Historical trends in the employment distribution for the aerospace industry. Note: employment in avionics and other categories is not shown.

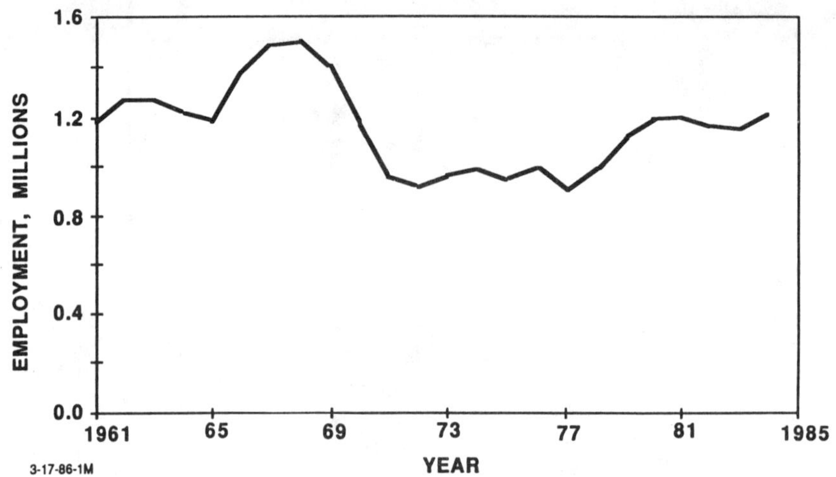

Fig. 12. Historical trends in the aerospace total employment.

conservatism reaches deep into the civilian economy. Not only do employees not like to change jobs rapidly, but furthermore, Members of Congress are concerned about what industries are located in their districts. So, advancing technology in weaponry not only challenges some tenets of military sociology, it threatens in some ways the whole national economy and social structure.

Figure 12 can be used to demonstrate how long it takes to change employment patterns. It is clear that the total employment in the aerospace industry has been surprisingly constant for many years. The curve shows the dropoff around 1970, toward the end of the Vietnam War, followed by the buildup after 1978 that was started during the Carter administration. It required about three years to accomplish the shift up to a significantly higher level. The slowness of the shift was not due to the fact that people did not want a rapid shift. In fact, a lot of the inflation in that industry, and the increasing prices that one reads about, were due to exactly this attempt to achieve a significant increase in the military expenditures. This illustrates that one cannot rapidly decide to change the way things should be done, because in military expenditures one is confronting the sociological and economic roots of society.

Actually it is not structural arguments that are made against advanced conventional weapons. The actually-presented arguments are many, and quite different. The first complaint is that the advanced technology weapons are too complicated. It can, however, be argued that "complicated" is in the eye of the beholder. Table 2 makes that point by means of a well-known, almost trite example, namely calculators. Clearly the more advanced contemporary calculator is really not more complicated; it certainly is more effective and much cheaper. Another criticism often raised about advanced technology weapons is that they are much more expensive to maintain as they get more complicated. Table 2 indicates that this is at the very least not always true.

Table 2. Practical problems with indicators of complexity, as shown by the trite example of calculators.

	Electro-Mechanical Calculator	Electronic Pocket Calculator
Size & Weights	Like a typewriter	2"x3.5"x0.25"; ~ 2 oz.
Number of Parts	6000 (electromechanical)	<100 (mech.); order of 10^5 gates
Performance	4 functions (11-place) (with variations in manipulation)	24-60 functions (depends on how counted (8-place))
Procurement Cost	$1000 in 1965	$15 in 1988
O&S Cost	$73/year in 1965 (maintenance contract)	$3.00 @ 2-4 years (batteries)

One would expect the cost of an avionics system to be roughly proportional to its complexity. The horizontal axis of Fig. 13 shows increasing complexity, if it be assumed that cost and complexity are directly related. The point made by this curve is that the mean time between failures decreases as complexity increases. However, as avionics went from the transistor generation to the integrated-circuit generation, there was a factor of five improvement in reliability. At first glance the next generation of avionics, which involves highly-integrated circuits, does not seem to have a similar improvement in reliability. But if one looks at those avionic systems, one finds that the

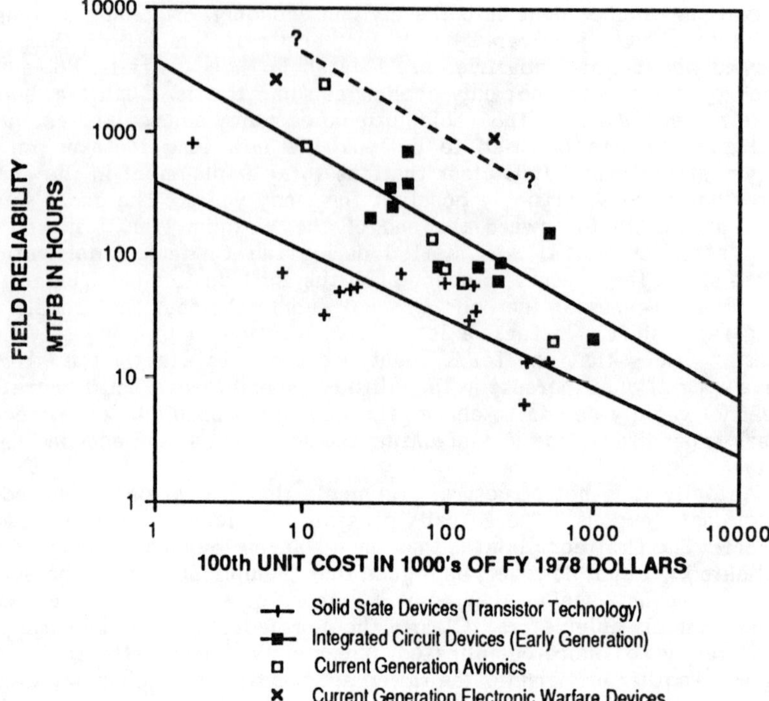

Fig. 13. Field mean-time-between-failure (MTBF) versus unit cost for selected items of avionic equipment.

mechanical parts and electronic tubes are causing the failures, not the solid-state electronics. Data on two purely solid-state warfare systems suggest that there may come to be a new plateau of reliability, which will be reached when ways are found to eliminate the mechanical failures in the current systems. Thus, complexity and unreliability do not necessarily go hand in hand.

Another argument against advanced technology weapons is that the increased expenditures for them have not led to sufficient increases in the capabilities. It is interesting to compare some new weapons systems with the old systems that they replaced, to calculate the ratio of improved performance to increased cost. In Fig. 14, systems whose performance doubled as the cost doubled would fall on the parity line; those systems that were costing more without improving their performance by an equivalent amount would lie below the parity line; and those systems that were costing more, but were gaining proportionally even more in performance, would lie above this line. Evidently, for most of the weapon systems compared in this analysis, the effectiveness improved more than the cost increased.

These data would seem to make the arguments against rapid deployment of advanced-technology weapons less than convincing. But perhaps a key part of the problem is with the ultimate question of how to pay for the more advanced-technology weapons. Indeed the acquisition budgets are very large, and the items are very expensive. Consider the costs of some typical military systems

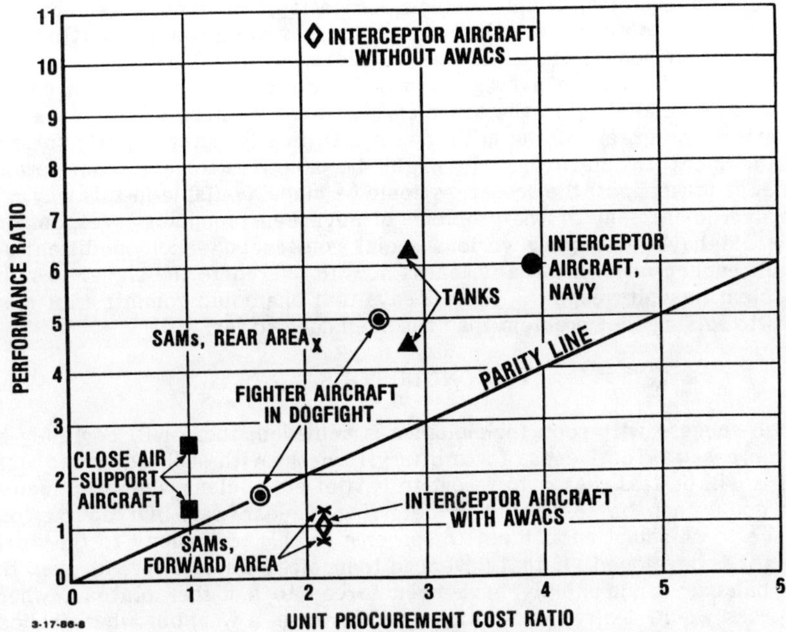

Fig. 14. Improvements in performance ratio versus increases in costs, displayed for selected advanced technology weapons.

as shown in Table 3. The way these costs are being compared here would drive "purist" cost analysts crazy, because all aspects of each weapon system acquisition have been aggregated into a single number, irrespective of how many items were bought. However, what it says in a very crude way is that the U. S. military has been willing to pay heavily for various complex weapons systems that involve advanced platforms, e.g. $43 billion for the F-14, $18 billion for the M-1 tank system. Consider now the relative cost of weapon systems and of

Table 3. Relative costs of platforms, combat support systems and guided weapons. Total program costs are given in billions of 1986 dollars.

Platforms		Guided Weapons		Combat Support Systems	
F-14	43.5	Harpoon	5.2	JTIDS	5.0
F/A-18	34.9	HARM	4.7	NAVSTAR	2.6
AV8-B	6.1	AMRAAM	9.4	KC-10A	4.4
CG-47	22.5	Tomahawk	11.0	C-17	23.7
DDG-51	15.0	Copperhead	1.6	AWACS	7.9
AH-64	9.0	Hellfire	2.3	EF-111A	2.0
M-1 Tank	18.1	JTACMS	1.9	LANTIRN	0.6
Bradley	10.3	Maverick	5.4	OTH-B	2.2
F-15	42.3				
F-16	47.8				
Average	24.9	Average	4.6	Average	6.3
				Avg. w/o C-17	3.5

combat support systems, which include navigation, electronic warfare, and airborne early warning--all considered expensive systems in the patterns of expenditure described earlier. If one neglects the C-17 transport aircraft program, one finds that an average weapon or ground-control system can be bought for a price that is on the order of 1/5 to 1/7 as much as the price of a major platform program. If the military were willing to defer slightly some of the advancement in platforms, it might be able to achieve considerable advancement in weapons; the resources could be made available in this way.

To overcome many of the problems of advanced-technology weapons one indeed would have to modify various social constructs. Such modifications might well happen once one really learns how to overcome the signal-to-noise ratio problem in such weapons. Table 3 says that one could commit a lot more resources to solving that problem than has been done so far.

VI. CONCLUSIONS

Once success with such technologies is achieved, there will certainly be greatly increased effectiveness for military forces. With such advanced weapons there will be less reason to resort to battlefield nuclear weapons; nuclear weapons could then be restricted to deterring nuclear war. Of course, one would have to be smart enough not to sell or give this technology to the world as fast as it is developed. If that advanced technology were to spread, then the military balance would simply have been forced to a higher plateau, where armed forces would destroy each other ever faster in a war, but where nobody would have gained any great military advantage over the opposition.

The situation with respect to advanced-technology weapons is that the movement has been along the path of a technological revolution, and the point of greatest difficulty, but also of the highest payoff, has now been reached. One could back away, as many people argue, including some political candidates who talk about a so-called reform movement in the Armed Forces. Or one can press ahead to complete the revolution, with all the attendant risks. Perhaps the worst choice is to argue and not reach any conclusion. That may well mean foregoing the revolution, or seeing the opponents gain it before the United States. Those are the choices and they certainly are not simple. I know the way I would make the choices.

CHAPTER 3

NUCLEAR ARMS RACE TECHNOLOGIES IN THE 1990s THE CASE OF INDIA AND PAKISTAN

Warren H. Donnelly, Congressional Research Service,
The Library of Congress, Washington, D.C. 20540

ABSTRACT

India and Pakistan continue to inch towards a capability to produce nuclear weapons, with India having the stronger industrial base of the two to supply the necessary nuclear materials. This trend challenges U.S. policy to discourage the further spread, or proliferation, of nuclear weapons. After providing background on the production of nuclear weapons materials, this paper briefly describes the nuclear industrial bases of India and Pakistan; reason for and against their acquisition of nuclear weapons, and related U.S. response; and suggests some options for action and for study.

I. DEFINING THE PROBLEM

Nuclear weapons abound by the thousands in the late 1980s. The United States and the Soviet Union have vast arsenals while the United Kingdom, France and China all have larger arsenals than did the United States for many years after 1945. India tested a nuclear explosive device in 1974 and today several other nations are suspected of having unannounced nuclear arsenals, or the ability to quickly make atom bombs. Ever since 1945, the United States has led the world in steps to prevent the further spread, or proliferation, of nuclear weapons to more countries. During the past four decades, most U.S. officials, policy analysts in think-tanks and public interest groups, and academicians have generally agreed upon the grave dangers of the further spread of nuclear weapons, although they sometimes disagree on the appropriateness of U.S. non-proliferation policy and its implementation. There is an overarching concern that if some non-nuclear weapons states were to acquire and use even a few atom bombs, their use could trigger a political chain reaction leading to nuclear war between the superpowers. Fears of widespread instant incineration have kept non-proliferation issues on government and academic agendas, even if not always as the top item.

For the 1990s, there are technological developments on the horizon which, if realized, would increase the risk of proliferation by making it easier to make or get materials needed to make atom bombs. Furthermore, some worst case analyses see strong reasons for some non-weapons states to reach for nuclear weapons, or at least to obtain some of the benefits of an ambiguous weapons status. Coupled with this is the continuing spread of improved aircraft and missiles capable of carrying nuclear weapons to distant targets.

Dealing with proliferation is one of many problems to be solved on the road to reaching the twin goals of assuring the reduction of the world's nuclear arsenals and avoiding the future use of nuclear weapons.

India and Pakistan are two states which continue to attract attention because it is widely thought in public circles, despite reassurances by their governments to the contrary, that both want some degree of military nuclear capability. This capability is likely to increase as their nuclear industrial bases gain in experence and facilities. India and Pakistan also are so tightly coupled by geography, culture, fears, tensions and competition that they cannot be considered separately.

II. THE MATERIALS FOR ATOM BOMBS

The fissionable materials for atom bombs are uranium-235 and plutonium-239. Typical amounts needed for well designed warheads are about 15 kilograms of uranium-235 or 5 kilograms of plutonium.[1]

A. Uranium-235

While uranium-235 exists in nature, it constitutes only 0.7 percent of natural uranium, the remainder is uranium-238 which will not sustain a chain reaction.[2] Although natural uranium can be used to fuel one type of reactor[3], its U-235 concentration is too small to fuel most of today's nuclear power plants, and is far too low to make nuclear explosives. Two technologies are well established today for separating uranium isotopes to produce fuel-grade materials for nuclear reactors, 3 to 4 percent U-235, or weapons grade materials, 90 percent U-235. The technology in common use diffuses uranium in gaseous form (uranium hexafluoride) through many porous barriers, with some separation occurring at each stage. This "gaseous diffusion" process is used by the United States, the Soviet Union and France.

A newer technology uses centrifugation of uranium hexafluoride. It is used commercially in England, the Netherlands and West Germany; Japan is building a commercial-scale gas centrifuge plant; and Pakistan has completed one small plant and is thought to be building another. South Africa, using a related separative process -- the jet nozzle -- has a pilot scale plant in operation and is completing a small commercial size unit.[4] A third promising technology for uranium isotope separation uses laser beams to excite atoms or molecules of uranium in ways that make the isotopes separable. The United States and other nations are far along in the development of laser isotope separation.

B. Plutonium

Aside from minor traces, plutonium is not found in nature. It is made by exposing atoms of uranium-238 to neutrons in a nuclear reactor, which transmutes some of the U-238 into various isotopes of plutonium. The irradiated uranium (spent fuel) is removed, chopped up, dissolved in acid and the plutonium and left-over uranium are separated from the intensely radioactive fission products that have come from the fissioning of some U-235 and plutonium atoms in the process. This separation process is called "reprocessing." Commercial reprocessing plants are operating in Great Britain and in France; West Germany has a pilot-scale reprocessing plant and is preparing to start construction of a larger unit; and Japan too is preparing to build one. Argentina has an almost completed reprocessing plant; India has two operating; Pakistan has one partially completed; and Israel is said to have a small one in operation. The United States and the Soviet Union have large reprocessing plants dedicated to the production of weapons-grade plutonium.[5]

The suitability of plutonium for nuclear weapons depends upon the proportion of two isotopes: Pu-239, which is the preferred isotope, and Pu-240 which, because of its high spontaneous fission rate, emits neutrons that complicate warhead design. The proportion of Pu-240 present depends upon how long uranium is irradiated with neutrons in a reactor: the longer the time, the greater the concentration of Pu-240 and the lower the quality. Plutonium produced for weapons has a purity of about six percent Pu-240, while plutonium recovered from spent fuel of nuclear power reactors, which may have been exposed for three years, can contain 20 percent Pu-240 (reactor grade).[6] Note, however, the Office of Technology Assessment cautioned in 1977 that:[7] ". . . militarily useful weapons with reliable nuclear yields in the kiloton range can be constructed with reactor-grade plutonium using low technology."

III. THE TECHNOLOGICAL SITUATION IN 1988

A. A Cautionary Note

Assessing today's nuclear weapons situation for India and Pakistan is a tricky business because there are few well established facts. Most of the information available about the ability of India and Pakistan to make nuclear weapons and about pressures and counter pressures affecting their attitude toward the nuclear option is vague, speculative, and sometimes hearsay. The world, however, does not enjoy the luxury of being able to wait until all of the facts are known before further political decisions have to be made about nuclear proliferation.

In the mid-1960s, the popular concern among non-proliferation analysts was that the industrial nations of Europe and Japan, as they moved vigorously towards the commercialization of nuclear power, would create nuclear industrial bases that could be quickly redirected to produce nuclear weapons. At that time, two decades into the nuclear age, the production of weapons materials was thought of as highly technological and therefore well beyond the capabilities of developing nations. Two decades later, the non-proliferation community now voices concern about the production of plutonium by India, largely through its own means, and of uranium-235 by Pakistan, with the help of surreptitious imports from several supplier states. This is a remarkable change in 20 years.

India and Pakistan differ substantially in their nuclear industrial bases, which will now be described.[8]

B. The Status of India's Nuclear Weapons Capability

India's test of a "peaceful nuclear explosive" in 1974 conclusively showed that it can make nuclear-fission explosives. Moreover, there are some suspicions that it may secretly have resumed design and fabrication of parts.[9] Indian Prime Minister Rajiv Gandhi said in 1985 that if India decided to become a nuclear power, it would require only a few weeks or months to do so.[10] However, during a visit to Washington in 1985, he also said that India had not continued its nuclear explosives program after the 1974 test.[11]

Canada cut off its nuclear assistance to India following the 1974 test. Given India's adamant resistance to external inspection and controls on its use of nuclear power, India has had to rely upon its own industrial base to provide much of the fuel, equipment and plants for the generation of electricity from uranium. During early Indian-Canadian cooperation, India adopted the Canadian type of heavy water moderated reactor fueled with natural uranium (the CANDU design), and after 1974 began an ambitious nuclear construction program. The subsequent cutoff of U.S. nuclear cooperation, because India would not agree to conditions of the U.S. Nuclear NonProliferation Act of 1978, added to India's desire for independence. Considering that it is a developing country, India has made astonishing progress in the design, construction and operation of nuclear power plants and of nuclear fuel factories.

India has many nuclear scientists and engineers, and many nuclear facilities. It has seven operating nuclear power plants with a total electrical generating capacity of 1,243 megawatts (MW), and six more under construction that will supply an additional 1410 MW. Another five are planned with a capacity of 1705 MW.[12] The initial overall goal for India's nuclear construction was 10,000 MW by the year 2000. However this goal has been pushed back to the year 2005 and now includes two 1,000 MW pressurized water reactors (PWRs) to be supplied by the Soviet Union.[13]

At present, India has no enrichment plants to produce weapons-grade uranium. Instead, it focuses on the production of plutonium-239, with some development work on the production of uranium-233 from thorium. Except for the Tarapur Atomic Power Plant's two light water reactors, all of India's nuclear power plants, operating and under construction, are heavy water plants. In principle, these can be operated to produce high quality plutonium. India also has six research reactors ranging in power from 1 to 100 megawatts thermal energy (MWt).[14] One of these, the 100 MWt Dhruva reactor, which began operation in August 1985, is reportedly capable of producing 25 kg of plutonium a year, enough for about three weapons annually assuming it was operated solely to produce plutonium.

One conspicuous weakness in India's nuclear base is the poor performance of its heavy-water production plants. India has five small plants to make heavy water and two more under construction. Their performance has been poor because of breakdowns and shortages of electricity.[15] This weakness has kept India under pressure to import heavy water. However, the Soviet Union, which is India's principal foreign supplier, insists that power reactors using Soviet heavy water be open to international inspection.[16] India has not published detailed information about any short-fall in its heavy water production.

As for reprocessing, India has operating plants at Trombay and Tarapur. The Trombay plant can reprocess 30 tonnes tons of spent fuel per year, the Tarapur plant 100 tonnes tons. Assuming the spent fuel contains one percent plutonium, these plants could separate yearly 0.3 tonnes and 1 tonne of plutonium respectively, if operated at full capacity. The Trombay and Tarapur plants are open to international inspection only when reprocessing spent fuel from one of India's safeguarded power reactors. India is also constructing a new reprocessing plant at Kalpakkam designed for 125 tonnes per year, which could separate up to 1.25 tonnes annually of plutonium.

Until recently, India had little, if any, plutonium that it could use in nuclear weapons without violating commitments made to either the IAEA or to the major supplier countries. However, according to Leonard Spector (senior associate at the Carnegie Endowment for International Peace), India has reprocessed spent fuel from its Madras reactors at the Tarapur reprocessing plant, and this material is free from any foreign controls on its use. The significance of this event depends upon the quality of the plutonium separated, whether it is weapons grade or reactor grade.[17] If it is weapons grade, the absence of inspection becomes more significant. Indian officials say the plutonium will be used in its breeder reactor program, although if it is of good quality, it could also be used to make nuclear explosives. As more of India's unsafeguarded power and research reactors become operational, India's supply of safeguards-free plutonium will grow substantially. According to one estimate, India could produce around 600 kg of plutonium annually from unsafeguarded plants by the early 1990s, enough for at least 60 bombs per year.[18]

C. The Status of Pakistan's Nuclear weapons Capability

Pakistan continues to inch towards an ability to make nuclear bombs, but so far appears not to have manufactured or tested them.[19]

President Reagan has certified to Congress (most recently on January 17, 1988) under Section 620E of the Foreign Assistance Act (as amended by Section 902 of P.L. 99-83) that Pakistan has no nuclear explosive devices. However, on February 16, 1987, U.S. Ambassador Hinton in Islamabad warned Pakistan about nuclear activities inconsistent with a purely peaceful program and cautioned that "it is open to question whether the President could so certify were he to conclude that Pakistan had in hand, but not assembled, all the needed components for a nuclear explosive device." Nuclear analyst Leonard Spector reported a consensus that Pakistan is at the nuclear-weapons threshold: "It either possesses all of the components needed to manufacture one or several atom bombs or else remains just short of this goal because its uranium enrichment plant at Kahuta has not yet produced a sufficient quantity of nuclear-weapons-usable highly-enriched uranium[20].

Some news accounts suggest that Pakistan may have already obtained many non-nuclear items needed to make a nuclear weapon:

--In 1982 Pakistan purchased a flash x-ray machine from Sweden, and in 1985 two employees of Pakistan's Atomic Energy Commission attempted to get training on how to operate the machine in the United States. These machines can take split-second x-ray images through solid materials that are needed to observe the behavior of an implosion device.[21]

-- On June 24, 1984, the New York Times reported allegations that China had provided Pakistan with the design of a nuclear weapon which China had exploded in one of its nuclear tests. There have also been unconfirmed reports that China allowed Pakistan to observe one of its nuclear tests.

--In July 1984, three Pakistanis were arrested for attempting to illegally export from the United States 50 high-speed electronic switches (krytrons). One use for krytrons is in nuclear weapon detonators.

--On July 11, 1985, ABC News, citing unidentified intelligence sources, reported that Pakistan had successfully tested the non-nuclear parts of a nuclear weapon trigger.

--On December 17, 1987, a Canadian, of Pakistani origin, was convicted for attempting to illegally export beryllium to Pakistan in violation of U.S. export regulations (the Pervez case).

Nonetheless, Pakistans' nuclear industrial base is limited and includes only one small nuclear power plant, the 125 MW CANDU type reactor. Pakistani efforts to solicit proposals to finance and build a large, 900 MW nuclear power plant have been unsuccessful. It continues to rely heavily upon foreign suppliers for nuclear equipment, and in some instances, is said to have obtained foreign items by subterfuge and in violation of U.S. nuclear export controls.

Pakistan's original plan to produce both plutonium and U-235 has not worked out. Construction of a reprocessing plant by France has stopped, tied up in a contract dispute over the chemical form for the produced plutonium. On the other hand, Pakistan has been more successful with uranium enrichment. Many in the non-proliferation community accept statements that Pakistan can produce weapon-grade uranium at its Kahuta enrichment plant.

Pakistan's enrichment plant at Kahuta was built with technology, equipment, and materials obtained secretly from a number of European countries and with plans and information obtained surreptitiously by a Pakistani scientist working in the Netherlands. Since Kahuta's operation is secret, there is speculation about how much enriched uranium it can produce and at what level of enrichment. If Pakistan could produce annually tens of kilograms of uranium enriched to 90 percent or more U-235, it would have a source of weapons grade material sufficient for a small nuclear arsenal. On March 27, 1986, The Economist Foreign Report said Pakistan had succeeded in enriching uranium beyond 30 percent. This report was not confirmed, and State Department officials were unwilling to comment publicly on it.

Again in November 1986, according to "authoritative sources," intelligence reports showed that Pakistan had enriched some uranium to 93.5 percent, although the report did not say how much had been produced.[22] The State Department declined to comment on these "alleged intelligence reports." More recently, on March 2, 1987, Dr. A.Q. Khan was quoted as saying Pakistan was producing weapons grade uranium. Three days later, Richard Perle, Assistant Secretary of Defense for International Security Policy, told the Senate Committee on Governmental Affairs that he had no definitive information that this was so. On January 15, 1988, President Reagan signed his determination that items involved in the Pervez case were to be used by Pakistan "in the manufacture of a nuclear explosive device."[23] Nonetheless, he waived the cutoff of U.S. economic and military aid required by the Foreign Assistance Act for countries that attempt to illegally export certain nuclear items from the United States. A more recent statement came before a joint hearing of two subcommittees of the House Committee on Foreign Affairs at which Deputy Undersecretary of State Robert Peck said:

> The Pakistan government has not modified its position that its uranium enrichment activities are strictly peaceful and that it will not enrich uranium above the 5 percent level, nor has it given any new assurances with respect to its enrichment activities. We continue to have very serious concerns about those activities.

As for plutonium, Pakistan reportedly is still trying to complete the construction of two reprocessing plants, one located at Rawalpindi and the other at Chashma. The plant at Rawalpindi would be capable of producing 10-20 kg of plutonium per year, theoretically enough for about two weapons annually. The Chashma plant is larger, with a reported output of 100 to 200 kg of plutonium per year. Construction of the latter was begun by a French company, which stopped work in 1978 under U.S. pressure. News reports indicated that Pakistan was trying in 1987 to get France to resume construction.[24] During a visit to Islamabad, the French Minister for Foreign Affairs said that once the dispute over the contract is settled, France would be ready to consider possible nuclear power cooperation with Pakistan.

D. The Nuclear Situation of India and Pakistan in the 1990s.

Looking ahead to the 1990s, if India's government continues its present level of support for nuclear power, India will have an even stronger nuclear industrial base, especially if its recent work in enrichment technology proves to be successful [25]. Moreover, India today is one of a handful of countries continuing to develop nuclear breeder technology. Success here would give India both another source of plutonium and also more reason to produce and use plutonium as a nuclear fuel. Indeed, India is pioneering in the transmuting of thorium into uranium-233, which is both a nuclear fuel and another fissile material that in theory could be used to make nuclear explosives. As suggested in Table I, by the early 1990s, India could have enough plutonium to produce a nuclear arsenal of perhaps 50 to 100 warheads. However, recent public criticisms of India's nuclear-power program could signal a loss of support that ultimately might lead to a cutback in government funding and a corresponding loss of technological momentum. On the other hand, the Indian government might decide to accelerate its nuclear program through expanded foreign assistance by agreeing to international inspections and the other requirements imposed by Canada, the United States and the Soviet Union as conditions for their nuclear assistance. That would enable India to import improved nuclear products and technologies, which would benefit both its nuclear power generation and the experience of its nuclear engineers and technicians.

Looking ahead to the 1990s, if Pakistan could operate its enrichment plant effectively and if it indeed were able to produce weapons grade uranium, then it could possibly accumulate enough material to make perhaps 15 to 50 warheads by the early 1990s. The additional main technological change into the mid-1990s would probably be improvement in Pakistan construction and operation of uranium centrifuges. However, there could be a technological surprise if Pakistan and France should reach agreement on resuming construction of Pakistan's partially-built reprocessing plant. Were that to be completed and operated to produce separated plutonium, there certainly would be increased concern whether Pakistan might decide to construct nuclear weapons with its output. [26]

IV. REASONS TO ACQUIRE NUCLEAR WEAPONS

The classic reason for wanting nuclear weapons is for protection, for a means to deter attacks by hostile neighbors. Another reason often given is that nuclear weapons appear to confer pre-

TABLE I
INDIA'S AND PAKISTAN'S CAPACITY TO PRODUCE
NUCLEAR WEAPONS MATERIALS:1988

Elements of production capacity	India	Pakistan
Uranium reserves, tonnes	46,090	?
Uranium mines	1	1
Uranium mills	1	1
Uranium conversion plants (UF_6)	0	1
Uranium enrichment plants:		
Operating	0	1
Planned/under construction	1	1
Plutonium producing reactors:		
Not open to IAEA inspection		
Research reactors (150 MWth each)	2	0
Power reactors:		
Operating (440 MWe each)	2	0
Under construction (880MWe each)		
Open to IAEA inspection		
Research reactors (5 MWth)	0	1
Power reactors (420 MWe each)	2	0
Power reactors (125 MWe each)	0	1
Reprocessing plants		
Industrial scale		
Operating	2	
Incomplete		1
Laboratory scale		1
Estimated production		
Uranium-235 (kg/yr)	0	21-63
Plutonium (kg/yr)	150	0
Stockpiles, estimated		
Uranium-235		
By 1987 (kg)		25-50
By 1992 (kg)		130-365
Plutonium		
By 1987 (kg)	100-200	
By 1992 (kg)	500-1000	

Sources:

Leonard S. Spector. Going Nuclear (Ballinger Publishing Company, Cambridge, MA., 1987), pp. 95-98.

Carnegie Task Force on Non-Proliferation. Nuclear Weapons and South Asian Security (Carnegie Endowment for International Peace, Washington, D.C., 1988), pp. 9, 11, 16.

stige upon states that have them. Both reasons, as well as others, apply to the South Asian situation. India is fearful of China, Pakistan is fearful of India, and factions in both countries see protection in nuclear arsenals.

Whether or not a nation can build a nuclear weapon now or in the future is a technical question involving the capabilities of the country's nuclear industrial base, including its engineers and scientists, and its nuclear facilities. Whether or not such a nation actually would construct a bomb is a political decision, based in part on its calculations of the costs and benefits of nuclear weapons for its security and status, and on its relations with other states. Domestic political considerations can also influence this decision.

One indication of a nation's nuclear weapons intentions can be found in its nonproliferation commitments made in treaties and public statements. In measuring this commitment, however, the direction of a nation's technical nuclear development can be as important an indicator of intention as are official statements.

A. India

Many Indians see their country as surrounded by hostile or potentially hostile neighbors and as unable to defend itself or to deter attack unless a nuclear weapons option is available. Jawaharlal Nehru, India's first Prime Minister, who presided over the start-up of India's nuclear power program, saw India as having been defeated and occupied by foreigners in the past due to its scientific and technological backwardness, especially in military technology. India's initial steps toward a nuclear weapon capability were probably a response to a perceived threat from China, although the founder of India's atomic energy program, Dr. Homi Bhabha, aspired to a nuclear weapons option for India as early as the 1950s. In 1962, China attacked and decisively defeated the Indian army in a series of major border clashes in the Himalayas. In 1964 China tested a nuclear explosive. These events led opposition and ruling parties in India to call for the development of nuclear weapons. It appears that the decision to test a nuclear explosive was not taken until around 1972, however, after the war with Pakistan.

While China's nuclear arsenal still provides a long-term reason for Indian interest in a nuclear deterrent, Pakistan's nuclear activities appear to be India's immediate security concern. Other developments in South Asia also are viewed in India as threatening to its security. These include the Islamic revolution in Iran, the Soviet military presence in Afghanistan, U.S. military assistance to Pakistan, the Iraq-Iran war, and the increased U.S. and Soviet military presence in and around the Indian Ocean. U.S. military aid to since the 1979 Soviet invasion of Afghanistan is also viewed by Indians as threatening, although the military balance of power continues to favor India. The 1985 annual report of India's defense ministry described Pakistan as India's "principal security concern."

Indian concerns over Pakistan's growing military capabilities are compounded by India's belief that Pakistan has been inciting Sikh unrest in the Punjab. In light of prospects that Pakistan might indeed soon test or produce nuclear weapons, it is widely argued by many Indians that their country should at least retain the option of producing its own nuclear weapons as a hedge against the appearance of a nuclear arsenal on the other side of their border.

Popular sentiment in favor of a nuclear capability has reportedly had a significant impact on India's nuclear program in the past, and is still a factor the government must consider. The 1962 war with China led to the first pro-bomb posture by any Indian political party (the Jana Sangh party). China's 1964 nuclear test stimulated public debate about India's defense and reinforced the pro-bomb sentiment. This sentiment was particularly strong in urban areas among the educated elite. The 1974 Indian test led to higher levels of support for its nuclear-weapons effort and increased the popularity of Indira Gandhi's government. In 1981, a public opinion poll in 15 cities showed that 70 percent of the urban residents questioned believed that India should develop nuclear weapons regardless of what Pakistan's nuclear stance or actions. A strong pro-bomb constituency remains today.

India's interests in the military applications of atomic energy extend beyond nuclear weapons. On December 5, 1987, India's Defense Minister K.C. Pant spoke of India's efforts to acquire nuclear powered submarines from the Soviet Union as part of the modernization of its forces.[27] Subsequently, in early February 1988, Prime Minister Rajiv Gandhi accepted the delivery on lease of a Soviet nuclear submarine (renamed the INS Chakra) for training purposes. While the submarine is not armed with nuclear missiles, Pakistan criticized the Soviet action as contrary to the goal of a nuclear weapons-free zone in South Asia, and described the deal as an alarming step on the road to nuclearization by India, because the submarine's nuclear reactor would not be under IAEA safeguards which do not apply to military facilities.[28] Under the Indian-Soviet arrangement, India apparently will ultimately receive four Soviet nuclear submarines. A spokesman for the Reagan Administration on February 18, 1988 described this as "an unfortunate development."

B. Pakistan

Pakistan's nuclear activities have been significantly influenced by its intense rivalry with India. Pakistan and India have fought three wars since 1947. Pakistan suffered a major defeat in the 1971 war, resulting in the creation of the separate nation of Bangladesh in territory that had been East Pakistan. Border skirmishes in the Kashmir region even occur today. When India began a nuclear program in the late 1950s, some Pakistani leaders said that Pakistan should match India's efforts. One reason Pakistan may wish at least to remain close to a nuclear weapon capability is that such a position provides a hedge against the Indian deployment of nuclear weapons. The view is widely held in Islamabad that it would be better if both Pakistan and India had nuclear weapons, than for only India to have them.

V. REASONS NOT TO ACQUIRE NUCLEAR WEAPONS

On the other hand, there are reasons not to acquire nuclear weapons, even for states that still have not taken a no-nuclear-weapons-pledge. Several of these applicable for India and Pakistan will now be described.

A. India

India has ratified the Limited Test Ban Treaty and is thereby committed not to test nuclear explosives in the atmosphere or oceans. However, India is not party to the Non-Proliferation Treaty (NPT), which commits non-nuclear weapon states to foreswear the acquisition of nuclear weapons and to accept the verification of this commitment through international inspection of all of their nuclear activities. India considers the NPT to be discriminatory because it allows the nuclear weapon states to keep their nuclear arsenals while denying them to other countries, and because the treaty requires international inspections of civil nuclear facilities on non-nuclear weapon states but not in nuclear weapons states. Moreover, from India's viewpoint, the NPT does not constrain China's nuclear forces. India remains one of the severest critics of the NPT, and has repeatedly said it will not sign the NPT unless the nuclear weapons states disarm.[29]

India has not supported the proposals for a regional nuclear weapons-free zone that Pakistan has periodically proposed at the United Nations since 1974. India, for its part, insists that China be included in such a zone. Although Pakistan has proposed mutual inspections of Indian and Pakistani nuclear facilities, India has said that such inspections would not be sufficient to verify that Pakistan is not making nuclear weapons. India is a member of the International Atomic Energy Agency (IAEA) and accepts IAEA inspection of imported, but not of domestically constructed, nuclear facilities to verify that they and the materials they produce are used only for peaceful purposes. However, only four of India's nuclear power reactors are under IAEA safeguards. India's powerful new research reactor is not safeguarded, nor are its reprocessing plants or its experimental enrichment facility, nor its four nuclear plants now under construction.

Over the years the attitudes of successive Indian prime ministers on whether India should have the bomb have altered notably, as shown below, changing from Jawaharlal Nehru's anti-bomb position to the current "open option" view of Rajiv Gandhi.

Table II.
Attitudes of Indian Prime Ministers Toward Nuclear Weapons

Prime Minister	Date of Office	Attitude Toward Nuclear Weapons
Jawaharlal Nehru	1947-64	Never
Lai Bahadur Shastri	1964-66	Not at present
Indira Gandhi	1966-77	Open option
Morarji Desai	1977-79	Against testing
Indira Gandhi	1980-84	Open option
Rajiv Gandhi	1984-	Open option

B. Pakistan

Pakistan has signed but never ratified the Limited Test Ban Treaty. It is not a party to the Non-Proliferation Treaty (NPT) and has neither pledged to foreswear acquisition of nuclear weapons nor to accept international inspection. Pakistan, however, has said it will sign the NPT if India does. Since 1974, Pakistan has proposed at the UN a nuclear-weapon free zone in South Asia. On June 16, 1987, Pakistan's foreign minister again called for a regional non-proliferation treaty.[30] Pakistan is a member of the IAEA and in 1986 was elected to its board of governors for a 2-year term. Pakistan has accepted IAEA inspection (safeguards) on its imported power and research reactors, but not for its enrichment plant at Kahuta. If Pakistan's reprocessing plants ever become operational, they will be safeguarded only when processing spent fuel from safeguarded reactors, but not otherwise. An agreement between Pakistan, France, and the IAEA to safeguard the Chashma reprocessing plant at all times has been concluded but is not yet in force. The smaller reprocessing plant, when it is completed, may also be covered under this agreement, although this is not certain. Pakistan has not agreed to safeguards for its enrichment plant. It has proposed mutual inspection of Indian and Pakistani nuclear facilities.

Pakistani officials have said many times that Pakistan is not attempting to develop nuclear weapons. For example, on October 21, 1985, President Zia was quoted in the New York Times as saying: "Pakistan has neither the resources nor the means nor the desire" to develop a nuclear weapon. In July 1984 President Zia stated: "Pakistan has the capability to enrich uranium but it would never use that capability for any purpose other than peaceful ones."[31] Dr. A. Q. Khan, head of Pakistan's enrichment program, reportedly said in May 1984 that it was "theoretically" possible for Pakistan not only to manufacture atomic bombs but also hydrogen bombs, but Pakistan's enrichment was purely for peaceful purposes.[32]

In sum, Pakistan's incentive to practice nuclear ambiguity is strong. By staying a short step away from making nuclear weapons while publicly disclaiming interest in them, Pakistan reaps several benefits. It maintains a hedge against an Indian move to manufacture a small nuclear arsenal and yet gives India no excuse to resume nuclear testing. It preserves some leverage over U.S. economic and military aid without causing a cutoff. It discourages conventional attacks from neighboring countries. And it enhances Pakistan's prestige in the Arab world.

VI. U. S. RESPONSES TO PAKISTAN AND INDIAN NUCLEAR ACTIVITIES

The aim of U.S. policy to keep India and Pakistan from developing nuclear weapons has had limited success. For example, the United States withholds cooperation in nuclear power from both countries because they have yet to agree to the U. S. statutory requirement for international inspection of all nuclear activities in non-nuclear weapons states. The Reagan Administration has warned Pakistan not to test or assemble a weapon, at least openly, because that could cause the United States to cut off economic and military aid. Pakistan is currently the fourth largest recipient of U.S. aid, after Israel, Egypt, and Turkey. Given Pakistan's current dependence on the United States for military aid, such a cutoff could have severe consequences for Pakistan's security.

Pakistan is the only country to have been cut off from U.S. economic and military aid because of its nuclear activities. In 1979 all such aid to Pakistan was terminated pursuant to Section 669 of the Foreign Assistance Act of 1961 because of Pakistan's attempts to import enrichment technology. However, the Soviet invasion of Afghanistan in December 1979 caused a shift in U.S. priorities and in December 1981, Section 669 was amended to authorize the President to waive the cutoff if he determined that to do so is in "the national interest of the United States." This was done on February 10, 1982, clearing the way for a six-year, $3.2 billion dollar assistance package to Pakistan.[33]

In August 1985 Congress further amended this legislation by Public Law 99-83, to require the President to certify to Congress that Pakistan does not possess a nuclear explosive device and that the proposed U.S. assistance will reduce significantly the risk that Pakistan will possess a nuclear explosive device before any aid to Pakistan can be furnished. The President must make this certification in each fiscal year in which aid is requested. The latest certification, for Fiscal Year 1988, was sent to Congress in mid-December 1987.[34]

Public Law 99-83 also added a new subsection to 670 of the Foreign Assistance Act (The Solarz amendment). In effect it authorizes the President to cut off U.S. economic and military aid to a country that illegally exports, or attempts to illegally export, nuclear items that would "contribute significantly" to the ability of a country to make a nuclear explosive if he determines that the items are to be so used. However, he can waive the cutoff if he determines and certifies to Congress that "termination of such assistance would be seriously prejudicial to the achievement of United States nonproliferation objectives or otherwise jeopardize the common defense and security." On January 15, 1988, President Reagan used this authority with respect to the conviction of a Pakistani, in the Pervez case, on charges of illegally exporting certain beryllium to Pakistan and waived the cutoff until April 1, 1990.[35]

VII. WHAT TO DO? OPTIONS FOR ACTION AND FOR STUDY

A. Proposals for Policy and Action

The 1970s and 1980s have produced many policies and ideas on what to do to prevent the further spread of nuclear weapons in general and particularly to India and Pakistan. These are well articulated in an extensive literature of periodicals, reports and books. Notable among these are policies and ideas actions to:

(1). Reduce reasons why India and Pakistan would want nuclear weapons. This could include reduction of tensions in South Asia, the offer of new or stronger security assurances to both states by the superpowers and by China, and deemphasis of the value of nuclear weapons through nuclear arms control and reduction.

(2). Strengthen world pressures not to acquire nuclear weapons. This could include more support for the Nuclear Non-Proliferation Treaty and steps to assure it is extended beyond 1995, negotiation of a South Asia nuclear weapons free zone, support for the world predisposition against nuclear weapons, and rewards for abstinence from nuclear weapons.

(3). Deny access to external nuclear suppliers by India and Pakistan until they agree to open all of their nuclear activities to international inspection and (ideally) abandon the development of plutonium as a nuclear fuel and the production of highly enriched uranium.

(4). Deny their access to external suppliers of missiles and aircraft capable of carrying nuclear warheads.

(5). Increase both U. S. and international sanctions against India and Pakistan if they persist in producing weapons-grade nuclear materials, or if they test nuclear explosives, or make nuclear weapons.

(6). Patiently and persistently keep diplomatic pressure on both governments to abandon and refrain from suspicious nuclear activities.

B. Subjects for Study

Even a cursory review of the literature, particularly the proceedings of congressional hearings, shows widely differing views on what U.S. non-proliferation policy should be and how it should be carried out. Such differences can be expected to continue well into the future. This provides reason for continued independent research, analysis and thinking about ways to avoid or to deal with the further spread of nuclear weapons and the means to deliver them. More analysis could be used on such questions as:

(1). Will nuclear weapons become obsolete for non-weapons states in the 1990s because of advances in missiles that can carry chemical or biological agents?

(2). What are the implications for U.S. security and world peace of more nuclear weapons states, ambiguous or overt, in the 1990s?

(3). What might be new ways to analyze the changing, dynamic balances between forces for and against nuclear weaponry in general, and in India and Pakistan in particular?

(4). What are some ways to reduce Indian-Pakistani tensions and hostilities?

(5). What are some ways for the United States, the Soviet Union and China to work together to reduce proliferation risks in South Asia?

(6). What are the limitations of technology denial as a non-proliferation measure, and ways to increase the effectiveness of present denial measures?

(7). How might Indian and Pakistani support be gained for the nonproliferation treaty and international inspection of peaceful nuclear activities?

REFERENCES

1. For a detailed recent primer on nuclear weapons, see Thomas B. Cochran, William M. Arkin and Milton M. Hoenig, Nuclear Weapons Databook. Volume I. U. S. Nuclear Forces and Capabilities. (Cambridge, MA.: Ballinger Publishing Company, 1984).

2. The nuclei of atoms of uranium-235 can absorb slow moving neutrons, which causes them to split apart, or fission, with a release of energy and also of several new neutrons. These, in turn, can cause more fissioning of U-235 atoms. This is called a chain reaction. If the process proceeds quickly in a uncontrolled manner, the enormous energy release of an atom bomb follows. If the chain reaction is controlled in a nuclear power plant, the same energy may be released over many years in the form of electricity.

3. Natural uranium can be used to fuel reactors moderated with graphite or heavy water, such as in the early power reactors in Great Britain or the Canadian heavy-water power reactors.

4. For details about the present enrichment situation in these countries, see Leonard S. Spector, Going Nuclear: The Spread of Nuclear Weapons 1986-1987 (Cambridge, MA: Ballinger Publishing Company, 1988).

5. For comparable information about the reprocessing situation in these countries, see, for example, Spector, ibid.

6. See, for example, Thomas B. Cochran, William M. Arkin and Milton M. Hoenig, op. cit. pp. 22-24.

7. U. S. Congress. Office of Technology Assessment Nuclear, Proliferation and Safeguard (New York:Praeger Publishers, 1977) p. 142.

8. For a recent detailed review of the nuclear industrial bases of Pakistan and India, see Spector, op. cit., Chapter III.

9. Spector, The New Nuclear Nations, p. 91. Also, two recently published reports indicate that India already has nuclear bombs. On March 20, 1988, UPI carried a story by Richard Sale that India had assembled a handful of low-yield atomic bombs that can be delivered by combat aircraft. More recently, a July 11, 1988 Newsweek article by Rod Norland stated that India has 12-20 nuclear devices.

10. Le Monde, June 5, 1985

11. Meet-the-Press, June 14, 1985

12. World Nuclear Industry Databook 1988, pp. 13-15.

13. Nuclear Engineering International, April 1988, pp. 6 and 7.

14. The power output of nuclear power plants is expressed in megawatts of electrical generating capacity (MW) which is roughly one-third of the heat or thermal output of their reactors -- the other two-thirds is waste heat rejected to the environment. Since research reactors do not produce power, their power output is expressed in megawatts of thermal energy produced (MWt).

15. For example, on March 13, 1988 an explosion and fire at India's Baroda heavy water plant further reduced India's heavy water production at a time when India is becoming more dependent on imports (Nucleonics Week, March 24, 1988, p. 5). Also, a report by India's

comptroller and auditor general said that the Tuticorin heavy water plant, which is the most successful of India's plants, had worked on average at only 20 percent of capacity since it was commissioned in 1978 (Nucleonics Week, May 5, 1988, p. 1).

16. India's heavy water production was further disrupted on March 18, 1988 by an explosion and fire at its Baroda heavy water plant. See Nucleonics Week, March 24, 1988, p. 5.

17. The suitability, or quality, of plutonium depends upon the ratio of the Pu-239 and Pu-240 isotopes. The higher the Pu-239 concentration, the better it is for weapons use. However, the higher the burn-up of nuclear fuel in a reactor, the greater is the concentration of Pu-240. To determine the quality of the plutonium produced in the Madras power plant would require information about the burn up of the individual fuel elements as they are discharged from its reactor.

18. Nucleonics Week, August 15, 1985, p. 1.

19. However, it should be noted that at least one recently published article claims that Pakistan has as many as four atomic weapons. Norland, Rod. Newsweek, July 11, 1988, pp. 42-43.

20. Leonard Spector. Going Nuclear (Cambridge, MA: Ballinger Publishing Company) 1987, p. 101.

21. Simon Henderson. Financial Times, August 14, 1985.

22. The Washington Post, November 4, 1986, p. A1; also November 5, 1986, p. A29.

23. Presidential Determination No. 88-5, January 15, 1988.

24. Nucleonics Week, June 4, 1987, p. 11; Nuclear News, July 1987, p. 58.

25. While heavy-water reactors can be run with natural uranium for fuel, a slight enrichment will greatly increase the working life of the fuel which brings economic advantages and reduces demands on spent fuel disposal.

26. Completion of the reprocessing plant would mean little unless Pakistan has a reactor not under international inspection that could produce spent fuel to be reprocessed. If Pakistan were to reprocess the fuel from its safeguarded Kanupp nuclear powerplant, it is already committed to have that reprocessing and the recovered plutonium open to IAEA inspection. This would make it difficult to secretly divert any of plutonium from Kanupp to weapons purposes.

27. Far Eastern Economic Review, December 24, 1987, p. 18.

28. Nucleonics Week, January 21, 1988, p. 12.

29. Prime Minister Raji Gandhi recently (June 9, 1988) addressed the third Special Session on Disarmament of the United Nation General Assembly and proposed an "Action plan" which would eliminate nuclear weapons in three stages over the next 22 years.

30. Nucleonics Week, July 2, 1987, p. 11.

31. The Pakistan Times, July 12, 1984, p. 1.

32. Lahore, Nawa-I-Waqt, February 10, 1984, p. F1.

33. Federal Register, Vol. 47, March 8, 1982, p. 9805.

34. Presidential determination no. 88-4, Federal Register, Vol. 53, January 13, 1988, p. 773.

35. Presidential determination no.88-5, January 15, 1988. In it the President said that he had:

-- determined that the items involved were to be used by Pakistan "in the manufacture of a nuclear explosive device";

-- determined and certified that not providing assistance to Pakistan would "be seriously prejudicial to the achievement of United States nonproliferation objectives and otherwise jeopardize the common defense and security";

-- determined that assistance to Pakistan was "in the national interest of the United States"; and

-- therefore waived the prohibitions of Section 669 through April 1, 1990.

CHAPTER 4

SEISMIC METHODS FOR VERIFYING TEST BAN TREATIES*

Paul G. Richards
Department of Geological Sciences, Columbia University, and
Lamont-Doherty Geological Observatory, Palisades, NY 10964

ABSTRACT

Seismic monitoring of underground nuclear explosions, for purposes of treaty verification, is a process that entails three separate steps: event *detection*, explosion *identification*, and *yield estimation*. For the 150 kiloton yield threshold treaties now in effect, the first two steps are easy. The third has been remarkably contentious and is the subject of new US – USSR negotiations, begun in November 1987, that have the potential both to resolve questions of Soviet non-compliance with current test ban treaties, and to give credibility to seismic methods of yield estimation that have relevance for verifying further restrictions on nuclear testing, for example a low yield threshold test ban.

INTRODUCTION

This paper is about a specialized part of seismology, and its relationship to a specialized issue in arms control, namely the discussion of nuclear test bans. Yet this specialized subject, the seismic monitoring of underground nuclear explosions, occasionally attracts wide attention as a key element of the debate on whether nuclear weapons systems should indefinitely be modernized, or whether modernization should be hampered by restrictions on nuclear testing.

Discussion of nuclear test bans has become contentious, not only over questions of policy. For example, on a historical issue, some say that the U.S. and the U.S.S.R. nearly concluded a Comprehensive Test Ban Treaty (CTBT) in the years 1961 to 1963 under President Kennedy, as a result of initiatives begun under President Eisenhower. But others disagree and say that a CTBT was not even close, and the real opposition to it was never seriously mobilized. As will be documented below, there have also been disagreements over technical issues, such as how the yield of an underground nuclear explosion may be estimated by seismic means. However, there appears to be general agreement on the basic issue that indeed the modernization of nuclear weapons requires a test program. In the words of Dr. Donald M. Kerr, former Director of the Los Alamos National Laboratory, "Nuclear weapon testing is ... a process intimately intertwined with the design of nuclear weapon systems."[1]

The history of nuclear testing is characterized in Figure 1 by a pair of histograms, giving the numbers of nuclear explosions carried out each year by the superpowers, beginning in 1945 with TRINITY. The explosions at Hiroshima and Nagasaki are included. The atmospheric explosions are shown plotted above the zero line, and those underground are plotted below. Note the big series of atmospheric tests in the 1950's and again in the early 1960's. The first underground nuclear explosion was named RAINIER and it occurred in 1957 at the Nevada Test Site. There is a gap for the years 1959 to 1960, associated with a moratorium on testing that began in 1958 and which ended in 1961 in a time of severe antagonism between the US and the USSR over nuclear testing issues, when the Soviets quite suddenly began a major

*Lamont-Doherty Geological Observatory Contribution Number 4334

test program of explosions in the atmosphere. Such atmospheric explosions gave us the mushroom cloud as a symbol of nuclear explosions; and of course the mushroom cloud presented a major environmental problem, the problem of radioactive fallout. But in 1963 this type of testing was banned by the Limited Test Ban Treaty. Nuclear explosions were prohibited in the atmosphere, in oceans and in space, but permitted underground.

US Nuclear Explosions

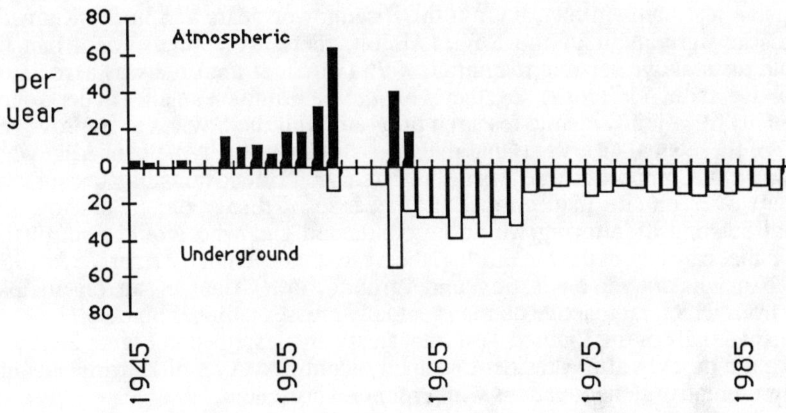

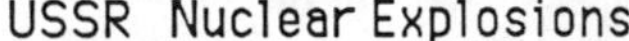

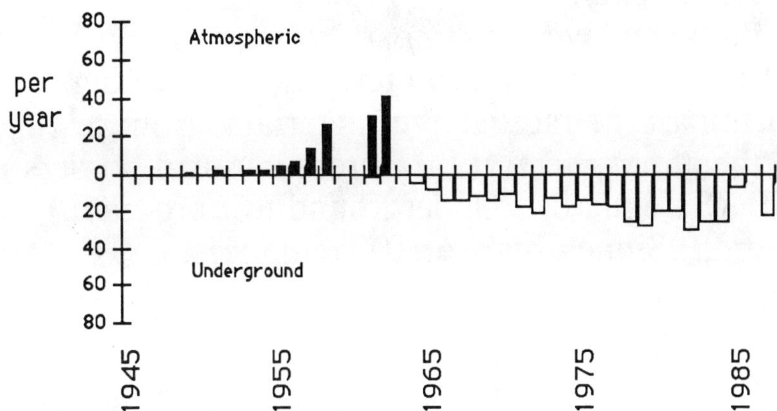

Fig. 1. Numbers of nuclear explosions, conducted by the US and USSR, to 1987.

In years subsequent to 1963, the US and the USSR have both carried out usually between 10 to around 30 underground nuclear explosions per year. In more recent years, Figure 1 shows the unilateral moratorium announced by the Soviet Union, beginning in 1985, with no tests in 1986; and then a resumption in 1987 when the Soviets executed 23 underground explosions. The numbers of tests shown in Figure 1 are derived from the Swedish Ministry of Defense, which publishes quite widely its conclusions on how many underground nuclear explosions occur each year. Of course there is testing by the United Kingdom, France and China; and India carried out one underground nuclear explosion in 1974. I shall return to some of the underlying political debate associated with these recent years, but note here that atmospheric testing was one of the leading political issues of the 1950's and '60's. And the debate which led in the end to the Limited Test Ban Treaty of 1963 was part of what at that time was a serious commitment by Presidents Eisenhower and Kennedy to conclude a comprehensive test ban. In fact, both in the Eisenhower years and in the Kennedy years, with the agreement of the Soviet Union, there were plans for a ban on underground tests above seismic magnitude 4.75 (which at the time was associated with tests above about 5 kilotons), together with a moratorium on smaller underground tests; and plans for a joint seismic research program with the Soviets to improve the monitoring of these smaller tests so that they too could eventually be banned in what would then be a Comprehensive Test Ban Treaty[2]. But, ostensibly because of disagreement over on-site inspections, and because of disagreement on how the operation of seismic monitoring would be conducted and who would control the organization that carried out the monitoring, the treaty that eventually emerged in 1963 placed no ban whatsoever on testing underground, other than a ban on nuclear explosions from which radioactive debris eventually crosses national borders.

The main details of the Limited Test Ban Treaty are described in Figure 2. It was negotiated quite quickly after President Kennedy identified Averell Harriman as his personal representative in negotiations with Premier Khruschev[3].

Banned nuclear testing in
atmosphere, outer space, and under water.
Negotiated quickly.
Not a comprehensive ban, in part because of
verification problems with underground testing.
Banned release of radioactive materials (from testing)
into the atmosphere, if they cross national borders.
Ratified. (Safeguards. Underground test program.)
Multilateral. Signed by over 100 countries.
Unlimited duration.

Fig. 2. Main points of the Limited Test Ban Treaty of 1963.

It was quickly ratified and has subsequently been signed by over 100 countries. As part of the political process of getting the advice and consent of the Senate, President Kennedy made a commitment under what are called the "safeguards" associated with this treaty, to promote a vigorous program of underground nuclear testing.

In the years which followed, the Johnson years, the debate on arms control mostly centered on the ABM treaty, and the question of non-proliferation of nuclear weapons. But attention returned to nuclear testing in the Nixon years when quite suddenly in the last few months of the Nixon administration the Threshold Test Ban

Treaty (TTBT) of 1974 was negotiated, and signed at a summit meeting in Moscow. The main points are shown below:

After March 1976, **banned underground weapons tests greater than 150 kilotons.**
Introduced a new technical issue – yield estimation.
Reagan Administration:
"There have been likely violations."
"Cannot be effectively verified."
"Need to negotiate extra verification measures."
Weapons' lab spokesmen (e.g. Dr. Batzel of Livermore):
"The Soviets appear to be observing a yield limit. Our best estimate of this yield limit is consistent with TTBT compliance."
Bilateral negotiation.
Unratified (as of July 1988), though submitted to the Senate in the Ford Administration and again in 1987.

Fig. 3. Main points and comments on the Threshold Test Ban Treaty of 1974.

This treaty introduced a new technical issue of importance in treaty monitoring, namely the question of yield estimation, because it banned underground nuclear weapons tests greater than 150 kilotons, so that in addition to the requirements of detecting these explosions, and identifying them as nuclear, it now became important to estimate how big they were. Some would say this is a fairly minor arms control treaty, but insofar as it was intended to terminate the practice of high-yield testing, it can be called successful. (Both superpowers had been carrying out weapons tests prior to 1974 at yield levels above a megaton.) Though not yet ratified, this treaty is in effect and has been at the focus of political debate on nuclear testing in recent years in the Reagan Administration, as I shall discuss later. The companion to this treaty is the Peaceful Nuclear Explosions Treaty (PNET) of 1976 which was concluded after 18 months of intensive discussion, and which has a very extensive protocol for verification. It too places a ban on underground nuclear explosions that exceed 150 kilotons.

The Carter Administration made a presidential commitment to achieve a comprehensive test ban; and, in trilateral negotiations that included the United Kingdom, agreement was in fact reached on four issues as shown here in Figure 4. But in the end, in deep political trouble over failure to ratify the SALT II Treaty, and with the problems of Soviet occupation of Afghanistan, these negotiations never reached their intended goal.[4]

The main feature of the Carter Administration's CTBT negotiations was the agreement by the superpowers to operate seismic instruments on the other side's territory. Note that to monitor treaties now in effect (LTBT, TTBT, PNET), there is basically no need to have access to "in-country" seismic data, because seismic signals for explosions of concern with respect to the 150 kiloton threshold are large enough to be picked up by seismometers all over the globe. But for a CTBT or a Low Yield

Threshold Treaty (LYTTBT), in-country data would be essential.

"Agreement" reached on:
(1) networks of seismometers deployed in the U.S.S.R. and the U.S., with data sent to International Data Centers;
(2) an unlimited number of "challenge-type" on-site inspections;
(3) a moratorium on Peaceful Nuclear Explosions;
(4) no dclay until agreement was reached with China and France.

Limited duration.
Trilateral (U.S./U.S.S.R./U.K.).
Dropped, after Afghanistan/Cuban Brigade/SALT II.

Fig. 4. Summary of Carter Administration attempts to negotiate a CTBT.

When President Reagan came into office in 1981 there was a question as to whether the trilateral negotiations on a Comprehensive Test Ban Treaty would resume. But on July 20 1982 a press briefing was given at the White House to convey the decision that these negotiations would not resume, and that there were problems with verification of the TTBT and PNET. Specifically it was stated that: "... on several occasions, seismic signals from the Soviet Union have been of sufficient magnitude to call into question Soviet compliance with the threshold of 150 kt., and it's because of the uncertainties of our yield estimation process that the United States cannot prove beyond any reasonable doubt that the Soviets have violated the Threshold Test Ban Treaty....The Soviets have always asserted, when challenged on this point, that they have not violated the agreement."

In recent years, the Reagan Administration has stated that a comprehensive treaty does remain a long term goal of the United States. But it has given a series of restrictions that govern the circumstances under which a CTBT would be favored, as listed in a March 7 1986 letter from the President to the then Majority Leader of the U.S. Senate:

> "A CTB remains a long-term goal of the U.S. However, it must be viewed in the context of achieving broad, deep and verifiable nuclear arms reductions, substantially improved verification capabilities, a greater balance in conventional forces and at a time when a nuclear deterrent is no longer as essential an element as currently for international security and stability."

From this brief summary of treaties now in effect, and recent contrasting positions of the last two U.S. administrations, let me turn to an introduction to the technical issues that arise in seismic monitoring.

The first line of Figure 5 lists three main components of seismic monitoring of underground nuclear explosions. The first is that of *detection*, and I would include location under this heading. Can we detect seismic signals from underground nuclear testing, and can we use them to locate the source of the signal? The second major component of a monitoring program is *identification*, sometimes called discrimination, which is the question of whether or not nuclear explosions, once detected, can be distinguished from earthquakes and whether too they can be distinguished from chemical explosions. And the third major issue identified here is that of *yield estimation.*

Detection, Identification, Yield Estimation

Source theory (earthquakes, explosions)
Propagation theory (different seismic waves, such as P, S, Rayleigh, Love, Lg, Pn, Sn, P-coda)
Soviet geology (distribution of dry alluvium, and salt)
Seismicity (where do earthquakes occur, and how often)
Earth noise, quarry blasts
Instrumentation, array design, data management
Network assessment
Decoupling (full and partial, theory and practice)

Fig. 5. Three aspects of seismic monitoring, and a list of technical elements.

Somewhat different conclusions can be reached on these three main components of seismic monitoring, as to whether our capabilities are good or poor. For example, I would say that detection capability in general is very good. Some examples to illustrate this capability are given in the next section. Our identification capability is also very good for events above a certain size, including those that raise questions of compliance with the 150 kiloton limit of the Threshold Test Ban Treaty (TTBT); is good even for events ten times smaller (i.e. down to 15 kt); and is adequate in most but not all situations down to around 1 kt. But there is lack of agreement on how well we can now identify nuclear explosions at very low magnitude, and whether we could, with access to seismic data from the Soviet Union, identify all underground nuclear explosions down to 1 kiloton. On the question of yield estimation, there is a series of debates, first on the question of whether yield estimation can be done accurately for purposes of monitoring the main treaty now in force, that is the TTBT of 1974; and then separately on the question of monitoring prospective new treaties that would allow testing only at low yields, with a threshold set perhaps at around 10 kilotons or maybe even set at one kiloton. As part of the debate on how well one would be able to estimate yield when yield is low, there is a need for discussion of what restraints on allowed testing environment might be negotiated, that would address the requirement of the monitoring side to be assured that a new low-yield threshold was not being exceeded.

Associated with each of these three main technical issues there is a set of questions about what can be done to evade or "spoof" the side doing the monitoring. Thus, one asks how well one could expect to hide a nuclear explosion in various environments, and again this debate has to be developed in terms of thresholds. What size nuclear explosion might go undetected, what size unidentified, and at what size can yield estimation be spoofed? Nuclear explosions can be conducted even down to such small sizes as pounds or even a few grams of TNT equivalent. Associated with these broad categories of technical questions are a series of subdisciplines as listed in the remainder of Figure 5. The most important evasion scenarios are those based on a limited ability to "decouple" the nuclear explosion by carrying out the test in an underground cavity.

In sections that follow I introduce some basic terminology and give examples of detection capability; describe briefly the recent progress and current problems of identification capability; and then at greater length review methods for estimating yield. Yield estimation is emphasized, because this is often stated to be a sticking point in current political debates on nuclear testing, and also because I believe that a correct description of capabilities to estimate yield, at low yield, may be the chief contribution the technical community can make to informed debate on options for new limitations on nuclear testing. The final section covers developments from 1985 to July 1988.

BASIC TERMINOLOGY, AND DETECTION CAPABILITY

Four different types of seismic waves are described in this review, namely

- body waves (specifically the P-wave),
- surface waves (specifically the Rayleigh wave),
- regional waves (specifically the Lg wave), and
- the P-wave coda.

Each of these may be made the basis of a seismic method for yield estimation. P-waves and Lg-waves are important for assessing detection capability.

As for any specialized technical field, jargon is pervasive. Thus, a **body wave**

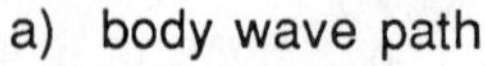

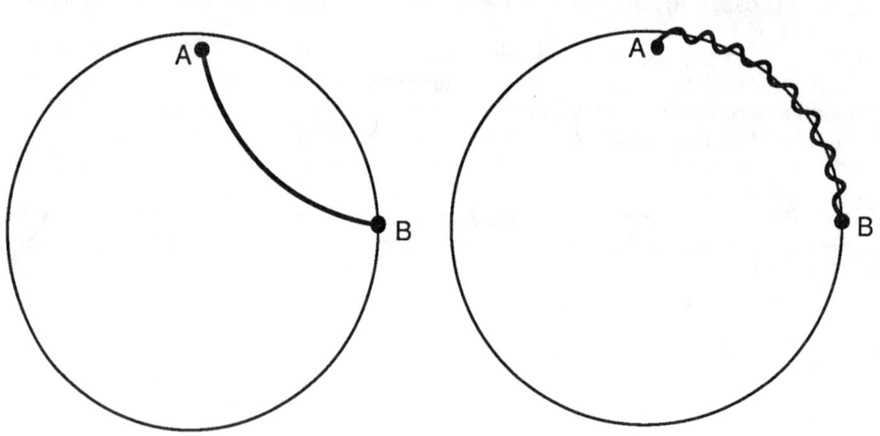

Fig. 6. a) This shows the path of a P-wave through the Earth, from a source at point A to a receiver at point B. This is a type of body wave, meaning that it travels through the Earth's deep interior. P-waves are a type of sound wave, traveling through solid rock. An explosion at A may be recorded on a seismometer at B.

b) This is a schematic for surface waves. A Rayleigh wave is shown, going from a source at point A to a receiver at point B. This wave travels around the Earth's surface, and is similar to the set of ripples spreading away from a disturbance, over the surface of a pond.

is a seismic wave that propagates through the body of the spherical Earth (see Figure 6a). A **surface wave** travels a completely different path (Figure 6b).

P-waves and Rayleigh waves (a type of surface wave discovered by Lord Rayleigh in 1887) have long been used to estimate the size of seismic sources. Discussion of the other two types of waves (Lg, P-coda), used to estimate explosion yield, is developing as a research subject.

The main Soviet sites for nuclear weapons testing are at Novaya Zemlya (an island north of the Eurasian land mass), and in Eastern Kazakhstan near the town of Semipalatinsk. The Eastern Kazakhstan site includes both Degelen and Shagan River as separate test sites. All explosions of concern with respect to the 150 kiloton Threshold Test Ban Treaty (TTBT) have occurred since 1976 at Shagan River.

Fig. 7. Location of the main Soviet nuclear weapons test sites, in Novaya Zemlya and Eastern Kazakhstan.[5]

A **seismogram** is a record of ground motion at a particular location. Here in Figure 8 is a typical seismogram showing a P-wave (which is always the first wave to arrive, traveling faster than, for example, surface waves):

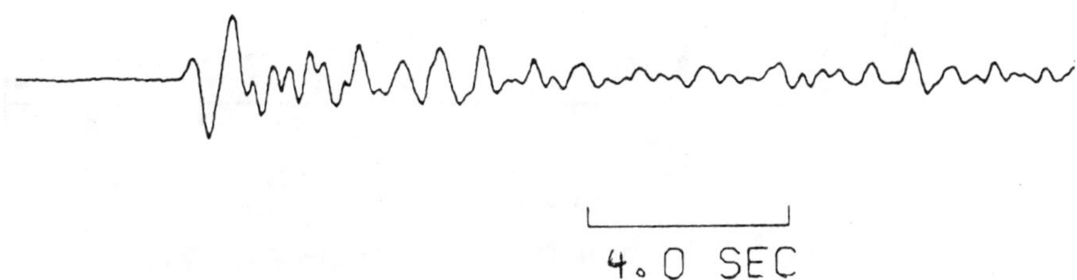

Fig. 8. A teleseismic P-wave.

This shows the vertical motion of the ground, recorded in Scotland (July 4, 1976) due to a Shagan River explosion. Note the time-scale, indicating that the period of the P-wave is roughly one second. A seismologist would call this a **short-period** wave. It is a "teleseismic" signal, in that the recording is made at substantial distance (more than a thousand kilometers) from the source. Note that the first motion of the ground is always upward, for a well-recorded explosion. In order to obtain the seismic magnitude, one measures the amplitude (A) of the maximum ground motion in the first few seconds, and also the period (T) of this motion. This leads into the concept of **body wave magnitude**, symbolized as m_b, for which one uses the logarithm of ground motion because of the wide range of sizes of observed seismic signals.

Definition of body wave magnitude, m_b

$$m_b = \log(A/T) + \text{distance correction}$$

where:

A is the maximum trace amplitude, read off a seismogram at a particular distance and converted to ground motion by allowing for the instrumental gain; and

T is the period (often around 1 second) of the wave at maximum trace amplitude.

The distance correction makes allowance for the distance between the receiver and the source of the seismic waves. For example, this is the distance between Scotland and Shagan River in the seismogram above. Use of m_b began in 1935, to study earthquakes. There is now 50 years of experience in working with this magnitude scale. The scale is directly related to signal levels. Thus, it is phenomenological – not a quantity with any fundamental physical interpretation.

Figure 9 illustrates how the amplitude of the P-wave falls off at great distances between source and receiver.

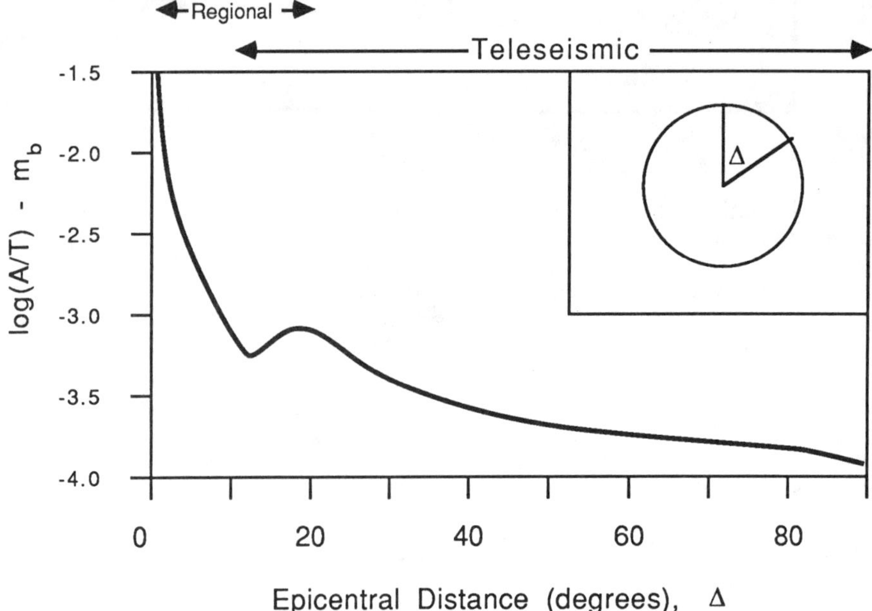

Fig. 9. Decay of P-wave amplitude with distance. A is measured in nanometers.

This is an important curve, showing several basic properties observed for P-waves. The ordinate, "epicentral distance," is conventionally taken as the angle (Δ) subtended by the source and the receiver at the center of the Earth (see inset in Figure 9). Thus, one degree (1°) represents about 110 kilometers measured over the Earth's surface. The abscissa, $\log(A/T) - m_b$, with A and T as noted above in the definition of seismic magnitude, varies little over the teleseismic distance range from about 30° to 85°. If A is measured as 0.5 nanometer at 30°, with T as 0.5 second, Figure 9 indicates m_b would be about 3.5. For a well designed seismic station located in a quiet site, such a signal would be above the "noise" of background Earth motions. The irregularity in the curve at around 20° is a result of focussing of rays arriving at this distance range, due to a region within the Earth's mantle at depths from about 350 to 700 kilometers in which the velocity of P-wave propagation increases anomalously with depth. The range from 0° up to about 20°, known as the "regional" distance range (i.e. distances of up to roughly 2000 kilometers) shows relatively strong amplitudes – and, amplitudes that change substantially with distance – and thus indicates the advantages in terms of stronger signal that "in-country" monitoring would have as compared to typical teleseismic monitoring (i.e, for distances over about 1000 kilometers).

In practice, the teleseismic P-wave signal from a one kiloton (1 kt) underground nuclear explosion corresponds roughly to about $m_b = 4$. Since signal strength is roughly proportional to yield and the magnitude scale is logarithmic, it follows that 100 kt has signal strength roughly about $m_b = 6$, and that factors of 2 in yield (up or down) correspond to about 0.3 magnitude units (plus or minus).

In practice the relationship between magnitude and yield is complicated by naturally occurring variability in both the efficiency with which nuclear yield couples into seismic signal at the shot point, and the efficiency with which seismic signal propagates to teleseismic distances. The degree of coupling depends on rock type at the shot point, on the amount of water saturation there, and on whether the explosive device is "tamped" (i.e., well packed in the emplacement hole), or whether it is placed in some type of cavity. The efficiency of propagation to teleseismic distances is thought to depend basically on temperatures of the solid material in the upper layers of the Earth within which the seismic wave propagates for part of its path – with resulting effects on wave attenuation that may be estimated for different propagation paths, using a wide variety of seismic data that may be quite separate from the data on the explosion under study. We shall return to this subject below.

It should be noted that in monitoring for small explosions under a new test ban regime, reliance would be placed not so much on teleseismic signals, propagating as illustrated in Figure 6a above, but on so-called regional waves (i.e. as recorded at distances less than about one thousand kilometers between source and receiver). There are several such seismic waves, guided by layering in the crust and upper mantle, and though for a given yield they are stronger than teleseismic waves, they are significantly harder to interpret because crustal and upper mantle properties vary from region to region. Thus, measures of regional wave amplitude often show much more scatter than is the case for teleseismic observations. This is important both for assessing detection and identification capabilities using regional waves, and for purposes of yield estimation.

Some examples of detection capability

The map in Figure 10 is for a part of Asia and Europe, showing the location in Norway of one of the most sophisticated sets of seismic monitoring instruments in the world. It is just north of Oslo, and is known as NORSAR (which is an abbreviation for "Norwegian Seismic Array").

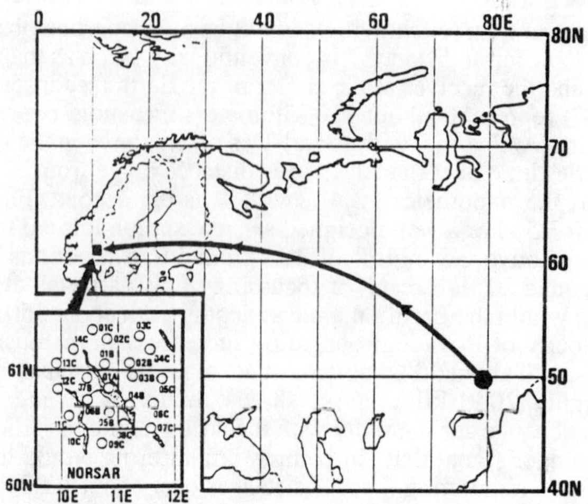

Fig. 10. Location of NORSAR, and (inset) map of stations[6].

Shown too, is the great circle path for seismic signals to NORSAR from the Soviet Union's main nuclear test site in Kazakhstan, as well as an inset showing how NORSAR was, as originally configured, spread out at 22 locations in an area about 200 kilometers across. (Only seven sites are now in operation.) Figure 11 is that inset map view again, but showing the seismograms of a Soviet nuclear explosion at each of the 22 original locations within NORSAR. These are recordings of the sound wave (the "P-wave"), that has propagated in solid rock through the Earth's deep interior to get from Kazakhstan to Norway in about seven minutes:

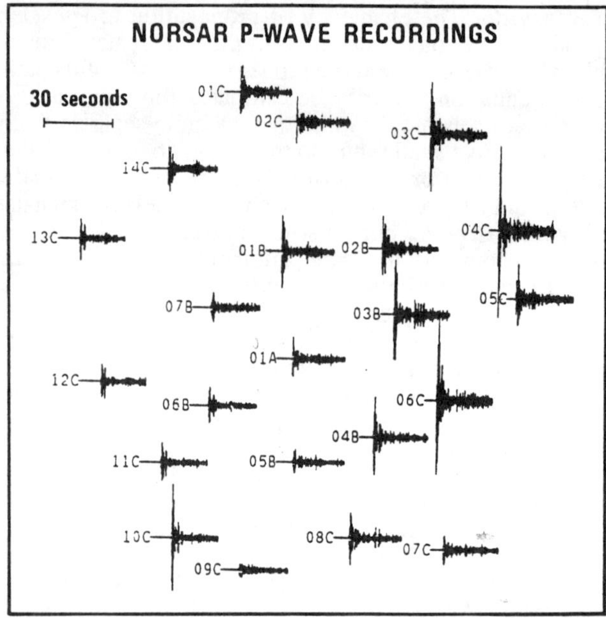

Fig. 11. Map view of NORSAR-recorded P-waves from a Soviet nuclear explosion[6].

One can see that at different sites the signal comes in with somewhat different strengths. At site 06C, it is a relatively large signal. At site 07B it is quite small. This variability points to a problem down the road if one intends to use seismograms to do yield estimation, but the present discussion concerns detection capability and Figure 12 gives a display of the excellent signal-to-noise characteristics of the NORSAR array:

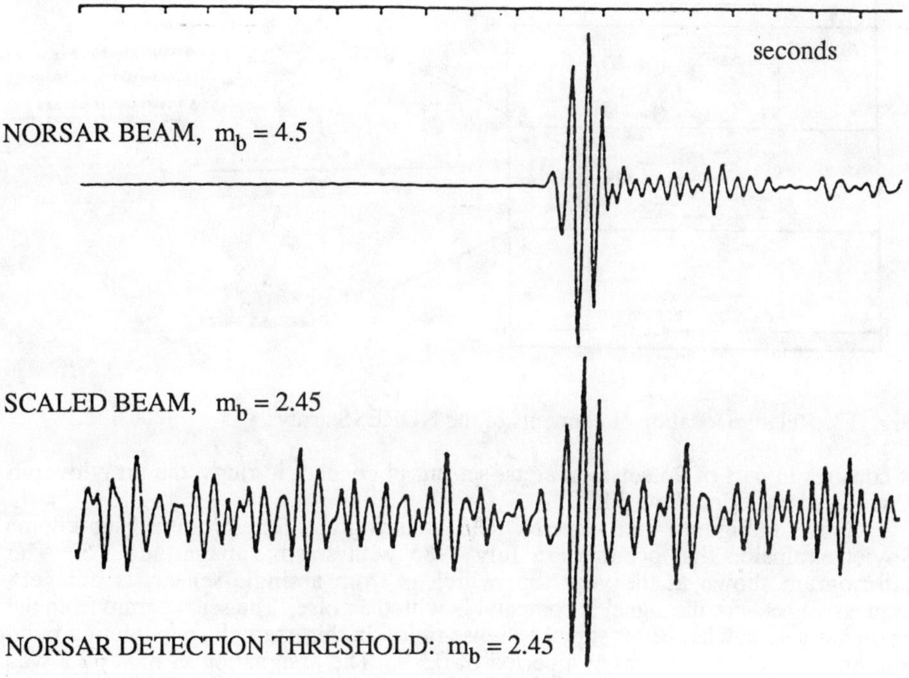

Fig. 12. Illustration of NORSAR detection threshold[6].

Shown here in the upper part of the picture is the outcome of adding up (or "stacking") all the NORSAR seismograms with an appropriate delay time, to bring the signals from each instrument coherently into the sum (or "beam"), and one can see that noise levels prior to the signal arrival are indeed very low. The seismogram appears as virtually a straight line prior to the signal arrival. This signal-to-noise comparison is for an actual Soviet nuclear explosion about 4000 kilometers away, with seismic magnitude 4.5, which probably corresponds to a yield somewhat less than 5 kilotons. The excellence of this signal-to-noise ratio makes it is natural to ask, of the display in the upper part of the Figure, how small might this signal be, before it becomes indistinguishable against a background of Earth noise. This is answered in the lower part of the picture by scaling up the noise rather than by scaling down the signal, making a seismogram in which both signal and noise are of the same order. It turns out that the ratio of signal to noise can degrade by a factor of about a hundred before the automatic detectors associated with this NORSAR instrumentation would run into any problem in detecting the signal. It follows that NORSAR can detect signals down to around m_b 2.5 in Kazakhstan (i.e., m_b 4.5, less a factor of 100 on a logarithmic scale), which corresponds to a very small explosion, perhaps just a few hundredths of a kiloton, even as far away from Soviet Central Asia as Norway!

Looking again at this same array, recall from Figure 11 that site 06C was for some reason quite favorable for picking up seismic signals from Soviet explosions.

Since 1985, a special small array known by another acronym, NORESS, has been deployed at this site:

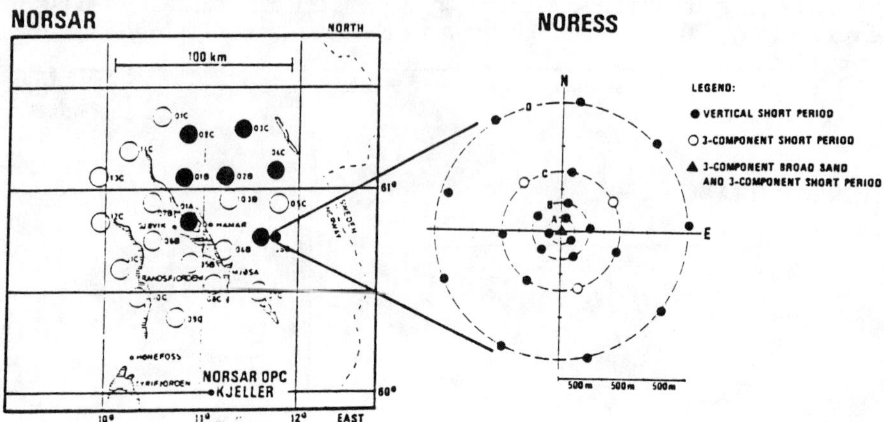

Fig. 13. Relative location of elements of the NORESS array[6].

It consists in part of 25 seismometers, set out in concentric rings, the array overall having a diameter of about 3 km.

Shown in Figure 14 are some NORESS seismograms for a presumed sub-kiloton Soviet explosion that occurred in July 1985 with seismic magnitude 3.5. The seismogram shown at the very top, which is from a single sensor, is not very impressive because the signal is comparable with the noise. The seismogram from the beam has a somewhat better signal-to-noise ratio. In this example, maximum ground motion was 1.7 nanometers at a period of 0.5 s. The assignation of m_b as 3.5 was made after making a station magnitude correction, to allow for the difference between NORESS signal strengths and those of a typical seismic station. Shown in the middle of Figure 14 is the spectrum for the beam, both of the signal and of the noise. There is a favorable signal-to-noise ratio from a little less than two Hertz up to about 8 Hertz, and if the signal is band-passed in this frequency range one gets the seismogram shown at the bottom. The outcome is a quite satisfactory signal, having about a 30 to 1 signal-to-noise ratio even though the ground motion at NORESS was less than 2 nanometers. Thus, for an actual example of a small explosion, namely a sub-kiloton explosion that occurs about 4,000 kilometers away, we can say there is an excellent detection capability. With in-country seismic stations in the Soviet Union such as were agreed to in the 1970's, detection capability would even further be enhanced.

IDENTIFICATION CAPABILITY

Moving on to the question of identification capability, we ask how well can one discriminate between seismic signals, to know whether they come from an earthquake; or, for an explosion, whether it was a chemical or nuclear explosion.

In the case of a located seismic event for which signals are large – that is, larger than could be ascribed to a chemical explosion – the only candidates are an earthquake or a nuclear explosion. Identification in this case is generally routine, and is based on an appreciation of the physical difference between earthquakes and underground nuclear explosions.

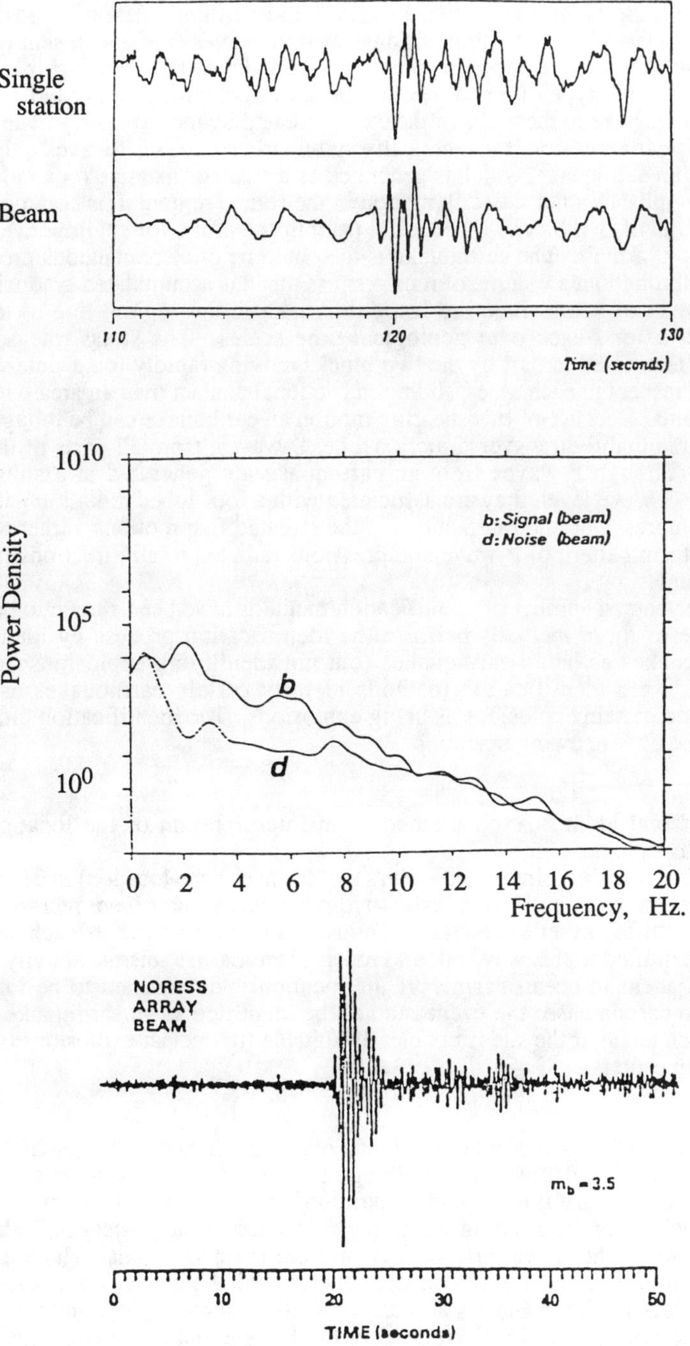

Fig. 14. NORESS signals for a sub-kiloton nuclear explosion in Kazakhstan[6,7].

Thus, from the monitoring perspective, an underground nuclear explosion is a highly concentrated source of compressional seismic waves (P-waves), sent out with approximately the same strength in all directions from the point of emplacement of the nuclear device. This type of signal occurs, because the explosion products apply a fairly uniform pressure to the walls of the cavity created by the explosion. The simple model of a nuclear explosion is a spherically symmetric source of P-waves only. This contrasts with an earthquake, which is generated as a result of massive rock failure that typically accomplishes a net shearing motion in the source region. It is common here[8] to think in terms of two blocks of material (within the crust, for shallow events) on opposite sides of a fault. The earthquake is the outcome of a spontaneous process of stress release throughout a volume of rock – stress that has accumulated gradually over tens or hundreds of years since the last earthquake in the region, due to tectonic processes that extend back over geological time scales. The stress release of an earthquake is expressed in part by the two blocks moving rapidly (on a time scale of seconds) with respect to each other, sliding in frictional contact over an area within the plane of the fault. Because of this shearing motion an earthquake can be thought of as radiating predominantly transverse motions, i.e. S-waves, from all parts of the fault that rupture. Though P-waves from an earthquake are generated at a substantial fraction of the S-wave level, they are associated with a four-lobed radiation pattern of alternating compressions and rarefactions in the radiated first motions, rather than the relatively uniform pattern of P-wave compressions radiated in all directions from an explosion source.

Over the years, a number of identification methods have been shown to be fairly robust. Some of these methods perform the identification process by identifying certain earthquakes as being earthquakes (but not identifying explosions as being explosions). Other identification methods identify certain earthquakes as being earthquakes and certain explosions as being explosions. The identification process is therefore a type of winnowing operation.

Location

The principal identification method is an interpretation of the location of a detected seismic source.

If the epicenter (the point on the Earth's surface above the location) is determined to be in an oceanic area, but no explosion hydroacoustic signals were recorded, then the event is identified as an earthquake. Thousands of seismic events each year can routinely be identified in this way, since so much of the Earth's seismic activity occurs beneath or adjacent to ocean basins. If the location is determined to be in a land region, then in certain cases the event can still be identified as an earthquake on the basis of location alone, if the site is not clearly suitable for nuclear explosions (such as near population centers).

Depth

With the exception of epicenter interpretation, interpretation of the seismic source depth is probably the most useful identification technique. A seismic event can be identified as an earthquake if its depth is determined to be well below 15 km.

The procedure for determining source depth is a part of the process of finding the event location using the arrival time of four or more P-wave signals. There are also certain body-wave signals caused by seismic energy which has traveled upward from the source, reflected off the Earth's surface above the source region, and which has then traveled down into the Earth, following the P-wave out to great distances. A depth estimate can be obtained by measuring the time-difference between the first arriving P-wave energy and the arrival time of these reflected signals. The analysis of broadband data through waveform modeling is particularly useful for detecting these reflected signals. Empirical methods can also be used, based on comparison with previously interpreted seismic events in the same general region as the event under study.

An advantage of depth as an identification method is that it is not dependent on magnitude – it will work for small events as well as large ones, provided the basic data is of adequate signal quality. It will not alone, however, distinguish between chemical and nuclear explosions. The further discrimination between types of explosion, chemical and nuclear, may require other than seismic data – for example, a photograph of ground surface deformation caused by an industrial chemical explosion in the vicinity of a suspicious seismic epicenter.

M_s: m_b

Underground nuclear explosions generate signals which tend to have surface wave magnitudes and body wave magnitudes that differ systematically from earthquake signals. We have already described the way the body wave magnitude, m_b, is assigned. For surface waves, the **surface wave magnitude, M_s** is defined by

$$M_s = \log (A/T) + \text{distance correction}$$

where, as for body waves, A is the amplitude of ground motion and T is the period at which this motion takes place. The main differences from the m_b scale stem from the two ways in which surface waves appear very different from body waves in a seismogram:

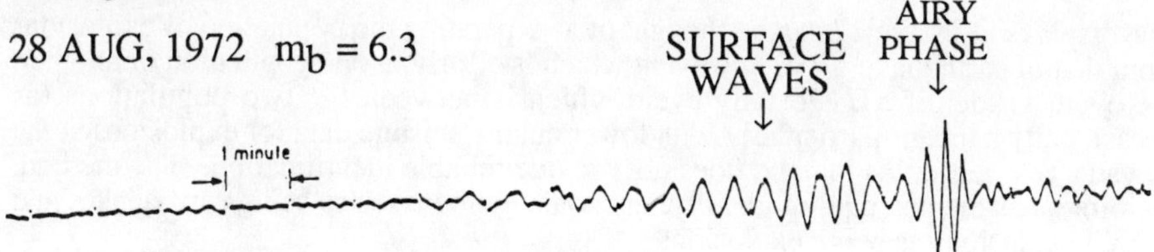

Fig. 15. Example of a surface wave, teleseismically recorded at Eilat, Israel, from an underground nuclear explosion at Novaya Zemlya.

Thus, surface waves are **dispersed**, meaning that different frequency components travel at different speeds. The effect is to make a surface wave spread out in time over several minutes. The second way in which surface waves differ from body waves is that they are a **long-period** signal. The period, T, is typically in the range 18 to 20 seconds, for the part of the well-dispersed wave train where amplitudes are at their maximum. Contrast this seismogram with that for a teleseismic body wave in Figure 8.

In measuring the maximum surface wave amplitude, it is important to avoid the "Airy phase", a portion of the wave train appearing late in the record, and corresponding to a poorly-dispersed part of the overall signal in which a range of frequencies (typically, with periods shorter than 20 seconds) all arrive at about the same time.

The fact that M_s as compared to m_b is different for explosions and earthquakes, is basically an expression of the efficiency of explosions in exciting body waves (short-period), and of the efficiency of earthquakes in exciting surface waves (long-period). The phenomenon is directly apparent in original seismograms (Figure 16), and examples are shown in terms of an M_s: m_b diagram in Figure 17 (see ref. 9 also).

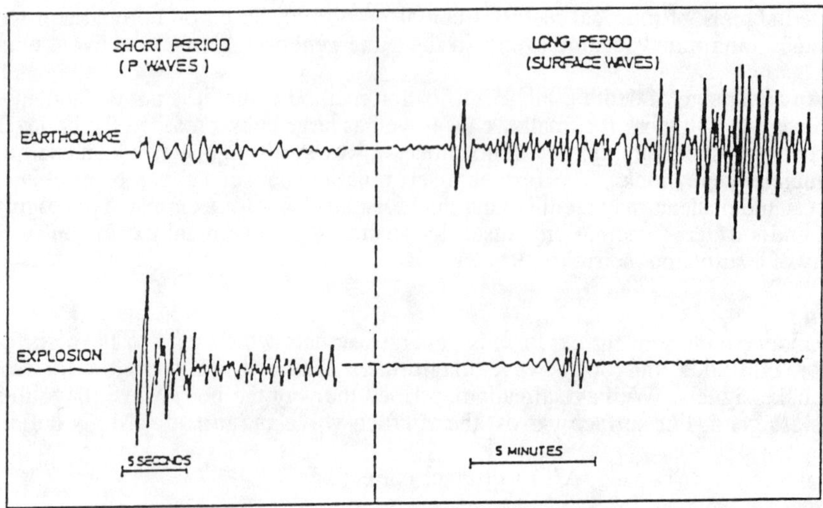

Fig. 16. Short period and long period seismograms, showing the different efficiency of excitation of P-waves and surface waves by earthquakes and explosions. (Note the use of different timescales, to show these different waves.)

The displays in Figure 17 can be thought of as separating the population of explosions from that of earthquakes. For any event which is clearly in one population or another, the event is identified. For any event which is between the two populations (as occasionally happens particularly with lower quality seismic data for explosions at the Nevada Test Site), this method does not provide reliable identification. The method, therefore, has the potential of identifying certain earthquakes as being earthquakes and certain explosions as being explosions.

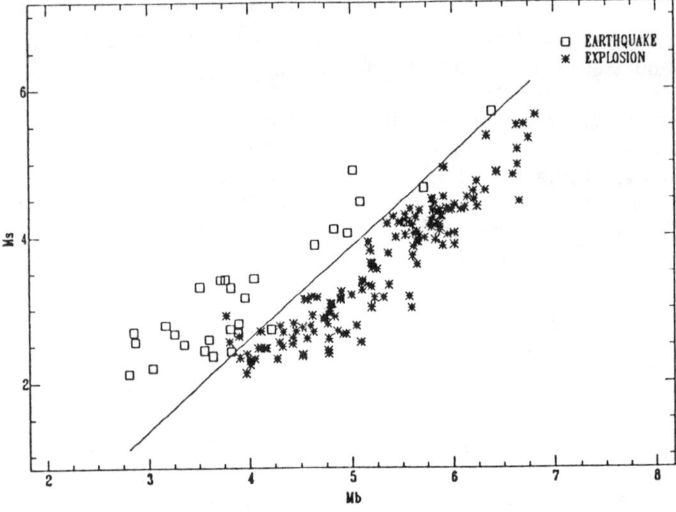

Fig. 17. Example of the M_s: m_b method of discrimination between earthquakes and explosions[10], using the station Elko in the Western U.S.

To use this identification method, both m_b and M_s values are required for the event. This is no problem for the larger events, but for smaller events (below 4.5 m_b) it can be very difficult to detect long-period surface waves using external (i.e., teleseismic) stations alone. This difficulty is present particularly for explosions since they excite surface waves less efficiently, with the result that the M_s: m_b method intrinsically works better for identifying small earthquakes than it does for identifying small explosions. Internal stations would provide important additional capability for obtaining M_s values of events down to m_b 3.0 and perhaps below.

It is known that M_s: m_b diagrams, with their separate populations as shown in Figure 17, tend in detail to exhibit somewhat different properties for sources occuring in different geophysical provinces. That is, the best identification capabilities are obtained, for a particular event of interest, if there is already a data base and an associated M_s: m_b diagram tailored for the general region of the Earth in which that event occurred.

The M_s: m_b method has proven to be the most robust identification technique available for shallow events. For events below m_b 4.0, the separation between the two M_s: m_b populations is found to be less in some studies. In the opinion of many seismologists, however, a useful separation (at low magnitudes) is still likely to be present if good quality data for such small events can be found. Again, in the context of monitoring for small underground explosions in the U.S.S.R., this would require in-country stations and empirical confirmation. (Empirical confirmation that the method is effective at low magnitudes could also possibly be judged from Soviet seismic data not currently available to the U.S.)

Spectral Methods

The basis of the success of M_s: m_b diagrams is that the ratio of low frequency waves to high frequency waves is typically different for earthquakes and explosions. As a way of exploiting such differences, M_s: m_b diagrams are quite crude compared to methods that use a more complete characterization of the frequency content of seismic signals. Thus, M_s is a measure of signal strength at around 20 sec (frequency 0.05 Hz), and m_b is a measure of signal strength at around 1 sec (frequency 1 Hz).

What then can be learned from analysis of earthquake and explosion signals across the whole spectrum of frequencies that they contain? This is the question that is driving much of the current R&D efforts in seismic monitoring.

Because M_s: m_b diagrams work so well when surface waves are large enough to be measured, the main contribution required of more sophisticated analysis is that it makes better use of the information contained in short-period P-wave signals. This offers the prospect of improved identification capabilities just where they are most needed, namely for smaller events. Thus, instead of boiling down the P-wave signals at all stations simply to a network m_b value, one can measure the amplitude at a variety of frequencies and seek a discriminant which works on systematic differences in the way earthquake and explosion signals vary with frequency. One such procedure is the variable frequency magnitude (VFM) method, in which the short-period body wave signal strength is measured from seismograms filtered to pass energy at two different frequencies, f_1 and f_2. Discrimination is based on a comparison of $m_b(f_1)$ and $m_b(f_2)$ that in many cases shows a clear separation of earthquake and explosion populations[11].

An analogy for the VFM discriminant would be a person's ability to tell by ear the difference between a choir with loud sopranos and weak contraltos, and a choir with loud contraltos and weak sopranos. The conventional M_s: m_b discriminant is like comparing basses and contraltos. For small events, VFM is complementary to M_s: m_b in several ways. Thus, VFM is improved by interference of the main P-wave with waves reflected from the surface above the explosion, while M_s: m_b is degraded;

VFM is more effective for hard rock explosions, while M_s: m_b will preferentially discriminate explosions in low-velocity materials; and VFM is insensitive to fault orientation[11]. VFM is more sensitive to noise and regional attenuation differences than M_s: m_b, but the main point is that simultaneous use of both methods may be expected to allow improved discrimination of small events.

High-Frequency Signals

The utility of high frequency signals in seismic monitoring is strongly linked to the question of what can be learned from seismic stations in the U.S.S.R. whose data is made available to the U.S. Such "in-country" or "internal" stations were considered in CTBT negotiations in the late 1950's and early 1960's, and the technical community concerned with monitoring in those years was well aware that seismic signals up to several tens of Hz could be observed to propagate from sub-kiloton explosions out to distances of several hundred kilometers. The seismic waves most prominent at these so-called "regional distances" (by convention, this means distances less than about 2000 km, contrasting with "teleseismic distances") are known as Pg, Pn, Sn and Lg. Pg propagates wholly within the crust; Pn and Sn travel at the top of the mantle, just below the bottom of the crust; and Lg, guided largely by the crustal layer, is often the strongest signal and is sometimes observed across continents even at distances of several thousand km. (See Figures 35-37 below, for further descriptive detail.)

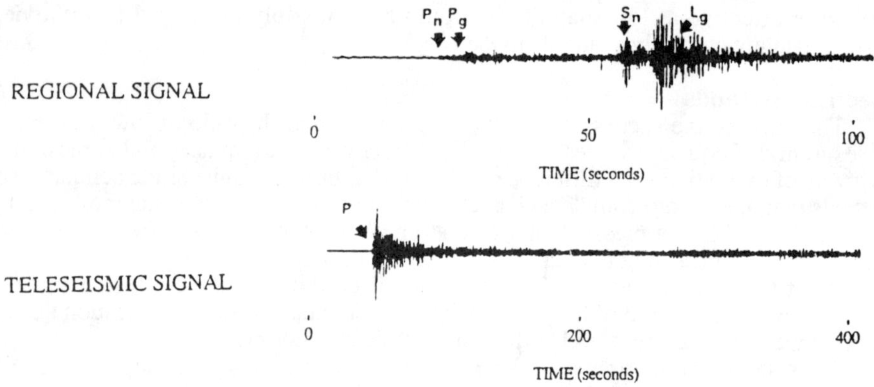

Fig. 18. Regional waves are more complex than waves observed teleseismically. Note the different time scales.

However, this early interest in regional waves lessened once the Limited Test Ban Treaty of 1963 was signed, because it was then recognized that allowed programs of large underground nuclear explosions conducted by different countries could be monitored teleseismically (i.e., at distances beyond 2000 km, so that internal stations were not needed). At teleseismic distances, seismic body wave signals are usually simpler, are not seen with frequency content much above 10 Hz (Figure 14), and have indeed proved adequate for most purposes of current seismic monitoring, even under treaty restrictions on underground nuclear testing that have developed since 1963. Thus, it is in the context of considering further restraints on testing, such as a CTBT or LYTTBT, that the seismic monitoring community is required to evaluate high frequency/regional seismic signals. The basis for the role of high frequency monitoring, in such a hypothetical new testing regime, is that the greatest challenge will come from very small nuclear explosions (i.e. with $m_b \leq 4$), and for them the most favorable signal-to-noise ratios will often be at high frequency. Internal stations, of high quality and at quiet sites, would be essential.

The fact that high-frequency seismic data from internal stations within the U.S.S.R. can lead to great improvements in detection capability, has been supported in 1986-1988 by the data recorded within that country as part of a non-governmental agreement which has allowed U.S. seismologists to deploy sophisticated monitoring equipment at sites within Soviet Central Asia. The daily recording of very small seismic events, made possible by such an improvement in detection capability, focusses attention on the discrimination problem for events in the magnitude range m_b 2.0 to m_b 4.0. In practice, the main use of in-country high-frequency signals, i.e. above 10 Hz, in improving identification capability, may lie in interpretation of the improved locations (including depth estimates) that would result.

To get a sense of the number of seismic events that will be detected by a good seismic monitoring network, and that would require identification in the context of a CTBT or low yield TTBT, it is salutory to review the basic statistics on how many earthquakes and explosions (chemical and nuclear) occur each year, at different m_b levels. The following Table gives two published estimates of the earthquake statistics:

Seismic magnitude	Numbers of earthquakes, worldwide, estimated by:	
	Ringdal (1984)	Lilwall and Douglas (1985)
4.5	2600	2300
4.0	7400	7100
3.5	21000	19000
3.0	59000	68000
2.5	166000	209000

Table 1. Two estimates[12, 13] of the numbers of earthqukes occurring per year, at and above various magnitude levels (m_b) of relevance for CTBT and low yield TTBT monitoring. About 2 or 3% of these events occur on Soviet continental territory[14].

For an explosion that is well coupled into the ground (a "tamped" explosion), m_b 2.0 can be achieved with about a hundredth of a kiloton, that is, about ten tons of TNT equivalent. At this level, a monitoring program would confront perhaps 1000 to 10,000 chemical explosions per year. Given the difficulty of discriminating between chemical and nuclear explosions, it is clear that seismic monitoring fails at some levels of low but still detectable yield, and thus that a truly comprehensive ban on all nuclear testing could not be strictly verified seismically. However, relatively few chemical explosions occur at the upper end of this m_b range (2.0 to 4.0), and, for such large events, discriminants between chemical and nuclear explosions do exist.

Summary of identification capability

It is difficult to reach precise but general conclusions on what future low yield levels could be monitored, because much would depend on the degree of effort put into new seismic networks, and data analysis. It would depend too on judgments about the level of effort that might be put into clandestine nuclear testing, and on judgements about the level of risk deemed acceptable by the side planning to execute such a test program.

The identification capability of any network will always be poorer than the detection capability for that network. Experience with teleseismic signals indicates that in this case the difference amounts to about 0.5 m_b units[16] (i.e., a detection threshold of about 3.5 indicates an identification threshold of about 4.0). However, for regional signals, for events with magnitude less than m_b 3.5, where there is less experience than with teleseismic data analysis, the difference may be larger than 0.5 m_b. A summary statement conveyed in a May 1988 Office of Technology Assessment Report[14] was as follows:

> "... there appears to be agreement that, with internal stations that detect down to m_b 2.0–2.5, identification can be accomplished in the U.S.S.R. down to *at least* as low as m_b 3.5. This cautious identification threshold is currently set by the uncertainty associated with identifying routine chemical explosions that occur below this level. Many experts claim that this identification threshold is too cautious and that with an internal network, identification could be done with high confidence down to m_b 3.0."

High-frequency seismic data would clearly be important in defeating some types of attempt to hide small nuclear explosions – most notably, in attempts to "decouple" the event by executing the explosion in a large cavity. Whereas decoupling can result in signal reduction by factors of around 70 at frequencies used for teleseismic monitoring, the reduction is lessened to a factor of about 10 at the higher frequencies accessible with regional (i.e. "in-country") monitoring equipment[15]. Thus, even a fully decoupled 1 kiloton nuclear explosion would be clearly detected by an in-country network if it was capable of picking up signals down to about m_b 2.5. Given that such a test would be difficult to execute (requiring clandestine construction of a cavity over 30 meters in diameter), and would attract some attention because it would generate a detectable seismic signal (albeit a signal not identifiable seismically as that from a nuclear explosion), and given also a role for on-site inspections, some seismologists have concluded that in practice a CTBT can be monitored down to 1 kiloton. Note, however, that such a conclusion is based on more than a purely seismological assessment.

On the general subject of capability to monitor explosions, it should be noted that the technical community can review technical issues, and can (if pushed) strive for consensus, but the standard of what is or is not an acceptable capability is ultimately a *political* question. An example of this arises in connection with yield estimation, which has been studied extensively in the last ten years from political and technical perspectives, as I next discuss.

YIELD ESTIMATION

The yield of a nuclear explosion is measured in kilotons. "1 kiloton" is an energy unit, originally defined as the energy of a thousand tons of TNT. The modern definition is:

$$1 \text{ kiloton} = 1{,}000{,}000{,}000{,}000 \text{ calories} = 10^{12} \text{ calories.}$$

If yield rises, then so does the size of the seismic signal. Since these signal amplitudes are reported in terms of seismic magnitude, which is related to the logarithm of wave amplitude, it is expected and indeed found that magnitude is linearly related to the logarithm of the yield. Because the most easily observed seismic signal from a large underground nuclear explosion is the P-wave, the commonest method of obtaining a yield estimate from interpreting seismic waves is that based on use of m_b. The method is described in this section, and use of surface waves and regional waves for purposes of yield estimation is also briefly reviewed. The section concludes with a discussion of the uncertainty of different methods of yield estimation.

Yield estimation using m_b

If Y = the yield in kilotons, it is expected that

$$m_b = a + b \log Y.$$

The constants a and b are first determined empirically by fitting a line to data points corresponding to explosions for which both yield and seismic magnitude are known. This gives a **magnitude-yield curve**. An example for U.S. nuclear explosions in Nevada is given in Figure 19.

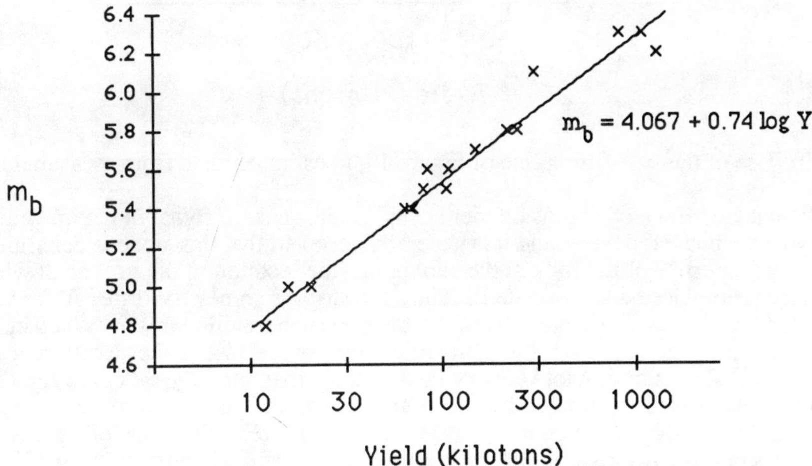

Fig. 19. The magnitude-yield curve for nuclear explosions in hard rock below the water table at the Nevada Test Site (data from ref. 17).

The seismic magnitudes here were obtained from the International Seismological Centre (an organization now based in England that gathers data for the research community from hundreds of seismic stations operated all over the globe). The yields for these 17 explosions were announced by the U.S. The yield is plotted here on a logarithmic scale. By fitting a line through these points, as shown, one obtains a "calibration curve" which can then be used at this test site to obtain an estimate of the yield for an explosion for which only the seismic magnitude is known. For example, for an explosion for which the seismic magnitude is measured as 5.7, the yield estimated in this case would be 161 kilotons, as shown below in Figure 20.

The question then arises: is this magnitude - yield curve "portable" to other test sites?

The answer, typically, is "No." If the curve is indeed known at one test site, where both seismic magnitudes and yields are available, it must usually be modified for application of the seismic method to a new area where explosions occur with unknown yield.

The slope of a magnitude-yield curve is a measure of the way in which seismic signal strength increases for an increase in the explosion yield. This is determined by

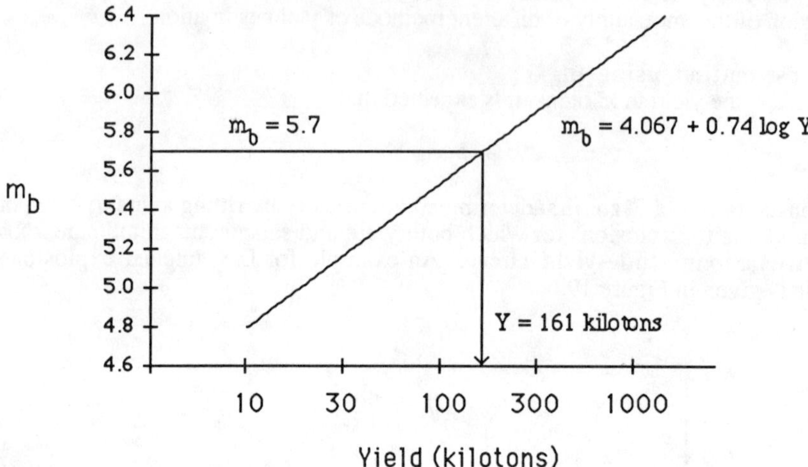

Fig. 20. Use of the best-fitting line of Figure 19, to estimate yield from measured m_b.

the efficiency with which explosion yield couples into seismic signal in the immediate vicinity of the nuclear device, and it is generally accepted that this slope is determined largely by properties of the rock at the shot point (the location in the ground at which the device is emplaced and detonated). Thus, the slope is somewhat different for very hard rocks (such as granite) than it is for soft rocks (such as alluvium)[18]. Whether the rock is water-saturated, or is dry with air-filled pore space, also makes a difference to the slope of the magnitude-yield curve. To develop a magnitude-yield curve for a site at which yield is to be estimated by the seismic method, it is therefore necessary to work from knowledge of the geology of the site and to take as the slope of the desired curve the slope from a magnitude-yield curve at another site with rock of similar hardness and water-saturation.

This gives the value of the constant b in the relation $m_b = a + b \log Y$. In practice, the value is somewhat less than unity. Thus, ref. 18 concludes that b lies in the range 0.8 to 0.9 for salt, granite, tuff/rhyolite, and shale.

The other constant, a, is also determined in part by the efficiency of coupling of explosion yield into seismic signal. It is also determined by how efficiently the seismic signal propagates along the paths between the shot point and receivers (seismometers) where the signal is recorded. There is evidence that this efficiency of seismic signal propagation is quite variable for different test sites. It can even vary for sub-regions of the same test site. This can slightly change the value of the constant, a, in the magnitude-yield formula. A common convention in describing the difficulty is to adopt a standard value for a, say a_0, and then to introduce a correction which is called the **bias**. Thus, $a = a_0 + \text{bias}$, and the magnitude yield relation becomes

$$m_b = a_0 + \text{bias} + b \log Y.$$

The constant a_0 has a simple interpretation: it is the seismic magnitude of a one kiloton explosion at a site that one has decided to take as the standard, for which the bias is zero. Note that if an m_b value is available for an explosion for which the yield is known (or is estimated from a method that does not use seismic body waves), then a is estimated directly from $a = m_b - b \log Y$, without the need to break a into the sum of two terms (a_0 + bias).

The fact that different test sites have different bias values can be seen from many lines of evidence. One of the most direct, is to compare two U.S. underground nuclear explosions in water-saturated hard rock: PILEDRIVER (announced as 62 kilotons in granite at the Nevada Test Site, with m_b about 5.5), and LONGSHOT (announced as 80 kilotons in crystalline rock in the Aleutians, with m_b about 5.9). The yield of these explosions is so similar (specifically: the term $b \log Y$ differs for the two events only by about 0.1 unit for choices of the slope b in the plausible range 0.7 to 1.0) that one might expect their seismic magnitudes to be about the same. But, they differ by about 0.4 units. This is a specific illustration of a difference in bias between the two sites, which in this example is about 0.3 units. Since 0.3 is the logarithm of 2, this difference in bias indicates that an explosion in hard rock at the Aleutian Test Site makes about twice the size of seismic signal as the same size explosion in the same type of rock at the Nevada Test Site. If the concept of bias between sites were ignored and experience from Nevada were applied without the necessary correction, then the seismic signal of an underground nuclear explosion in the Aleutians would indicate an Aleutian explosion about twice as large as it really was.

Bias, as introduced here, has an explanation in terms of a region of attenuating material that underlies some test sites (including Nevada), as shown in Figure 21:

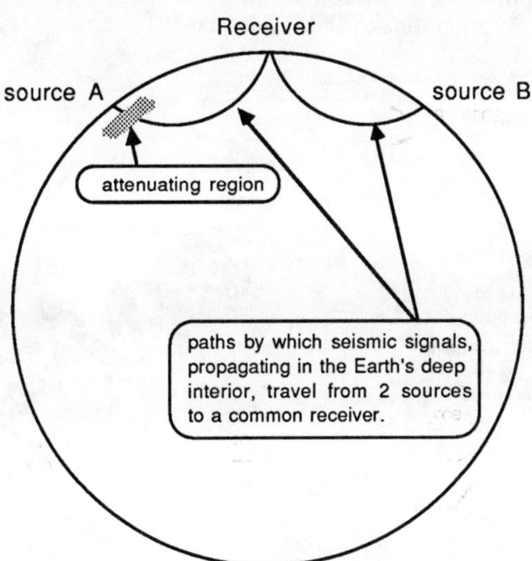

Fig. 21. An explanation for unequal amplitudes, from explosions with the same yield but at different test sites.

Seismologists readily adopt the above picture, in which there is a **source region bias** because of attenuation below some sources but not others, because it is expected from what is known about a more easily studied phenomenon in which attenuation underlies a *receiver*, rather than a test site. The picture in this case is shown in Figure 22.

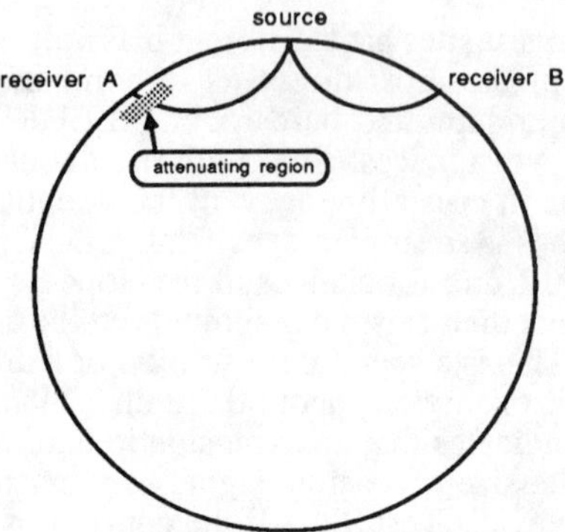

Fig. 22. In this diagram, a receiver at point A typically reports anomalously low m_b for all distant earthquakes and explosions, an effect which is referred to as **station bias**. This type of bias may be quantified from observations at every well-calibrated station, and may even be mapped for areas of dense station coverage.

Maps of station bias indicate broad regions of the western U.S. that have attenuated seismic signals. In the following maps for North America (Figure 23) and Eurasia (Figure 24), note that a station on the Nevada Test Site (OB2) has bias – 0.2, whereas a station at Semipalatinsk (SEM) has bias 0.1.

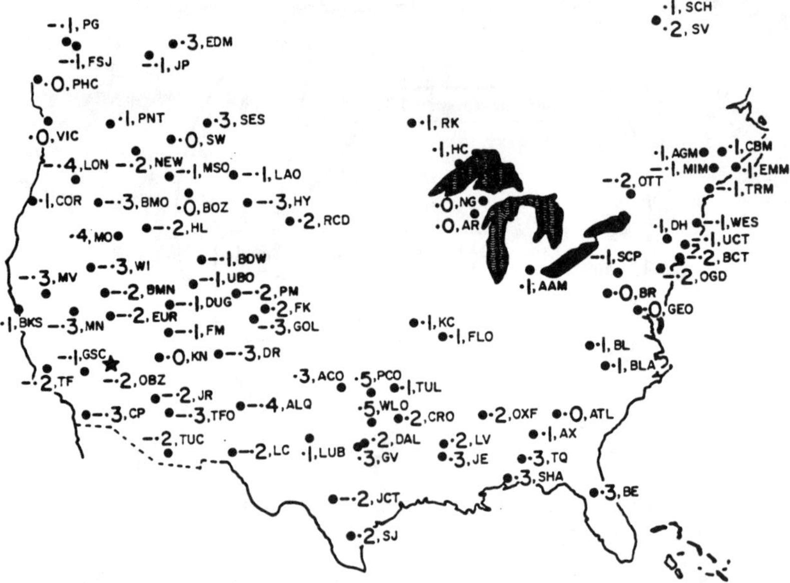

Fig. 23. Station bias in the continental United States and parts of Canada[19].

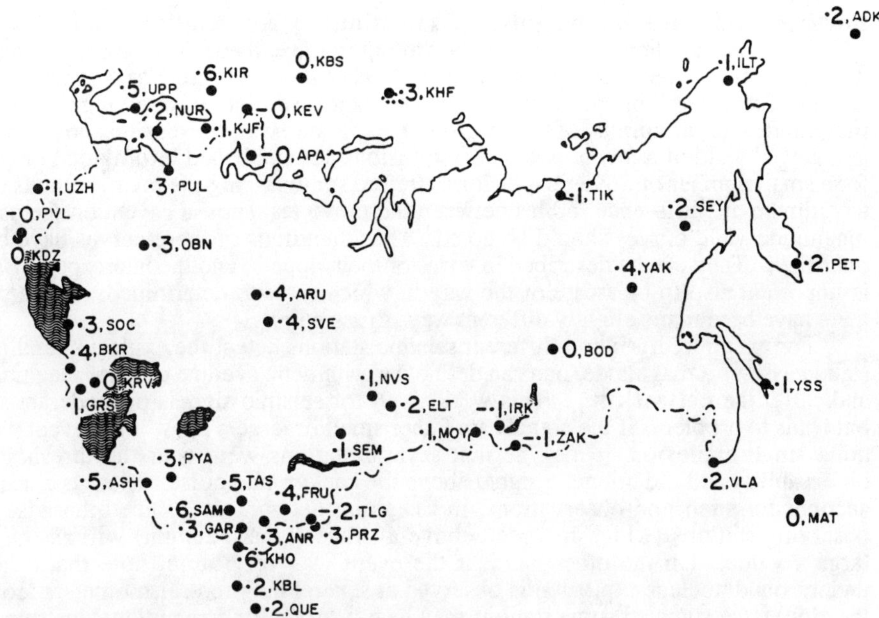

Fig. 24. Station bias in and near the U.S.S.R.[19]

The explanation of bias in both source and receiver regions can be given in terms of the temperature of the Earth at depth, below the site of interest. Seismic waves are attenuated by propagation through rock near its melting temperature. It is known that volcanoes have been active near the Nevada Test Site in the last few million years, and that this is not enough time for cooling at depth to have occurred. A comparison of several pertinent geophysical features for the Nevada Test Site, and for other regions that in geological terms have been relatively inactive over the last several hundred million years, is summarized in Figure 25.

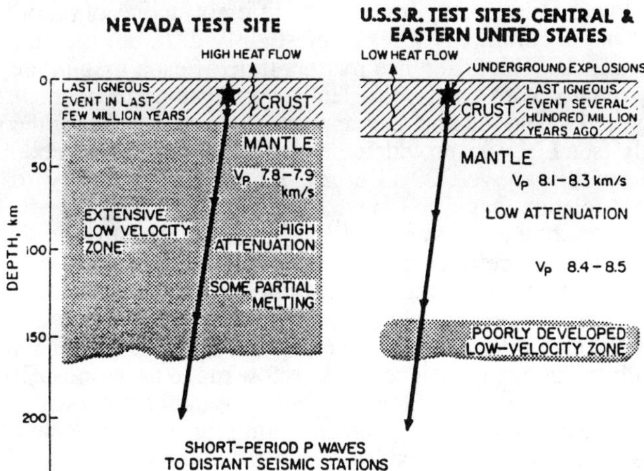

Fig. 25. Upper mantle and crustal structures for the Nevada Test Site, and for other test sites, as indicated by a variety of geophysical observables[5].

Further comments on magnitude, magnitude-yield relation, and bias

There is an extensive literature, beginning in the early 1970's, on estimating the difference in bias between the Shagan River Test Site in Eastern Kazakhstan and the Nevada Test Site. It draws on the many open data bases that exist, giving the seismic magnitudes of hundreds of underground explosions at these sites; and on the announced yield of a few tens of U.S. explosions and published information on yield for a small number of Soviet explosions. Before summarizing the main methods used to estimate the difference in bias between these two test sites, a caveat on the use of magnitude-yield curves should be noted. Thus, hundreds of these curves have been published. They can be described in terms of their slope *b*, and the intercept *a*. But it is important also to be aware of the way in which m_b was determined. In practice, there have been many slightly different ways to assign m_b.

For example, if several different seismic stations detect the explosion and each station assigns a magnitude, one can define the arithmetic average of these magnitudes and call it the **network m_b**. This works well for seismic signals of moderate size, but leads to problems if the signals are either small or large. Thus, if the event was a quite small explosion, it may be that several stations were near the threshold of observability and did not get a signal above the background noise. There is a need to account for such non-observations, in assigning network m_b, since otherwise the observing stations (which recorded above-average signal strength) will assign too large a value. On the other hand, if the event was very large (note that a large underground nuclear explosion is observed at several hundred seismometers around the globe), the signal at some stations may be too strong for conventional instruments to give a correct measure of signal strength. Commonly, such overloaded instruments can report only a maximum reading: they are said to be "clipped", and if the actual signal strength exceeds the maximum that some instruments in the network can report, the average of all station magnitudes will be too small. In recent years, these types of problems with network m_b have been recognized. Many workers have developed, and made openly available, statistically-based corrections applied to conventional network m_b to obtain what is known as the **maximum likelihood m_b**[12, 20].

It is therefore essential that, to use a published value of m_b to estimate yield for a particular explosion, a magnitude-yield curve is used that was developed with m_b assigned in the same way as for the event of interest. The fact that slightly different magnitude-yield curves are often produced in different studies of purportedly the same set of explosions, is commonly a result of slightly different m_b determinations by different institutions. The difference matters little, if each magnitude-yield curve is used correctly (that is, with a self-consistent set of seismic magnitudes).

The magnitude-yield relation does appear to be a straight line if the yield is plotted logarithmically (see Fig. 19), provided the yield does not exceed about 300 kilotons. However, for yields above 300 kilotons, it may be necessary to work with a magnitude-yield relation that is no longer a straight line, but is curved. The reason for this is that estimates of explosion size should be made from seismic signals at periods long enough for the source to be perceived as effectively a point in space and time. (This permits some averaging over the radiation put out by the source, preventing the scatter in measurement which would result if signals were used at periods short enough to suffer from interference between different parts of the exploding source region.) Recalling that m_b is assigned in a narrow range of periods, fixed at around 1 second, it follows that seismic amplitude will be roughly proportional to explosion strength, and hence that magnitude will increase in proportion with the logarithm of the yield, provided one restricts the explosions to those for which 1 second is a period long enough for the explosion to appear as a point source. For explosions above 300 kilotons, this last restriction no longer holds. Such an explosion is so big, that a seismic signal at one second is affected by slightly destructive interference of signal coming from different parts of the large source region. The signal at a period of one second (and hence m_b), for such large explosions, is typically not as big as it would be if all the initial P-wave radiation were coming in coherently to a seismometer. The result is that at high yield the magnitude increases with log(yield) at a lower rate than it does at low yield. Thus, the slope decreases at high yield, and it follows that magnitude - log(yield) curves have lower slope at high yield.

A caveat is also needed concerning the definition of the bias of a particular test site. It is unfortunate that there are at least three definitions in common use:

that given above, in terms of a difference between a and a standard value, a_0;

the bias in comparison to a scale based on NTS (i.e. the bias at NTS is taken by definition as zero), the comparison being made at 1 kiloton; and

the bias in comparison to a scale based on NTS, the comparison being made at 150 kilotons.

In general, these slightly different definitions will result in three different values. (They would all be the same, if magnitude - log(yield) curves had the same slope, and the constant a_0 were appropriate for a zero bias at NTS.) As we review some of the different ways that bias may be estimated, a detailed approach reveals that slightly different bias definitions are often used in the different methods. To obtain a common standard, it is then necessary in the most accurate studies to correct the different bias estimates to make them conform to a common definition. However, though necessary for some purposes, such detail obscures the overall presentation of the physical basis for concluding that significant bias does exist and is different for different test sites. Therefore, in the following, I shall not adopt a detailed approach.

Methods of estimating test site bias

The quick negotiation of the 150 kiloton Threshold Test Ban Treaty of 1974 focussed attention on a technical problem that, to that date, had received little attention. Namely, the question of how the yield of an underground nuclear explosion should be estimated. We shall find in this section that several different seismic waves can be made the basis of a yield estimate, but it was soon recognized that there would be considerable advantages for a method based on the teleseismically-observed P-wave, because this wave is relatively simple to detect and measure. It was also soon recognized widely that use of the P-wave and the associated m_b scale, for purposes of yield estimation, boiled down to obtaining values for the constants a_0 and b, and then estimating the correct bias value for the test site of interest.

There have been hundreds of studies of test site bias conducted since the early 1970's. Many of them have been published in the reviewed journals of professional societies, and many more are known as "grey literature" studies, i.e. they have not been reviewed, and are of variable quality, but they are openly available. These many studies are of four types:

- those based on an overall assessment of geological and geophysical observables;
- those based on detailed properties of P-wave signals (for example, their spectra);
- comparisons of P-wave-based yield estimates with estimates obtained from other seismic waves; and
- comparisons of P-wave-based yield estimates with yield estimates available from non-seismic methods (for example, from Soviet-supplied information on certain explosions).

The principal features of these methods are as follows:

Overall assessment of geological and geophysical observables

The gross characteristics of continental geology that influence the way seismic waves propagate in a particular continental region are those that are reflective of the temperature of rocks throughout the uppermost 200 kilometers of Earth structure in the region. If there have been active volcanoes in the area (e.g. the Western U.S. in the vicinity of the Nevada Test Site), temperatures at depth can be presumed to be high. The opposite extreme is a continental shield region, such as found in parts of Canada and the Eastern U.S., Scandinavia, Australia, India, and parts of the USSR. In shields, temperatures are thought to be relatively low: there has been no mountain-building activity or volcanism for hundreds of millions of years. The main Soviet test site near Semipalatinsk in Eastern Kazakhstan is thought to be in geology of a type that, if not a shield region, at least is much closer to being a shield than to a geologically active area.

Within these broad categories, geologically active and inactive, certain types of observable features can be measured and found to have values roughly diagnostic of high or low temperatures. The list of observable quantities includes heat flow itself (is it high or low); the speed of teleseismic waves influenced by propagation in the region of interest (do these waves arrive earlier or later than the global average); the value of station bias (is it negative or positive – see Figs. 23 and 24); and the speed of seismic waves (Pn) that propagate horizontally in the upper mantle directly below the region. Thus, just from knowledge of gross properties, one can expect to estimate the bias of a test site to within roughly ±0.2 magnitude units. An explosion on a shield region would have high bias - perhaps as high as 0.5 relative to NTS. An explosion in a tectonically active area such as the East African rift or part of the USSR near Lake Baikal would be expected to have a low bias – perhaps, about the same as NTS.

On the basis of what is known about the gross geology of Kazakhstan, a positive bias for this region would be expected, relative to NTS.

Detailed properties of the P-wave signal

The P-wave from an explosion, observed teleseismically (i.e., at more than 2000 kilometers from the source), is attenuated during propagation from source to receiver. An observable feature of this attenuation is the way the higher frequencies are preferentially removed. For example, we may compare an NTS explosion and a Semipalatinsk explosion as follows:

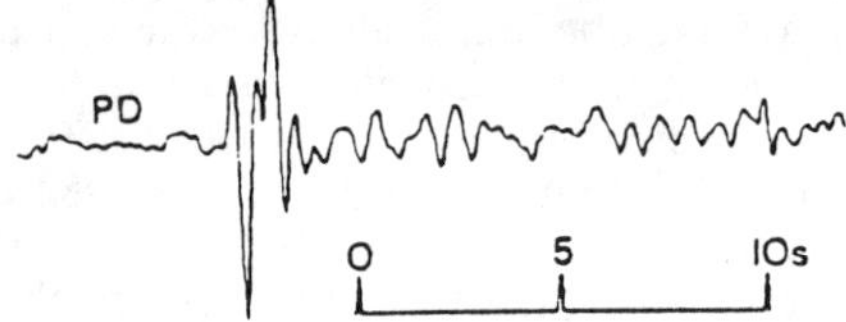

NTS explosion PILEDRIVER (PD)

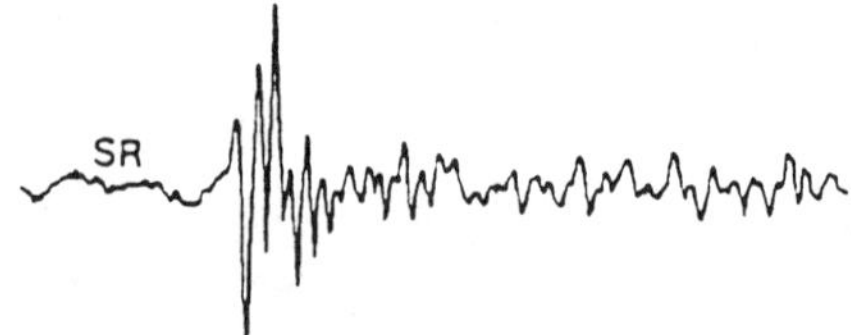

Semipalatinsk explosion of June 30, 1971, at Shagan River (SR)

Fig. 26. Seismograms at the same station, from explosions at different test sites.

These two recordings were made at the same seismic station, at Eskdalemuir in Scotland. Inspection of Figure 26 reveals that the first signal (PD) appears to have undergone attenuation that has removed high frequencies compared to the second (SR). It is presumed that the cause is the attenuating upper mantle beneath the Nevada Test Site (see Figs. 21 and 25).

Seismologists characterize attenuation along a whole propagation path from source to receiver by a parameter labelled as t^*. Attenuation is then quantified by the presence of a factor

$$e^{-\pi f t^*}$$

in the observed spectrum; f here is the frequency, in Hz. It can be difficult to identify this factor uniquely, because the spectrum is usually influenced by other frequency-dependent effects that are more or less unknown. But, fortunately, it is only the **differential attenuation**, Δt^*, that is needed in obtaining the bias between two test sites, one of which may be NTS.

Thus, the extra amount of attenuation in the PD seismogram, above, can be quantified by multiplying the PD spectrum by a factor

$$e^{+\pi f \Delta t^*}$$

to restore the high frequencies, and doing this with various choices for Δt^* to find a value such that the resulting seismogram looks like the SR seismogram. If a value of 0.2 seconds is used for Δt^*, the result[21] is

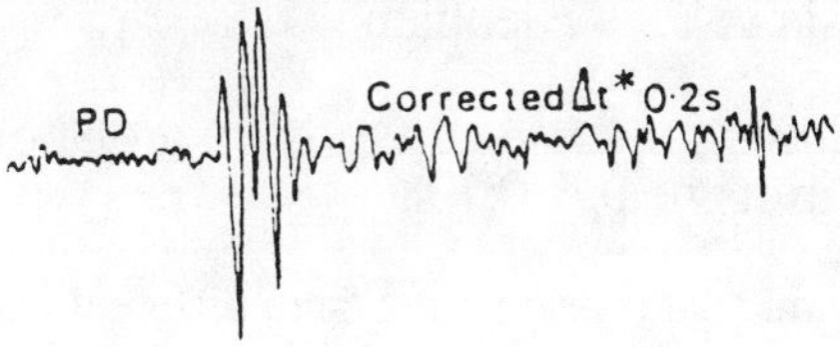

Fig. 27. The filtered PD seismogram now looks much more like that for SR.

In this way, Δt^* can be measured between two test sites. Many studies of this type have been done, using sources and receivers all over the world to refine the basic method.

There is a simple relationship between Δt^* and the bias between these two sites:

$$\Delta m_b = \text{site bias} = \log_{10}(e^{+\pi \Delta t^*}).$$

This follows, because seismic magnitude is obtained from taking a logarithm (to the base 10) of amplitude at a frequency near to 1 Hz. The right-hand side of the above equation simplifies, giving

$$\Delta m_b = 1.36\, \Delta t^*.$$

If, in practice, the seismic magnitude is obtained from the somewhat stronger signals that may be present at frequencies somewhat higher than 1 Hz, then a more representative relationship may be

$$\Delta m_b = 1.5\, \Delta t^*.$$

A value of 0.2 for Δt^* then translates to a bias of 0.3 between NTS and Shagan River, on the basis of an interpretation of the difference in frequency content, visible in the seismograms discussed in Figure 26.

A summary of the results of this method, in application to several different explosion locations, may be tabulated as follows[22]:

Test Site	Region	t^*
Pahute Mesa	Nevada	0.34 *
Climax Stock	Nevada	0.34 *
Yucca Flats	Nevada	0.34 *
Faultless	Nevada	0.32
Gasbuggy	New Mexico	0.37 *
Salmon	Mississippi	0.12
Shoal	Nevada	0.33 *
Novaya Zemlya	USSR	0.08
Degelen Mtn	Kazakhstan	0.06 +
Shagan River	Kazakhstan	0.08 +
Hoggar	Algeria	0.35 +
Rajahsthan	India	0.24
Mururoa	Tuamotu Archipelago	0.35 +
Lop Nor	Sinkiang	0.10
Amchitka Island	Aleutians	0.10

* 95% confidence limits for these values of t^* are about $\pm$ 0.05 s;
+ 95% confidence limits near $\pm$ 0.1 s.

Table 2. t^* for teleseismic paths originating from different nuclear explosions.

As a use of this Table, for example, to get a bias for Shagan River compared to the Yucca or Climax Stock areas of the Nevada Test Site, note that

$\Delta t^* = t^*$ for Yucca or the Climax Stock (0.34), minus t^* for Shagan (0.08),

giving Δt^* as 0.26. Multiplying by 1.36, gives 0.35 for the bias; multiplying by 1.5, would give a bias of 0.39.

Comparison of P-wave-based yield estimates, and yield estimates derived from other seismic waves

This sub-heading introduces a large subject, namely the use of seismic surface waves and regional waves, each of which may be made the basis of a yield estimate. The methods are broadly similar to that which has been developed for yield estimates based on m_b. Thus, a logarithmic magnitude scale is defined for each new wave type; its magnitude - yield relation is found from knowledge of yield for a group of explosions whose seismic magnitude is also measured; and the properties of this magnitude - yield relation are explored, in particular its portability to test sites other than that for which it was developed.

To the extent that seismic waves other than the P-wave can successfully be used to estimate yield, these estimates can then be used for explosions at a test site for which an estimate of the m_b bias is desired. Knowing the m_b for these explosions, the yield estimate derived from other seismic waves can be used to measure the bias.

Note, this program can be carried out by special study of a small group of explosions whose body waves and surface waves are studied in detail. Thereafter, the resulting m_b bias value can be applied to P-waves from all explosions at the site of interest.

What, then, are the main steps and pitfalls of such an approach to estimating m_b bias? We shall answer this question here with a look at the use of surface waves.

Recall that the surface wave magnitude is denoted by M_S. It is found empirically that the surface wave magnitude - yield relation is a straight line (when M_S is plotted against the logarithm of yield). This is also expected on theoretical grounds. A representative example is

$$M_S = 2 + \log Y,$$

which was proposed several years ago by British seismologists[23].

In general, surface waves (in comparison to body waves) appear to have one great virtue and one great failing. The virtue is that these waves appear to propagate in a fashion that is less influenced by regional geology, than is the case for body waves. This might be expected, from the long-period nature of surface waves. The failing, is that a large underground nuclear explosion can often trigger a concurrent release of earthquake-like energy in the source region, so that the observed seismogram is a superposition of waves proportional to the explosion strength, plus waves from the triggered earthquake. The earthquake component is a phenomenon sometimes called **tectonic release.**

It took about ten years, beginning in the mid-1970's when the presence of tectonic release was recognized in surface wave observations, for a method to be developed that correctly identified the explosion component, allowing it to be used for yield estimation.

There are in fact two types of surface wave, known as Rayleigh waves and Love waves. Surface waves are observed at stations all around in azimuth, as shown in the disposition of stations in Figure 28, for an explosion in Eastern Kazakhstan:

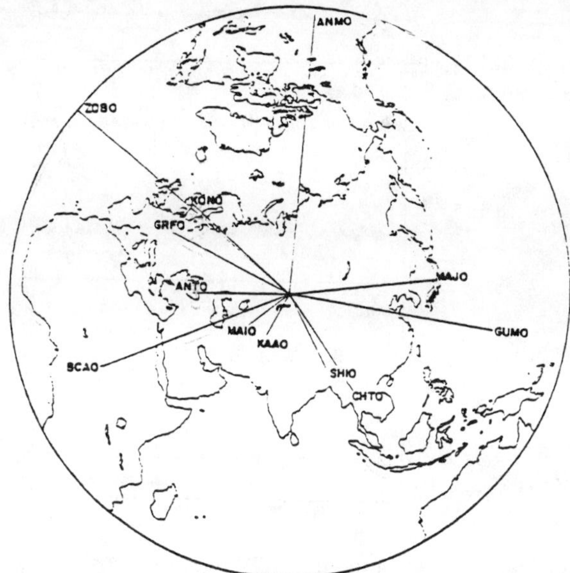

Fig. 28. Stations from a global digital network, shown at different azimuths from an explosion source located in Eastern Kazakhstan.

The question of whether the contaminating effect of tectonic release is present, in surface waves from an underground explosion, is answered by examining how the surface wave signals vary in azimuth at stations around the source. This variation is systematically different, for the earthquake component (i.e., that associated with tectonic release), than it is for the pure explosion component of the total signal. Thus, the Rayleigh wave from the explosion should ideally leave the source with the same strength at all azimuths. But for an earthquake there is diagnostic variability with azimuth. The pure explosion component ideally does not generate Love waves, whereas Love waves are efficiently generated by earthquakes.

A Love wave is identifiable because it is a horizontal motion in the direction transverse to the path between source and receiver. A Rayleigh wave is identifiable because it is a motion in the vertical plane containing the path between source and receiver. (In several respects, a Rayleigh wave is similar to the motions set up on the surface of a pond by throwing in a stone, as noted in Fig. 6b.) The basis for recognizing these waves in seismograms is that a seismic station records the motion of the ground in all three directions: the vertical direction, denoted on seismograms as the Z direction; and two horizontal directions, from which the motion in the transverse direction of Love waves can be measured. This latter direction is denoted on seismograms as the T direction. Thus, Rayleigh waves may be measured from motions in the Z direction; and Love waves from motions in the T direction.

In Figure 29 are shown some surface wave seismograms for an explosion that exhibits only minor contamination from tectonic release: the Love waves (T) are small and the Rayleigh waves (Z) turn out to have little azimuthal variability.

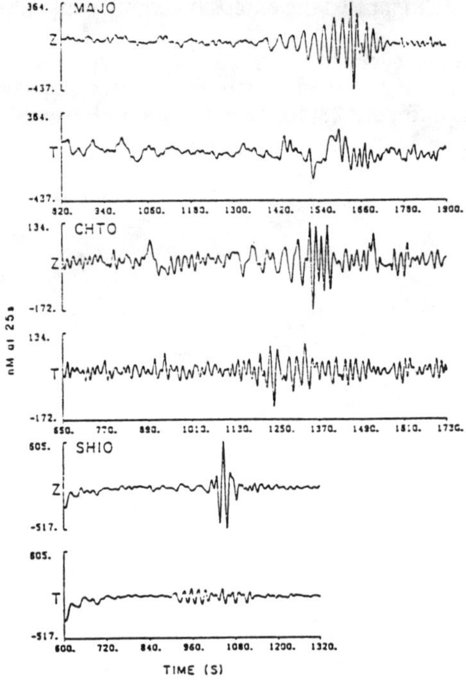

Fig. 29. Vertical (Z) and tangential (T) component digital seismograms from the Shagan River explosion of 2 December 1979. The Rayleigh waves were large compared to the Love waves at all azimuths for this event, suggesting little non-isotropic source contribution.

In contrast, the seismograms in Figure 30 show surface waves from an explosion that has substantial contamination. This contamination must be removed, if these records are to be used for accurate estimation of explosion yield:

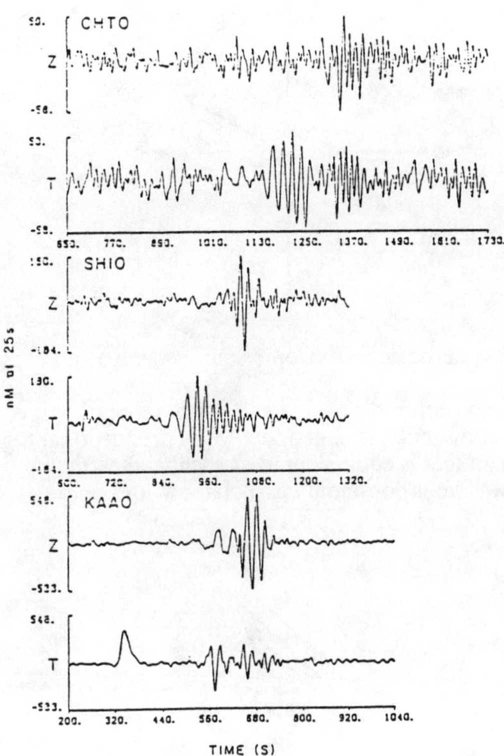

Fig. 30. Digital seismograms from the Shagan River explosion of 18 August 1979. The Love waves from this event are energetic and the Rayleigh waves show strong azimuthal variation when compared to the event of Figure 29.

The way in which the effects of tectonic release are now quantified has entailed a new approach to understanding the way a seismic signal is initiated at the source. In particular, it has led to routine use of a new concept, called **seismic moment**, that is a fundamental departure from use of traditional seismic magnitude scales. The seismic moment is another way to describe the **size** of a source of seismic waves. (For extensive discussion of this modern approach to quantitative seismology, see chapters 3, 4, 7, 8, 9 of ref. 8.)

The use of "seismic moment" for a particular source gives a direct description of the force system, acting in the Earth, that would make seismic waves of the size and shape we actually record for that particular source.

Thus, the seismic moment of a seismic source is something that has a well defined (albeit abstract) meaning. This contrasts with the purely phenomenological definitions of seismic magnitude. Different seismic magnitude estimates for the same explosion often cannot be compared, because they are based on different techniques for processing the records.

For an idealized explosion, the equivalent force system is one that pushes out equally in all directions:

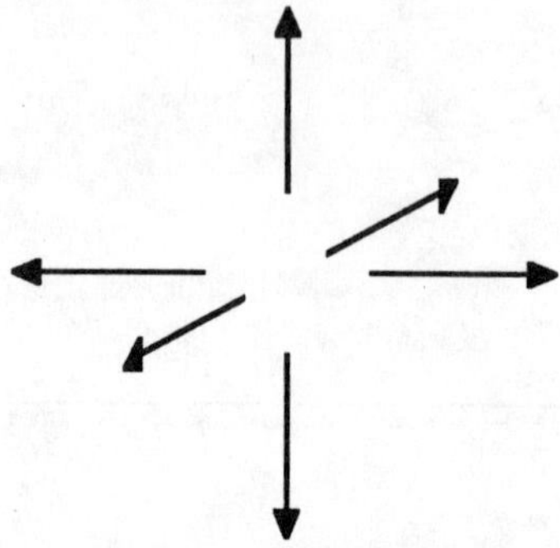

Fig. 31. The force system for an isotropic source, such as an explosion.

It is a sum of three vector dipoles, each with strength given by the **isotropic moment**, M_I. It is commonly assumed that M_I is proportional to yield.

The force system that is equivalent to an idealized earthquake is quite different.

Thus, an earthquake is commonly associated with shearing motions across a fault plane:

Fig. 32. The relative motions that occur across a fault in a simple model of an earthquake.

The equivalent force system, which would generate the same seismic motions, can be thought of as two pairs of opposing forces, one pushing in and one pushing out:

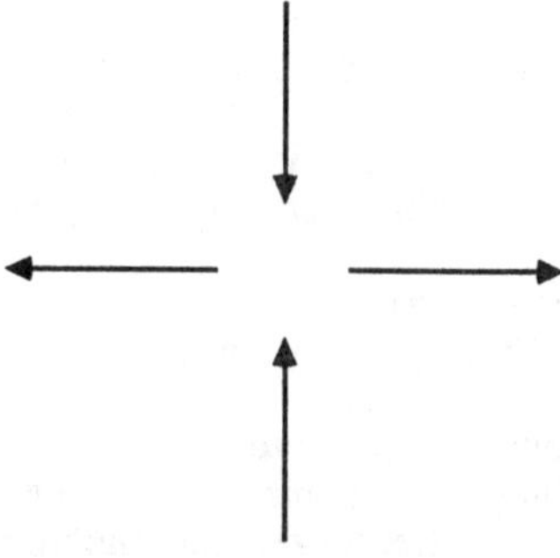

Fig. 33. The force system, acting at an earthquake source, that generates seismic waves equivalent to those generated by the motions shown in Figure 32.

A force system that gives exactly the same mechanical results is a pair of couples, of equal moment but acting in opposite directions:

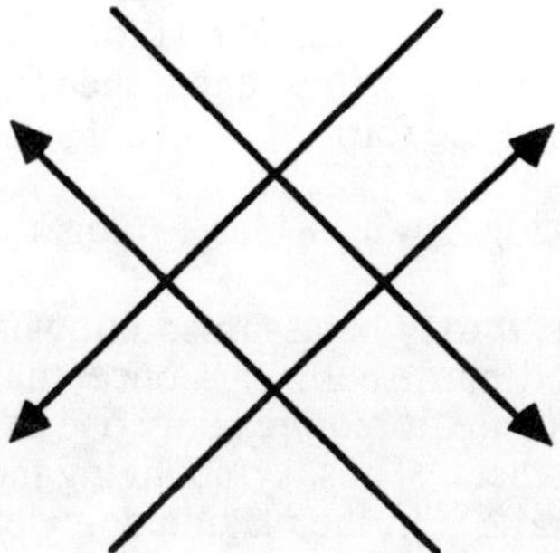

Fig. 34. A force system, equivalent to that shown in Figure 33, that also generates seismic waves as if from an earthquake.

This is known as a "double couple", with each component having strength M_0.

Thus: explosions are modelled with M_I (three equal and mutually orthogonal vector dipoles), earthquakes are modelled with M_0 (two equal and opposite couples).

We often observe seismic waves from an explosion that appear to come from a mixed system, M_0 and M_I. It appears as if the explosion has triggered release of tectonic stress.

For purposes of yield estimation, M_0 has contaminated the signal. M_0 must be estimated, and removed, resulting in a correction so that the pure explosion signal can be properly measured. Part of the exercise of estimating M_0 is estimating the orientation of the double couple in three-dimensional space.

For a "strike slip" type of tectonic release, such as prevails at the Nevada Test Site (NTS), the orientation of the double couple is such that there is little net effect on the surface wave yield estimate because the effect at each station averages out (provided the observing network of seismic stations has a fair coverage in azimuth). But for a "thrusting" type of tectonic release, thought to be typical of the Shagan River Test Site, the double couple orientation is such that a typical station observes a reduced surface wave amplitude due to contamination with M_0. The amount of the reduction is variable from one explosion to another at this test site, but is about 0.2 surface wave magnitude units for the typical event [24]. So, correction for the thrusting type of tectonic release apparently occurring at Shagan River has the effect of raising the surface wave estimate of explosion yield.

One use of such corrected surface wave yield estimates is an evaluation of the body wave magnitude bias. There have been many studies on this general subject. The steps typically might include: removal of the double couple component, from NTS and Shagan River explosion seismograms; development of a relation between log M_I and yield based on NTS data; use of this relation to estimate yields at Shagan River; and a comparison of NTS and Shagan River m_b values at the same M_I. The m_b bias estimated in this way lies around 0.3 to 0.4[25].

An estimate of the m_b bias can also be made using yield estimates based on regional seismic waves. (As with surface waves, regional waves can directly be used as the basis of a yield estimate, provided there is a way of calibrating the relevant magnitude - yield relation.) Some discussion of regional waves is given below, but next I comment on the fourth and final type of method for estimating bias.

Comparisons of P-wave-based yield estimates with yield estimates available from non-seismic methods

A fourth way to estimate yield at a test site of interest (from P-waves observed teleseismically) is to calibrate the m_b – log Y relation with yield information obtained non-seismically for one or more explosions at that specific test site.

Recall that this relation has the form

$$m_b = a + b \log Y = a_0 + \text{bias} + b \log Y.$$

For many practical purposes, b may be assumed known. (It has little variability for explosions in water-saturated hard rock[18].) Conceptually, if m_b is measured for a suite of explosions at that test site, it requires only one explosion with known yield to estimate the yield of all the others. Thus, symbolizing the magnitude and yield of the calibration event as m_b^{calib} and Y^{calib}, the magnitude - yield relation is

$$m_b = m_b^{calib} + b \log (Y / Y^{calib}),$$

from which Y may be estimated from knowledge of m_b. This estimate is insensitive to the value of b, for explosions with yield not more than a factor of ten different from that of the calibration event. The method is equivalent to determining a bias value, since $a_0 + \text{bias} = m_b^{calib} - b \log Y^{calib}$.

There are conceptually several ways in which such calibration yield(s) may be obtained.

One way, is in the context of a ratified Threshold Test Ban Treaty. The treaty protocol stipulates that data will be exchanged on two tests per test site, for calibration purposes. However, whether to accept such data for calibration entails a political judgement. It was presumably deemed acceptable in 1974 when the treaty was signed by President Nixon, and in July 1976 when it was first submitted to the Senate by President Ford for advice and consent to ratification. Note, however, that the treaty, as signed, does not specify any technical method for independently verifying the accuracy of data delivered as part of such a data exchange. Dr. Milo Nordyke has suggested that Soviet PNE's, monitored on site by the U.S., be used for this purpose (testimony before the Senate Foreign Relations Committee, January 15, 1987).

Another way is to have (say) a U.S. nuclear device, of presumably known yield, exploded underground at a Soviet weapons test site. This has been proposed on a reciprocal basis by the U.S.S.R. (In the early 1960's, the Atomic Energy Commission approved the release to the Soviets, of blueprints for certain obsolete designs of nuclear device to be used in seismic research. The U.S. also proposed that designs for the devices to be used for all peaceful nuclear explosions be released to the Soviets on a reciprocal basis[3].)

Another way to obtain calibration yields, which beginning in the summer of 1988 may make a major contribution to the technical and political debates on Soviet yields, is one based on measuring the speed of the shock wave sent out into the rock from the exploding nuclear device. The so-called CORRTEX method is a particular way of making such a shock wave measurement, and the relevant equipment must be deployed right on the test site itself (unlike typical seismic methods), so it is a relatively intrusive technology. Initially the shock wave propagates at a speed faster than the usual speed of sound in solid rock. But this occurs only over a scale of a very few tens of meters from the shock point, and after the shock wave has propagated such a distance it slows down somewhat and eventually propagates the rest of the way through the Earth at the sound speed (as a P-wave). The CORRTEX technology as planned for deployment at the Shagan River test site entails drilling into the ground in the vicinity of the shock point itself, and making many measurements over a time period lasting about a hundredth of a second, to see how quickly the shock wave spreads in space and time. These measurements of the changing speed of the shock wave are diagnostic of the yield of the nuclear explosion. The method is important, because at the May 1988 summit meeting in Moscow between President Reagan and General Secretary Gorbachev, use of CORRTEX in a "Joint Verification Experiment" was agreed to for a Soviet underground nuclear explosion to be conducted at Shagan River in the yield range 100 to 150 kilotons. Several plane loads of U.S. equipment have been dispatched to make this non-seismic yield measurement in the vicinity of the shot point, for an explosion that will of course also be recorded teleseismically.

Finally, a method of calibration is available from one of the earliest Soviet "peaceful nuclear explosions" (PNE's), which was conducted right on the Shagan River Test Site. For this explosion, conducted on January 15 1965, the U.S.S.R. has released technical information from which the yield may be estimated non-seismically as around 120 kilotons. This PNE made a substantial crater, and because of its shallow depth of burial its observed m_b (which was 5.88) is thought to have been somewhat less than it would have been if the explosion had been fired at the normal depth for a weapons test. Making a correction for the different interaction with the free surface, one may conclude that 150 kilotons, exploded at Shagan River at a depth typical for a weapons test, would have an m_b around 6.17. (Note that the largest tests at this site, of concern with respect to TTBT compliance, have had m_b values around 6.2[5, 14, 26].)

Use of regional waves and P-wave coda

Previous sub-sections have covered the use of body waves and surface waves to estimate explosion yield. These waves can be detected by National Technical Means (NTM), and do not usually require a network of seismometers near the source. This contrasts with what seismologists call "regional waves", which are a whole class of waves that are neither body waves nor surface waves. Examples of such waves are known as Pn, Pg, Sn, and Lg. They are seismic waves that travel at relatively high frequencies within the Earth's crust or outer layers, and typically they are observed only at distances less than 2000 km. (Lg is an exception – it may be observed across continental paths as long as 4000 km from large explosions.) Therefore, they are associated with seismic stations that, for application to monitor Soviet nuclear explosions, would typically have to be located within the territory of the USSR. Such stations are called "in-country" or "internal" stations, and, as noted in the above sections on detection and identification, they would be of use in monitoring small explosions, as well as providing yield estimates for explosions small and large. A comparison of teleseismic and regional seismic signals appears above in Figure 18. The more complicated paths of regional waves are sketched in Figs. 35–37.

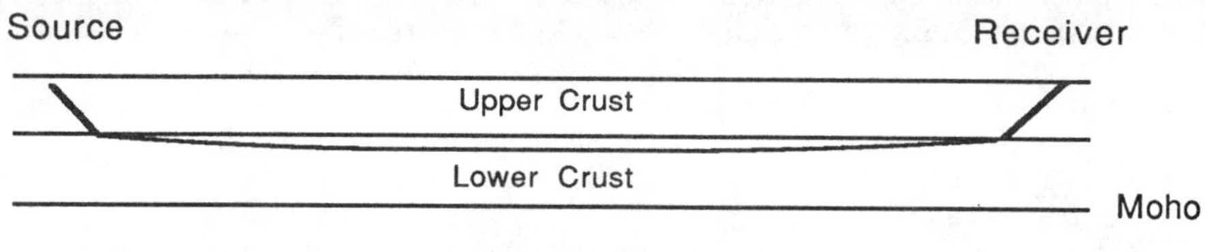

Fig. 35. Pg is a wave that can be the first to arrive at regional distances. It travels the path between source and receiver wholly within the Earth's crust. It is thought to be guided by an interface within the crust. The **Moho** is the name for the interface between the crust and the mantle.

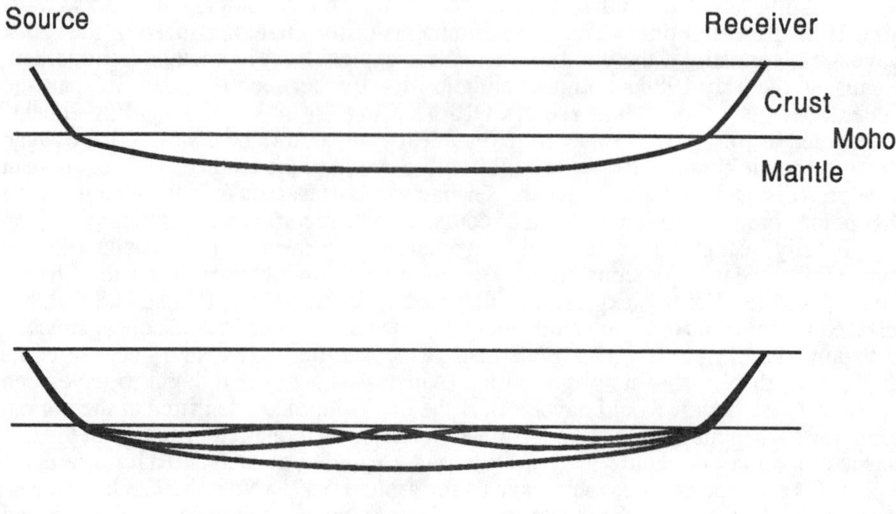

Fig. 36. Pn is a wave that at "regional distances" is usually the first wave to arrive on a seismogram (though sometimes it is preceded by Pg). It is a wave that goes down through the crust of the continent on which the source is located, then travels mostly horizontally at the top of the upper mantle (in contrast to the usual body wave, which goes deeply within the mantle), and finally travels upward through the crust to where the receiver is located. This path is shown in the upper part of the above figure. In the lower part, are shown some of the multiple bounces that also contribute to the Pn-wave.

Sn is a wave that travels a path very similar to that of Pn, but arriving later because Sn is composed of shear waves, which propagate slower than the P-waves that make up Pn. In a shear wave, particle motion is in a direction perpendicular to the direction in which the wave propagates.

Lg can be the most important of the regional waves, for purposes of explosion monitoring, because it is typically the largest wave observed in a seismogram at a regional distance. Lg is a type of shear wave that has most of its energy trapped in the Earth's crust. In fact, this wave can be so strong, that (contrary to what is implied by its inclusion in the class of "regional" waves) Lg for a large explosion can be observed even at teleseismic distances, i.e. well in excess of 2000 km. However, the observation of Lg out to teleseismic distances can occur only across large continents. This happens, because the crust of continents is typically much thicker than the crust of the Earth underneath oceans. Thus, Lg fails to propagate across even short distances beneath an ocean.

A schematic explanation for the path of propagation of Lg is given in Figure 37. This shows a seismic source within the Earth's crust. Energy departing downwards will be partially reflected back into the crust, and partially transmitted down into the mantle. However, for waves that depart more toward the horizontal direction, as shown in the middle part of the figure, energy cannot get into the mantle and is wholly reflected back into the crust. The type of reflection occurring here is the total internal reflection that is similar to that occurring within an optical fiber. For a fixed source

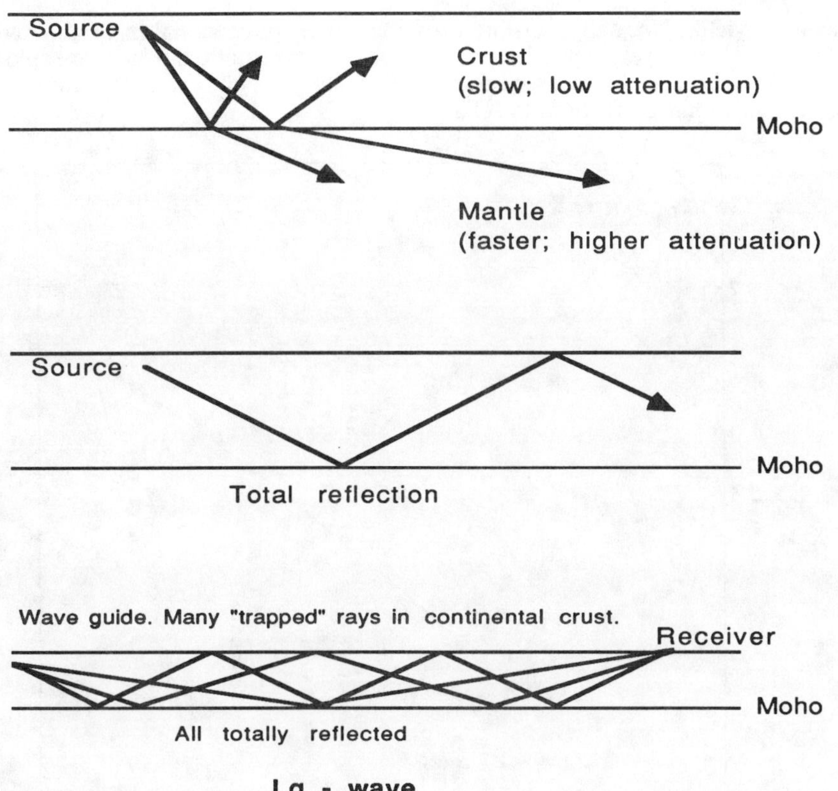

Fig. 37. Schematic illustration of the crustally-guided Lg-wave.

and receiver, there may be many reflection paths, all totally reflected and thus trapped within the crust. This is shown at the bottom of the figure. The Lg-wave is composed of the whole family of trapped waves. The low velocity crust acts as a waveguide. Because the crust is commonly composed of materials that support seismic wave propagation with little attenuation loss, Lg may be observed to propagate over thousands of kilometers. It is stopped by oceanic structures, because for them the crust is very much thinner.

For most of the last thirty years, regional waves for purposes of seismic monitoring have been thought of in the context of monitoring small explosions. (Small explosions may not be detected teleseismically.) However, in recent years Lg has come to be recognized as potentially a useful seismic wave for obtaining yield estimates, even for large explosions, using a method that is independent of the more conventional body wave and surface wave methods. A magnitude-yield curve for Lg, that indicates some very useful properties of this wave, published by Otto Nuttli in 1986, is shown in Figure 38.

Nuttli's curve is for 22 Nevada Test Site (NTS) nuclear explosions in hard rock, and it shows very little scatter in the data. It is remarkable that he obtained this result using only 2 or 3 stations. Thus, he documented a method of obtaining a seismic amplitude from the Lg wave, took the logarithm to obtain a seismic magnitude that he calls m_b(Lg), and found that there is very little scatter when plotted against the announced yield. Normally, seismic magnitude (the conventional m_b, measured on teleseismic body waves) is not deemed reliable (i.e., does not relate well to explosion yield) unless the data is processed from many stations. Nuttli worked with stations that were all within 1000 km from NTS.

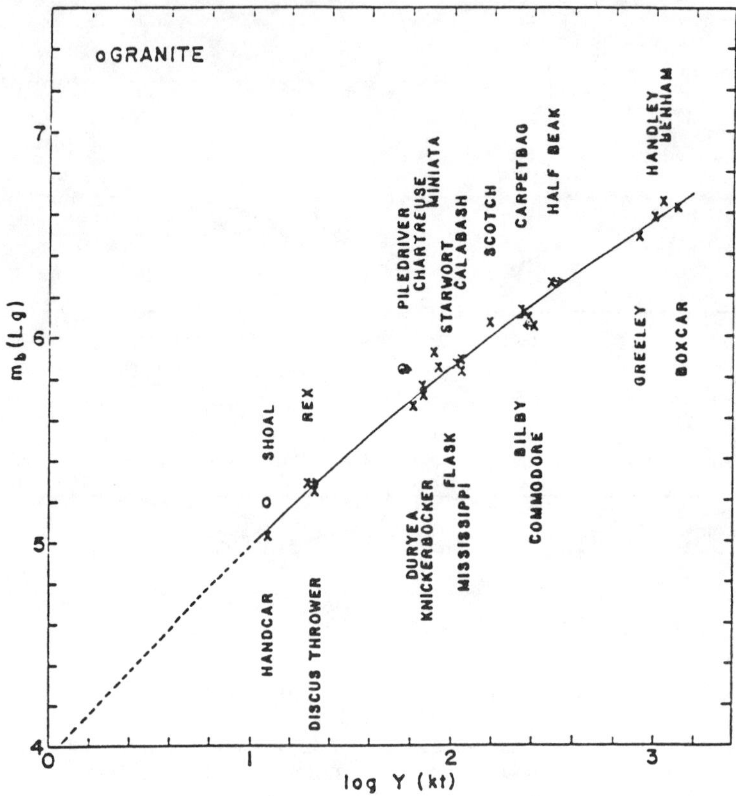

Fig. 38. Relationship between Lg magnitudes and the logarithm of explosion yield for NTS explosions of announced yield in water-saturated hard rock[17].

Unfortunately, he did not use the best values for announced yield. The most recent list of announced yields gives different numbers for most of these 22 events. For four of them (REX, PILEDRIVER, HALF BEAK, COMMODORE), the change is more than 10%. These four changes all are in the direction that improves the fit, reducing the scatter in the data. The standard deviation is around 0.06 magnitude units, indicating that a yield estimate based on the Lg wave can be almost as accurate as what has been claimed for the CORRTEX method in application to explosions on Soviet test sites (see below).

Others who have examined the scatter for Lg magnitude against yield, for a larger number of explosions at NTS, have found a somewhat larger scatter in which the standard deviation is nearer 0.1 magnitude unit. An example, supplied by the Defense Advanced Research Projects Agency (DARPA) to the Congressional Research Service, is a m_b(Lg) – yield curve for NTS as follows:

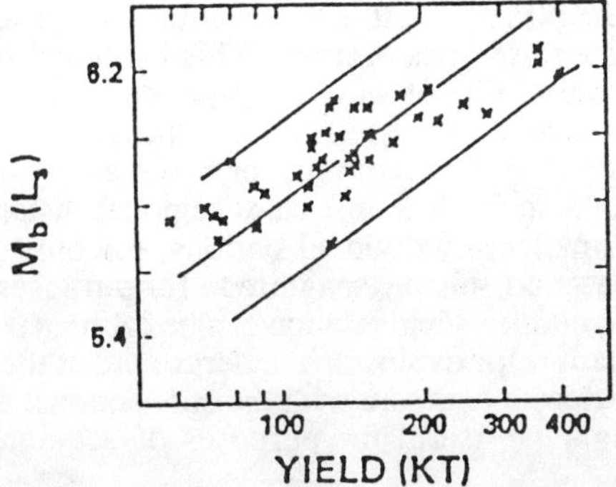

Fig. 39. A second study of Lg magnitudes versus yield, at NTS.

Presumably, this larger scatter, for Lg when a larger population of events is used at NTS, is associated with variability in the efficiency of coupling of explosion energy into seismic wave energy, and is an expression of the range of geology and degree of water saturation for different explosion locations at this site. It appears that Nuttli worked with 22 NTS events drawn from a narrow range of explosion coupling. Therefore, because NTS is such an unusual area geologically (and other test sites may have a lesser degree of geological variability), it is interesting to go on with these 22 events studied by Nuttli, which may be representative of what might expected at sites more homogeneous than the full range of NTS variability, to look at their P-wave magnitude and see how much this scatters when plotted against yield.

Interestingly, one finds, even with m_b taken directly from the International Seismological Centre (these m_b values are given to only 2 significant figures, and are based on data of variable quality) that there is a very high correlation with announced yield, as shown in Figure 40. The RMS residual to a best-fitting straight line (not shown) is only 0.07 magnitude units.

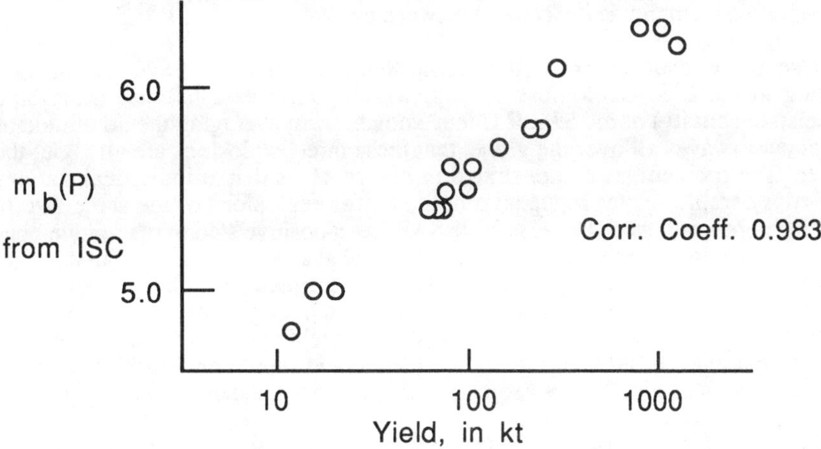

Fig. 40. The conventional P-wave m_b – yield relation, for events at NTS with a narrow range of explosion coupling.

Thus, the picture for the Nevada Test Site is that for events with a narrow range of coupling, both P and Lg exhibit little scatter. This is not circular reasoning, but rather a demonstration that variability of wave propagation may be less of a problem, in achieving accurate yield estimates by seismic methods, than has been supposed. Thus, the scatter due to propagation is reduced, for P-waves, by using many tens of stations all over the globe. And for Lg, at least at regional distances, the scatter in amplitudes is intrinsically smaller at individual stations, and only 2 or 3 high quality stations are needed to get a good seismic magnitude for purposes of accurate yield estimation, provided the magnitude - yield relation can be calibrated.

What, then, is the situation for explosions underground at the Shagan River test site? Can Lg be used, how does it compare with the conventional m_b, and what other seismic measurements might be useful for purposes of accurate or precise yield estimation?

In answering these questions, we can begin with a reminder of the basic problem, looking in Figure 41 at the way scatter presents itself in the conventional P-wave method:

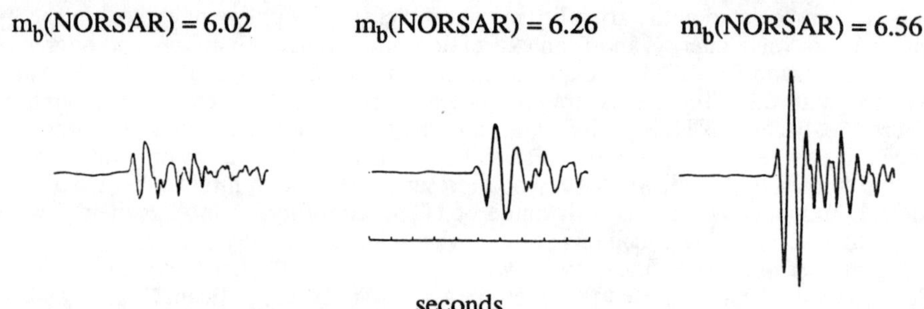

Fig. 41. NORSAR P-wave recordings (instrument 01A06) of three Semipalatinsk explosions plotted to the same amplitude scale. All three events have m_b 5.9, when m_b is averaged for a large network. Note the positive station m_b bias of NORSAR, and the significant amplitude differences between events[6].

These three nuclear explosions near Semipalatinsk all have magnitudes (according to the U.S. Geological Survey, which publishes a bulletin describing global seismic activity) of m_b 5.9. It is thus known, from averaging the amplitude of conventional P-waves all over the globe, that these three explosions are all about the same size The problem, as demonstrated in Figure 41, is that at individual stations there is considerable scatter in signal amplitude, for explosions of the same size in slightly different locations. (Because NORSAR has a positive station magnitude bias, each station magnitude in the figure exceeds the global average of 5.9.) Can one find ways to reduce this scatter, to get a precise seismic estimator of relative event size? Even better, can one get independent estimators, such as globally-averaged P and continentally-averaged Lg, so that joint estimation of yield will improve precision (and perhaps accuracy) even further? Preferably, the method should minimize the number of seismic stations needed to meet some pre-determined standard of accuracy in the resulting yield estimate.

To address this last criterion, it is of interest to introduce a fourth measure of seismic source size that may be made the basis of a yield estimate. (The first three were body waves, surface waves, and Lg.) This measure is based on the sequence of waves that are seen on a seismogram in the seconds and minutes immediately following the first P-wave. This sequence is known as the **P-wave coda**, and is under study as a research project. An example may seen in the "tail" of the teleseismic P-wave in Figure 18. Instead of working from the amplitude of the largest part of the signal in the first few seconds (this is the basis of conventional m_b), the P-wave coda amplitude is a measure of signal strength averaged over some tens of seconds, beginning just after the first arrival.

Studies of P-coda amplitude at NORSAR have shown that there is less variability at individual stations than is the case for conventional P-wave amplitudes, which are based on the maximum ground motion[6, 27]. Published values of P-coda magnitude for 23 Semipalatinsk explosions[27] are shown in Figure 42, compared with an m_b re-worked from ISC data[28], and the correlation is excellent.

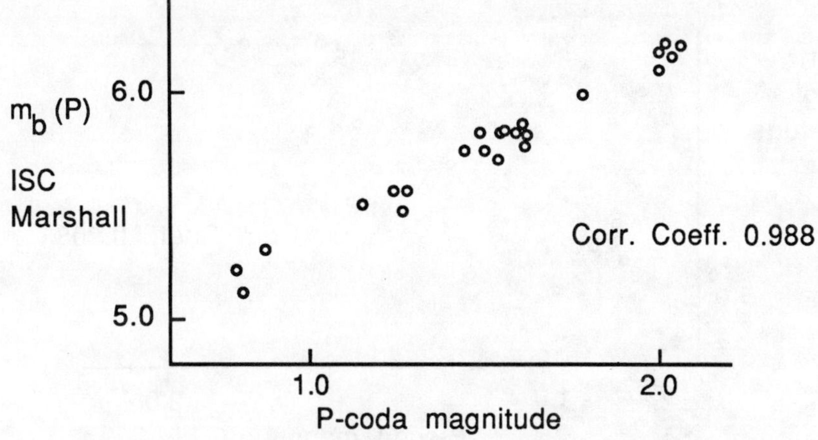

Fig. 42. Comparison of m_b and P-coda magnitude for Shagan River explosions. (Coda magnitude here is based on a logarithmic scale, with arbitrary absolute level.)

The P-coda magnitude is obtained here at a single teleseismic station. The excellence of the correlation validates both seismic estimators of explosion size – that is, both the m_b measure, based on the hundreds of stations reporting to the ISC, and the P-coda magnitude measured from one station. Either of these may be made the basis of a method of estimating the relative yields for a suite of explosions. Once an absolute calibration is obtained, for example with CORRTEX, either of these methods may be used to supply absolute estimates of yield.

Unfortunately, the Lg data for the same 23 Semipalatinsk explosions, as studied by Nuttli[29], do not exhibit the same degree of internal consistency when compared with the P-coda method, as shown in Figure 43.

The problem, quite plausibly, is with the quality of the data Nuttli was obliged to use. Lg is a regional wave, yet he had to work with stations at least 2000 km and in some cases more than 3000 km from the source. This contrasts with the data underlying Figure 38, which was obtained for NTS with seismic stations that were less than 1000 km from the explosions.

The overall conclusions of this review of Lg and P-wave coda, for purposes of yield estimation, are therefore:

(1) that Lg, which works so well for NTS, is not yet showing the self-consistent properties one should require for making precise relative yield estimates at Shagan River; and

(2) that the P-wave coda method has promise.

In order for Lg to work better, as an independent seismic method for estimating yields of nuclear explosions at Shagan River, it appears necessary to improve signal-to-noise ratios for this wave. This could be achieved by using in-country seismic stations. If a few calibration yields became available, either from CORRTEX measurements or from the exchange of yield information on a few past explosions (such as would follow upon TTBT ratification), then experience with regional waves from NTS explosions indicates that Lg signals would be competitive with the conventional P-wave signal, for purposes of accurately estimating explosion yield.

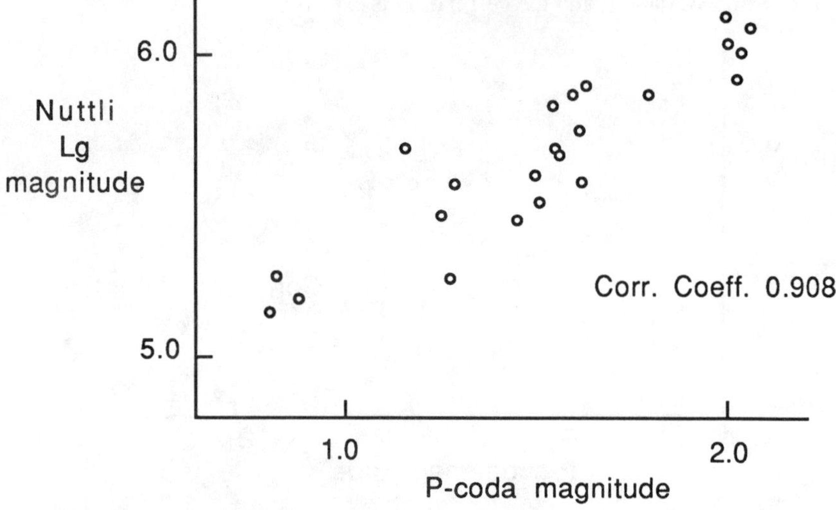

Fig. 43. Comparison of Lg magnitudes[29], and P-coda magnitude[27], for 23 Shagan River explosions, showing a worse correlation than between m_b and P-coda.

What is meant by "uncertainty" in yield estimation?

Once the yield of an underground nuclear explosion has been estimated, the next step to be taken is to try and characterize the accuracy of the estimate. This technical subject comes close to some political issues, because the Threshold Test Ban Treaty has been characterized as "not effectively verifiable" by the Reagan Administration. To discuss this subject we need to understand what is meant by the so-called "factor of uncertainty", which is conventionally used to characterize yield estimation accuracy. The term may be explained in conjunction with Figure 44, which presents an example in which the yield estimate is centered on 150 kilotons.

Thus, when one gives a yield estimate as, say, 150 kilotons, one is not saying 150 kilotons is the certain value. Instead, one is saying the actual yield is somewhere near 150 kilotons, and the question has become: "what is the uncertainty that one associates with the yield estimate?" The Figure illustrates an answer to this question, in the case where the "factor of uncertainty" is 2, which is a value that in recent years has often been stated as characterizing the uncertainty of seismic methods of yield estimation, for Soviet underground nuclear explosions. It is important first to understand what is meant by the "factor of uncertainty"; and then to appreciate why its value, in application to yield estimates based on teleseismic data, is currently thought by seismologists to be better than (i.e., smaller than) the often-quoted value of 2.

If we say the estimate basically is "150 kilotons with a factor of 2 uncertainty", it means that the range created by starting with 150 kt, and multiplying and dividing by 2 (i.e., from 75 to 300 kt), has a 95% probability of including the actual yield. So, the "factor of uncertainty", symbolized as F, is simply the multiplier needed to specify the range of yields, around the central value, corresponding to the 95% confidence interval.

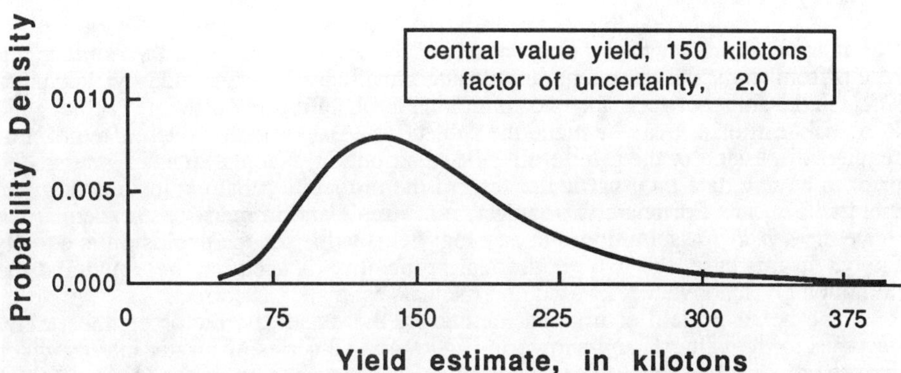

Fig. 44. Depicts the uncertainty of yield estimation. The probability that actual yield lies in the yield range from A to B is given by the area under the curve from yield A to yield B. The curve is correctly scaled, in that the total area under the curve is exactly unity. Half the area lies to the left of 150, and half to the right. This is a conditional probability, in the sense that it depends on the correctness of assumptions used to generate the probability density. For example, it depends on whatever arguments went into conclusions that the central value yield is unbiassed; and that logarithms of yield estimates are normally distributed. It is assumed before the event that any magnitude for the explosion is equally likely.

The basis for an appreciation of this definition of F is that, for explosions of fixed yield, seismic magnitudes show scatter with a normal (gaussian) distribution. Since m_b is linearly related to log(yield) and this is the route to providing a yield estimate (see Fig. 20), it follows that yield estimates are log normally distributed for explosions of fixed yield. "One sigma" expressed in magnitude units (e.g. $\sigma = 0.07$, as in Fig. 40) must be divided by the slope b of the m_b – log(yield) relation to obtain the relevant sigma value for the normal distribution of log(yield). It is this sigma value that, when multiplied by 1.96, gives the 95% confidence limits on estimates of log(yield). It follows that F is related to the standard deviation of a magnitude determination, σ, by

$$F = 10^{1.96\,\sigma/b},$$

giving, for example, an F of 1.53 for Fig. 40 if the slope b is taken as 0.74 (as in Figs 19, 20).

Note first that this relationship between the standard deviation of magnitude and the factor of uncertainty of the resulting yield estimate presumes the availability of a magnitude - yield curve that is correctly calibrated for the test site on which the explosion of interest is conducted. Uncertainty in the constants of this curve (intercept a, often broken down into a reference, a_0, plus the bias; and slope b) translates into greater uncertainty in F; and the wrong choice of constants will in general lead also to systematically biassed values of the central value yield [i.e., the yield corresponding to the best estimate of log(yield)].

Note also that the definition of F, given above, with its reference to "95% probability of including the actual yield", has applied in reverse the interpretation of scatter in observed magnitudes that is the basis for our use here of gaussian distributions. We can determine experimentally the scatter in *estimates* of log(yield) (i.e., in m_b), for a set of actual yields that are fixed at a particular value. Thus, for explosions at fixed yield, we can find by repeated experiment, conceptually if not in practice, the 95% confidence interval of *estimated* yield, given the *actual* yield. It is based on the conditional probability distribution,

P(estimate | actual).

But, in application, we want the probability of the *actual* yield, given the *estimate* that results from an m_b measurement (and its interpretation via a magnitude - yield curve). The relationship between the two conditional probabilities, P(estimate | actual) and P(actual | estimate), can be made the subject of a Bayesian approach. In turn, this requires discussion of the prior probability distribution, P(actual), i.e. an assessment, prior to having data on a particular test, of the probability that explosions of given yield will occur. Fortunately, however, it is simple and in many cases adequate to make the *a priori* assumption that any log(yield) value for the explosion is equally likely. In this case, the two conditional probability distribution are equal and the definition given above for *F*, based on P(actual | estimate), is correct.

The better a yield estimation method is, the closer the factor of uncertainty becomes to the value 1, and shown in Figure 45 is a series of master curves which convey graphically a sense of the increasing confidence one has in the "central value" yield estimate of 150 kilotons, as the factor of uncertainty drops from 2 down to 1.3.

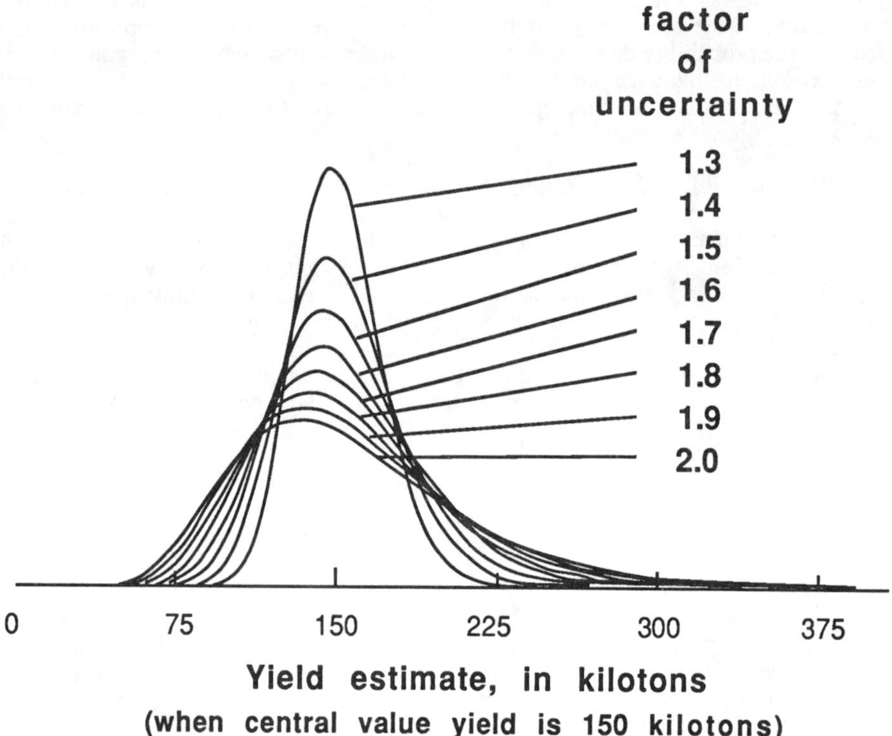

Fig. 45. Probability distributions for yield estimates, with different factors of uncertainty.

The factor of uncertainty for the CORRTEX method, if it were to be applied to Soviet explosions, is claimed to be 1.3[30]. How does this accuracy compare with seismic methods? This question is addressed in Figure 46, for the Nevada Test Site, using three different types of seismic waves (P-waves, Lg-waves, and Rayleigh waves), which separately are shown to have factors of uncertainty amounting to 1.45, 1.74, and 2.13, the number in each case being a characterization of the amount of scatter on the relevant magnitude-yield curve. It is possible then to combine the yield estimates, derived from the three types of seismic wave measurements, into a weighted average called the "unified seismic yield estimate". This turns out to have a factor of uncertainty of only 1.33 in this Figure 46 study, done by the Defense Advanced Research Projects Agency, of the Nevada Test Site.

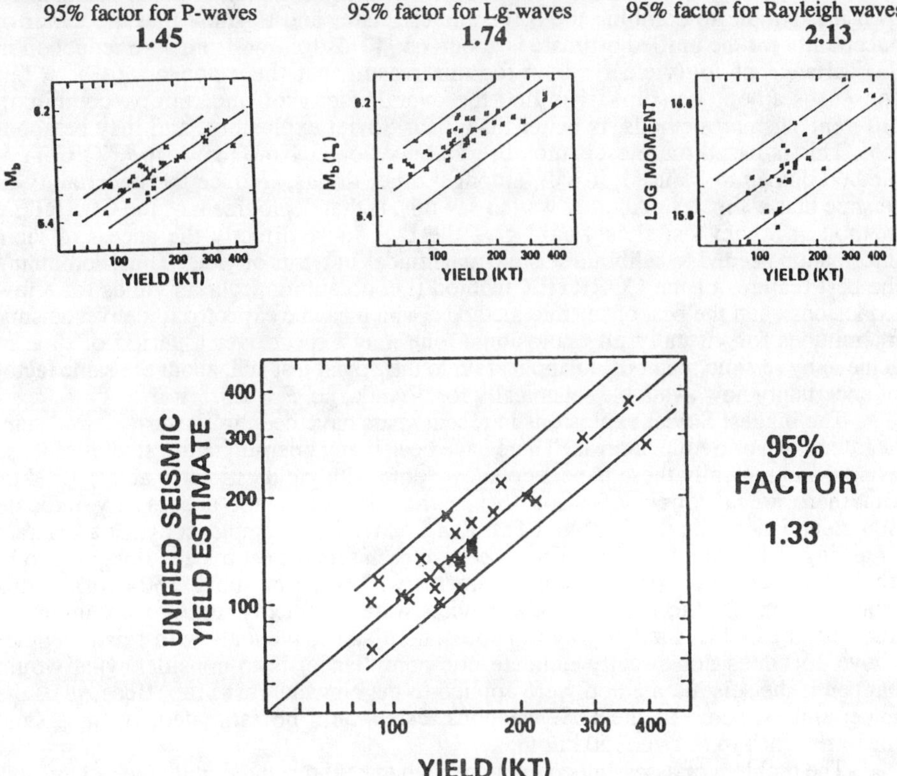

Fig. 46. Data scatter, and related factors of uncertainty, pertinent to seismic methods of yield estimation at the Nevada Test Site[14].

Note that there are two basic reasons for uncertainty in estimating yields at test sites where one does not have direct access to a calibrated magnitude-yield curve. There is the type of uncertainty which arises simply because of scatter in the data that shows up in magnitude-yield diagrams, when one has access to yield information (as the U.S. has in the case of the Nevada Test Site). And then there is additionally the systematic uncertainty which arises as a result of having to make an estimate of the bias. This latter type of uncertainty is particularly important in estimating the size of

Soviet explosions. Figure 46 indicates that the factor of uncertainty for nuclear explosions at the Nevada Test Site is about 1.5 using P-waves alone, and is about 1.3 overall if P, Lg, and Rayleigh waves are all used and the results combined into a unified estimate. But there is argument on what the overall factor of uncertainty is using seismic methods for the Soviet test site, because at present it is necessary to estimate the bias of a magnitude-yield curve derived elsewhere in order to use it for the Soviet explosions. Allowing for this uncertainty in the bias, the factor of uncertainty here is usually taken to be 2 for the main Soviet test sites using P-waves alone[30]. If the factor is also taken as 2 for independent estimates based on Lg data and Rayleigh wave data (and note that surface waves can be observed somewhat better for explosions at Semipalatinsk than in Nevada, because of the better azimuthal distribution of seismic stations observing over purely continental paths in Eurasia), then it is simple to combine the 3 seismic estimates and to show that the factor of uncertainty for the unified estimate is about 1.5. [This follows from a $\sqrt{3}$ reduction in the variance of log(yield).] It is for this reason that there appears to be a fair consensus among seismologists that the overall factor of uncertainty, combining different seismic methods, is better than 2 for Soviet explosions, and may be about 1.5. This is based on teleseismic observations alone. Note that if the CORRTEX method comes to be used, it will, amongst other things, reduce the uncertainty in seismic bias estimates. Another way to say this, is that deployment of the CORRTEX method at Soviet test sites would give the U.S. more directly the access to yield information needed to calibrate seismic magnitudes in terms of yield. Thus, combining the best features of the CORRTEX method (i.e. obtaining accurate yields for a few explosions), and the best of seismic methods (which can be expected to deliver seismic magnitudes for virtually all explosions), one may expect over a period of time to achieve by seismic methods, in application to the Soviet test site, about the same factor of uncertainty now available seismically for Nevada, i.e. $F = 1.3$.

The biggest Soviet explosions in recent years have been in Eastern Kazakhstan, near the town of Semipalatinsk. There have been many open-literature studies of these events, and typically these papers do show some with yield estimates above 150 kt.[5] But there are a couple of ways to see that this does not necessarily indicate non-compliance with the Theshold Test Ban Treaty. For example, consider a program of testing at 149 kt every year. Then, because each seismic estimate is not going to be absolutely accurate, some seismic estimates will come in above 150. In fact, this would be *expected* to happen, if the Soviets were repeatedly testing right up to the allowed threshold. Another way to appreciate that a few events with seismic yields above 150 does not directly indicate non-compliance, is to consider what would happen if the seismic method were applied to the Nevada Test Site. Because of the uncertainties, some of the U.S. weapons tests would be estimated, from seismic measurements, to be over 150 kilotons.

The problem of assessing compliance with a yield threshold will always have this difficulty because of technical uncertainties. The technical uncertainty is not unique to seismology, nor to 150 kilotons. If a method that had a 1.1 factor of uncertainty were used, and testing was routinely taken up to the threshold but not above, then there would still be a nearly 50-50 probability that the "central value" estimate would exceed that threshold for the largest events.

Further comments on yield estimation

A key question, separate from the compliance issue, is whether the uncertainty is large enough that undetected cheating could have military significance (i.e., so that the Soviets could test significantly above the 150 kiloton threshold without U.S. detection that such violations were occurring). This issue was addressed in congressional testimony before the House Foreign Affairs Committee on May 8 1985, when

Congressman Stephen Solarz asked Dr. Donald M. Kerr, then Director of the Los Alamos National Laboratory, the following question: "Dr. Kerr, assuming ... there is a 30-percent variance here, do you believe that a test at 195 kilotons in comparison to a test of a device with a yield of 150 kilotons would give the Soviet Union any potentially military significant advantages which we would not have if we ... didn't test over 150 kilotons; and, if so, could you ... give us some sense of what those advantages might be?" Dr. Kerr replied: "Fortunately I don't have to deal with the 'if so' because I don't think the difference between 195 and 150 would be militarily significant. The sort of thing that I think would be, is if the Soviet Union, for example, could test at a level like 300 kilotons."

The reason why accuracy in yield at the level of ± 30% is not needed, from the point of view of military application, is that other characteristics such the accuracy of hitting a target are more important.

In looking at recent debates on TTBT verification, it appears that discussion is conducted on two levels. On the one level, there is a debate within the expert seismological community on what precise value should be used for the bias term in the equation relating seismic magnitude and explosion yield; and on what the factor of uncertainty is for the resulting yield estimate. It appears that the range of opinion extends roughly over factors from 1.5 to 2.0[14]. But there is another level of debate in which clearly some policymakers have reached a conclusion basically that the seismological method as a whole may be seriously flawed. This was apparent throughout the many 1985-87 congressional hearings on whether the TTBT should be ratified. An important reason why many policymakers have been sceptical of seismology is contained in the following quotation from the present Director of Los Alamos, Dr. Siegfried Hecker, writing early in 1987 to the Senate Foreign Relations Committee[31]: "The teleseismic yield estimate commonly is stated to be uncertain by a factor of 2 at a 95% confidence level ... Since this uncertainty first was attributed to teleseismic yield estimates, always at the 95% confidence level, actually official estimates of the yields of Soviet tests have retroactively been reduced several times; by now the cumulative reduction near the TTBT limit is almost a factor of 5."

This is important, obviously, as a commentary on past official estimates of Soviet yields. Dr. Hecker is apparently saying that for a Soviet nuclear test now officially estimated as near 150 kt, the seismic signals would at one time have resulted in an official estimate around 700 kt! Whether or not official estimates are now in line with the best that seismology can do, it is clear that changing them several times has led many policymakers to doubt the efficacy of seismology as a verification technology. Note, however, that for several years following negotiation of the TTBT it was evident from the technical literature that seismic bias estimates covered a wide range and that factors of uncertainty were up to 2 or 3[32, 33]. The situation today, in the technical community, after years of intensive research on yield estimation, is characterized more by the consensus described in the following excerpt from the executive summary of a recent Office of Technology Assessment (OTA) report[14]: "Most seismologists feel that if new [seismic] methods [incorporating the use of surface waves and Lg waves] were applied, the resulting uncertainty for measuring explosive yields in the range of 150 kt at the Soviet test site would be closer to a factor of 1.5 than a factor of two. The uncertainty of this comprehensive approach could be further reduced if calibration shots were performed and testing were restricted to areas of known geologic composition. **It is estimated that through such measures, the uncertainty in seismically measuring Soviet tests could be reduced to a level comparable to the uncertainty in seismically measuring U.S. tests. An uncertainty factor of 1.3 is the current capability that seismic methods are able to achieve for estimating yields at the Nevada test site.**"

On the question of Soviet compliance, note the view expressed for example by

Dr. Roger Batzel, who when Director of Lawrence Livermore National Laboratory adopted the view of seismologists at his institution in making the following careful statement (ref. 31, pp 220-1):

"With regard to [the correct relationship between seismic magnitude and yield for the Soviet test sites], a magnitude-yield relationship has been developed for explosions at the Nevada Test Site. However, virtually all seismic studies have shown that, because of geological differences, the relationship is different for the Soviet test sites by an amount known as the "bias". Because the effect of the bias is to shift the entire body of Soviet yield estimates to lower values, the assumed value of the bias has a major impact on one's view of the question of Soviet compliance with the yield threshold. Ultimately, it could affect one's view of the configuration of the Soviet arsenal.

"With regard to [random errors in the seismic magnitude measurements], the ... variations ... can make the estimated yield of any explosion appear to be either smaller or larger than its actual yield, even if the bias is known. For example, given a series of explosions with actual yields of 150 kt, half of them on the average will have estimated yields greater than 150 kt, and half smaller. Thus, the issue of compliance must be dealt with in a statistical framework ...

"Based on our own assessment of the relationship between yields and seismic magnitudes for the Soviet test sites and the patterns of Soviet testing, we have concluded that the Soviets appear to be observing some yield limit. Livermore's best estimate of this yield limit, based on a probabilistic assessment, is that it is consistent with TTBT compliance. However, because of the statistical uncertainty of teleseismic yield estimates and the uncertainty in extrapolating U.S. experience to Soviet test sites, we cannot rule out the possibility that a few Soviet tests may have exceeded the limit. ... I believe that it is impossible to state with certainty that the Soviets have or have not violated the TTBT."

RECENT DEVELOPMENTS, AND CONCLUSIONS

The consideration of underground nuclear testing, and whether it should be subject to further limitations, has seen increased activity in political and technical arenas in recent years. Thus, the U.S. House of Representatives passed a non-binding resolution in February 1986, requesting that the TTBT be submitted to the Senate for advice and consent to ratification. The U.S.S.R. maintained a unilateral moratorium on nuclear testing from August 1985 to February 1987. In August 1986, the House of Representatives voted to delete funding in 1987 for all U.S. test larger than 1 kt, provided the U.S.S.R. also desisted and accepted a satisfactory U.S. monitoring network. The Senate did not concur, but as part of an agreement resolving the whole federal budget for FY1987 the Reagan Administration agreed to submit the TTBT to the Senate. In 1987 and also in 1988, the House again voted to cut off funds for U.S. testing above 1 kt, and in each case the Senate acted to restore this funding.

Table 3 shows the funds allocated for nuclear testing within the Department of Energy and within the Department of Defense. They amount to several hundred million dollars each year. These numbers were supplied by DoE in 1986 as actual numbers for the prior ten years, and they are numbers for planning purposes only for the subsequent five years. The figures in Table 3 for DoD are primarily in support of special nuclear explosions conducted in horizontal tunnels in Rainier Mesa, on the Nevada Test Site, to test for the effects of nuclear weapons on military hardware.

In 1986, an extraordinary private agreement was reached between the Natural Resources Defense Council (a non-governmental U.S. group primarily working on environmental problems) and the Soviet Academy of Sciences, which has allowed U.S. seismologists to operate monitoring equipment within the Soviet Union. Tests with small chemical explosives have been carried out, showing that seismic signals up to several tens of Hz do propagate efficiently in Kazakhstan up to several hundred kilometers. However, the monitoring equipment has been deactivated on every occasion, to date, when the U.S.S.R. has carried out a nuclear explosion. In 1988, the Soviet Academy has also signed agreements with a consortium of U.S. graduate schools teaching seismology, known as IRIS (Incorporated Research Institutions in Seismology), and with researchers in France and the U.K. These agreements will permit the operation of several high-quality broadband seismometers on Soviet territory beginning in 1988, the data becoming openly available for general research. Agreements of this nature, pioneered by the N.R.D.C., are important in supplying the kind of data that is needed for future purposes of designing an in-country seismic monitoring network to meet pre-determined requirements.

($ in Millions)

	DoE Nuclear Testing	DoD Nuclear Testing*
1977	$229.1	$38.2
1978	240.0	27.7
1979	221.0	26.6
1980	213.0	28.6
1981	288.0	30.6
1982	360.0	66.2
1983	408.3	94.2
1984	475.9	94.1
1985	532.0	105.9
1986	496.1	89.8
1987	633.8	93.1
1988	682.0	110.8
1989	734.0	117.1
1990	790.0	127.9
1991	850.0	118.6

*Data supplied by Defense Nuclear Agency.

Table 3. Annual expenditures in support of the U.S. nuclear test program.

In 1987 the TTBT and PNET were formally submitted to the Senate for advice and consent to ratification, but President Reagan stated that he had reservations on the need for improved verification and that he would not proceed with ratification even if Senatorial consent were forthcoming. Several Senators concluded that this was an unfortunate constitutional precedent because they were being asked to give advice and consent to a pair of encumbered treaties that in fact were not on track to ratification, because a new verification agreement was being sought that, if concluded, would require a second stage of approval. In Senator Sarbanes words, the Senate was being asked to engage in "an empty act." Consent to the treaties has yet to be debated on the Senate floor.

Each year since 1984, including 1988, the Reagan Administration has reported to the U.S. Congress on Soviet non-compliance with arms control treaties, and has found that: "Soviet nuclear testing activities for a number of tests constitute a likely violation of legal obligations under the Threshold Test Ban Treaty."

In September 1987, reversing a decision reached early in the Reagan Administration, a U.S.-Soviet joint statement announced the resumption of negotiations on limitations on nuclear testing with the following words: "**In these negotiations the sides as the first step will agree upon effective verification measures which will make it possible to ratify the US-USSR Threshold Test Ban Treaty of 1974... and [will] proceed to negotiating further intermediate limitations on nuclear testing.**" What is important here, is the commitment by the U.S. to move on to consider new limitations on testing. This commitment, whether one deems it desirable or not, lends added importance to the outcome of new verification provisions for the TTBT. Negotiations began in Geneva in November 1987, and have centered on U.S. use of CORRTEX monitoring equipment for large Soviet tests. In part as a demonstration of this technique, a pair of "Joint Verification Experiments" (JVE's), agreed to under procedures signed at the May 1988 Summit Meeting in Moscow, are to be conducted in summer 1988, one JVE at the Nevada Test Site and one about a month later at Shagan River.

As part of the JVE agreement, several scientists associated with the negotiating teams have visited the other country's test site. To deliver the sophisticated equipment needed for execution of the CORRTEX procedure in the JVE, seven C5A planes of the U.S. Air Force have flown into Eastern Kazakhstan, into what hitherto has been one of the most sensitive parts of the U.S.S.R. Most of this equipment is for the specialized drilling needed to meet two requirements of the so-called "satellite hole", located near the hole in which the nuclear device itself is emplaced. Thus, the nuclear device will be exploded at a depth of around half a kilometer. The two requirements are that the satellite hole be steered while drilling so that it ends up about ten meters from where the nuclear device will be; and that the actual location of the bottom of the satellite hole be known, with respect to the nuclear device, to about one meter. There will also be CORRTEX equipment in the emplacement hole to serve as a standard measure of the yield, and the JVE will be a demonstration of the capability of CORRTEX as used in the satellite hole. In the forty years of nuclear weapons development by the superpowers, these JVE's are a remarkable event. They will, of course, be recorded seismically all over the world, and are intended to be about the same yield – a little under 150 kt. To a seismologist, they constitute a multi-million dollar experiment to check the concept displayed in Figure 21, and will perhaps supply as direct an estimate of the bias between test sites as could be conceived. More importantly, of course, they will enable an independent and direct calibration (for purposes of yield estimation), of seismic signals that have been recorded from explosions at the Shagan River test site for over 20 years.

As part of the agreement to conduct the JVE's, the superpowers exchanged yield information in June 1988 about the yield, depth of burial, measured seismic magnitudes, and geological setting of five large underground nuclear weapons tests conducted by each side at Shagan River and Nevada in the period 1978-1987. This is a more extensive exchange than that called for in the TTBT as originally negotiated – the agreement then being that such information on only two previous explosions (by each side, per test site) would be exchanged, once the treaty was ratified. The exchange had not taken place until newly negotiated in 1988, because of course the treaty has not been ratified. The new information has not as yet been openly released, because it is regarded as part of the ongoing negotiations. It is a political question, whether to use such Soviet-supplied data on yield, for the technical exercise of calibrating the Shagan River magnitude - yield curve. However, there is a sense in which this data too can potentially be validated by the JVE's themselves.

While the JVE's alone will not settle the question of what the bias value is to two or more decimal places (perhaps no experiment could ever do this, for at such a level of accuracy the bias value varies within a test site), their seismic signals will be looked at with more care and with more interest than perhaps any underground explosion since nuclear explosion seismology began with RAINIER in 1957. However, to Spurgeon Keeny, Jr., a specialist in arms control who has been close to nuclear test ban issues since he worked in the 1950's for President Eisenhower on negotiating a CTBT, the Shagan River JVE is "a monument to the incredible lengths to which the United States is prepared to go to achieve technical solutions to unimportant or even non-existent problems."[34]

Once the yield estimation issues related to the TTBT are resolved, and attention turns to the second item on the declared agenda of current U.S.-U.S.S.R. nuclear testing negotiations, the decision will have to be made on what "further intermediate limitations on nuclear testing" should be considered. (Recall that a CTBT is acknowledged as an eventual goal, for both the U.S. and U.S.S.R.) At least from the perspective of verification, and given the difficulty if not the impossibility of monitoring a truly comprehensive test ban, it appears that there is merit in reducing the allowed yield threshold in stages.[35, 36] Quoting again from the 1988 OTA report[14], such an approach "would begin with a limit that can be monitored with high confidence using current methods, but would establish the verification network for the desired lowest level. The threshold would then be lowered as information, experience, and confidence increase."

REFERENCES

1. D.M. Kerr, Nuclear Weapons Tests: Prohibition or Limitation? (edited by J. Goldblat and D. Cox, Oxford University Press, N.Y., 1988), p. 43.
2. Committee on International Security and Arms Control, Nuclear Arms Control - Background and Issues (National Academy Press, Washington, D.C., 1985), chap. 7.
3. G.T. Seaborg, Kennedy Khruschev and the Test Ban (Univ. of Calif. Press, Berkeley, 1981), p. 239.
4. H.F. York, Making Weapons, Talking Peace (Basic Books Inc., New York, 1987), chap. 14.
5. L.R. Sykes, and S. Ruggi, in Nuclear Weapons Databook, Volume IV: Soviet Nuclear Weapons (eds. T. B. Cochran, W.A. Arkin and J.I. Sands, Ballinger Publishing Company, Cambridge Massachusetts, in press for 1988), chap. 10.
6. NORSAR Semiannual Technical Summaries (Royal Norwegian Council for Scientific and Industrial Research, Kjeller, Norway, since 1982).
7. R.W. Alewine III, Defense 85, 11 (December 1985).
8. K. Aki and P.G. Richards, Quantitative Seismology – Theory and Methods, (W.H. Freeman and Co., San Francisco, California, 1980).
9. P.D. Marshall and P.W. Basham, Geoph. J. R. A. S., v 28, 431 (1972).
10. S.R. Taylor, M.D. Denny and E.S. Vergino, UCID-20642 (Lawrence Livermore National Laboratory, Jan 1986).
11. J.L. Stevens and S.M. Day, J. Geophys. Res., 3009 (1985).
12. F. Ringdal, in The VELA Program: A Twenty-Five Year Review of Basic Research (ed. A.U. Kerr, Executive Graphic Services, 1985), p. 611.
13. R.C. Lilwall and A. Douglas, AWRE Report O 23/84 (U.K. Atomic Weapons Research Establishment, 1985).
14. U.S. Congress, Office of Technology Assessment, Seismic Verification of Nuclear Testing Treaties, OTA-ISC-361 (U.S. Govt. Printing Office, Washington, DC, 1988).

15. J.F. Evernden, C.B. Archambeau, and E. Cranswick, Revs. of Geophysics, 143 (1986).
16. W.J. Hannon, Science, v 227, 251 (1985).
17. O.W. Nuttli, J. Geophys. Res., 2137 (1985).
18. J.R. Murphy, in Identification of Seismic Sources – Earthquake or Underground Explosion (ed. E.S. Husebye and S. Mykkeltveit, D. Reidel Publ. Co., 1981), p. 201.
19. Maps of station bias supplied as a personal communication to Richards in 1985, from Peter Marshall of the United Kingdom Atomic Weapons Research Establishment.
20. P.D. Marshall, R.C. Lilwall, and J. Farthing, AWRE Report No. O 21/86 (U.K. Atomic Weapons Research Establishment, 1986).
21. A. Douglas, Bull. Seism. Soc. Amer., 270 (1987).
22. Z. Der, T. McElfresh, R. Wagner, and J. Burnetti, Bull. Seism. Soc. Amer., 379 (1985), and erratum in Bull. Seism. Soc. Amer., 1222 (1985).
23. P.D. Marshall, A. Douglas, and J.A. Hudson, Nature, v. 234, 8 (1971).
24. J.W. Given and G.R. Mellman, Source Parameters of Shagan River East Kazakh USSR Events Using Surface Wave Observations, paper presented in Basic Research in the VELA Program, a symposium held in Santa Fé, New Mexico, May 7–9, 1984.
25. J.W. Given and G.R. Mellman, "Source Parameters for Nuclear Explosions at NTS and Shagan River from Observations of Rayleigh and Love Waves", paper presented at 7th Annual DARPA/AFGL Seismic Research Symposium, U.S. Air Force Academy, Colorado Springs, Colorado, May 6–8, 1985.
26. L.R. Sykes and I.L. Cifuentes, Proc. Nat. Acad. Sci. USA, v. 81, 1922 (1984).
27. I.N. Gupta, R.R. Blandford, R.A. Wagner, J.A. Burnetti, and T.W. McElfresh, Geophy. J. R. A. S., v. 83, 541 (1985).
28. P.D. Marshall, T.C. Bache, and R.C. Lilwall, AWRE Report No. O 16/84 (U.K. Atomic Weapons Research Establishment, December 1984).
29. O.W. Nuttli, Bull. Seism. Soc. Amer., 1241 (1986).
30. United States Department of State Special Report No. 152, "Verifying nuclear testing limitations: possible U.S. - Soviet cooperation" (August 14, 1986).
31. U.S. Congress, S. Hrg. 100-115, p. 378 (1987).
32. Defense Advanced Research Projects Agency, A Technical Assessment of Seismic Yield Estimation, 3 vols (January 1981).
33. T.C. Bache, Bull. Seism. Soc. Amer., p. S131 (1982).
34. S.M. Keeny, Jr., Arms Control Today, 2 (June 1988).
35. P. Doty, Foreign Affairs, 750 (Spring 1987).
36. P.G. Richards, in Verification and Compliance: a Problem-Solving Approach (eds. M.Krepon and M. Umberger, Macmillan, UK, in press for July 1988), chap. 4.

CHAPTER 5

MONITORING YIELDS OF UNDERGROUND NUCLEAR TESTS USING HYDRODYNAMIC METHODS

F. K. Lamb

Program in Arms Control, Disarmament, and International Security
and Departments of Physics and Astronomy
University of Illinois at Urbana-Champaign
and
Center for International Security and Arms Control
Stanford University

ABSTRACT

The yields of nuclear explosions can be estimated using hydrodynamic methods. The approach that has been proposed by the United States for nuclear test ban monitoring makes use of the fact that the initial speed of the expanding shock wave produced by an underground explosion increases with the yield. Several techniques have been developed in the United States to measure the speed of the shock wave, of which the so-called CORRTEX technique is the most recent and best. A variety of algorithms have been used to derive yield estimates from shock wave radius vs. time measurements. Although more intrusive than seismic methods, current hydrodynamic methods could be used to monitor the Threshold Test Ban and Peaceful Nuclear Explosion treaties, provided that appropriate changes in these treaties are negotiated and that adequate cooperative arrangements are made to assure accuracy. Significant engineering, operational, and analysis problems need to be solved before these methods could be used to monitor with confidence a low-threshold test ban. The methods are not relevant to a comprehensive test ban.

I. INTRODUCTION

Hydrodynamic methods have long been used to estimate the yields of nuclear explosions, both in the atmosphere[1] and underground.[2] All such methods are based on the fact that the strength of the shock wave produced by an explosion increases with the yield. Hydrodynamic methods were introduced as a treaty-monitoring tool in the Protocol of the Peaceful Nuclear Explosions Treaty (PNET) of 1976, which explicitly established such methods as among those that could be used to monitor the yield of any salvo of explosions with a planned aggregate yield greater than 150 kilotons (kt).[3] Hydrodynamic methods have recently become a focus of attention as a result of controversy over monitoring of the Threshold Test Ban Treaty (TTBT).

The TTBT was signed by the United States and the Soviet Union in 1974 and banned underground nuclear tests with yields greater than 150 kt after March 31, 1976.[3] Although neither party has ratified the TTBT, both have separately stated that they will respect the 150 kt limit. From that time to the present, the U.S. government has relied primarily on yield estimates based on teleseismic body wave magnitudes measured outside the Soviet Union to monitor the TTBT limit.[4] According to the U.S. Department of State, this method has an uncertainty of approximately a factor of two at the 95% confidence level for Soviet tests with yields near 150 kt.[5,6] This statement means that 95 times out of 100 the estimated yield of a 150 kt explosion will lie between 75 kt and 300 kt. Put another way, there is only one chance in 40 that a single explosion with a yield of 300 kt would appear to have a most likely yield of 150 kt or less. The probability that two such explosions would appear to be treaty compliant is 1 in 1600, and so on.[7] According to recent studies, the best seismic methods now available are expected to have an uncertainty of a factor of 1.5–1.6 at the 95% confidence level for explosions with yields above 50 kt, if measurements are made only outside the Soviet Union and the Soviet test site is not calibrated, or a factor of 1.3, once the Soviet test site is properly defined and calibrated.[4]

Prior to the Reagan administration, the precision of remote seismic methods was considered adequate. However, the Reagan administration has stated that "the remote seismic techniques we must rely on today to monitor Soviet nuclear tests do not provide yield estimates with the accuracy required for effective verification of compliance."[5] As an alternative to these techniques, the Reagan administration has since 1983 strongly advocated routine use of hydrodynamic methods to monitor the TTBT.[5,6] It believes that it has identified in hydrodynamic yield estimation methods an approach "which will reduce the uncertainty in yield measurement to an acceptable level and will do so without danger of compromising other sensitive information about the nature or performance of the nuclear device whose yield is to be measured."[6]

The present article reviews hydrodynamic yield estimation methods and their application to test ban monitoring. §II provides a brief overview of the development of an underground nuclear explosion and the variety of yield estimation methods currently in use. §III summarizes some relevant material properties of rock and describes the evolution of the shock wave produced by an underground nuclear explosion. In §IV we explain the technique advocated by the Reagan administration to measure the evolution of the shock wave. In §V we describe the algorithms that are currently used to derive yield estimates from these measurements. Technical issues related to the use of hydrodynamic methods to monitor test ban agreements are considered in §VI. In §VII we briefly compare hydrodynamic and seismic yield estimation methods, discuss the current status of U.S.-Soviet test ban negotiations, and mention several public policy issues raised by the U.S. drive to gain Soviet acceptance of hydrodynamic methods for routine monitoring of the TTBT. Our conclusions are summarized in §VIII.

II. MONITORING UNDERGROUND NUCLEAR EXPLOSIONS

Currently, about 90% of U.S. nuclear tests are conducted in vertical shafts at depths of 250–650 m; the remainder are conducted in tunnels.[8] In preparing for a test that will be conducted deep underground, the vertical hole that will contain the nuclear charge (the so-called emplacement hole) is first drilled, a process that typically takes 8–10 weeks. The charge and diagnostic equipment are then placed in canisters, which can be as much as 15 m in length, and lowered into the emplacement hole. Depending on the nature of the test, a variety of diagnostic pipes and cables may lead upward from the canisters to the surface (current U.S. nuclear weapon tests involve anywhere from ~50 to ~250 diagnostic pipes and cables). After the canisters are in place, the hole is stemmed with sand, gravel, and plugs.[9]

In the following two subsections we first describe what happens when a nuclear charge is exploded deep underground and then explain briefly the methods that are used to estimate the yields of such explosions.

A. Phases of an Underground Explosion

For present purposes the time development of the explosion may be divided into the following three somewhat simplified phases:[10,11]

Initial phase.—The release of nuclear energy is accompanied by emission of nuclear radiation, fission fragments, and thermal electromagnetic radiation. The temperature in the nuclear charge rises steeply, reaching 10^7 K within a microsecond or so. At the very earliest times, the energy of the explosion is carried outward by the expanding weapon debris and radiation. Soon, however, the vaporized nuclear charge and nearby rock form a bubble of hot gas in which the initial pressure is hundreds of Mbar. The enormous pressure in the bubble causes it to expand rapidly, creating a cavity and driving a shock wave into the surrounding rock. The final radius R_c of the cavity depends somewhat on the depth of the explosion and the composition of the surrounding rock, as well as the yield. For a burst of yield W, a useful appproximate expression is[12]

$$R_c \approx 14\,(W/1\,\mathrm{kt})^{1/3}\,\mathrm{m}\,. \qquad (1)$$

The cavity reaches its final radius in about $90\,(W/1\,\mathrm{kt})^{1/3}$ ms.

Hydrodynamic phase.—The shock wave initiated by the expansion of the hot gas propagates outward at a speed that is initially much greater than the speed of sound in the surrounding, undisturbed rock. At this early time, the stress produced by the shock wave greatly exceeds the critical stress at which the rock becomes plastic, so that to a good approximation the rock can be treated as a fluid. This phase is therefore referred to as the "hydrodynamic" phase. (In defining the hydrodynamic phase, we emphasize the prefix *hydro* and simply require that the shocked rock behave like a fluid. Other authors emphasize instead the root *dynamic* and require not only that the shocked rock behave like a fluid, but also that the speed of the shock wave greatly exceed the speed of sound in the rock; this second usage is common in the Soviet Union.) As the shock wave continues to expand, it weakens. Eventually, the strength of the rock can no longer be neglected. This marks the end of the hydrodynamic phase.

Final phase.—Even after the compression wave is no longer hydrodynamic, the rarifaction wave that follows is still strong enough to fracture rock. Intense fracturing typically occurs out to a radius $\sim 3R_c$.[12] Beyond this point, the degree of fracturing caused by the expanding shock wave drops dramatically until, at $\sim 5R_c$, fracturing essentially stops. (Rarifaction waves caused by reflection of the shock wave from the surface or collapse of the roof of the cavity may cause fracturing beyond this radius.) The shock wave then continues to expand nearly elastically, eventually evolving into the leading wave of a train of elastic (seismic) waves. These waves, which typically carry $\lesssim 5\%$ of the energy of the explosion, propagate through and around the earth and can be observed at points thousands of kilometers from the site of the explosion.

B. Monitoring Methods

Three different types of methods are commonly used to estimate the yields of nuclear weapon tests. These types make use of phenomena that occur during the three phases of the explosion identified above.

Radiochemical methods make use of the nuclear reactions that occur during the first phase of the explosion. By knowing the relative abundances of various nuclides in the original nuclear device and by determining the relative abundances of fission fragments and fusion products after the explosion, the yield of the explosion can be estimated. However, at present there are several barriers to using radiochemical methods to monitor test ban treaties. First, the monitoring party must be able to make a variety of measurements at the test site. Second, to achieve high precision with this method some knowledge of the design of the nuclear charge may be required. Third, in addition to the yield, radiochemical methods can provide other information about the design and performance of nuclear devices, which may be considered sensitive. For these reasons, radiochemical methods are not usually considered for treaty monitoring.[13]

Hydrodynamic methods make use of the fact that the strength of the shock wave produced by an explosion increases with the yield, other things being equal. As a result, the peak particle velocity, pressure, and density are greater at a given radius for explosions of greater yield. By comparing measurements of these quantities with a model of the evolution of the shock wave based on knowledge of the nature and structure of the geologic media in which the explosion occured, the yield of the explosion can be estimated. In order to use hydrodynamic methods to monitor test ban treaties, the monitoring party must have access to the test site in order to determine the relevant properties of the geologic media there before the test and to measure the evolution of the shock wave during the test. To assure high accuracy, constraints on the test geometry are also required. Hydrodynamic methods are not part of current TTBT verification provisions.

Seismic methods make use of the ground motions caused by the elastic waves that propagate through and around the earth during the third stage of an underground nuclear explosion. Some seismic waves (such as body and surface waves) propagate to so-called teleseismic distances (≥2,000 km) from the explosion. Yield estimation methods based on these waves can be used with measurements of ground motion made at stations outside the country in which the test occurs. Other seismic waves (such as L_g waves) typically propagate only to regional distances (<2,000 km). Yield estimation methods based on these latter waves may require data from in-country stations. To assure high accuracy, knowledge of the geologic media at the test site and the way in which the earth near the site transmits seismic waves is required. As noted earlier, the United States routinely uses seismic data taken at stations outside the Soviet Union to monitor the TTBT. In-country monitoring stations and independent access to data on the seismic properties of the test site are not part of the current verification provisions of the TTBT. For recent reviews of seismic methods, see refs. 4 and 14–18.

It has been claimed several times in recent Congressional hearings on TTBT and PNET verification[19–23] that hydrodynamic methods are "direct" whereas seismic methods are not. From a scientific point of view there is no such distinction. All three methods just described involve (1) production of a signal by the exploding charge, (2) propagation of the signal to locations more or less remote from the detonation point, and (3) detection of the signal by sensors at those locations. Important questions are how the size of the signal varies with yield, how well the propagation of the signal is understood, and how accurately and precisely the sensors can measure the signal.

It has also been asserted[22] that use of hydrodynamic methods in and of itself eliminates the possibility of systematic error or "bias". Obviously it does not. All three methods are subject to both systematic and random errors. Relevant questions are the expected sizes of the errors, and whether they are so large as to be of concern.

III. SHOCK WAVE EVOLUTION

In hydrodynamic methods of yield estimation, the size of the explosion is estimated by fitting a model of the evolution of the shock wave, which depends parametrically on the yield, to measurements of the motion. Shock waves in rock behave differently from shock waves in air primarily because the atoms in rock are close together and interact strongly.[24] Therefore, in discussing the application of hydrodynamic methods to underground explosions it will be helpful to have in mind some relevant material properties of rock as well how shock waves produced by underground explosions evolve in rock.

A. Rock Properties

The strength of a shock wave can be characterized by the peak pressure that it produces. Weak shock waves and acoustic waves in rock propagate at a constant speed, the so-called elastic wave speed[25]

$$c_\ell = \left(\frac{K_0 + \frac{4}{3}G_0}{\rho_0}\right)^{1/2} . \tag{2}$$

Here K_0 and G_0 are the bulk and shear modulii, respectively, of the rock in its standard state, and ρ_0 is the mass density. For granite, $K_0 \approx 360\,\mathrm{kbar}$ and $G_0 \approx 320\,\mathrm{kbar}$,[26] giving $c_\ell \approx 5\,\mathrm{km\,s^{-1}}$.

Shock waves that are strong enough to produce stresses in excess of the critical shear stress p_{crit} cause the rock to lose its firmness and to become plastic (for granite, p_{crit} is about 40 kbar for high strain rates[26]). Such waves are called plastic waves. The speed of a plastic wave increases with its strength. Thc weakest such waves propagate at the low-pressure plastic wave speed[25]

$$c_0 = \left(\frac{K_0}{\rho_0}\right)^{1/2} , \tag{3}$$

which is determined by the compressibility of the rock in its standard state. Since only the bulk modulus contributes to c_0, it is necessarily less than c_ℓ. For solid granite, $c_0 \approx 4\,\mathrm{km\,s^{-1}}$.

In the hydrodynamic regime, conservation of mass, momentum, and energy across the shock front imply[27]

$$\epsilon(p, V) - \epsilon_0(p_0, V_0) = \frac{1}{2}(p_0 + p)(V_0 - V) , \tag{4}$$

where ϵ_0 and ϵ, p_0 and p, and V_0 and V are, respectively, the internal energies, pressures, and specific volumes ahead of and just behind the shock front. By analogy with the equation relating the initial and final pressures and volumes during adiabatic compression of a fluid, this relation, which is of the form

$$p = H(V, p_0, V_0) , \tag{5}$$

is called the *shock adiabat* or *Hugoniot.*

Conservation of momentum across the front of a hydrodynamic shock wave implies

$$p = p_0 + \rho_0 D u , \tag{6}$$

where D is the speed of the shock front, measured in the rest frame of the undisturbed rock, u is the particle speed just behind the shock front, and we have assumed that the rock in front of the shock front is at rest. Thus, the Hugoniot may be expressed as a relation between D and u, that is

$$D = H(u) . \tag{7}$$

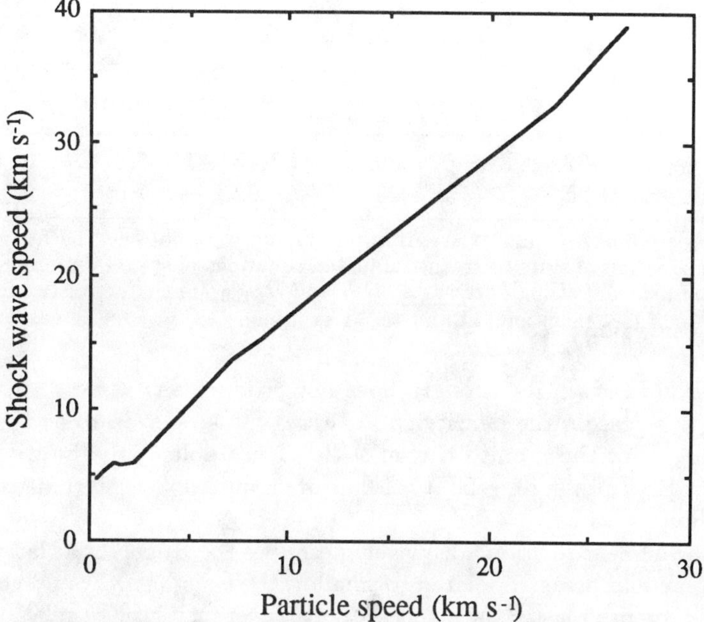

Fig. 1.—Relation between shock wave speed D and particle speed u just behind the shock front for granite (ref. 29). Note the approximate linearity of the Hugoniot at large u. The step in the curve at $u \approx 2\,\mathrm{km\,s^{-1}}$ reflects a phase transition that occurs at about 400 kbar.

Figure 1 shows a recent Hugoniot for granite expressed in this way. In general, the Hugoniot depends on the chemical composition, porosity, gas-filled porosity, fracture pattern, degree of liquid saturation, and other properties of the rock.

For some rocks, the Hugoniot may be adequately represented *in the hydrodynamic regime* by a single linear relation of the form[27,28]

$$D = A + Bu\,, \tag{8}$$

which implies the Hugoniot curve[27]

$$p = \frac{A^2(V_0 - V)}{(B-1)^2 V^2 \left[\frac{B}{B-1} - \frac{V_0}{V}\right]^2}\,. \tag{9}$$

Table 1 lists values of A, B, and ρ_0 for granite and wet tuff that were derived by fitting a Hugoniot of the form (9) to recent high-pressure equations of state for these materials.

For Hugoniots of the form (9), the ratio ρ/ρ_0 of the material density immediately behind the shock front to the material density ahead of the shock front increases with the strength of the shock wave until it reaches a certain value $(\rho/\rho_0)_{\max} = (V_0/V)_{\max} = B/(B-1)$. Once the wave has become this strong,

TABLE 1
Rock Equations of State[a]

Rock	ρ_0 (g cm^{-3})	A (km s^{-1})	B	$R_{t,1}$ (m)	$R_{t,150}$ (m)
Granite	2.67	2.80	1.45	3.8	20
Wet tuff	1.95	1.45	1.62	7.0	37

[a]The parameters ρ_0, A, and B are from ref. 30 and were obtained by fitting a Mie-Grüneisen equation of state to recent tabulated equations of state for granite (ref. 29) and wet tuff (ref. 31) at high pressures. $R_{t,1}$ and $R_{t,150}$ are characteristic shock wave transition radii (see text) for 1 kt and 150 kt explosions.

any further increase in its strength does not produce any increase in the ratio ρ/ρ_0. For this reason, the density ratio $(\rho/\rho_0)_{\max}$ is referred to as the limiting density ratio. For the granite Hugoniot listed in Table 1, the limiting density ratio is ~3. Peak pressures ~10–100 Mbar are required to achieve density ratios near the limiting value.

If a single linear relation adequately describes the Hugoniot at large u and if this relation could be extrapolated to small u, the constant A would correspond to the low-pressure plastic wave speed c_0. However, the large-u relation usually is not valid for small u, and hence A usually does not equal c_0. In granite, for example, A is about 3 km s^{-1} whereas c_0 is about 4 km s^{-1}.

Even if the Hugoniot is not linear over the range of u that is of interest, a curve consisting of piece-wise linear segments of the form (8) may serve as a practical approximation to $H(u)$ for many purposes.[28]

B. Simplified Model

To understand the evolution of shock waves produced by underground nuclear tests, it is helpful to consider first the shock wave that would be produced by release of a large amount of energy in an infinitesimal spherical volume (a so-called point explosion) in homogeneous rock. A qualitative understanding of the development of the shock wave produced by such an explosion can be gained from the following relatively simple model. Suppose we make the *ansatz* that throughout the motion the particle speed u just behind the shock front is related to the yield W of the explosion by the expression[30,32]

$$u(t) = \left(\frac{3fW}{4\pi R^3 \rho_0}\right)^{1/2}, \qquad (10)$$

where R is the radius of the shock front and f is a constant dimensionless factor that describes how the energy of the explosion is partitioned between kinetic energy of bulk motion and internal energy, and how the velocity, density, and internal energy of the shocked material vary with position. The factor f is constant for a self-similar shock wave (see below) but is not expected to remain constant as the shock wave evolves.[32] Nevertheless, shock wave radius and particle speed data from actual underground nuclear tests as well as computer simulations of such tests indicate that relation (10) with $f \approx 0.5$ is satisfied fairly well for granite and wet tuff until relatively late times.[30,32]

Let us now assume, for the sake of illustration, that the Hugoniot of the medium can be adequately represented by a *single* linear relation of the form (8) *over the whole range of* u *that is of interest.* Then $c_0 = A$. Therefore, in the following discussion we refer to A as the low-pressure plastic wave speed.

Given the *ansatz* (10), the Hugoniot can be rewritten as[32]

$$D \equiv \frac{dR}{dt} = A + B\,u = A\left[\left(\frac{R_t}{R}\right)^{3/2} + 1\right], \tag{11}$$

where

$$R_t \equiv \left(\frac{3fWB^2}{4\pi\rho_0 A^2}\right)^{1/3} \tag{12}$$

is the characteristic transition radius that separates the region where $D \propto R^{-3/2}$ from the region where $D \approx A$. Typical values of R_t for 1 kt and 150 kt explosions in granite and wet tuff are listed in Table 1 for the values of A and B given there. In dry alluvium, R_t can sometimes exceed 60 m for a 150 kt explosion.

Equation (11) can be integrated to obtain a simple, closed expression for $R(t)$, from which one can calculate $D(t)$, $u(t)$, $p(t)$, and $\rho(t)$.[32,33] This model shows in a qualitative way how the evolution of the shock wave depends on the yield of the explosion and the equation of state of the rock. As an example, the peak pressure, peak density, and radius of the shock front at various times are listed in Table 2, for 1 kt and 150 kt explosions in granite. The pressure p_0 of the overburden is ~200 bar for a depth of 1 km and hence can be neglected for all depths and times of interest.

The motion of the shock front given by equation (11) can be divided into three different intervals:

Strong Shock Interval.—Initially, the speed of the shock front is much greater than the speed of sound in the undisturbed rock, the pressure behind the shock front is predominantly thermal pressure, and the ratio of the density immediately behind the shock front to the density ahead of the front is close to its limiting value. This is the strong shock interval.[24]

For a spherically-symmetric point explosion in a homogeneous medium, the motion of the shock wave in the strong shock interval is self-similar.[34–36] In such a motion, the distributions with radius of the pressure, density, and particle velocity evolve with time in such a way that only their scales and the radius of the shock front change, while the shapes of the distributions remain unaltered. The evolution of such a shock wave is only weakly dependent on the properties of the medium. Thus, simple models can be used to estimate the yield if there is an interval of self-similar motion and if data from this interval are available.

TABLE 2
Shock Wave Evolution in Granite[a]

Pressure (kbar)	Density (ρ_{max})	1 kt Explosion Time (μs)	1 kt Explosion Radius (m)	150 kt Explosion Time (μs)	150 kt Explosion Radius (m)
70,000	0.9	4	0.5	20	3
10,000	0.8	10	0.9	80	5
4,000	0.7	40	1.4	200	8
1,500	0.6	90	2	500	11
500	0.5	200	3	1,200	17
150	0.4	600	5	3,000	30

[a] For the model of a spherically-symmetric, point explosion described in the text. The Hugoniot (8) was used, with the values of A and B given in Table 1. The phase transition that occurs at ~400 kbar (cf. Fig. 1) has been neglected. The unit of density, ρ_{max}, is the limiting density of granite (see text), which is 9.4 g cm^{-3} for this equation of state.

For example, in the evolution given by equation (11), the motion of the shock front radius R during the strong shock interval ($R \ll R_t$) satisfies

$$R(t) \approx W^{1/3}\left(\frac{75fB^2}{16\pi\rho_0}\right)^{1/5}\left(\frac{t}{W^{1/3}}\right)^{2/5} \propto W^{1/3}\left(\frac{t}{W^{1/3}}\right)^{2/5} . \qquad (13)$$

Thus, the exponent of time is independent of the properties of the medium. Moreover, the evolution is only affected by ρ_0 and B.

Unfortunately, in actual underground nuclear tests self-similar motion does not have time to develop, given current testing practices and the yields of interest, as explained below.

Transition Interval.—As the shock wave expands, it weakens and slows, and the peak pressure and density drop. When the shock front reaches a certain radius R_1, the peak density ratio is 0.8 times the limiting value and we say that the shock wave has entered the transition interval. (The motion of the shock wave changes only gradually and so the point at which it is said to enter the transition interval is purely conventional. Throughout the present article we use the convention that the transition interval begins when the peak density ratio falls to 80% of its limiting value.) For an explosion in granite, this occurs when the peak pressure has fallen to ~10,000 kbar (cf. Table 2). For a 1 kt explosion in granite R_1 is ~1 m, whereas for a 150 kt explosion R_1 is ~5 m.

Over most of the transition interval, the thermal pressure just behind the shock front is not much greater than the cold pressure of the compressed rock, although the speed D of the shock front is still much larger than the low-pressure plastic wave speed A. In this interval, the motion of the shock wave is more sensitive to the properties of the medium than it is in the strong shock interval, depending on A as well as B and ρ_0 even in the simple model of equation (11). Consequently, more knowledge of the ambient rock is required in order to make accurate yield estimates using data taken in this interval.

Plastic Wave Interval.—As the shock wave expands and weakens further, the thermal pressure behind the shock front becomes a small fraction of the total pressure and the shock speed D approaches the low-pressure plastic wave speed A. At a certain radius R_2 ($\sim R_t$), the shock speed is 1.2 times the low-pressure plastic wave speed and we say that the shock wave has entered the plastic wave interval. (Again, the motion of the shock wave changes only gradually and so the point at which it is said to enter the plastic wave interval is purely conventional. Throughout the present article we use the convention that the plastic wave interval begins when the shock speed falls to 1.2 times the low-pressure plastic wave speed.) For an explosion in granite, this occurs when the peak pressure has fallen to ~150 kbar, corresponding to a peak density ratio ~0.4 times the maximum (cf. Table 2). For a 1 kt explosion in granite R_2 is ~5 m, whereas for a 150 kt explosion R_2 is ~30 m.

For the model given by equation (11), the plastic wave interval corresponds to $R \geq 3\,R_t$. In this regime

$$R \approx \text{const.} + A\,t\,, \qquad (14)$$

where the constant is determined by the motion in the strong shock and transition intervals.

The evolution of the shock wave in the plastic wave interval is sensitive to the equation of state and the constitutive relations describing the rock. For actual rocks, the evolution in this interval is complex, as described below. Hence relatively detailed modeling is required in order to obtain accurate yield estimates from data taken in this interval.

C. Other Effects

The evolution of the shock wave produced by an actual underground nuclear test is more complex than the evolution just described, for several reasons. First, the actual shock wave is not produced by a spherically-symmetric point explosion. The emplacement holes currently used in U.S. tests have radii R_e as large as 1.5 m[10] and emplacement holes with larger radii are planned for the future. Moreover, the nuclear charge and diagnostic canisters may be many meters in length. As a result, the source of the shock wave is vapor and radiation filling a volume with a dimension of meters. Moreover, the explosion is usually not spherically symmetric, causing the expanding shock wave to be aspherical initially. The energy flows produced by explosions in tunnels may be even more complex.

The motion of the shock wave cannot become self-similar until the shock wave has enveloped a mass of material much greater than the mass of the nuclear charge and casing, and energy transport by radiation is negligible. The radius R_0 at which this occurs is necessarily much greater than the radius R_e of the emplacement hole and depends on the design of the nuclear charge and surrounding diagnostic equipment. Unless there is a range of radii satisfying $R_0 \ll R \ll R_1$, the shock wave will not have time to become self-similar before entering the transition interval. The data in Table 2 show that no such range exists in granite for current emplacement practices, even for explosions as large as 150 kt. Thus, the simplicity of estimating yields from an interval of self-similar motion cannot be realized, given current testing practices and the yields of interest.

Other complications arise from the complexity of the motion in the transition and plastic wave intervals in actual rocks. As the shock wave expands and weakens, one or more phase transitions may occur. We have already noted that granite undergoes a phase transition when the peak pressure falls to ~400 kbar. At a somewhat lower peak pressure (~200 kbar in granite), the shock speed becomes less than the elastic wave speed c_ℓ. At this point the shock wave splits into an elastic wave followed by a plastic wave.[25,37] The initial pressure jump in the elastic wave is p_{crit}, which is ≈ 40 kbar for granite.[26] Since $c_0 < c_\ell$, the weakening plastic wave cannot overtake the elastic wave and the two-wave structure is stable. As a result, the plastic wave propagates through rock that has already been compressed and accelerated to a velocity ~1–10 m s^{-1} by the elastic wave. (Since c_0 is several km s^{-1}, for most purposes the acceleration of the medium by the elastic wave may be neglected and the plastic wave may be taken to propagate with a speed c_0 relative to the undisturbed ambient medium, as was done in eq. [6]). Finally, when the peak pressure in the plastic wave is no longer much greater than the critical shear stress p_{crit}, the shear strength of the rock can no longer be neglected. In granite, for example, p_{crit} is ≈ 40 kbar, and the hydrodynamic approximation therefore begins to fail when the peak pressure falls below ~150 kbar.

In addition to these complications that occur in a homogeneous medium, theoretical models and experimental data show that the evolution of shock waves in actual geologic media can be affected by voids, layering, and other geologic structures that may vary from one test site to another.

IV. MEASURING SHOCK WAVE EVOLUTION

A variety of sensors have been used at different times in the United States to measure the evolution of shock waves produced by underground nuclear explosions. These include strain gauges, particle velocity gauges, accelerometers, and pressure sensors.[26] In the United States, recent efforts to develop instrumentation for test ban monitoring have focused on radius vs. time (RVT) techniques, which measure the radius of the shock front as a function of time.

A. Using Sensing Cables

One RVT measuring technique that has been used in the United States since the early 1960s is based on the fact that the pressure peak near the front

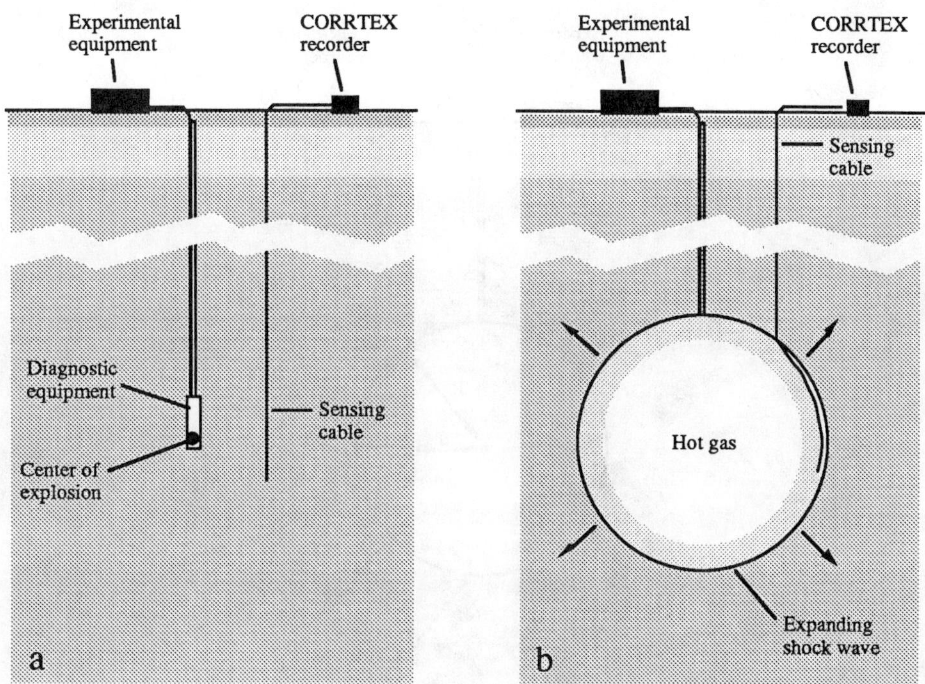

Fig. 2.—Schematic drawings illustrating (a) placement of the shock front sensing cable in a satellite hole and (b) progressive shortening of the cable by the expanding shock wave produced by a nuclear explosion.

of the shock wave can crush an electrical cable. In this approach, a coaxial sensing cable is lowered into the emplacement hole before it is backfilled or into one or more other holes (so-called satellite holes) that have been drilled nearby specifically for this purpose. The satellite-hole geometry is shown in Figure 2a. If the sensing cable is strong enough that it is not crushed by the elastic precursor or other unwanted signals but weak enough that it *is* crushed by the pressure peak in the hydrodynamic shock wave, the cable will be electrically shorted or its impedance substantially changed near the point where the hydrodynamic shock front intersects it. As the shock wave expands with time, the length of cable from the measuring equipment to the shallowest point at which it has been crushed is measured by electrical equipment attached to the cable and located above ground, as shown in Figure 2b. If the time at which the explosion began and the path of the cable relative to the center of the explosion are known and if the explosion is spherically symmetric, the radius of the shock front as a function of the time since the beginning of the explosion can be calculated from a record of the changing length of the cable.

As a concrete example, suppose the sensing cable is placed in a vertical satellite hole displaced laterally a distance d from the emplacement point. Then if the length L_0 of the sensing cable from the surface to the point where the cable

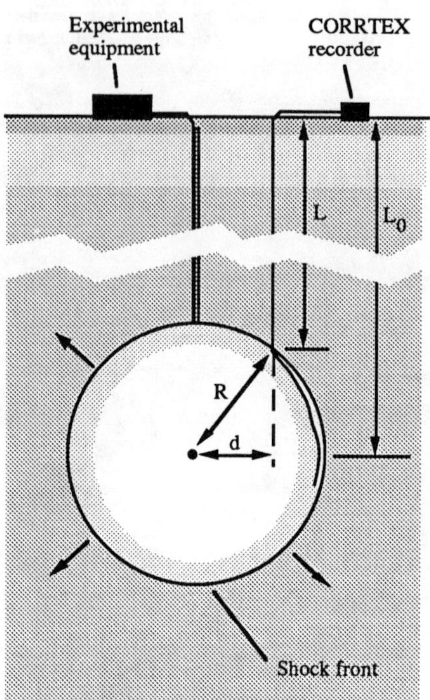

Fig. 3.—Schematic drawing illustrating how the radius R of the shock wave at time t is related to the lateral displacement d of the sensing cable from the center of the explosion, the length L_0 of the sensing cable from the surface to the point where the cable is first crushed, and the length L of the cable.

is first crushed is known, measurements of $L(t)$ can be converted into estimates of the radius R of the shock front as a function of time t using the Pythagorean relation

$$R(t) = \left[d^2 + \left(L_0 - L(t)\right)^2\right]^{1/2} . \tag{15}$$

The geometrical meaning of the quantities appearing in equation (15) is shown in Figure 3. In practice the sensing cable usually is not perfectly vertical and more complicated calculations must be performed to convert $L(t)$ to $R(t)$.

As discussed in §V, an error of 1 m in the measured distance of the crushing point from the center of the explosion will cause an error of about 50 kt in the yield estimate, for yields near 150 kt. Thus, accurate surveys of the emplacement and satellite holes and an accurate knowledge of the path of the sensing cable within the satellite hole are required in order to make an accurate yield estimate. If the cable wanders within the hole and this is not taken into account, the length of the cable crushed by the shock wave will be greater than the linear distance traveled by the shock front, causing the the speed of the shock wave and therefore the yield of the explosion to be overestimated. In the United States, hole surveys are currently made with special laser or gyroscopic equipment. In some yield estimation algorithms, the lateral displacement of the satellite hole from the center of the explosion can be treated as one of the unknowns in estimating the yield.

If the explosion is not spherically symmetric, due to the effect of the canister, the design of the test geometry, or inhomogeneities in the ambient medium, interpretation of the sensing cable data becomes more complicated and could be ambiguous or misleading under the conditions likely to be encountered in treaty monitoring. Potential problems of this kind could be addressed by cooperative agreements, as discussed in §VI.

Sensing cables with a variety of crushing strengths, ranging from as little as 30 bar to as much as 30 kbar, have been used.[38] However, even the latter cable can be crushed by the elastic precursor in granite. Thus, once the shock wave has split, the length of cable to the crushing point may reflect the position of the elastic shock front rather than that of the hydrodynamic shock front which follows.[39,40] If so, the sensing cable cannot provide data about the motion of the hydrodynamic shock wave.[26] Worse, if the motion is misinterpreted as that of the hydrodynamic wave, the estimated yield of the explosion will be erroneously high. As a result, interpretation of data in this interval requires great care.

The Reagan administration has strongly advocated adoption of the satellite-hole geometry for monitoring the TTBT using hydrodynamic methods. In order to collect ground shock data in the hydrodynamic region from explosions with yields near 150 kt, the satellite hole must be slightly deeper than the emplacement hole, which is typically ~650 m deep for such explosions, and the deepest portion of the satellite hole must be no further than ~30 m from the center of the explosion. In order to use the particular hydrodynamic yield estimation algorithm advocated by the administration (insensitive-interval scaling, see §V), the sensing cable must cover the algorithmic interval, which is a more demanding criterion and requires that the bottom of the satellite hole be no more than ~10 m from the center of the explosion, for yields near 150 kt.

In order to apply hydrodynamic methods to monitoring a 10 kt low-threshold test ban, the bottom of the satellite hole would have to be well within ~10 m of the emplacement point, which is typically ~250 m deep for charges with yields near 10 kt. In order to apply insensitive-interval scaling, the satellite hole would have to be within ~4 m of the center of the explosion.

Use of the satellite-hole geometry requires sophisticated drilling capabilities in order to make sure that the satellite hole maintains the proper separation from the emplacement hole.[41] In addition, conversion of cable length to shock wave radius is more complicated if the cable is in a satellite hole than if it is in the emplacement hole. On the other hand, the satellite-hole geometry reduces the intrusiveness of the method and eliminates the disturbing effects of jetting within diagnostic pipes and other phenomena that can crush or short the sensing cable ahead of the hydrodynamic ground shock. In the discussion that follows, we shall assume that the sensing cable is in a satellite hole, unless otherwise stated.

B. Measuring the Position of the Crushing Point

During the 1960s and 1970s, the position of the crushing point was measured in the United States using a technique called SLIFER (which is an acronym for Shorted Location Indicator by Frequency of Electrical Resonance).[26,42] In this approach, the cable is used as the inductive element of a resonant oscillator. As the cable is progressively crushed, the frequency of the oscillator changes. By knowing the propagation velocity of electromagnetic signals in the cable and the frequencies of the oscillator that correspond to $L = 0$ and $L = L_0$, one can convert measurements of the change in oscillator frequency during the explosion to estimates of the change in the length of the cable.

In the late 1970s, an improved approach to measuring the length of sensing cables, called CORRTEX, was developed (CORRTEX is an acronym for Continuous Reflectometry for Radius versus Time Experiments).[39,40,43,44] In this approach, a sequence of electrical pulses is sent down the cable. At the crushing point, these pulses are reflected back up the cable to recording equipment. By knowing the speed at which the pulses propagate down and up the cable, the round-trip travel time of each pulse can be converted into an estimate of the length of the cable at the time the pulse was reflected.

Current (CORRTEX III) equipment can store up to 4,000 data points. Pulse separations from 10 μs to 90 μs can be selected, giving a record of the changing cable length that is 40 ms to 360 ms in length. The pulses typically propagate down and up the sensing cable at about 2×10^5 km s^{-1}. A typical uncertainty in the round-trip travel time during a nuclear explosion is 500 ps, corresponding to an uncertainty of about 10 cm in the measured length of the cable or about 5 cm in the distance to the crushing point. The cable length measurements can be checked by creating fiducial loops in the cable at predetermined points, which will create downward jumps in the cable length as the crushing point passes over them. Using these jumps, the cable length data can be adjusted for systematic errors. The time at which the explosion begins is estimated by recording the time at which the electromagnetic pulse (EMP) from the explosion arrives at the CORRTEX recorder. The CORRTEX technique is less affected by disturbing signals produced by the explosion than were earlier techniques and is the technique that has been advocated by the Reagan administration for monitoring the TTBT.

RVT techniques are also used to estimate the yields of underground nuclear tests in the Soviet Union. The Soviets use two different sensing techniques, called Mis (or Miz) and Contactor.[45,46] The current Soviet approach reportedly uses a sensing system with switches that are sequentially destroyed by the shock front.[47]

V. YIELD ESTIMATION ALGORITHMS

Once measurements of the length of the sensing cable have been converted to estimates of the radius of the shock front as a function of time, the yield of the explosion is estimated by applying some algorithm, by which we mean a particular procedure for comparing the RVT data with a particular model of the motion of the shock front. Because hydrodynamic methods of yield estimation are evolving as research aimed at increasing our understanding of underground explosions and improving yield estimation methods continues, the description of yield estimation algorithms in the present section is necessarily a status report.

Although the algorithms used by different individuals or groups can (and usually do) differ in detail, most of the algorithms currently in use are of four basic types: insensitive interval scaling, similar explosion scaling, semi-analytical modeling, and numerical modeling. These are the algorithms that are discussed here. In order to simplify matters, we will assume at first that the explosion is spherically symmetric and that the ambient geologic medium is homogeneous. Some of the complications that can arise if the explosion is aspherical or the medium is inhomogeneous are described at the end of this section.

A. Insensitive Interval Scaling

The simplest algorithm currently in use is insensitive interval scaling. This is the algorithm that the Reagan administration has proposed to use in analyzing CORRTEX data as a new routine method of monitoring Soviet compliance with the TTBT.

Insensitive interval scaling assumes that the radius of the shock wave produced by an explosion of given yield does not depend on the rock in which the explosion occurs during a certain interval in time and radius called the "insensitive interval". This assumption is based on empirical evidence that the radius of the shock front *is* relatively insensitive to the medium during a certain interval in time and radius toward the end of the transition interval, for the particular geologic media for which the United States has good experimental data or theoretical models. These media include the dry alluvium, partially saturated tuff, saturated tuff, granite, basalt, and rhyolite at the nuclear test sites the U.S. has used. These media are mostly silicates and almost all are located at the Nevada Test Site. The radius of the shock front appears to depend only weakly on the medium despite the fact that phase transitions and shock wave splitting occur in some of these media during the insensitive interval. The radius of the shock front in one medium approaches and then deviates from that in another gradually, so that the insensitive interval is not sharply defined.

Although the reason for the existence of an insensitive interval for this collection of media is currently not well understood from a fundamental physical point of view, it appears to stem from a particular correlation among the relevant properties of these media. It *is* known that the relevant properties of other geologic media are *not* correlated in this way, so that the radius of the shock front in these media during the "insensitive interval" is very different from the radius of the shock front in the silicates cited above. Thus, the existence of an insensitive interval must be established by test experience or modeling, and is only assured for certain geologic media.

In using insensitive interval scaling, the shock wave sensing cable must be placed close enough to the center of the explosion that it samples the insensitive interval. Yield estimates are then derived by fitting a simple empirical formula, called the Los Alamos Formula, to the RVT data in this interval.[48] If the radius R of the shock front is expressed in meters, this formula is

$$R(t) = a\,W^{1/3}(t/W^{1/3})^b\,, \tag{16}$$

where W is the yield of the explosion in kilotons, t is the elapsed time since the beginning of the explosion in milliseconds, and a and b are constants. According to the assumption on which the algorithm is based, the values of a and b do not depend on the medium (because of this, eq. [16] has sometimes been referred to as the Los Alamos "universal formula"). The values of the constants a and b are typically determined by fitting equation (16) to a selected interval of RVT data from a collection of nuclear explosions. Different individuals and groups have found different values of a and b at various times. Even the values used by a single group have changed with time by amounts that have caused estimated yields to change by tens of percent. The values of a and b used here are 6.29 and 0.475.[48]

The Los Alamos Formula is a simple power law that approximates the actual RVT curve during the insensitive interval. This is illustrated in Figure 4, which compares the Formula with a model of the evolution of the shock wave produced in granite by a spherically-symmetric point explosion with a yield of 62 kt. In practice, the Los Alamos Formula is usually first fit to a broad interval of radius vs. time data that is thought to include the insensitive interval. The result is a sequence of yield estimates. Due to the departure of the Formula from the actual RVT curve at both early and late times, the sequence of yield estimates typically forms a U-shaped curve. This is illustrated in Figure 5, which shows the sequence of yield estimates obtained by applying the Formula to the relatively high-quality SLIFER data from the *Piledriver* explosion in granite (yield estimation using CORRTEX data cannot be illustrated here because at present all CORRTEX data remain classified). If the assumptions on which the algorithm is based are satisfied, the yield estimates near the bottom of the curve should approximate the actual yield of the explosion.

In the usual form of the algorithm, only the RVT data that fall within a certain predetermined interval chosen on the basis of previous experience (the so-called *algorithmic interval*) are actually used to make the final yield estimate. Because both the location and the extent of the algorithmic interval depend on the yield W of the explosion (both are proportional to $W^{1/3}$), an iterative procedure must be followed in estimating the yield of an explosion whose yield is initially unknown. Table 3 lists the algorithmic interval for several yields.

The sensitivity of an individual yield estimate to an error in the inferred location of the shock front depends on the position of the data point within the algorithmic interval and the yield of the explosion. For example, the sensitivity

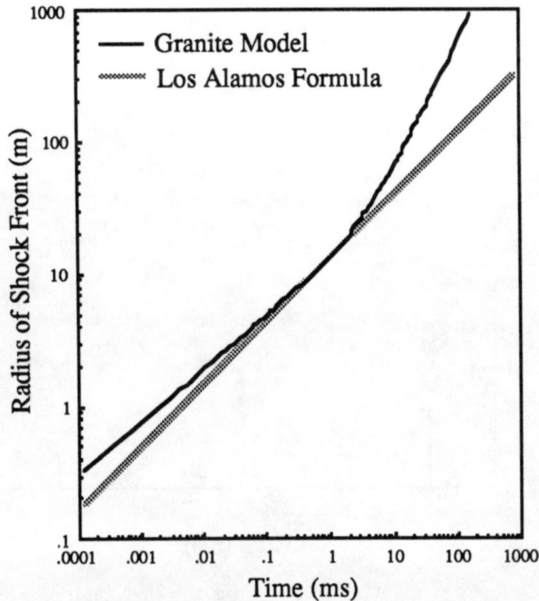

Fig. 4.—Comparison of the Los Alamos Formula with a model of the evolution of a shock wave in granite produced by a spherically-symmetric point explosion with a yield of 62 kt, showing the agreement during the transition interval and the departure of the Formula from the model at earlier and later times. The effect of the phase transition at 400 kbar has been included in the model.

varies from $13\,\mathrm{kt\,m^{-1}}$ at the beginning of the interval to $7.4\,\mathrm{kt\,m^{-1}}$ at the end of the interval, for a 10 kt explosion, and from $77\,\mathrm{kt\,m^{-1}}$ to $18\,\mathrm{kt\,m^{-1}}$, for a 150 kt explosion.

The algorithmic interval for a yield of 62 kt is indicated in Figure 5 by the two vertical bars at the bottom of the figure. In this example the assumptions on which the algorithm is based appear to be satisfied and the average of the yield estimates that lie within the algorithmic interval is very close to the announced yield of 62 kt.

The insensitive interval algorithm does not work as well if the assumptions on which it is based are not satisfied. This is illustrated in Figure 6, which shows the yield estimates obtained by fitting the Los Alamos Formula to good-quality SLIFER data from a typical low-yield explosion in alluvium. In this example the RVT data have been scaled in the manner described below in the subsection on similar explosion scaling, so that the derived yield should be 1 kt (the actual yield is classified). The yield estimates given by the Los Alamos Formula are systematically low, ranging from 30% to 82% of the actual yield, and do not form a U-shaped curve. The average of the yield estimates that lie within the algorithmic interval is about 60% of the actual yield. The overall appearance of the yield vs. time curve shows that the assumptions of the algorithm are not satisfied.

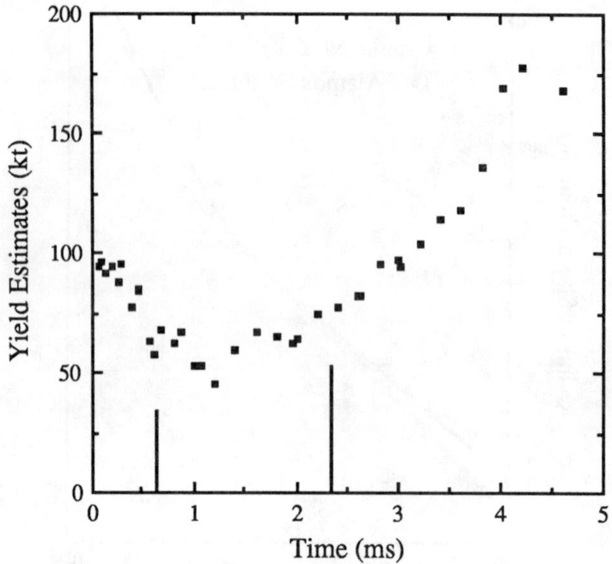

Fig. 5.—Yield estimates derived by applying the Los Alamos Formula to SLIFER data from the *Piledriver* explosion in granite, which had a nominal yield of 62 kt. The so-called algorithmic time interval is the interval between the two vertical bars. The average of the yield estimates within the algorithmic interval is in good agreement with the nominal yield. Note the departure of the yield estimates from 50–70 kt at earlier and later times, which is caused by the deviation of the Formula from the actual radius vs. time behavior of the shock wave outside the algorithmic interval (cf. Fig. 4).

TABLE 3
Algorithmic Intervals[a]

Yield (kt)	Time Interval (ms)	Radius Interval (m)
1	0.10 — 0.5	2.1 — 4.5
10	0.21 — 1.1	4.5 — 9.7
50	0.37 — 1.8	7.7 — 17
100	0.46 — 2.3	9.8 — 21
150	0.53 — 2.7	11 — 24

[a]The intervals used by various individuals and groups vary. Throughout this article the algorithmic interval is taken to be from $0.1\,W^{1/3}$ ms to $0.5\,W^{1/3}$ ms after the beginning of the explosion, where W is the yield of the explosion in kilotons.

A common misconception has been that the algorithmic interval lies within the strong shock region and that the relative insensitivity of yield estimates to the properties of the medium stems from this.[49] This misconception apparently

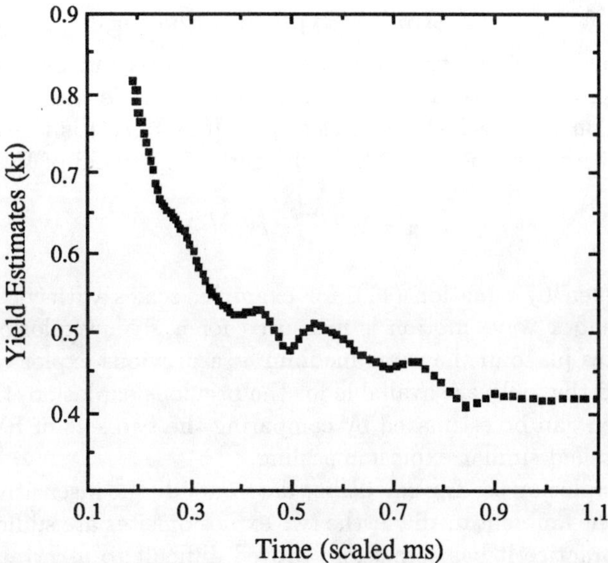

Fig. 6.—Yield estimates derived by applying the Los Alamos Formula to SLIFER data from a low-yield explosion in alluvium (note that the vertical axis is offset from zero). The data have been scaled so that the estimated yield should be 1 kt. Difficulties in applying the insensitive interval algorithm to low-yield explosions in alluvium are not uncommon.

has arisen at least in part because the interval formerly used to estimate the yields of nuclear explosions in the atmosphere using hydrodynamic methods *is* within the strong shock region. As explained earlier, the relative insensitivity of the radius of the shock front to the medium in the algorithmic interval would be explained if the algorithmic interval were within the strong shock region and if the motion were self-similar. The formula for the radius of the shock front would then be a power-law function of time like the Los Alamos Formula, except that the exponent of t would be 0.4. However, in reality the shock wave motion is not self-similar during the algorithmic interval, for current test geometries and the yields permitted by the TTBT. In fact, the shock wave is not even strong during this interval, since the shock speed is only a few times the low-pressure plastic wave speed while the peak pressure is much less than the pressure required to achieve the limiting density ratio. Indeed, the exponent of time usually used in the Los Alamos Formula, 0.475, is significantly greater than the exponent 0.4 characteristic of a strong, self-similar shock wave. The power-law Los Alamos Formula is not a theoretical result like the power-law formula for the radius of a strong, self-similar shock wave. Rather, it is an approximate, empirical relation, which was obtained by fitting a simple power-law expression to a selected interval of RVT data from a collection of nuclear explosions in several different media.

B. Similar Explosion Scaling

As long as the final state of the shocked rock is independent of the rate of compression and explosions take place in the same medium, the RVT curve depends only on the yield of the explosion.[11] If $R = f(t)$ is the RVT curve for an explosion with a yield of 1 kt, then the curve for a yield of W kt is

$$R = W^{1/3} f(t/W^{1/3}) . \tag{17}$$

The model given by equation (11), for example, scales with yield in this way. Thus, if the shock wave motion is measured for a given explosion of unknown yield that takes place in the same medium as a previous explosion, and if both RVT data and the yield are available for the previous explosion, the yield of the given explosion can be estimated by comparing the two sets of RVT data. This algorithm is called similar explosion scaling.

Similar explosion scaling can utilize data outside the insensitive interval and works well if the ambient media at the two explosion sites are sufficiently similar. However, in practice it has sometimes proved difficult to ascertain whether the relevant properties of the media are similar enough to give the desired accuracy. As a result, application of this algorithm has sometimes led to unexpected errors in the estimated yield.

If hydrodynamic yield estimation is adopted for TTBT verification, similar explosion scaling would presumably be used where possible to check the results of insensitive interval scaling.

C. Semi-Analytical Modeling

Another approach that is useful for studying the evolution of shock waves in geologic media and for estimating yields is semi-analytical modeling.[30,32,50–52] In this approach both the properties of the ambient medium and the motion of the shock front are treated in a simplified way that nevertheless includes the most important effects. The result is a relatively simple, semi-analytical expression for the radius of the shock front as a function of time. If the required properties of the ambient medium are known and inserted in this expression, the yield of an explosion can be estimated by fitting the expression to RVT data.

Semi-analytical algorithms can make use of data over a more extensive interval than the interval used in the insensitive interval algorithm. For example, application of equation (10) to particle velocity data taken 8 m from an explosion with a nominal yield of 10.4 kt gave an estimated yield of 10.3 kt.[53] Even data taken at a radius of 13.5 m, well outside the algorithmic interval, gave a yield of 7.9 kt. Semi-analytical models can also be used to estimate the uncertainty in the yield caused by uncertainties in the properties of the ambient medium.

D. Numerical Modeling

If a treatment that includes the details of the equation of state and other properties of the ambient medium is required, if it is desired to utilize RVT data in the region where the shock wave has split, or if the explosion is not spherically symmetric, modeling of the motion of the shock front using numerical codes may be necessary.[54] In principle, such simulations can provide RVT curves that extend over much of the shock wave evolution, making it possible to base yield estimates not only on data from the transition interval but also on data from later phases of the shock wave evolution. In these later phases, the shock wave is no longer hydrodynamic, so yield estimation methods that make use of such data are not hydrodynamic methods. Yield estimates made using data from later phases are fairly sensitive to the equation of state and constitutive relations that characterize the ambient rock. Since these are accurately known only for a few rocks, accurate yield estimates are possible using late-time data only for explosions in a few geologic media.

E. Complications

In order to use hydrodynamic methods, RVT data must be taken within meters of the center of the explosion (cf. Tables 2 and 3). At such small distances, the arrangement of the nuclear charge and the canister or canisters containing it and the diagnostic equipment can introduce asymmetries in the expansion of the shock wave that will affect the yield estimate:

- A long canister can produce a shock wave that is initially cylindrical and hence crushes the sensing cable in such a way that only part of the total yield is sensed over most of the interval sampled, as shown schematically in Figure 7a. The physical size of canisters and diagnostic lines-of-sight tend to pose more of a problem for tests of nuclear directed-energy weapons than for tests of traditional nuclear weapons.[55]
- Explosions of nuclear charges in tunnels may be accompanied by complicated (and unanticipated) energy flows and complex shock wave patterns. If significant energy reaches the sensing cable ahead of the ground shock and shorts or destroys it before the ground shock arrives, the CORRTEX data will describe that flow of energy and not the motion of the ground shock. Moreover, the ground shock itself may become sufficiently distorted that the RVT data are confusing or misleading.

In complicated geometries, disturbing effects like these are difficult to analyze and correct for using data from a single sensing cable, since it senses only the depth of the shallowest point where a pressure wave first crushes it, at a single azimuth and lateral displacement from the explosion. As a result, unambiguous interpretation of the data may become difficult or impossible. The disturbing effect of a canister and emplacement hole of given size is less for higher-yield than for lower-yield explosions, since the hydrodynamic region extends further from the canister and emplacement hole for a higher-yield explosion. In addition, higher-yield charges are not usually exploded in tunnels.

In addition to the potential disturbing effects of the nuclear test design, any errors in characterizing the surrounding geologic media will introduce errors

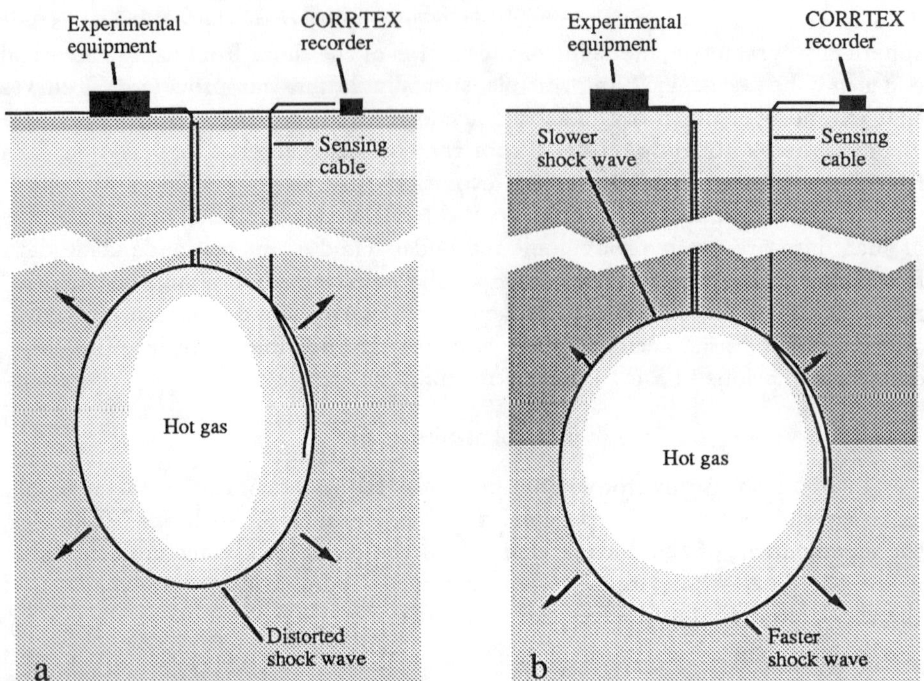

Fig. 7.—Schematic drawings illustrating disturbance of hydrodynamic yield estimates by (a) a nuclear test design that produces a cylindrical shock wave during the algorithmic interval and (b) slowing and weakening of the upward traveling portion of the shock wave when it encounters a much denser geologic stratum.

in yield estimates derived using hydrodynamic methods. Incorrect assumptions about the average properties of the rock surrounding the emplacement hole may bias the yield estimate, decreasing its accuracy, while small-scale variations in the rock will cause scatter in the RVT record, decreasing the precision of the yield estimate. In addition, geologic structures like that shown in Figure 7b can affect the yield estimate. At the Nevada Test Site, for example, the alluvial deposits are weakly consolidated erosion products of the surounding mountains with physical properties that vary widely. Layers of gravel, the residues of ancient stream beds, are often encountered in drilled holes. While most RVT records follow the expected behavior, an occasional event will produce an RVT record whose irregular behavior defies simple explanation. Such behavior has been attributed to spatial variations in the ambient medium.[26]

F. Summary of Yield Estimation

Shock waves produced by an underground nuclear explosion propagate differently in different media and different geologic structures. As a result, knowledge of the rock and geologic structure within $\sim 10\,(W/1\,\mathrm{kt})^{1/3}$ meters of the center of the explosion is required in order to make accurate yield estimates using hydrodynamic methods. The evolution of the shock wave can also be affected by the test geometry. Knowledge of the test geometry and/or limitations on the aspects that could disturb the RVT data are therefore required in order to make accurate yield estimates. Several different yield estimation algorithms have been developed. Like the algorithms used to estimate yields from seismic data, these algorithms involve some complexity and require sophistication to understand and apply correctly.

Hydrodynamic yield estimation methods have not yet been studied as thoroughly or as widely as seismic methods. Although approximately 100 tests have been carried out with the CORRTEX sensing cable in the emplacement hole, only a few have been carried out with the cable in a satellite hole[5,6] (in addition, SLIFER data from sensing cables in satellite holes are available for several tens of earlier explosions). Moreover, no systematic and comprehensive review of the scientific evidence concerning the accuracy and precision of U.S. hydrodynamic yield estimation methods has yet been carried out. The U.S. Department of State has asserted that the methods are accurate to within 15% (at the 95% confidence level) of radiochemical yield estimates for tests with yields greater than 50 kt in the geologic media in which tests have been conducted at the Nevada Test Site.[5,6] According to these same reports, the algorithm is expected to have an uncertainty of a factor of 1.3 at the 95% confidence level if used in a treaty-monitoring context at the Soviet test sites near Shagan River to monitor explosions with yields greater than 50 kt.[5,6] However, some scientists familiar with the methods believe that the uncertainty could be somewhat larger.[4]

Some key terms that have been introduced in this discussion are listed and explained in Table 4.

VI. APPLICATION TO MONITORING TEST BAN TREATIES

A. Assuring Accuracy

Ambient medium.—As explained in §III and §V, the physical properties and geologic structure of the media around the emplacement point affect the evolution of the shock wave produced by an underground nuclear explosion. Thus, it is important in test ban monitoring to gather information about the types of rock present at the test site and their properties, including their chemical composition, bulk density, porosity, gas-filled porosity, and degree of liquid saturation, as well as the speed of sound in the rock and any specific features of the local geologic structure that could affect the yield estimate. Such data are easily gathered by the party conducting the explosion. Availability of the required data to the monitoring party could be assured by appropriate negotiated cooperative measures.

Some information about the geologic medium at the test site can be obtained by examining the contents of the hole drilled for the CORRTEX sensing cable. Verification could be improved by cooperative arrangements that would also allow observation of the construction of the emplacement hole, removal and examination of rock core or rock fragments from the wall of the emplacement

TABLE 4
Glossary of Hydrodynamic Yield Estimation Terms

Term	Explanation
Algorithmic interval	The interval, in time usually 0.1–0.5 scaled ms after the beginning of the explosion, from which data is selected in applying insensitve interval scaling
CORRTEX technique	A technique for measuring the position of the shock front expanding away from an underground explosion by determining the round-trip travel time of electrical pulses sent down a sensing cable placed in a hole in the ground near the explosion
Insensitive interval scaling	A yield estimation algorithm in which the Los Alamos Formula is fit to measurements of the position of the expanding shock front as a function of time during the algorithmic interval
Los Alamos Formula	The empirical formula used in insensitive interval scaling to make yield estimates by fitting to shock front radius vs. time data
Plastic wave interval	The interval in radius and time in which the speed of the weakening hydrodynamic shock wave is less than 1.2 times the low-pressure plastic wave speed
Similar explosion scaling	A yield estimation algorithm in which shock front radius data are compared with similar data from a previous explosion in the same medium
SLIFER technique	A technique for measuring the position of the shock front expanding away from an underground explosion by determining the resonant frequency of an electrical circuit that includes a sensing cable placed in a hole in the ground near the explosion
Strong shock interval	The interval in radius and time in which the density just behind the shock front is close to its limiting value
Transition interval	The interval in radius and time in which the density just behind the shock front is less than 80% of its limiting value but the speed of the hydrodynamic shock wave is still greater than 1.2 times the low-pressure plastic wave speed

hole, examination of any logs or drill core from existing exploratory holes, removal and examination of rock core or rock fragments from the walls of existing exploratory holes, and if necessary, construction of new exploratory holes.

There is precedent for such cooperative arrangements in the Protocol to the Peaceful Nuclear Explosions Treaty (PNET), which explicitly established that hydrodynamic methods could be used to monitor the yield of any explosion with a planned aggregate yield greater than 150 kt and which specified verification measures like those described here.[3]

Test geometry.—As explained in §V, the arrangement of the nuclear charge and the canister or canisters containing it and the diagnostic equipment can introduce asymmetries in the expansion of the shock wave which disturb the yield estimate. In using hydrodynamic methods to estimate the yields of one's own tests, the design and placement of the nuclear charge and related equipment are known and can be taken into account. This information may not be available when monitoring the nuclear tests of another party. Cooperative agreements to ensure satisfactory placement of the sensing cable and to exclude nuclear test geometries that would significantly disturb the yield estimate are therefore required.[55–58]

Such agreements could, for example, limit the length of the canister containing the nuclear charge and the cross-sectional dimensions of the emplacement hole, and mandate stemming of the emplacement hole with certain types of materials. Such agreements could also provide for observation of the emplacement of the nuclear charge and the backfilling of the emplacment hole, confirmation of the depth of emplacement, and limitations on the placement of cables or other equipment that might interfere with the CORRTEX measurement. For test geometries that include ancillary shafts, drifts, or other cavities, additional measures, such as placement of several sensing cables around the nuclear charge emplacement point, may be required to assure an accurate yield estimate. For tunnel shots, sensing cables could be placed in the tunnel walls or in a special hole drilled toward the tunnel from above. Again, there is precedent for such cooperative measures in the PNET.[3]

The restrictions on the size of canisters and diagnostic lines-of-sight that would be required even with the sensing cable placed in a satellite hole would cause some interference with the U. S. nuclear testing program at NTS. However, these restrictions have been examined in detail by the U.S. nuclear weapon design laboratories and the Department of Energy, and have been found to be manageable for the weapon tests that are planned for the next several years.[55,58] For the more distant future, the disadvantages of the required restrictions on testing must be weighed against the potential contributions to treaty monitoring that could be made by hydrodynamic methods.

In summary, the accuracy that could be achieved using hydrodynamic methods to monitor test ban treaties depends on the amount of information about the ambient medium that can be gathered by the monitoring party and the nature and extent of cooperative arrangements to limit disturbing effects. Some tests and simulations to identify troublesome configurations have been carried out, but such configurations have not been thoroughly explored. Given the possibility that hydrodynamic yield estimation methods may have to be used to monitor treaty compliance in an adversarial atmosphere, the possibility of deliberate efforts to introduce error or ambiguity, and the tendency for worst-case interpretations to prevail, additional research to improve understanding of the accuracy and precision of these methods and to reduce further the chances of confusion, ambiguity, spoofing, or data denial is very important.

B. Minimizing Intrusion

Hydrodynamic yield estimation methods are more intrusive than seismic methods for several reasons:

- Drilling of the emplacement and satellite holes and emplacement of the nuclear charge and diagnostic equipment typically take 8–10 weeks. In order to monitor these operations, personnel from the monitoring country would have to be present at the test site of the testing country for ten weeks or so before as well as during each test, and would therefore have an opportunity to observe test preparations. The presence of these personnel would pose some operational security problems.[55,58,59]
- The exterior of the canister or canisters containing the nuclear charge and diagnostic equipment must be examined to verify that the restrictions necessary for the yield estimate to be valid are satisfied. For tests of nuclear directed-energy weapons, this examination could reveal sensitive design information unless special procedures are followed.[58]
- Sensing cables and electrical equipment will tend to pick up the electromagnetic pulse (EMP) generated by the explosion. A detailed analysis of the EMP would reveal sensitive information about the design and performance of the nuclear device being tested.[60]

Intrusiveness can be minimized by careful attention to monitoring procedures and equipment. For example, the electrical recording equipment can be designed to avoid measuring sensitive information about the nuclear devices being tested. CORRTEX equipment has been designed in this way, and the United States could insist that any Soviet equipment used at the Nevada Test Site be similarly designed. Placement of the recording equipment in Faraday cages and other measures could be mandated in order to minimize the chance that sensitive information concerning the test could be picked up by the monitoring party.

The security problems posed by opportunities to observe test preparations are more severe for nuclear directed-energy weapon tests, since they tend to have larger and more complex diagnostic systems and canister arrangements which, if fully revealed to the Soviets, might disclose sensitive information.[55,58] The United States has determined that the Soviet personnel and activities that would be required at the Nevada Test Site to monitor U. S. tests would be acceptable both from a security standpoint and from the standpoint of their effect on the U.S. test program. Detailed operational plans have been developed to accommodate such visits without adverse impact on operations.[55]

C. Specific Applications

Threshold Test Ban Treaty.—As noted earlier, hydrodynamic yield estimation methods have been proposed by the Reagan administration as a new routine measure for monitoring the sizes of nuclear tests, in order to verify compliance with the 150 kt limit of the TTBT. These methods can be used only if appropriate changes in the TTBT are negotiated. In addition, adequate cooperative arrangements would need to be made in order to assure the accuracy of the resulting yield estimates. To reduce the cost and intrusiveness of verification, use of hydrodynamic methods could be restricted to tests with planned yields greater than some threshold that is an appreciable fraction of 150 kt. As an alternative to routine use, hydrodynamic methods could be used to help calibrate seismic yield estimation methods by measuring the yields of one or more nuclear calibration explosions at test sites,[57,61] provided that the medium in which the explosion takes place is within U.S. test experience and the necessary procedures are followed to assure the accuracy of the hydrdodynamic yield estimate.

Peaceful Nuclear Explosions Treaty.—As it stands, the PNET does not provide for use of hydrodynamic yield estimation except for salvos in which the planned aggregate yield is greater than 150 kt.[3] Thus, if the TTBT is modified to allow hydrodynamic yield estimation for all weapon tests with planned yields above a certain value, the purpose of the modification could in principle be circumvented by carrying out weapon tests as "peaceful" nuclear explosions of "planned yield" less than or equal to 150 kt, unless the PNET is also modified to close this loophole.

Low-Threshold Test Ban Treaty.—Tamped underground nuclear explosions as small as a few kilotons produce shock waves that evolve in the same way as those produced by explosions of larger yield. However, because the algorithmic interval for such explosions is much closer to the nuclear charge and diagnostic canisters than the algorithmic interval for tests with much larger yields, the disturbing effect of the canisters is generally more serious for such explosions. Moreover, low-yield tests can be and often are set off at shallow depths in softer material, such as alluvium, or in tunnels. As a result, the propagation of the shock wave can differ markedly from the models on which standard hydrodynamic yield estimation methods are based, causing confusion or error in the yield estimate. There can also be significant variations in the motion of the shock wave from explosion to explosion under these conditions.

In addition, serious practical, operational, and engineering problems arise in trying to use hydrodynamic methods to estimate the yields of explosions with yields of a few kilotons. For one thing, the sensing cable must be placed very close to the nuclear charge. Drilling a satellite hole within 2–5 meters of the emplacement hole to the depth at which the nuclear charge is placed (typically $\sim$250 m even for low-yield explosions in vertical shafts), which would be required in order to use hydrodynamic methods to monitor explosions with yields in the 1–10 kt range, is at or beyond the capabilities of current drilling techniques.[41] The need for such close placement would also necessitate more stringent restrictions on the maximum sizes and orientations of the nuclear charge and the diagnostic canisters. Such restrictions might be deemed an unacceptable interference with test programs. The alternative of using small canisters with numerous diagnostic lines-of-sight to the detonation point could disturb the shock wave and introduce errors in the yield measurements. Finally, because the shock wave radii to be measured are much smaller at low yields, survey errors become much more important.

It is possible that some of these difficulties could be circumvented by developing models and algorithms that would allow routine use of RVT data at radii beyond the hydrodynamic interval, as discussed in §V. Others could be alleviated by conducting tests with simple geometries to define and calibrate test sites so that insensitive interval scaling could be accurately applied. However, these and other potential solutions to the problems that would be encountered in monitoring low-yield tests using hydrodynamic methods have not yet been carefully and thoroughly studied. Thus, at the present time hydrodynamic yield estimation methods could not be used with confidence to monitor compliance with threshold test bans in which the threshold is less than several tens of kilotons.

Comprehensive Test Ban Treaty.—Hydrodynamic methods can only be used with shock wave data taken within meters of the center of the explosion. Moreover, their use requires advance notification of tests, extensive preparations in advance at the test site, and the presence of monitoring personnel during the test. Hydrodynamic methods were not intended and clearly are not able to detect or identify unannounced, remote, or clandestine nuclear tests. As a result, they cannot contribute to monitoring a comprehensive test ban.

VII. DISCUSSION

Hydrodynamic methods could be a potentially valuable component of a cooperative program to improve monitoring of the TTBT and PNET. Hydrodynamic yield estimates are more affected than seismic estimates by local features that disturb the evolution of the shock wave, such as canisters or local geologic structures. For explosions larger than ~10 kt, seismic yield estimates are less affected by local features but are affected by how the shock wave evolves into seismic waves. Thus, yield estimates based on hydrodynamic methods would complement yield estimates based on seismic methods.

Apparently, a systematic and comprehensive review of the scientific evidence concerning the accuracy and precision of hydrodynamic yield estimation methods has not yet been conducted in the United States. Nor has the United States any experience as yet in applying these methods at Soviet test sites. As noted in §V, the U.S. Department of State has asserted that yield estimates based on CORRTEX data analyzed using the insensitive interval algorithm are expected to have an uncertainty of a factor of 1.3 at the 95% confidence level, for explosions larger than 50 kt at the Soviet test sites near Shagan River. If achieved, such an uncertainty would be slightly smaller than the uncertainty of a factor of 1.5–1.6 that according to recent studies[4] could probably be achieved using the best current seismic methods, if the test site is not calibrated and measurements are made only outside the Soviet Union.

Hydrodynamic methods could also be used to help calibrate seismic methods. According to the results of a recent study, application of the best current seismic methods to measurements made outside the Soviet Union is expected give yield estimates that have an uncertainty of a factor of 1.3 at the 95% confidence level, the same precision reportedly expected for hydrodynamic methods, once the Soviet test site is properly defined and calibrated.[4]

In its first years, the Reagan administration declined to resume the U.S.-Soviet negotiations on a CTBT that had been adjourned in 1980. However, in the summer of 1986 both parties agreed to begin new, low-level talks on nuclear testing.[60,62] The Soviet Union reportedly pressed for resumption of negotiations on a CTBT and emphasized seismic verification methods whereas the United States sought an agreement allowing use of hydrodynamic methods as an additional verification measure for the TTBT and PNET.[62] Then, in April 1987, the Soviet Union suggested experiments at the U.S. and Soviet test sites as a way to advance the negotiations and to compare verification methods.[63–65] As a result of further discussions[66,67] the United States and the Soviet Union announced on September 17, 1987, that they had agreed to renew full-scale negotiations on nuclear testing limitations.[68,69] Formal negotiations began in November 1987 in Geneva.

The parties agreed as a first step to negotiate verification measures for the TTBT and PNET. In these negotiations, the sides agreed to make no changes in the treaties themselves but to negotiate a new protocol to each treaty, which would specify verification provisions.[69] Once these two treaties have been ratified, the parties have agreed to begin negotiations on ways to implement a step-by-step program to further limit nuclear tests.[69]

At the Washington Summit in December 1987, President Reagan and General Secretary Gorbachev agreed to conduct a joint verification experiment (JVE) at each other's nuclear test sites.[69] In this experiment, one nuclear charge with a yield near the 150 kt limit will be exploded at the Nevada Test Site; another will be exploded at the Soviet test site near Semipalatinsk. The purpose of the experiment is to compare the verification techniques advocated by the two sides in the Geneva negotiations.

In preparation for the JVE, 20 nuclear testing experts from each side visited the other side's nuclear testing site in January 1988 to familiarize themselves with how each side conducts tests and to provide a basis for the design and conduct of the JVE.[69,70] The second round of the current U.S.-Soviet test ban negotiations began in February 1988, and remained in continuous session until June 30, 1988.[47] In March, drilling of the necessary holes began.[69]

An agreement on the conduct of the JVE was signed on May 31, 1988, as part of the Moscow Summit.[71] The agreement, which with its annex totals 105 single-spaced pages, specifies that each party will have the opportunity to estimate the yield of each of the JVE explosions using both hydrodynamic methods (with sensing cables in both the emplacement hole and a satellite hole) and teleseismic methods (using recordings made at five designated seismic stations of each side).[69,71] Each party will also have the opportunity to make yield estimates using its national seismic network. To assist the teleseismic measurements, on June 28, 1988, each side shared with the other data on five of its high-yield underground tests conducted after January 1, 1978, but before January 1, 1988, recorded at its five designated seismic stations located at teleseismic distances from its nuclear test site. Associated geological and geophysical information and estimates of the yields of the five events were also exchanged.[69]

Originally, the Soviet Union had suggested that each side furnish the nuclear charge to be exploded at the other side's test site, so that the monitoring side would be able to compare the yields estimated using both seismic and hydrodynamic methods with a yield estimate based on knowledge of the design of the nuclear charge.[64–66] However, the JVE is actually being carried out with a U.S. nuclear charge at the U.S. test site and a Soviet nuclear charge at the Soviet test site.[71] As a result, the monitoring party will not have available an estimate of the yield of the explosion that is independent of both seismic and hydrodynamic methods. In order to solve this calibration problem, the sides have agreed to accept the hydrodynamic yield estimate derived from the data taken with the sensing cable in the emplacement hole as the yield of the explosion for the purposes of the experiment.

The JVE test at the Nevada Test Site, code-named *Kearsarge*, was completed as scheduled on August 17, 1988. The Soviet Union reportedly gathered hydrodynamic data on the explosion using sensing cables in the emplacement and satellite holes, regional seismic data from the seismic stations that have been installed nearby as part of the NRDC-Soviet Academy cooperative verification project,[72,73] and teleseismic data from its own national seismic network.[46,72] The Soviets will also be able to use data from a variety of seismographic stations in the United States that routinely publish such data. The United States reportedly gathered hydrodynamic data on *Kearsarge* and will also be able to use data from its national seismic network as well as published data. Data on this explosion were exchanged in Geneva on August 30, 1988, at the start of the third session of the current round of negotiations.

According to press reports following *Kearsarge*,[74] the nuclear charge was expected to have a yield of about 140 kt. However, one U.S. hydrodynamic measurement reportedly gave a yield estimate slightly more than 150 kt while the other gave a yield estimate slightly more than 160 kt. In contrast, data from the U.S. seismic network reportedly confirmed expectations that the yield of the explosion would be about 140 kt. If the yield of the explosion *was* 140 kt and if the uncertainty of the hydrodynamic yield estimation algorithm used by the United States is a factor of 1.15 at the 95% confidence level at the Nevada Test Site, as asserted by the U.S. Department of State,[5,6] then the probability that a single yield estimate would be larger than 160 kt is only about one in 40, while the joint probability that one yield estimate would be more than 150 kt *and* the other more than 160 kt is about one in 300. As a result of this outcome, the U.S. Department of Energy has reportedly decided to make a radiochemical yield estimate in order to better assess the yield of *Kearsarge*.

As this is being written, preparations for the explosion at the Semipalatinsk Test Site, which is currently scheduled for September 14, 1988, are in progress. Because of its strong advocacy of hydrodynamic methods, the United States has chosen not to gather regional seismic data on this explosion but will gather hydrodynamic data using CORRTEX sensing cables in the emplacment and satellite holes.[75] The United States will also have available teleseismic data from its national seismic network. Regional seismic data on the test at Semipalatinsk may be recorded by seismic stations that have been installed nearby as part of the NRDC-Soviet Academy cooperative verification project. If they are, these data would presumably also be available to the U.S. government. Thus, the JVE will provide the United States with a variety of information with which to improve its seismic estimates of the yields of past as well as future nuclear tests at Semipalatinsk.

The U.S. drive to negotiate hydrodynamic monitoring of the TTBT and PNET raises several important issues concerning the interaction of technology and public policy. From a purely scientific point of view, gathering of additional data on U.S. and Soviet test yields and achievement of better precision in estimating yields are advances. However, the principal purpose of TTBT and PNET monitoring is presumably to increase the security of the United States and the Soviet Union, rather than to advance scientific studies.

From a security perspective, the purpose of requiring a precision of a factor of 1.3 at the 95% confidence level for monitoring the 150 kt limit of the TTBT appears unclear. Such a precision means that there is only one chance in four that a single explosion with a yield of 164 kt would appear to be in compliance with the TTBT. The administration has not explained why such a high precision is needed, nor has it explained why current seismic methods, which according to recent studies could achieve a similar precision using measurements made only outside the Soviet Union, are still considered inadequate. Was the requirement for such high precision established after a careful review of the purposes of the TTBT and PNET and the costs and benefits of requiring various levels of precision? Or was it adopted simply because it was the highest precision that was thought to be achievable with hydrodynamic methods at the time the policy advocating such methods was adopted?

In assessing whether this degree of precision is necesssary, one must bear in mind the fact that if tests producing yields of 150 kt are performed, then approximately half of unbiased yield estimates will be above 150 kt (just as approximately half will be below 150 kt), no matter how precise the estimates.

Moreover, both parties to the TTBT recognized at the time it was signed that the yield of a nuclear test cannot always be predicted accurately in advance, so that even if the planned yield is 150 kt, some variation from this value may occur.[3]

Whatever the justification for requiring a precision of a factor of 1.3 at the 95% confidence level in monitoring the 150 kt limits of the TTBT and PNET, the considerations that enter an assessment of the precision that would be optimal in monitoring possible future lower-threshold test ban agreements are different, and depend on the threshold chosen. For some choices of threshold, an uncertainty much greater than a factor of 1.3 might be adequate. However, the emphasis on achieving a precision of a factor of 1.3 at 150 kt could establish this precision in the public mind as a requirement that must also be met in monitoring any future limitations on nuclear testing. Such a development could impede progress toward achieving such limitations.

Similarly, the emphasis on hydrodynamic monitoring of the TTBT and PNET might also establish this verification technology as a required part of any low-threshold test ban agreement. Given the difficulties that would have to be overcome to apply these methods with confidence to monitoring low-yield tests, such a requirement could become an additional impediment to achieving a low-threshold test ban.

The current test ban negotiations and the Joint Verification Experiment have led to increased openness about the U.S. and Soviet nuclear testing programs and increased cooperation between the national security establishments of the United States and the Soviet Union. According to participants, both sides have approached the JVE as a useful joint venture, in which it has been possible to make enquiries, receive answers, and resolve concerns.[69] In addition to these benefits, the JVE will also provide useful data on the seismic properties of the U.S. and Soviet test sites that may help both parties to resolve past concerns over compliance and to advance toward ratification of the TTBT and PNET.

VIII. CONCLUSIONS

Hydrodynamic methods were developed primarily as a tool to estimate the sizes of explosions in the U.S. and Soviet nuclear test programs and are inherently more intrusive than seismic methods. From a scientific point of view hydrodynamic methods are no more "direct" than seismic methods. Nor are hydrodynamic methods necessarily free of systematic error. The use of hydrodynamic methods at Soviet test sites would give estimates of the yields of Soviet tests larger than 50 kt that are slightly more precise than could be achieved using the best currently available seismic methods, if (1) the cooperative arrangements needed to assure the accuracy of hydrodynamic methods are successfully negotiated and followed, (2) U.S. hydrodynamic methods applied at Soviet test sites prove to be as precise as expected by the U.S. Department of State, (3) the Soviet test sites are not calibrated for seismic measurements, and (4) seismic data are collected only outside the Soviet Union. On the other hand, the precision of the best available remote seismic methods is expected to be comparable to the precision that could be achieved using hydrodynamic methods if the Soviet test sites are appropriately defined and calibrated. Significant engineering, operational, and analysis problems need to be solved before hydrodynamic methods could be used to monitor with confidence a low-threshold test ban. The methods are not relevant to a comprehensive test ban.

Possible advantages of an agreement allowing use of hydrodynamic yield estimation methods for monitoring the TTBT and PNET include the increased openness about nuclear testing programs to which this would lead and the ongoing cooperation between the national security establishments of the United States and the Soviet Union that would occur, as well as the additional deterrence against cheating and the additional assurance of compliance that use of these methods would provide. Such an agreement would also establish the principle of on-site inspection at nuclear test sites, would make available more accurate data on the yield distribution of large nuclear tests, and might indirectly limit nuclear testing in the future.

Possible disadvantages of seeking an agreement allowing routine use of hydrodynamic methods for test ban monitoring include the potential difficulty of successfully negotiating such use, the restrictions that such use would impose on nuclear test designs, which could adversely affect nuclear testing programs in the future, and the operational security problems at test sites that would be created by the continuous presence there of dozens of monitoring personnel from the other country. Adoption of hydrodynamic yield estimation methods for routine test ban monitoring might also create an impediment to further progress in limiting nuclear testing. Many of the possible advantages of hydrodynamic methods could be gained and many of the possible disadavantages of routine use avoided by an agreement allowing limited use of such methods, such as for calibration of seismic methods.

It is a pleasure to thank T. Lomelino for help in analyzing SLIFER data and rock equations of state and I. Fushiki and G. Miller for help in simulating ground shock propagation. I also thank D. Barr, D. Bernstein, S. Drell, D. Eilers, A. Gancarev, R. Geil, D. Hafemeister, J. Hannon, M. Heusinkveld, R. Hill, R. Jeffries, S. Keeny, D. Metzger, W. Moss, M. Nordyke, W. Panofsky, J. Parmentola, P. Richards, R. Scribner, J. Sullivan, G. van der Vink, and D. Westervelt for helpful comments. I am grateful to Professor Sidney Drell and the other staff members for the warm hospitality extended to me at the Center for International Security and Arms Control, Stanford University, where part of this work was completed, and to the University of Illinois Program in Arms Control, Disarmament, and International Security and the Carnegie Corporation of New York for support. Some of the analysis reported here was carried out for the Congressional Office of Technology Assessment study of seismic verification of nuclear testing treaties and for a study of test ban monitoring sponsored by the Institute for Defense Analyses.

NOTES AND REFERENCES

1. Fermi estimated the yield of the first nuclear test *Trinity* within seconds after the explosion, using a simple hydrodynamic technique. See R. Rhodes, *The Making of the Atomic Bomb* (New York: Simon and Schuster, 1988), p. 674. A model of the shock wave produced by a nuclear explosion in air was published by G. I. Taylor, *Proc. Roy. Soc.*, **A 151**, 421 (1950).
2. See, for example, G. W. Johnson, G. H. Higgins, and C. E. Violet, *J. Geophys. Res.*, **64**, 1457 (1959) and J. H. Nuckolls *Rep. UCRL-5675* (Livermore, Cal.: Lawrence Radiation Laboratory, 1959).
3. *Arms Control and Disarmament Agreements* (Washington, D.C.: U.S. Arms Control and Disarmament Agency, 1982).
4. For an excellent introduction to the scientific and policy issues related to nuclear test ban verification, see U.S. Congress, Office of Technology Assessment, *Seismic Verification of Nuclear Testing Treaties*, OTA-ISC-361 (Washington, D.C.: U.S. Government Printing Office, May 1988).
5. U.S. Department of State, Bureau of Public Affairs, *U.S. Policy Regarding Limitations on Nuclear Testing* (U.S. Department of State Special Report No. 150, August, 1986).
6. U.S. Department of State, Bureau of Public Affairs, *Verifying Nuclear Testing Limitations: Possible U.S.-Soviet Cooperation* (U.S. Department of State Special Report No. 152, August 14, 1986).
7. The uncertainty at the 95% confidence level is usually cited simply because it is conventional to cite the uncertainty at this confidence level in seismology. It is important to note, however, that this confidence level may not be appropriate in the context of test ban monitoring and that the uncertainty is quite sensitive to the choice of confidence level. For example, for a normal distribution a factor of two uncertainty at the 95% confidence level is the same as a factor of 1.27 uncertainty at the 50% confidence level. This means that there is only one chance in four that a single explosion with a yield of 190 kt would appear to have a most likely yield of 150 kt or less and only one chance in 16 that two such explosions would appear to be treaty compliant.
8. In order to prevent seepage of radioactive gases to the surface, the depth of burial (DOB) must be at least $120\,(W/1\,\mathrm{kt})^{1/3}$ m. This requires a DOB of at least 650 m for a 150 kt explosion. When the DOB given by this relation (the so-called scaled depth of burial, or SDOB) would be relatively small, and in media with a substantial water content, the actual DOB is increased in order to achieve containment of radioactive gases. Thus, the actual DOB of an explosion at the Nevada Test Site is normally not less than 250 m. See S. Glasstone and P. J. Dolan, *The Effects of Nuclear Weapons* (Washington, D.C.: U.S. Government Printing Office, 1977), p. 261.
9. For an example of a stemming plan, see H. D. Glenn, T. F. Stubbs, and J. A. Kalinowski, and E. C. Woodward, *Rep. UCRL-89410* (Livermore, Cal.: Lawrence Livermore National Laboratory, August 1983). For a shorter account,

see H. D. Glenn, T. F. Stubbs, and J. A. Kalinowski, in *Shock Waves in Condensed Matter*, ed. Y. M. Gupta (New York: Plenum, 1986), p. 639.
10. S. Glasstone and P. J. Dolan, *op. cit.*, pp. 61–63 and 260–262.
11. H. L. Brode, *Ann. Rev. Nuclear Sci.*, **18**, 153 (1968).
12. R. W. Terhune, H. D. Glenn, D. E. Burton, H. L. McKague, and J. T. Rambo, *Nucl. Tech.*, **46**, 159 (1979).
13. W.K.H. Panofsky, in *Threshold Test Ban Treaty and Peaceful Nuclear Explosions Treaty, Hearings Before the Committee on Foreign Relations, United States Senate* (Washington, D.C.: U. S. Government Printing Office, 1987), at pp. 66 and 247.
14. P. G. Richards, this volume.
15. L. R. Sykes, J. F. Evernden, and I. Cifuentes, in *Physics, Technology, and the Nuclear Arms Race, AIP Conf. Proc. No. 104* (New York: American Institute of Physics, 1983).
16. W. J. Hannon, *Science*, **227**, 251 (1985).
17. J. F. Evernden and C. B. Archambeau, in *Arms Control Verification* (New York: Pergamon-Brassey's, 1986).
18. J. F. Evernden, C. B. Archambeau, and E. Cranswick, *Rev. Geophys.*, **24**, 143 (1986).
19. R. B. Barker, in *Threshold Test Ban Treaty and Peaceful Nuclear Explosions Treaty, Hearings Before the Committee on Foreign Relations, United States Senate* (Washington, D.C.: U. S. Government Printing Office, 1987), at pp. 8, 19, and 89–90.
20. D. A. Vesser, in *Threshold Test Ban Treaty and Peaceful Nuclear Explosions Treaty, Hearings Before the Committee on Foreign Relations, United States Senate* (Washington, D.C.: U. S. Government Printing Office, 1987), at p. 94.
21. S. R. Foley, in *Threshold Test Ban Treaty and Peaceful Nuclear Explosions Treaty, Hearings Before the Committee on Foreign Relations, United States Senate* (Washington, D.C.: U. S. Government Printing Office, 1987), at p. 11.
22. J. H. McNally, in *Threshold Test Ban Treaty and Peaceful Nuclear Explosions Treaty, Hearings Before the Committee on Foreign Relations, United States Senate* (Washington, D.C.: U. S. Government Printing Office, 1987), at pp. 27 and 99–101.
23. H. A. Holmes, in *Threshold Test Ban Treaty and Peaceful Nuclear Explosions Treaty, Hearings Before the Committee on Foreign Relations, United States Senate* (Washington, D.C.: U. S. Government Printing Office, 1987), at pp. 5 and 108.
24. See Ya. B. Zel'dovich and Yu. P. Raizer, *Physics of Shock Waves and High-Temperature Phenomena* (New York: Academic Press, 1967 [English Translation]), pp. 685–705.
25. Ya. B. Zel'dovich and Yu. P. Raizer, *op. cit.*, pp. 741–746.
26. F. Holzer, *J. Geophys. Res.*, **70**, 893 (1965).
27. Ya. B. Zel'dovich and Yu. P. Raizer, *op. cit.*, pp. 49–50 and 705–710.
28. W. R. Perret and R. C. Bass, *Rep. SAND 74–0252* (Albuquerque, N.M.: Sandia

Corp., 1975).
29. J. D. Johnson and S. P. Lyon, *Rep. LA-10391-MS* (Los Alamos, N.M.: Los Alamos National Laboratory, 1985).
30. W. C. Moss, *J. Appl. Phys.*, **63**, 4771 (1988).
31. J. D. Johnson, unpublished.
32. F. K. Lamb, *ACDIS WP-2-87-2* (Urbana, Ill.: University of Illinois Program in Arms Control, Disarmament, and International Security, February 1987).
33. One can also obtain a closed expression for $R(t)$ for any Hugoniot $H(u)$ that is piece-wise linear.
34. Ya. B. Zel'dovich and Yu. P. Raizer, *op. cit.*, Chap. I and XII.
35. L. I. Sedov, *Similarity and Dimensional Methods in Mechanics* (New York: Academic Press, 1959 [English Translation]).
36. G. Barenblatt, *Similarity, Self-Similarity, and Intermediate Asymptotics* (New York: Consultants Bureau, 1979 [English Translation]).
37. See, for example, Fig. 2 of R. C. Bass, in *Shock Waves in Condensed Matter*, ed. Y. M. Gupta (New York: Plenum, 1986), p. 633.
38. G. C. Schmitt and R. D. Dick, *Rep. LA-UR-85-231* (Los Alamos, N.M.: Los Alamos National Laboratory, 1985).
39. C. F. Virchow, G. E. Conrad, D. M. Holt, and E. K. Hodson, *Rev. Sci. Instrum.*, **51**, 642 (1980).
40. R. G. Deupree, D. D. Eilers, T. O. McKown, and W. H. Storey, *Rep. LA-UR-80-3382* (Los Alamos, N.M.: Los Alamos National Laboratory, 1980).
41. During preparations for the 1988 U.S.-Soviet Joint Verification Experiment discussed in §VII, the Soviets stated that they did not have the technology to drill satellite and emplacement holes to the required depth (presumably $\sim$650 m) while maintaining a horizontal displacement within the tolerance (presumably $\sim$10 m) required by the U.S. As a result, the United States flew its drill rig and crew to the Soviet test site. See C. P. Robinson, Testimony before the Subcommittee on Arms Control, International Security, and Science, House Committee on Foreign Affairs, June 28, 1988.
42. M. Heusinkveld and F. Holzer, *Rev. Sci. Instrum.*, **35**, 1105 (1964).
43. W. H. Storey, D. D. Eilers, T. O. McKown, D. M. Holt, and G. C. Conrad, *Rep. LA-UR-82-558* (Los Alamos, N.M.: Los Alamos National Laboratory, 1982).
44. Public information sheet on CORRTEX (Los Alamos, N.M.: Los Alamos National Laboratory, April 1986).
45. R. B. Barker, in *Threshold Test Ban Treaty and Peaceful Nuclear Explosions Treaty, Hearings Before the Committee on Foreign Relations, United States Senate* (Washington, D.C.: U. S. Government Printing Office, 1987), at p. 16.
46. S. Blakeslee, *New York Times*, August 15, 1988, p. A4.
47. Robert C. Toth, *Los Angeles Times*, March 23, 1988.
48. R. C. Bass and G. E. Larsen, *Rep. SAND 77-0402* (Albuquerque, N.M.: Sandia National Laboratories, 1977).
49. For example, according to refs. 5 and 6 "The accuracy of the method is believed to be relatively, but not wholly, independent of the geologic medium, provided

the satellite hole measurements are made in the 'strong shock' region ..."

50. M. Heusinkveld, *J. Geophys. Res.*, **87**, 1891 (1982).
51. M. Heusinkveld, *Rep. UCRL-52648* (Livermore, Cal.: Lawrence Livermore National Laboratory, 1979).
52. R. A. Axford and D. D. Holm, *Proc. Nuclear Explosion Design Conference* (Los Alamos, N.M.: Los Alamos National Laboratory, October 1987).
53. W. C. Moss, private communication (1987).
54. For a list of some of the numerical codes currently in use, see F. K. Lamb, "Monitoring Yields of Underground Nuclear Tests", in *Threshold Test Ban Treaty and Peaceful Nuclear Explosions Treaty, Hearings Before the Committee on Foreign Relations, United States Senate* (Washington, D.C.: U. S. Government Printing Office, 1987), p. 359.
55. Sylvester R. Foley, Jr., Assistant Secretary for Defense Programs, Department of Energy, letter to Edward J. Markey, Congressman from Massachusetts (March 23, 1987).
56. S. S. Hecker, in *Threshold Test Ban Treaty and Peaceful Nuclear Explosions Treaty, Hearings Before the Committee on Foreign Relations, United States Senate* (Washington, D.C.: U. S. Government Printing Office, 1987), at pp. 50–54 and 233–235.
57. M. D. Nordyke, in *Threshold Test Ban Treaty and Peaceful Nuclear Explosions Treaty, Hearings Before the Committee on Foreign Relations, United States Senate* (Washington, D.C.: U. S. Government Printing Office, 1987), at pp. 70–71 and 283–284.
58. R. E. Batzel, in *Threshold Test Ban Treaty and Peaceful Nuclear Explosions Treaty, Hearings Before the Committee on Foreign Relations, United States Senate* (Washington, D.C.: U. S. Government Printing Office, 1987), at pp. 56 and 221–224.
59. Seven C5 military transport aircraft were required to ferry drilling equipment and personnel to the Soviet test site for the 1988 Joint Verification Experiment. See C. P. Robinson, Testimony before the Subcommittee on Arms Control, International Security, and Science, House Committee on Foreign Affairs, June 28, 1988.
60. See R. J. Smith, *Washington Post*, September 4, 1986.
61. It has been suggested that the nuclear calibration charges to be detonated at the test sites could be provided by the monitoring country. If they were, a hydrodynamic estimate of the yield might not be needed, since the yields of certain types of nuclear charges are accurately reproducible. Knowledge of the surrounding rock and geologic structure would still be needed to provide assurance that the coupling of the explosion to seismic waves was understood. However, some way would have to be found to provide assurance that no sensitive information about the design of the nuclear weapons of either country would be revealed.
62. M. R. Gordon, *New York Times*, July 10, 1986, p. 1.
63. M. R. Gordon, *New York Times*, April 7, 1987, p. 1.

64. M. R. Gordon, *New York Times*, April 18, 1987, p. 1.
65. W. Pincus, *Washington Post*, April 18, 1987, p. 1.
66. M. R. Gordon, *New York Times*, June 4, 1987, p. 3.
67. A. Rosenthal, *New York Times*, September 1, 1987, p. 3.
68. W. F. Burns, Testimony before the Subcommittee on Arms Control, International Security, and Science, House Committee on Foreign Affairs, June 28, 1988.
69. C. P. Robinson, Testimony before the Subcommittee on Arms Control, International Security, and Science, House Committee on Foreign Affairs, June 28, 1988.
70. Associated Press, *Los Angeles Times*, January 16, 1988.
71. *Agreement Between the United States of America and the Union of Soviet Socialist Republics on the Conduct of a Joint Verification Experiment* (Washington, D.C.: U.S. Department of State, May 1988).
72. S. Blakeslee, *New York Times*, August 18, 1988, p. A3.
73. W. Sweet, *Physics Today*, November 1987, p. 83.
74. M. R. Gordon, *New York Times*, September 11, 1988, p. 9.
75. R. B. Barker, News Conference in Geneva, Switzerland, November 20, 1987 (see *Arms Control Today*, December. 1987, p. 26).

CHAPTER 6

PUBLIC VERIFICATION: THE SPOT SATELLITE TECHNOLOGY

Pierre Bescond
SPOT Image Corporation
1897 Preston White Drive
Reston, VA 22091-4326

ABSTRACT

The successful launch and operation of the SPOT satellite system has raised significant interest from independent arms-control and third-party treaty-verification organizations. SPOT is the first system to make detailed satellite images of any location on earth available to anyone on an unrestricted commercial basis. The operation of the satellite and the commercial distribution of the imagery follow the United Nations' resolution referred to as "open skies, open access," which recognizes space as an open resource, the peaceful use of which is openly available to everyone.

For the first time ever, SPOT combines an unrestricted image acquisition and distribution policy with advanced capabilities including: 10-meter resolution panchromatic images; the ability to acquire repeat images of any area several times per month; and the production of stereoscopic imagery which can be viewed and analyzed in three dimensions. SPOT's characteristics have significantly decreased the minimum object size which can be detected, recognized, identified and described by interpreters of civilian satellite imagery.

SPOT imagery has already been used several times by the news media to investigate and portray various strategic facilities around the world which are significant to arms control and treaty verification activities, including the Chernobyl nuclear plant, the Krasnoyarsk radar facility, the Semipalatinsk nuclear testing area, and other locations. The real significance of this usage is that, for the first time ever, detailed images of these types of facilities are openly available to anyone, and are not restricted solely to military intelligence organizations.

In addition to direct observation of facilities, mapping and monitoring of environmental conditions relevant to nuclear-test monitoring can also be performed. Examples include accurate mapping and location studies for instrumentation siting, and analysis of environmental conditions which could affect monitoring activities and instrumentation results.

Because of its commercial distribution policy and advanced capabilities, SPOT represents an excellent opportunity for governments, corporations and individuals to develop the potential for monitoring arms control, treaty verification and geopolitical situations.

I. THE SPOT SYSTEM

A. The Space Segment

1. The orbit of the satellite

The SPOT satellite is placed in a sun-synchronous orbit with the following characteristics:

- o The whole globe can be fully covered with a revisit capability of 26 days over 369 orbits (14+5/26 periods per day).
- o The distance between two adjacent orbit tracks is 108.4 km at the equator and only 76 km at a 45° latitude.
- o The satellite altitude above the earth is 822 km at the equator and 832 km at 45° latitude.
- o The satellite velocity with respect to the ground is 6.6 km/s.
- o As a consequence of its sun-synchronous orbit, the satellite always flies at the same local time over areas located at the same latitude. The selected time is 10:30 a.m. for the descending node (which is the intersection of the south-going ground track with the equator). This means 10:15 a.m. for a 45° latitude.

Daytime passes, which are the only ones that are favorable for imaging, are always in the descending orbit. However, the sun's elevation above the horizon of the zones to be observed must exceed 20°.

2. The satellite

There is now one satellite in orbit, SPOT 1, which was launched on February 22, 1986. Its designed life in orbit was two years; this has already been exceeded, with at least three years expected. SPOT 2 is ready for launch when needed with an anticipated launch date early in 1989. SPOT 3 is under construction. SPOT 2 and 3 are identical to SPOT 1, while an upgraded satellite, SPOT 4 (with an extended life and an infrared observation band), is under design, with an option for a SPOT 5 identical to SPOT 4. The continuity of the system is therefore assured well beyond the year 2000.

The SPOT satellite weighs approximately 1,750 kg at the start of its life and is launched by the ARIANE launcher. The dimensions of the body are 2 m x 2 m x 3.5 m, while the overall length of the deployed solar panels is 15.6 m. The satellite consists of the normal two main parts, the bus which one could also call the vehicle, and the payload which performs the actual mission.

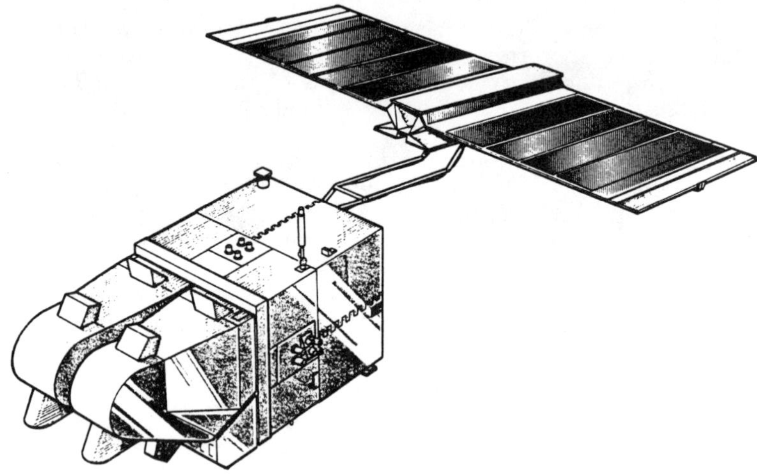

Fig. 1. The SPOT satellite.

The SPOT bus is a standard multipurpose platform with the following essential functions:

o Precision control of the orbit
o Three-axis stabilization
o Electrical power supply
o Housekeeping telemetry transmission
o Command reception
o Monitoring and programming of the payload by an on-board computer loaded through ground control.

The SPOT payload is mounted on one of the side panels of the bus; it consists of two identical HRV instruments (high-resolution telescopes operating in the visible) and a telemetry package consisting of two magnetic-tape data recorders and a telemetry transmitter.

3. The HRV instrument

The image is collected by an instrument of the push-broom type: a complete image line is impressed at the same time on a linear array of 6,000 equidistant CCD (charge-coupled device) detectors; this line is perpendicular to the satellite's track; the successive lines being read at regular intervals, the satellite's motion provides scanning in the direction of the track without any mechanical motion.

The HRV telescope has a field of view of 4.13°, and is designed to operate in either of two modes, in the visible or the infrared portions of the spectrum.

o The panchromatic (black and white, or Pa) mode: the array operates within the wave-length range of 0.5-0.75 μm. Each detector (13 μm x 13 μm) records the image of a ground element (pixel) of about 10 m x 10 m, so that there are 6,000 pixels per line. The image recorded by the array is sampled every 1.504 ms, i.e. every ten meters on the ground.
o The multispectral (color, or XS) mode: the incoming light is split into three bands, XS1, XS2 and XS3, operating simultaneously, respectively within wave-length ranges of 0.5-0.59 μm, 0.61-0.69 μm, and 0.79-0.9 μm. In this mode, two consecutive lines and two consecutive pixels per line are aggregated together so as to obtain a pixel representing a 20 m x 20 m ground element. There are then 3,000 pixels per line and per spectral band and a sampling time internal of 3.008 ms.

The basic design decisions were dictated by the small-scale subdivision of much of the agricultural land in many parts of the world, and also by the requirements of mapping applications. The SPOT optical resolution is much better than in any other existing satellite remote-sensing system. It introduces a great number of novel opportunities into the processing possibilities, as has already been shown by two years of operational experience.

3. Nadir and off-nadir viewing

Another interesting and new feature of the SPOT system is that each telescope is equipped with a plane mirror, steerable by ground control, with 91 positions (45 on either side of the central position). This allows the central direction of the instrument to be moved between +27° and -27° off the vertical, off the perpendicular to the satellite track. This provides coverage of areas located as much as 450 km from either side of the orbit track. The length of the lines pictured on the ground can then reach 80 km, but of course for the same number of pixels.

This configuration of the instruments offers a great number of very advantageous possibilities. For nadir (vertical) viewing, one instrument allows coverage of a 60-km wide ground strip, but the two instruments can be used in the so-called "joint" mode. When coupled with a +/- 1.8° overlap (i.e. 3 km on the ground), they provide a total swath width of 117 km. Since the distance between adjacent ground tracks at the equator is approximately 108 km, complete coverage of the earth can be obtained with this setting of the instrument field of view.

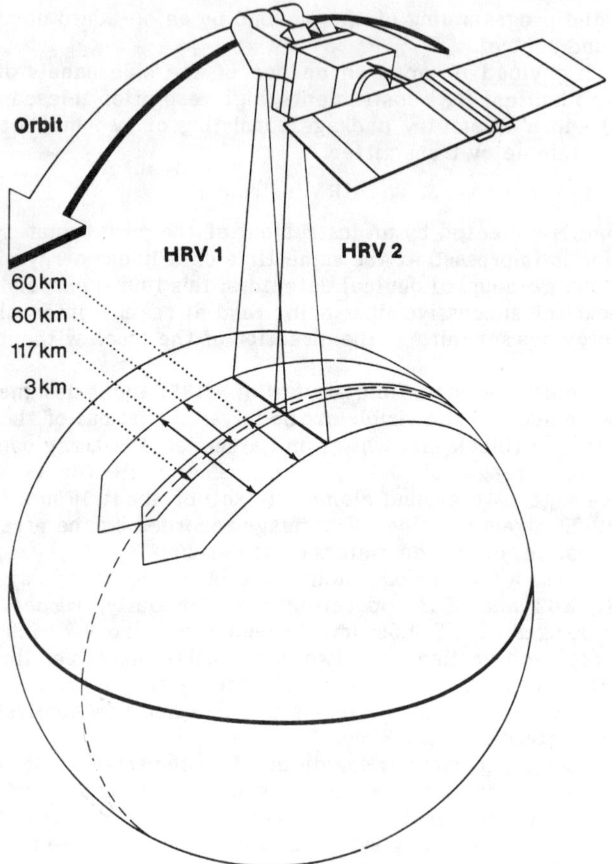

Fig. 2. Nadir viewing by HRV instruments.

The number of possibilities offered by off-nadir (lateral) viewing is quite interesting.

- o The revisit capability for a given area decreases from 26 days (for vertical viewing), to 5 days by increasing the viewing angle to 27°. Specifically, a given region can be revisited on dates separated alternately by one and four (or occasionally five) days. And of course the high-latitude regions can be revisited much more often.
- o The programming of observations along the track of the satellite can be very flexible: a sequence of recorded images may include on- and off-track pictures, and this with either or both operation modes (panchromatic and/or multispectral). The only constraint is that a minimum downtime of 15 seconds (no image acquisition) must be made available between two such pictures, to allow for mirror rotation and reconfiguration.

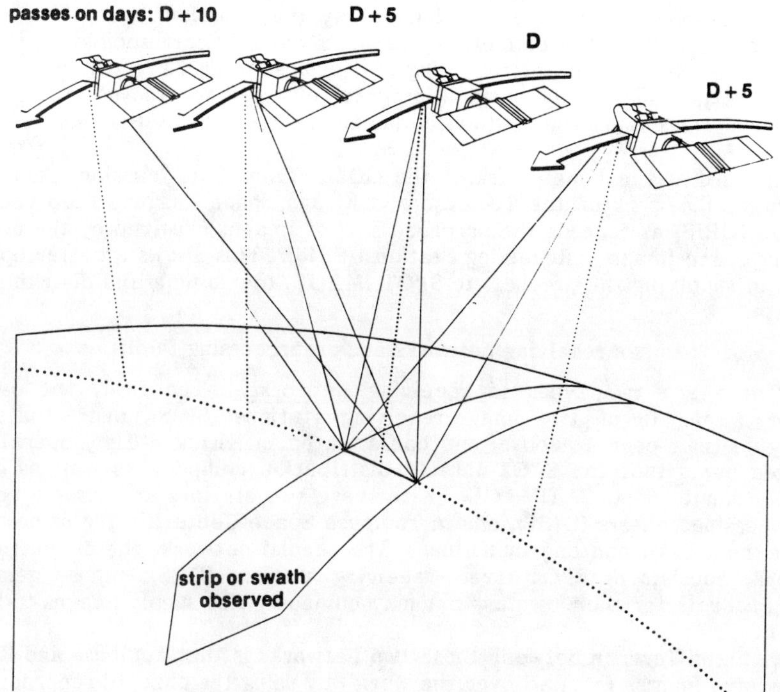

Fig. 3. Revisit Capabilities.

o Since two observations can be made on successive days such that the two images correspond to equivalent pointing angles on either side of the vertical, this gives a stereoscopic viewing capability, which is of great interest for photo-interpretation, photogrammetry, topographic cartography, geology, etc.

B. The Ground Segment

1. Satellite control and payload programming

The SPOT Mission and Control Center (CCM) in the Toulouse Space Center, located in the southwest of France, is responsible for satellite control, payload programming and ground facilities coordination. The master satellite control station is located in Aussaguel-Issus, near Toulouse, and performs, using the 2-GHz frequency band, such functions as satellite control, ranging measurements and satellite housekeeping telemetry reception during its passes in sight of the station. It is directly connected to the Control Center which processes the telemetry and the ranging measurements, then defines the various maneuvers to be performed to maintain the satellite under operational conditions, and eventually initiates the relevant commands.

The Control Center may be connected to the other 2-GHz CNES (Centre National d'Etudes Spatiales, The French Space Agency) TTC (Telemetry, Tracking & Command) network stations of other networks: NASA and NASDA for instance, as is the case for the launch in an early orbit phase. It can also be connected to the 2-GHz Swedish control station of Kiruna.

Payload programming is carried out by the "mission" part of the CCM control center. The CCM collects the requests from its correspondents, namely SPOT IMAGE and all direct receiving stations, and carefully studies them in order to establish a daily optimal program of work for the satellite. This program of work is transmitted every day to the satellite by the "control" part of the CCM.

A third and last main task of the CCM, again of its "mission" part, is to coordinate the work of the Toulouse and Kiruna Space Imagery Preprocessing Centers (CRIS) as regards the preprocessing of images received by the corresponding Space Imagery Receiving Stations (SRIS). This allows a better optimization in supplying the products to SPOT IMAGE, the commercial distributor of the data.

2. Image receiving networks and preprocessing facilities

There are two types of receiving networks. The first, the central network, consists of two image-receiving stations (SRIS), one situated in Aussaguel-Issus near Toulouse, and one situated in Kiruna. They operate for the sole benefit of the SPOT IMAGE distribution company as well as of its Swedish counterpart, SATIMAGE. With these two stations are associated two preprocessing centers (CRIS), one in Toulouse Space Center (in the same building as the CCM), and one in Kiruna. The second network, the decentralized network, consists of SPOT direct-receiving stations (DRS): these generate SPOT products for their own use but may obviously also supply images to SPOT IMAGE.

A big difference between these two networks is that Toulouse and Kiruna can receive images from all over the world, by using the onboard recorders and the recorded playback telemetry mode, whereas DRS can only receive direct imaging telemetry within their own range. The coverage of these stations is a circle of about 2,500-km radius centered on the station.

Reception of the SPOT imaging telemetry is limited to the rate of 50 Mbit/sec, transmitted in the 8-GHz frequency band (8.025-8.4 GHz), with a band width of 100 MHz. The satellite pass lasts from a few minutes to a maximum of thirteen minutes. It is thus necessary to have the station antenna automatically track the satellite, and to record the data on high-density magnetic data tapes (HDDTs) for the purpose of further preprocessing.

It is very important to recall that all the above specifications, as well as the radiated power, have been chosen to ensure commonality in receiving-station design with other earth-observation systems, especially LANDSAT 5. As a result, receiving equipments exist that is compatible with both SPOT and LANDSAT 5, and there are now a number of stations that can receive signals from both SPOT and LANDSAT 5.

The SPOT system capacity can be described approximately as follows. The satellite payload is capable of handling 900,000 scenes per year (a scene being the amount of image data for an area 60 km in length--i.e. along the ground track, and 60 to 80 km in width--i.e. in the cross-track direction, depending on the instrument viewing angle. The limiting factor is not the imaging capacity of the satellite itself but the preprocessing and processing capacities of the stations. Each of the Toulouse and Kiruna stations is equipped to receive and process 250,000 scenes per year. That leaves nearly half of the satellite capacity to be received and processed by the foreign stations. It is estimated that judiciously distributing the direct-receiving stations (DRSs) throughout the world would mean an optimum number ranging between 10 and 15. The combination of the direct images received by the DRSs, and both direct and recorded images received by Toulouse and Kiruna, should quickly lead to worldwide coverage.

The nature of the preprocessing of the SPOT images is defined by each receiving station. However, in order to standardize the SPOT products and optimize their production and distribution, the stations may be required to produce various basic product types. The following is a list of standard levels of processing as defined today in the SPOT system, which are performed by the two main CRIS preprocessing centers.

- o Inventory/film level. It is the high-density magnetic tape (HDDT) processing operation which allows the scenes to be incorporated into the SPOT reference grid and which, for each image, produces small subsampled images on a 70-mm film, called quick-looks, which make it possible, in particular, to determine the cloud cover.
- o Level 1A, in which the relative detector calibration is taken into account.
- o Level 1B. At this level the images undergo additional radiometric corrections to take account of satellite motion, and geometric corrections for the earth's rotation and curvature, and for the viewing angle. Resampling at a mesh of 10 or 20 meters is also done.
- o Level 2. Bearings on ground control points are taken on a reference map and a distortion model is applied, so that the scene can be presented in a given cartographic projection.
- o Level S is comparable to Level 2, in the sense that it gives a product which can be superimposed on a reference image. Each scene in Level 1, 2, or S is made available either on CCT (computer-compatible magnetic tape) or on 241-mm precision film, black and white, or color, in the standard scale of 1:400,000.

In operating the SPOT system, CNES is responsible for control and programming of the satellite, reception and preprocessing of the data in France. So its task ends in CRIS Toulouse.

3. The Commercial Distribution of Data

As described above, the SPOT system has been designed for the mass production of data. CNES, a public agency, realized very early in the program that efficient commercial user-oriented distribution of these data could only be achieved by a private company. CNES therefore created SPOT IMAGE, in which it holds 39% of the shares. Other shareholders include:

- o Government agencies, such as IGN (Institut Geographique National, the French mapping agency), BRGM (Bureau de Recherches Geologiques et Minieres), and IFP (Institut Francais du Petrole);
- o Private firms such as MATRA and SEP (Societe Europeene de Propulsion), two leading French aerospace companies heavily involved in the SPOT program: MATRA as prime contractor, SEP in the ground segment;
- o International partners in Sweden (SSC, the Swedish Space Corporation) and Belgium (Societe Nationale d'Investissements);
- o A consortium of banks.

SPOT IMAGE established a worldwide network of distributors (over 50), each covering a specific geographical area. For example, Sweden created a similar company, SATIMAGE, which was given exclusive distribution rights for Scandinavia. For the U. S. community, recognized as the largest and most advanced in the world, SPOT IMAGE created a subsidiary, SPOT Image Corporation (SICORP), based in Reston, Virginia. SICORP is responsible for:

- o Building up an archive of the U. S. data which are acquired through the two Canadian stations of Price Albert and Gatineau;

o Manufacturing the standard products for the customers through a preprocessing center which can produce 40 computer-compatible magnetic tapes (CCTS) per day;
o Marketing and selling the data to the U. S. customers.

Other distributors include licensees, dealers, agents and brokers; operating under a variety of agreements which clearly illustrate the flexibility that the commercial approach has brought into the market place.

This type of organization for distribution ensures that services and products can be tailored to the specific needs of the users in a specific area. It also helps in turning the technology from a research activity into a commercial enterprise.

II. TWO YEARS OF OPERATIONAL AND COMMERCIAL EXPERIENCE

SPOT 1 was launched on February 21, 1986 and was commissioned on May 6, 1986. It has already gone beyond its nominal orbital lifetime of 2 years, and is expected to remain in good working order for at least 3 years. This is based on the observed performance of critical systems such as power generation (see Fig. 4 below), orbit control and fuel consumption, and internal temperature control. These are operating well within specifications.

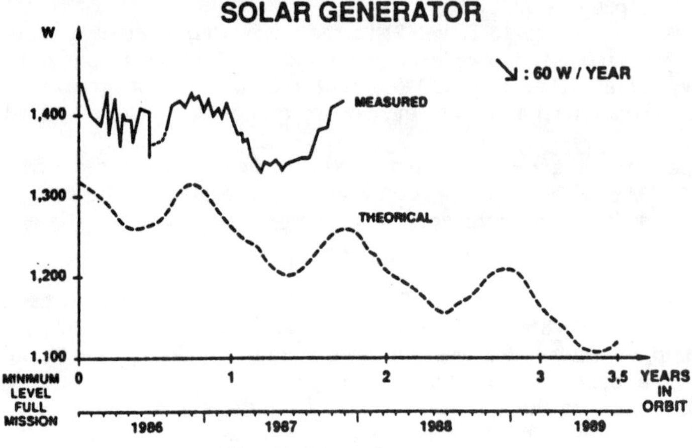

Fig. 4. Solar Generator.

A. Technical Performances

More than 440,000 SPOT scenes have been acquired by the two principal stations located in Kiruna, Sweden and near Toulouse, France. At the same time, more than 180,000 SPOT scenes have been collected by the two Canadian stations (Prince Albert, Saskatchewan and Gatineau, Quebec) and later archived either in Washington (U.S. data) or in Prince Albert (Canadian data). The Indian receiving station in Hyderabad started collecting data in May 1987.

The central catalog of all collected SPOT image includes the reference images of more than 600,000 scenes, out of which 25% have less than 10% cloud cover and an additional 5% have a cloud cover of less than 25%. Among these, 35% have been acquired in the panchromatic mode and 65% in the multispectral mode. More than 20,000 SPOT images have been processed by the facilities in

Toulouse, Kiruna and Reston, 85% at Level 1B, which is the most common processing level.

The quality of these data is excellent; specifications are exceeded most of the time. The following performance levels were achieved during commissioning tests and have been regularly achieved since.

1. Noise

The noise observed in the images has two different origins: one type of noise is generated by the electronic devices themselves, the other results from residual errors in the detector-normalization preprocessing.

Signal-to-Noise Ratio (SNR): The SNR describes the noise in the columns of images (along the track). It is measured at different gain levels, for each HRV telescope, and for each spectral band. It has exceeded specifications in the multispectral mode, even under the extreme conditions of high-reflectance terrain with a low sun angle (i.e. snow at high latitudes). It is close to specification in the panchromatic mode, except under extreme conditions, for example as described above just after switching on the electronics of the HRV.

Table 1. Signal-to-noise ratio (SNR) along the track.

	HRV Telescope 1				HRV Telescope 2			
	PA	XS1	XS2	XS3	PA	XS1	XS2	XS3
Specifications	200	210	200	270	190	190	200	270
Results	330	320	220	370	210	330	230	360

Detector Normalization: Variations in sensitivity among the detectors introduce line (cross track) noise which cannot be totally eliminated even after equalization processing. Normalization coefficients to be applied to the detectors (3,000 in the XS mode, 6,000 in the Pa mode) are measured through the analysis of uniformly-snowy landscapes (Greenland and Antarctica) approximately every 3 months. Residual errors are estimated statistically both for "local noise" (statistical fluctuations over a large number of 10 pixel x 10 pixel subimages) and "global noise" (over the full image). Table 2 shows that normalization is always accomplished within the specifications.

Table 2. Detector normalization (NΘΔL(10x10)).

	HRV Telescope 1				HRV Telescope 2			
	PA	XS1	XS2	XS3	PA	XS1	XS2	XS3
Specifications	1.3	1.05	1.05	0.8	1.3	1.05	1.05	0.8
Results	0.85	0.6	0.7	0.45	0.85	0.6	0.7	0.45

2. Absolute Calibration

Absolute calibration gives a proportionality coefficient A_k between the scene radiance L_k and the normalized output x_k. To calculate these coefficients $x_k = A_k \cdot L_k$ (one per spectral band), atmospheric and reflectance measurements are made over the White Sands test site while onboard calibration is performed through fiber optics which project the solar radiance (which is accurately known) onto the detectors. The 10% specified accuracy is met: it was 8% during in-flight commissioning. It is 6% today. Intercalibration between HRV1 and HRV2 is excellent (1%).

3. Modulation Transfer Function

The Modulation Transfer Function (MTF) allows a comprehensive measure of the signal degradation as it passes through all the components (optics, sensors, electronics) of the instruments. Using a catalog of high-resolution aerial photographs of cities of Southern France which have been digitized and then degraded through known MTFs, a qualitative assessment of the MTF in the panchromatic Pa mode has been made. It confirms the visual conclusion that in the Pa mode the telescope HRV2 performs less well than telescope HRV1. Both however nominally meet or exceed specifications, and no in-flight refocusing of the instruments has been necessary.

Quantitative comparison have also been made between the two instruments, which show that differences in the multispectral bands are minimal.

4. Level-1B Geometric Quality

Image Localization: This parameter measures the accuracy with which the actual position of a given point can be determined on a geometrically corrected scene. As shown in Table 3, the actual accuracy is much better than the specification.

Table 3. Localization accuracy.

Specification:	
Vertical Viewing	1500 m (rms)
Oblique Viewing	1800 m (rms)
Measured:	
In-Flight Assessment Period	832 m (rms)
One Year (August 1986/August 1987)	510 m (rms) [from 220 m minimum to 899 m maximum]

Length Distortion: In a given image, the length distortion is the relative error in the distance between two points. The specification for length distortion, given for two points separated by more than 500 m on the ground, again is very largely exceeded by the measurements reported in Table 4.

Table 4. Length alteration.

Specifications	0.01 (rms)
In-Flight Assessment Period	0.0015 (rms)
Follow-On Quality Control	0.0005 (rms)

Anisomorphism: At a given point, anisomorphism gives the difference between the length distortion along two perpendicular directions. The following Table 5 shows that this specification for anisomorphism is met.

Table 5. Anisomorphism

Specifications	0.001 (rms)
In-Flight Assessment Period	$0.0008<A<0.0016$ (rms)
Follow-On Quality Control	0.0007 (rms)

5. Registration

Multispectral registration plays a very important role in the quality of color images since it determines on how well the 3 color bands are superimposed. The specification for this registration was 1/3 of a pixel. Measurements show a very good registration of 0.15 pixel.

Multidate registration measures the ability to superimpose the same scene pictured at different dates, and is a quality criterion for the Level-S product. The specification of 1/2 of a pixel is met for this registration.

Cartographic registration measures the ability to superimpose a SPOT scene on a given cartographic projection (a map) and is a quality criterion for the Level-2 product. The specification was 50 rms (root-mean-square). The actually-measured registration was much better, at 20-m rms.

6. Elevation measurement accuracy

The specification in extracting the altitude of a given point from a stereoscopic pair of scenes called for an accuracy of 10-m rms. Measurements were made by the French Mapping Agencies (IGM) during the in-flight commissioning tests, using two areas in Southern France which were equipped with control and check points before launch. For a +27°/-27° stereo pair the altitude accuracy was 3.5 m; for a 0°/+27° stereo pair the altitude accuracy was 7.0 m.

Other organizations worldwide confirmed those findings during 1987. The excellent results described in II.A.4-6, both planimetric and in elevation, confirmed the adequacy of SPOT data for topographic cartography. SPOT data can thus be used--and are used--for creating and updating maps at the scales of 1/50,000 and even 1/24,000. Furthermore, some geologists have recognized their potential usefulness for working in the field at scales as precise as 1/12,000.

B. Commercial Experience Since Launch

Commercial distribution of SPOT data expanded very rapidly in 1987, with 10,000 products being distributed compared to 3,000 in 1986. The proportion of digital products continually increases for markets in developed nations, while photographic products are still more commonly requested in less developed markets.

The market share is also very strongly characterized by large sales in Europe (51%), North America (30%) and the Asia-Pacific region (12%). A significant proportion of the distributed SPOT data (30%) results from specific satellite imaging request, while the rest corresponds to orders of data already in the archive.

The satellite-based remote sensing industry has matured significantly since it was first conceived as a government research and development program. The technology is no longer the province of a relatively small group of researchers and scientists, but has become a growing commercial industry with widespread applications. Global recognition of its value is constantly increasing.

SPOT has provided the technology, ground systems and a commercial operation which are designed to succeed as a commercial enterprise. The SPOT system has tremendous potential for proving the worth of commercial remote sensing, and for guiding its development. The future of the industry now depends on successful marketing of the available technology.

The formula for industrial success involves everyone from data users, both traditional and newly developing, to hardware and software manufacturers, to processing (value-added) companies. In addition, there is a growing group of "indirect" clients who never see an image but who use the information that the images provide.

SPOT has established very strong relations with several value-added firms. They are developing new services, analytical methods and products and are helping to identify new markets within the expanding user communities. New applications that are having increasing potential thanks to the SPOT approach include urban planning, topographic cartography, generation of Digital Terrain Models (DTM) and as a consequence terrain simulations. SPOT data are also very well adapted to become an input to Geographic Information Systems (GIS). And last, but not least, SPOT has become a valuable resource to news gathering, especially as visual and information support for stories by the media. This in turn has opened up the way for new applications related to monitoring of geopolitical situations, as well as arms control and international treaty verifications.

III. SPOT AND THIRD PARTY VERIFICATION

Recent advances in arms control by the world superpowers, and in fact by all nations, are becoming increasingly important in this era of complex geopolitics and advanced weaponry. What was once the province of government officials and military organizations is now becoming more and more a forum for public involvement. While the signing of treaties by government officials is characterized by political posturing on all sides, the power behind such a document can only be solidified with successful verification procedures.

Private sector organizations are increasingly involved in researching and supporting arms-control activities. Non-intrusive means of observing any area in the world, a necessary technique for treaty verification, was once available only to military intelligence groups in the United States and the Soviet Union. Now, with SPOT, public interest groups and government agencies have a tool which provides a level of information never before publically available.

A. SPOT Capabilities for Verification

The SPOT satellite system has three unique attributes which are of particular significance to arms control and treaty verification applications: (i) highest resolution available from a civilian satellite system; (ii) the ability to acquire imagery of any location in the world (open skies); (iii) an operational policy of commercially distributing this imagery to any interested party in the world on an unrestricted basis (open access).

Because of these capabilities, SPOT is being used to study and report on strategic facilities around the world, particularly in those countries where information is unavailable due to political or physical limitations.

B. Resolution and Image Interpretation

When using satellite imagery there are four general levels of interpretation accuracy related to studying objects on the earth's surface: detection, recognition, identification, and description. These levels and the numbers associated with them provide general guidelines for understanding the application capabilities of SPOT data.

Detection is the ability to see the presence of an object on an image. The minimum object size detectable on SPOT imagery is 10 to 20 meters.

Recognition refers to the ability to determine the type of object (building, vehicle, airplane) one is observing on an image. The minimum object size capable of being recognized on a SPOT image ranges from 20 to 60 meters.

Identification is the ability to classify a recognized object, for example, to identify a building as a munitions plant or a ship as an oil tanker. Minimum object size for identification ranges from 60 to 180 meters.

The final level is description--the ability to analyze an identified object, or to assess such characteristics as age and physical condition. Minimum object size for this level of interpretation is 400 to 600 meters.

SPOT has the unique capability of producing stereoscopic imagery that can be viewed and analyzed in three dimensions. By adding this third dimension, one increases the ability to interpret objects on an image and the minimum object size stated above for each level is generally reduced by one half.

These four levels serve as basic guidelines for the potential of SPOT imagery to reveal information about objects of interest. There are several other techniques used in image interpretation such as analyzing the terrain, environment, and associated facilities around an object, and looking at multidate images of an object for change detection purposes. Also the use of multispectral imagery adds radiometric information which can reveal significant other characteristics and conditions on the earth's surface.

Since SPOT images exist in digital form, they can be processed as either computer-compatible tapes or as photographic products. The range of hardware/software image processing systems, and the rapid advances in this technology add greatly to the value of the digital imagery as a primary source of information. Manipulation and special processing of the digital imagery can be used to extract information and highlight features for closer study.

C. Open Skies and Open Access

SPOT operates according to the U.N. resolution which declares space as an open resource over which no country has sovereignty. The HRV sensors can be programmed for image acquisition over any area on the earth's surface. In fact, because of the ground receiving station network and onboard tape recorders, SPOT is the only civilian remote sensing system which can acquire images of any location on earth.

Once image data are received by a station, they are processed, and then enter SPOT's commercial distribution system. One of the principal tenets of SPOT's commercial system is providing unrestricted access of image data of any area to anyone in the world. This is essential to establishing remote sensing as an internationally accessible technology, and to developing its full potential as a tool for monitoring worldwide arms-control efforts.

Open access eliminates the ability of an organization or country to control and limit distribution of information on other areas or geopolitical situations. Effective verification requires open availability of necessary tools to all interested parties.

D. SPOT's Strategic Observations

The significance of the new earth observation capabilities introduced by SPOT was first exposed to the world community shortly after the launch of SPOT 1 in February of 1986. The explosion at the Chernobyl nuclear plant in the Soviet Union held the world in the grip of tension and anxious speculation, exacerbated by the fact that information and photographs of the disaster were unavailable. Shortly after the accident, SPOT 1 acquired a detailed image of the crippled plant, and the next day the image was seen on international news broadcasts and in the print media.

Much has happened since this occurrence, involving further growth and development of the SPOT system, and the increasing experience of image interpreters and the news media as they explored the capabilities presented by this technology. With the open availability of such a powerful technology comes the responsibility for its proper use. While the news media have been pioneering users of SPOT imagery for reporting on events of global significance, their use of the imagery has also revealed the importance of proper analysis and information-extraction from the imagery. While the imagery is a primary and invaluable source of information, it is sometimes only one piece of an information gathering system needed to study and understand events of global significance. Often corroborating information must be used to support conclusions drawn from image analysis and to ensure that those conclusions reflect reality.

The news media have been instrumental in revealing the power of this technology for significant information gathering and reporting on events of concern to the world community. Much of the work they have done has revealed the potential and laid the foundation for arms-control and treaty-verification studies using SPOT imagery.

The Krasnoyarsk radar facility in the Soviet Union had long been studied by U.S. government agencies using military satellites leading to the conclusion that the facility was in violation of an arms-control treaty between the United States and the Soviet Union. Few, if any, individuals outside these agencies were aware of the facility or its significance to the treaty. SPOT images of the facility were analyzed by trained interpreters. These images became the basis of a news story which was broadcast internationally, giving the public their first view of this facility, and exposing the controversy surrounding it. A few months later, in an unprecedented move, the Soviets invited civilian representatives from the United States to tour the facility and see what was actually there.

A year prior to that, the Soviets had invited Western journalists to inspect the Semipalatinsk nuclear test area, only months after SPOT images of the site (see Fig. 5) appeared in news broadcasts with claims by reporters that the Soviets may have been preparing for a new round of nuclear testing in the area.

Use of SPOT imagery has not been limited to the Soviet Union. Multidate images have been analyzed to detect changes on the ground, and then displayed during news broadcasts to illustrate the build-up of military facilities in the Persian Gulf region. This again gave the public its first real view of the strategic geography in this region, as it pertains to the current crisis there.

IV. CONCLUSION

These and several other examples have demonstrated the effectiveness of SPOT for visual support and information gathering related to events of global significance. SPOT is the only satellite-based remote sensing system which operates under the U.N. policies of open skies/open access and has the ability to acquire worldwide coverage. These are both prerequisites to effectively supporting public involvement with arms control efforts. The Carnegie Endowment for International Peace, which is actively involved in these efforts, is undertaking a large study to assess the benefits of commercial remote sensing in three specific areas: cross-border conflicts; U.S.-Soviet strategic weapons development; and slowing the spread of nuclear weapons.

For the first time ever, detailed images are openly available to the public and are helping to increase the flow of vital information necessary for understanding complex geopolitical situations, information which was previously available only to a limited few.

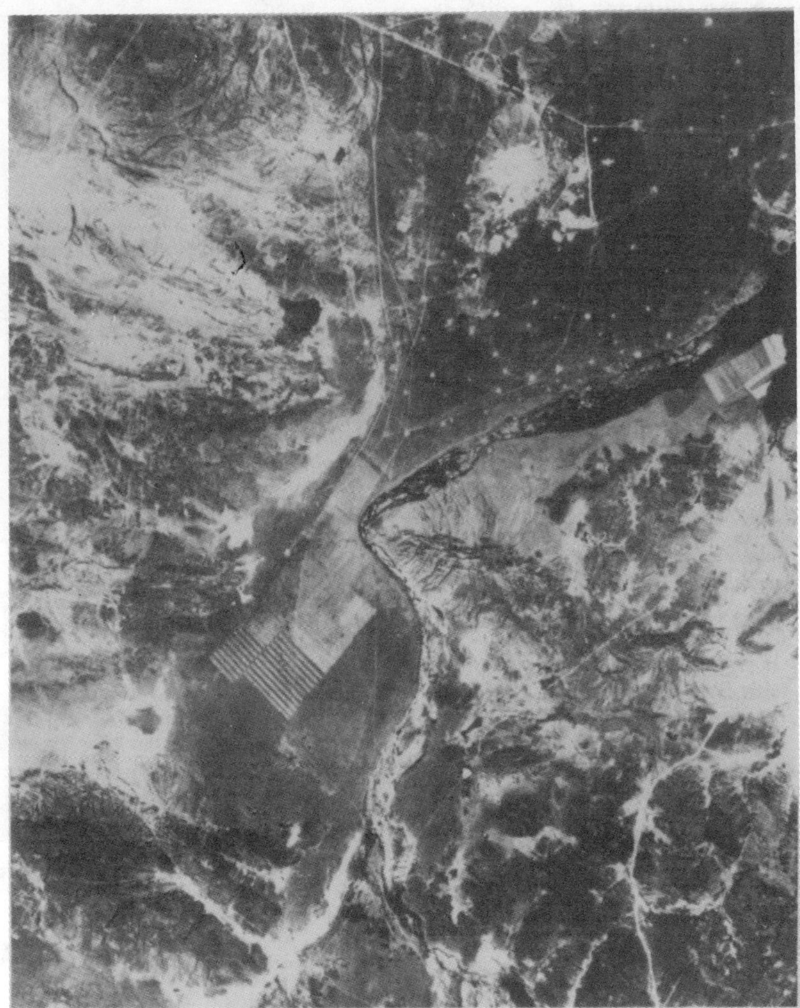

Photo credit: Copyright © 1987 CNES. Provided courtesy of SPOT Image Corp., Reston, VA

Fig. 5. Soviet nuclear testing facility, Semipalatinsk, U.S.S.R. Visible are cable scars and access roads connecting with drill holes. This ten-meter panchromatic image was taken by the French SPOT satellite.

V. BIBLIOGRAPHY

- o From the Paris Symposium of November 1987 on "SPOT 1 Image Utilization, Assessment, Results":
  1. G. Brachet, "SPOT Data Acquisition and Distribution."
  2. G. Begni, P. Henry, M. Leroy, M. Dinguirard, "La Qualite des Images SPOT."
  3. G. Cales, "Bilan du Fonctionnement de SPOT 1."
- o Various presentations by P. Bescond on SPOT, mainly during the last two years in the United States and Canada.

LARGE PHASED-ARRAY RADARS

Dr. Eli Brookner
Raytheon Company
Wayland, MA 01778

ABSTRACT

Large phased-array radars can play a very important part in arms control. They can be used to determine the number of RVs being deployed, the type of targeting of the RVs (the same or different targets), the shape of the deployed objects, and possibly the weight and yields of the deployed RVs. They can provide this information at night as well as during the day and during rain and cloud covered conditions. The radar can be on the ground, on a ship, in an airplane, or space-borne. Airborne and space-borne radars can provide high resolution map images of the ground for reconnaissance, of anti-ballistic missile (ABM) ground radar installations, missile launch sites, and tactical targets such as trucks and tanks. The large ground based radars can have microwave carrier frequencies or be at HF (high frequency). For a ground-based HF radar the signal is reflected off the ionosphere so as to provide over-the-horizon (OTH) viewing of targets. OTH radars can potentially be used to monitor stealth targets and missile traffic.

1.0 INTRODUCTION

Figure 1 shows the trajectory of a missile launched from Tyuratam to Kamchatka. Radar phased arrays can be used during the boost phase to determine the down-range capability of the ICBM system. By observing the post-boost phase, during which time the RVs may be deployed from the post-boost vehicle (PBV), the radar can determine the number of RVs being deployed and the type of targeting being used for the RVs. For example, one could determine if multiple reentry vehicles (MRVS) or multiple-independently targeted RVs (MIRVs) are being fired. For the former the RVs are deployed in a fixed relative close pattern (over for example a silo field). For the latter the RVs can be independently deployed to targets separated widely, like 100 nmi (180 km). By imaging the PBV and RVs it may be possible to also determine the shape and potentially the weight and payload of the RVs. Such observations made during the mid-course phase could potentially provide the same information as obtainable during the PBV phase. If observations are made during the reentry phase it is possible to obtain more and better

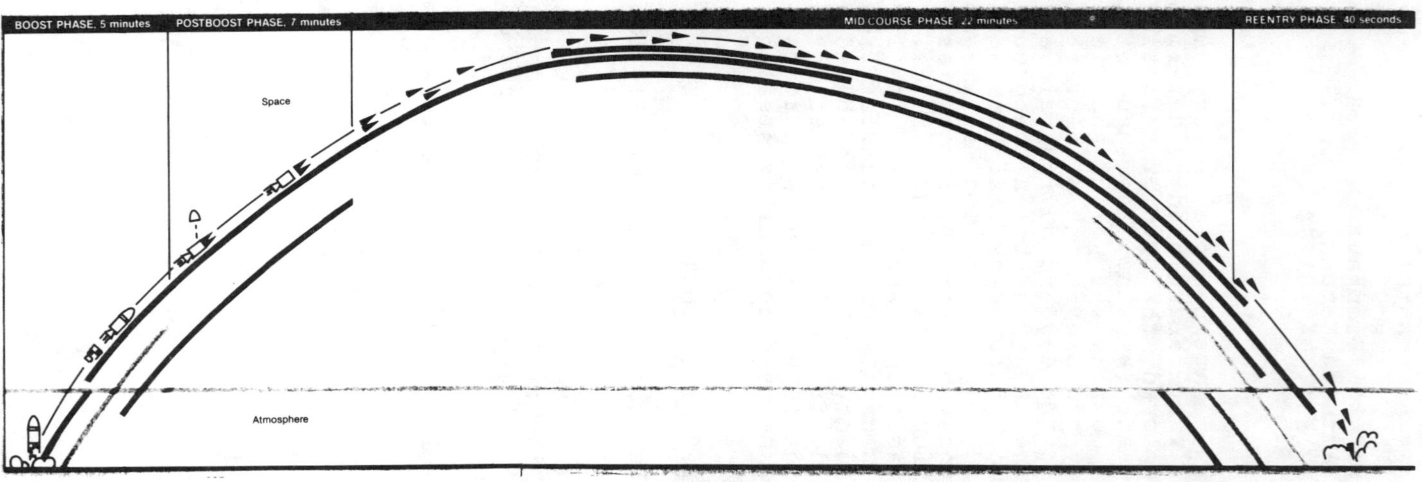

Figure 1. Missile Trajectory: Tyuratam to Kamchatka. (From Zorpette, G., IEEE Spectrum, July 1986. Copyright 1986 by IEEE. Used with permission.)

information than when observing the PBV or mid-course phase. In addition to being able to determine the number of RVs being deployed, their targeting, their shape and possibly weight and payload, it is also now possible to determine if the RVs maneuver, that is, if maneuvering reentry vehicles (MaRVs) are used. Because of the interaction of the RVs and decoys with the atmosphere, it is possible to make a better determination of the weight of the RVs and decoys. As a result a better estimate of the payloads of the RVs is obtainable. In addition, it is also possible to determine the effectiveness of the decoys. These large phased arrays used for this purpose can be either ground-based, airborne, ship-borne or space-borne. Their carrier frequencies should be in the microwave regime, specifically, at L-band or above; the higher the carrier frequency, the better the imaging.

Radars operating at the HF carrier frequencies between 5 and 30 MHz can potentially provide over-the-horizon (OTH) coverage of stealth bombers and of missiles. For these carrier frequencies, the ionosphere acts like a mirror for the radar signal.

Airborne and spacecraft-borne synthetic aperture radars (SAR) can provide day-night reconnaissance, and mapping of anti-ballistic-missile (ABM) ground radar sites, missile launch facilities, and tactical targets such as trucks and tanks.

Section 2.1 gives examples of some large ground-based phase-phase steered array radars, (that is, radars which are phase steered in azimuth and elevation). Section 2.2 will describe the future potential of large ground-based radars to do target imaging. Section 2.3 will discuss over-the-horizon radars that can be used potentially for monitoring stealth targets and missiles. Section 3.0 will discuss the use of airborne and space-borne radars to do synthetic aperture radar (SAR) high resolution imaging.

2.0 LARGE GROUND-BASED PHASED-ARRAY RADARS

2.1 EXAMPLE SYSTEMS

2.1.1 COBRA DANE RADAR (AN/FPX-108, REFS. 1-12)

The COBRA DANE radar, shown in Fig. 2, is a 95 ft diameter, 15.4 MW peak power and 0.92 MW average power L-band radar deployed in 1977 on the next to the last island in the Aleutian chain of Alaska, providing close proximity to U.S.S.R. for observing the conformance of test firings of Soviet intercontinental ballistic missiles (ICBM) systems as agreed to under the SALT I treaty. Because of its location, the COBRA DANE radar can observe the mid-course phase of the U.S.S.R. test firings of ICBM Systems in Kamchatka. In addition to its function of intelligence gathering on these firings for purposes of determining possible treaty violations, it is used for space track support and ICBM early warning.

Figure 2. COBRA DANE (AN/FPS-108). (From Brookner, E., Aspects of Modern Radar, Artech House, 1988, [1].)

Phased-array technology such as used in the COBRA DANE radar provides many advantages over conventional mechanically scanned radar systems. It can track more targets, more accurately and at greater speed. The phased-array system incorporates thousands of small radar antenna elements coordinated by a large-scale Control Data CYBER computer. It directs the energy emitted by the antenna elements in precise, controlled patterns, allowing the radar to detect and track objects at very high rates. While the conventional dish-shaped radar may take up to a minute to mechanically swing from one area to another, the phased-array radar can electronically change its pointing direction in microseconds. COBRA DANE's off-the-shelf computers are programmed to steer the beam, store and display the data, and perform the post-mission data reduction and analysis. The COBRA DANE radar can simultaneously track hundreds of targets.

A significant feature of this radar is its ability to provide high range-resolution observations of targets at ranges of the order of 1,000 nmi (1850 km). These wideband measurements are used to obtain target size and shape data. A 200 MHz waveform, which can provide a 2.5 ft range-resolution, is used to provide the high range-resolution measurements. Normally it is not possible to obtain such a high range-resolution measurement from a 95 ft aperture antenna because of the time dispersion across the antenna. For example, when pointing up at a target which is 22° off boresite in elevation, the signal from the lower part of the aperture arrives at the target 95 ft $\times \sin(22^{\circ})=35.6$ ft after the signal from the top of the array arrives at the target. As a result the 2.5 ft pulse is spread out over 35.6 ft in space which is equivalent to 36 nsec in time, the velocity of electro-magnetic waves being 1 ft/nsec, making it impossible to achieve the desired high range-resolution. To circumvent this problem, the 95 ft diameter array is broken up into smaller subarrays. These subarrays are sized so that the time dispersion across them is small compared to the desired range resolution. Furthermore, the signal to each subarray is delayed by the amount necessary to make the signals from all subarrays arrive at the target simultaneously. Thus, for the example given above, the top subarray unit receives a time delay of about 36 nsec relative to the bottom subarray unit. This process is called time-delay steering. Phase steering is then used over each subarray. The 95 ft aperture is divided into 96 subarrays, with each subarray having 160 active radiating elements that are phase steered.

Figure 3 is a simplified block diagram of the system which shows the time delay steering unit (TDU) used for each subarray and the phase steering within the subarray. There are a total of 15,360 (160x96) active elements in the array face. Each subarray is fed by a Raytheon ring bar QKW 1723 160 kW peak power traveling wave tube (TWT).

Another interesting feature of the radar is the use

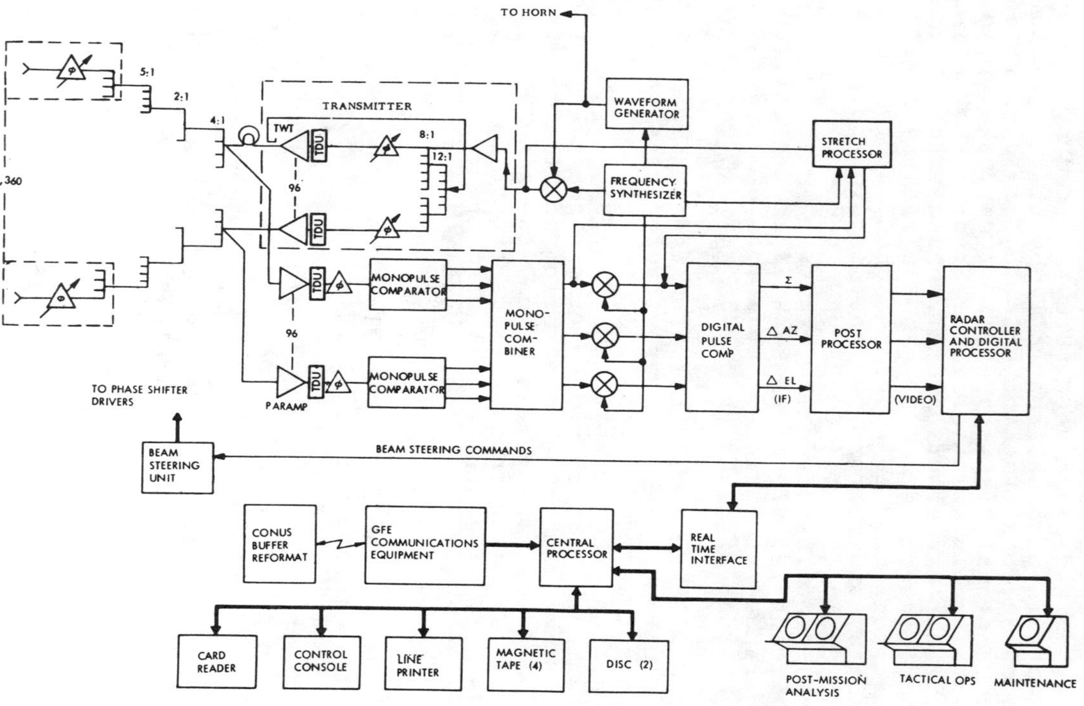

Figure 3. COBRA DANE block diagram. (From Filer and Hart [8], IEEE EASCON Proc., 1976.)

of array thinning [1,13,14]. Figure 4 illustrates the concept of array thinning. The black dots in the figure indicate positions in the array at which active elements are placed. Where there are no black dots, non-radiating elements (called dummy elements) are used. These elements are terminated into a dummy load whose impedance matches that of the active element loads. Notice that most of the elements near the center of the aperture are active with the density of active elements decreasing toward the edge of the aperture. This density variation effectively provides amplitude weighting of the aperture illumination on transmit. As a result low near-in sidelobes are achieved. The density tapering was designed to provide a 35 dB tapered-Taylor weighting on transmit. Array thinning also results in a narrower mainlobe beamwidth than would have been possible if a full array having the same number of active radiating elements had been used. The total number of dummy elements is 19,403 out of a total of 34,769 elements in the array. If desired, it is possible at a future date to increase the system's sensitivity by 9 dB. This would be done by increasing the number of active elements by a factor of 2, with 15,360 of the dummy elements being replaced by active elements, with another 96 TWTs feeding these additional active elements. Figure 5 shows a row of 1 to 160 subarray power divider units as seen looking approximately along the major diameter of the backside of the array face.

Dual-polarized circular-waveguide radiating elements are used in the array; currently these are connected for vertical polarization [15]. Growth featuring dual polarization is a system option.

Detailed parameters for this radar, as well as other radars to follow, are given in Table 1 of Chapter 1 in Reference 7 and in the Radar Catalogue in the Appendix of Chapter 2 in Reference 1.

2.1.2 COBRA JUDY (AN/SPQ-11) [16]

The COBRA JUDY ship-borne phased-array radar (AN/SPQ-11) is an Air Force detection and tracking system for collecting data on foreign strategic ballistic missile tests. The COBRA JUDY high priority, multi-million dollar surveillance system provides Department of Defense planners with information pertinent to current international treaties, as well as future Strategic Arms Reduction Talks. The radar has been acquired by the Air Force Systems Command's Electronic Systems Division, Hanscom Air Force Base, Mass. Raytheon Company, through its Equipment Division in Wayland, Mass., is the prime contractor for the COBRA JUDY radar. This ship-borne sensor will operate in the Pacific Ocean and be based out of Pearl Harbor, Hawaii.

The Observation Island, on which the radar is installed, is crewed by the Navy's Military Sealift Command. The ship is 563 ft long, 76 ft wide, weighs

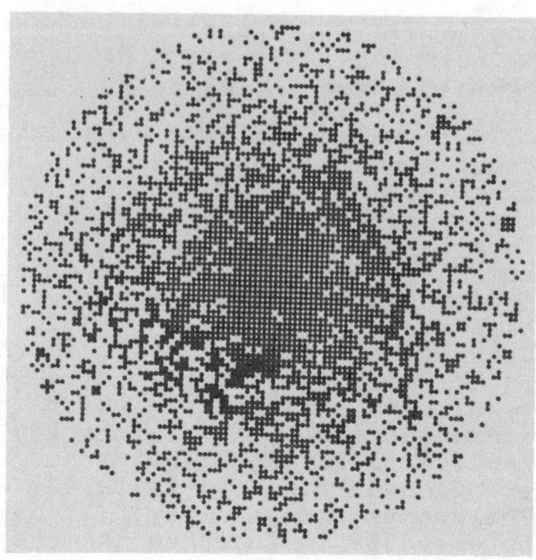

Figure 4. Illustration of thinned circular array (not COBRA DANE). (From Skolnik, et al. [14], IEEE Trans. on Ant. and Prop., July 1969. Copyright 1969 by IEEE. Used with permission.)

Figure 5. View of 1:160 power divider units along backside of array. (From Brookner, E., Aspects of Modern Radar, Artech House, 1988, [1].)

approximately 17,000 tons and has a speed of about 20 knots.

Installed on the stern of the ship, the sophisticated radar weighs 250 tons and stands four stories high. The steel turret containing the sensor face can mechanically rotate. The octagonal array, 22.5 ft in diameter, makes up one wall of the pyramid-like structure and contains 12,288 antenna elements; see Fig. 6 and 7. Its transmitter consists of 16 TWTs shown in Fig. 8.

2.1.3 PAVE PAWS RADAR (AN/FPS-115) [1-7,11,17,18]

The PAVE PAWS radar (see Fig. 9) is used to detect and provide warning of a sea launched ballistic missile (SLBM) attack on the Continental United States (CONUS). It will also furnish attack characterization data to the North American Air Defense Command's (NORAD) Cheyenne Mountain Complex (NCMC), the Strategic Air Command (SAC) and the National Command Authority (NCA). The hundreds of satellites that orbit the earth are also tracked by the PAVE PAWS radars. There are four such radars: one at Ottis Air Force Base, in Cape Cod, Massachusetts (completed in 1979; see Figure 9); one at Beale AFB, California (completed in 1980); one at Robins AFB, Georgia (completed in 1986); and one near Goodfellow AFB, Texas (completed in 1987).

The impressive advancement made in large UHF phased-array radars over the past decade is exemplified by the PAVE PAWS system. The first large phased-array radar was the UHF AN/FPS-85 which became operational in 1968 [7]. For this radar it was less expensive to use separate antennas for transmit and receive than to compensate for the duplexer insertion loss and pay the cost of the duplexer. The power generation was supplied by tetrode tubes driving each transmit radiating element. In contrast, the PAVE PAWS system transmits and receives on the same antenna. In addition, the transmitter is all solid state, using bipolar silicon UHF transistor power amplifiers with each active element of the array driven by a transceiver (T/R) solid-state module, which contains the transmitter and receiver as well as the phase shifter needed per radiating element for electronically steering the antenna beam. These modules receive the exciter signal which is first passed through a phase shifter on transmit and then through a solid state power amplifier. On receive the signal comes from a radiating element and passes through a low noise receiver in the transceiver module after which it is switched through the same phase shifter before going to a beamformer. The PAVE PAWS radar has two array antenna faces instead of one. The two faces provide continuous coverage for two 120° azimuth regions for a total coverage of 240° in azimuth. The use of a solid-state transmitter leads to better reliability and ease of maintenance over that obtainable with a tube transmitter. The PAVE PAWS radar represents the first

Figure 6. COBRA JUDY on Observation Island ship. (Coutesy of Raytheon Co.)

Figure 7. Closeup of COBRA JUDY array face. (From Brookner, E., Aspects of Modern Radar, Artech House, 1988, [1].)

Figure 8. View of some of the COBRA JUDY traveling wave tubes. The hoses are for water cooling the TWTs. (Courtesy of Aviation Week & Space Technology.)

Figure 9. Multifunction UHF phased-array radar AN/FPS-115115 (PAVE PAWS). (From Brookner, E., Aspects of Modern Radar, Artech House, 1988, [1].)

phased-array radar capable of scanning in 2-dimensions (azimuth and elevation) that uses a solid-state transmitter.

Figure 10 shows a simplified block diagram of the radar. As in the case of the COBRA DANE radar, the array is divided into subarrays. Each subarray consists of 32 active modules feeding 32 radiating elements. There are 56 such subarrays. These subarrays are in turn driven by modules identical to the array element modules except that they do not use the unneeded receiver portion. A predriver drives the 1-to-56 divider which supplies the subarray signals. This predriver also uses a module identical to the array-element module, except that again it does not use the receiver portion. Thus, to build the PAVE PAWS array, a large number of the UHF modules are proliferated.

Figure 11 shows a view of the backside of the array comparable to that of Fig. 5 for the COBRA DANE system. A subarray driver module is visible on the left mounted on the girder. The units standing in the middle of the floor are the power supply units for the array modules. What is striking when comparing Fig.11 with Fig. 5 is how much less hardware is needed for the PAVE PAWS solid-state array as compared to the TWT COBRA DANE array. The large 1-to-160 subarray divider units and 3-inch high-power coax cable of the COBRA DANE system are not needed. The high-voltage modulators and high-voltage (42 kV) power supplies (not shown) are also eliminated.

Figure 12 shows the modules in the backside of the array face near the center of the array. The subarray driver modules can also be seen. These are mounted on the girders. Array thinning is used similar to that in the COBRA DANE radar. There are 885 dummy elements and 1792 active elements spread over a 72.5 ft diameter circle. Provision has been made to allow the active area of the antenna to grow to a 102 ft circle if desired at a future time. This would permit the system sensitivity to increase by 10 dB. The number of active elements would be increased by slightly more than a factor of 2 in this case. This 420 to 450 MHz radar has a peak power of about 600 kW and an average of about 150 kW per face. Its detection range is 3,000 nmi for a 10 m^2 target. Figure 13 shows the eight-inch high tilted crossed-dipole radiating elements of the array. These are bent downward to alter electromagnetic interactions of adjacent elements that can at certain steering angles prevent energy from radiating (the energy instead going back into the radar) which would have the result that the target could not be seen at these steering angles. The radar is thus called blind at these angles. This is referred to as the phased array radar blindness phenomenon [7]. The metal stubs interspersed among the antenna elements aid in producing a circularly polarized beam, a characteristic that helps to maximize the system's sensitivity.

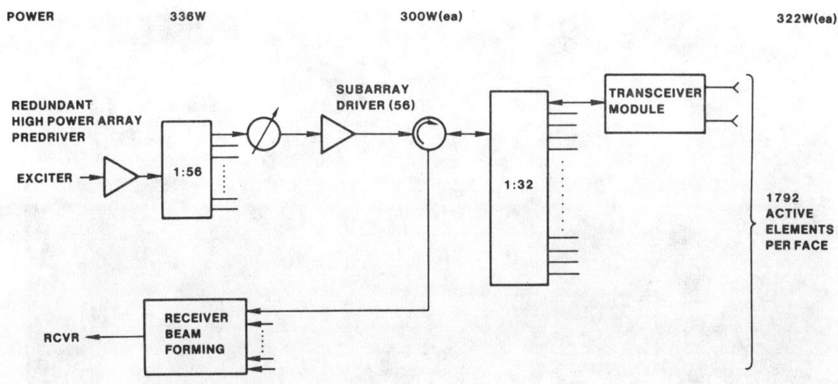

Figure 10. Simplified block diagram of PAVE PAWS system. (From Campbell and Hoft [17].)

Figure 11. Backside of PAVE PAWS along major diameter of the array. (From Brookner, E., Aspects of Modern Radar, Artech House, 1988, [1].)

Figure 12. View of backside of PAVE PAWS array near center of the array. Modules and driver modules shown. (From Brookner, E., Aspects of Modern Radar, Artech House, 1988, [1].)

Figure 13. View of front face of PAVE PAWS array showing bent dipoles. (From Brookner, E., Aspects of Modern Radar, Artech House, 1988, [1].)

The USAF can grow one of the PAVE PAWS radars by 6 dB in sensitivity. A sensitivity growth of 10 dB is also planned in the future at one of the southern sites in order to perform the SPACETRACK function and thereby replace the aging AN/FPS-85 radar system at Eglin AFB, Florida.

For a further discussion on the PAVE PAWS see Reference [1].

2.1.4 BMEWS RADAR UPGRADE (AN/FPS-123(V)5) [1,19]

The aging BMEWS radar in Thule, Greenland has been upgraded with a Raytheon solid-state 2-faced UHF phase-phase steered array radar using solid-state modules which are identical to those of the PAVE PAWS radar; see Fig. 14 [1,7]. Each face has an array 84 ft in diameter, containing 2,560 active elements and 1,024 dummy (inactive) elements. Similar upgrades are planned for the radars at Fylingdales Moor, England and Clear, Alaska. The Fylingdales system is to have 3 faces for 360° coverage.

2.1.5 TERMINAL IMAGING RADAR (TIR [20])

The Raytheon Company has a contract for the construction of an experimental phased-array radar for the Strategic Defense Initiative (SDI) program. This radar, shown in Figure 15, is called the Terminal Imaging Radar. This X-band radar is capable of providing high-resolution imaging of targets for the purpose of discriminating between RVs and decoys during the terminal phase. It will have about 40,000 phase shifters. The radar is to be installed at the Army's Kwajalein Atoll by the end of 1990 with demonstration/validation testing beginning in October 1991.

The Raytheon Company has proposed a variant of this ground-based radar as an alternative for the SDI system space-borne mid-course discrimination sensors. This ground- based radar (GBR) has the advantages over a space-borne system of lower risk (being based on proven radar phased-array technology), lower cost (costing less than $5 billion to acquire and maintain over its lifetime), and the ability to provide quick response to a rapid, unexpected deployment of Soviet anti-ballistic missiles. The radar would be used in conjunction with passive infra-red sensors. This would make it very difficult for the U.S.S.R. countermeasure designers to deceive the U.S. sensors because simultaneously building countermeasures for both the radar and the passive IR sensor is very difficult. The principles by which imaging can be done with a high-resolution microwave radar is described in Section 2.2. That section also illustrates the present state-of-the-art of microwave imaging through some actual examples.

Figure 14. UHF solid-state BMEWS radar system in Thule, Greenland. (Courtesy of Raytheon Co.)

Figure 15. Model of Raytheon X-band phased-array Terminal Imaging Radar (TIR). (Courtesy of Raytheon Company.)

Figure 16. The U.S.S.R.'s new anti-ballistic missile radar located at Pushkino (artist's concept).

2.1.6 U.S.S.R. LARGE PHASED-ARRAY RADARS [1]

The U.S.S.R. has recently developed two very impressive huge four-faced phase-phase steered array anti-ballistic missile radars providing 360° of coverage [21]. The radar at Pushkino is housed in a four-faced structure 120 ft high and 500 ft wide (see Fig. 16), while that at Abalakovo is even larger. The U.S.S.R. is also constructing a large radar (30-story receiver, 11-story transmitter) at Krasnoyarsk; see Fig. 17. This high-power (~20 MW [22,23,35]) radar has an estimated carrier frequency of 180 MHz.

2.2 FUTURE POTENTIAL FOR HIGH RANGE-RESOLUTION MICROWAVE RADARS TO PROVIDE IMAGES OF RVs

It is potentially possible to obtain high range-resolution images of tumbling and/or spinning RVs with a microwave radar. Such images would provide the shape of the RV. Section 2.2.1 explains the basis for obtaining such images with a microwave radar. Section 2.2.2 gives actual radar images obtained of an aircraft and ship with high range-resolution radars.

2.2.1 THEORETICAL BASIS FOR OBTAINING HIGH RESOLUTION IMAGES OF AN RV WITH A MICROWAVE RADAR

The basis for such imaging is explained using the tumbling RV shown in Fig. 18. Figure 19 shows the two-dimensional image obtained for this target. It was assumed that the radar has a range resolution of 0.5 ft. This is obtainable with a bandwidth of 1 GHz. The image in Fig. 19 shows the return from the RV tip, joins and base. These are successively further down range from the radar as expected, the tip being closest to the radar while the echos from the base are 10 ft further back, the RV length being 10 ft. What is also observed is that the echo from the top edge of the base and from the bottom edge of the base are separated out in a direction perpendicular to the line of sight to the radar. These echos are separated by a distance of 8 ft, which is equal to the RV base diameter. Thus, the image not only provides the length of the target, but also gives the base diameter of the target. In addition, the image shows where the cylindrical part of the RV starts and ends, as well as the diameter of the cylindrical part of the RV, which must be 6 ft because the echos from the joins of the cylinder are 6 ft apart in the cross-range direction, that is, the direction perpendicular to the radar line of sight (LOS).

We shall now explain how the radar achieves its resolution in the cross-range direction. To obtain this resolution in the cross-range direction the radar uses pulse doppler processing. Specifically, it coherently processes the return pulse doppler echos for a time long

Figure 17. The U.S.S.R.'s Krasnoyarsk phased-array radar in Siberia. On the right is the 30-story transmitter; on the left is the 18-story receiver.

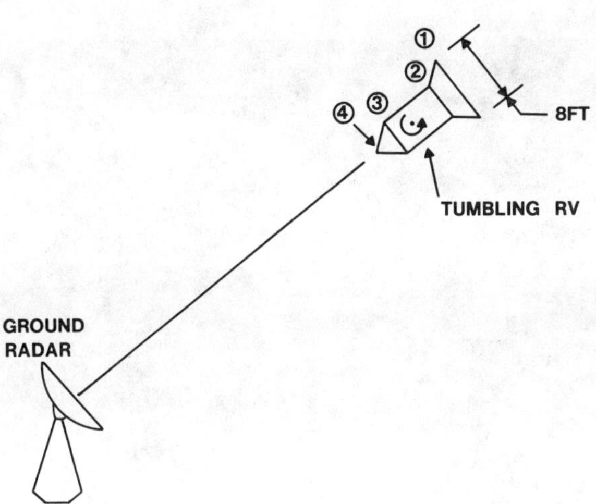

Figure 18. Tumbling reentry vehicle (RV): Nose-on view. (From Brookner, E., "Radar Imaging for Arms Control," Chapter 11 in Arms Control Verification, The Technologies That Make It Possible, Edited by K.T. Tsipis, D.W. Hafemeister and P. Janeway, Pergamon-Brassey's, 1986. Used with permission [24].)

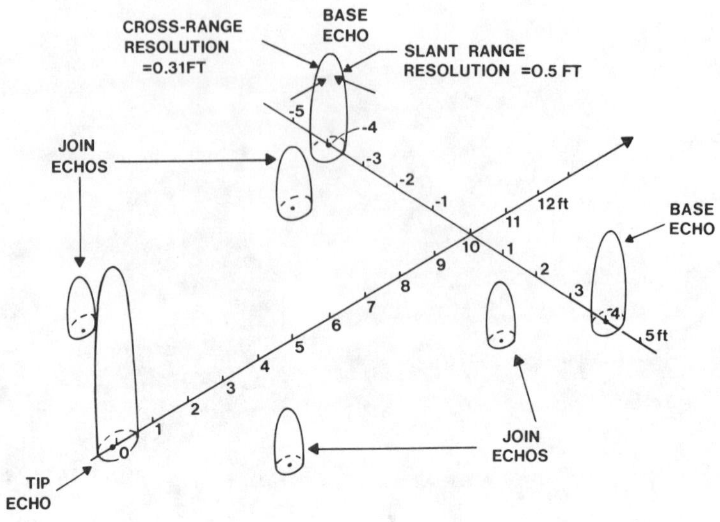

Figure 19. Two-dimensional image of tumbling RV: nose-on viewing aspect. (From Brookner, E., "Radar Imaging for Arms Control," Chapter 11 in Arms Control Verification, The Technologies That Make It Possible, Edited by K.T. Tsipis, D.W. Hafemeister and P. Janeway, Pergamon-Brassey's, 1986. Used with permission [24].)

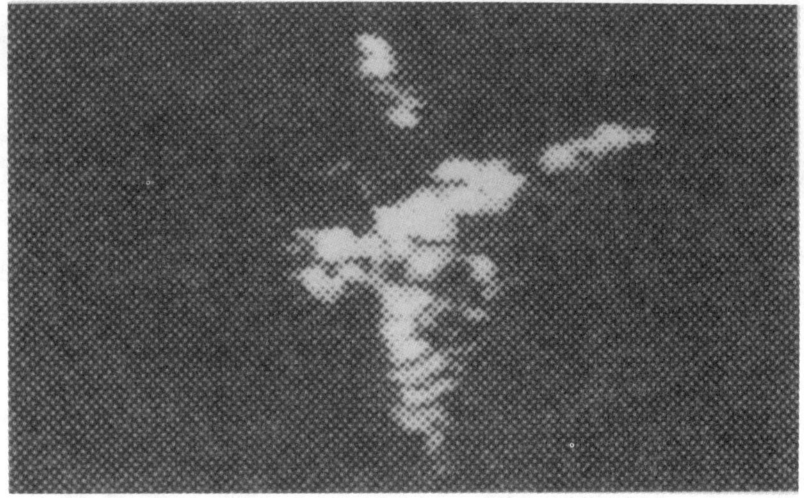

Figure 20. Aircraft ISAR image: 20 sec coherence time. (From: "Target-Motion-Induced Radar Imaging" by C. Chen and H.C. Andrews, 1980, IEEE Trans. on Aerospace and Electronic Systems [25]. Copyright 1980 by IEEE. Used with permission.)

enough to provide sufficiently high doppler resolution, doppler resolution equating to cross-range resolution.

To see how this comes about, assume for simplicity that the RV's center of gravity is in the base of the RV. Under this simplifying assumption, the RV will tumble about the center of the base of the RV. Assume furthermore that the RV is in the position shown in Fig. 18; that it is tumbling at a rate of ω rad/sec.; and that the base diameter is D. Then the doppler velocity from the top of the base of the RV will have a doppler velocity of $\omega D/2$ while the echo from the bottom edge of the RV base will have a doppler velocity of $-\omega D/2$. Thus the width of the doppler return from the base of the RV is ωD. Assume the tumble period of the RV is 2 sec.; by simple substitution the doppler velocity from the top edge of the base is +3.83 m/sec (+12.6 ft/sec) while that from the bottom edge is -3.83 m/sec. If the radar carrier frequency is 10 GHz then the doppler frequency shift f_D from the top edge of the base is 255 Hz (because $f_d=2v_d/\lambda$, where v_D is the doppler velocity of the target, f_D is the resulting doppler shift seen by the radar and λ is the wavelength of the radar). Thus the total doppler velocity across the base is 510 Hz. Further assume that the coherent integration time of the radar receiver is 0.05 sec. The doppler resolution of the radar is one over the coherent integration time or 1/0.05 sec=20 Hz. Thus the 510 Hz doppler spread across the base of the RV is broken up into 510 Hz/20 Hz=25.5 doppler resolution cells. Equivalently, the 8 ft base of the RV is broken up into 25.5 cross-range resolution cells so that the cross-range resolution of the radar along the base is 8 ft/25.5=0.31 ft, the value given in Fig. 19. A different physical explanation of how the radar achieves its cross-range resolution is given in Reference 24.

An image such as that shown in Fig. 19 is generally called an Inverse Synthetic Aperture Radar (ISAR) image. For the standard Synthetic Aperture Radar (SAR) images obtained with microwave radars, the antenna is on a moving platform while the ground being imaged is stationary; see Sect. 3.0. In the case of ISAR imaging, the radar is stationary while the target is moving; see Reference 24. That is the reason for calling it an inverse SAR (ISAR) image. The two, SAR and ISAR, are really equivalent.

2.2.2 EXAMPLE AIRCRAFT AND SHIP HIGH-RANGE TWO-DIMENSIONAL IMAGES OBTAINED WITH MICROWAVE RADARS

Figure 20 shows the image of a maneuvering aircraft target obtained with a high range-resolution ground-based microwave radar [25]. Fig. 21 shows a high range-resolution ISAR radar image of a ship [26,36]. The ship image is upside down in Fig. 21 because at the instant at which the image was obtained, the returns from the top of the ship had a negative doppler shift whereas those from

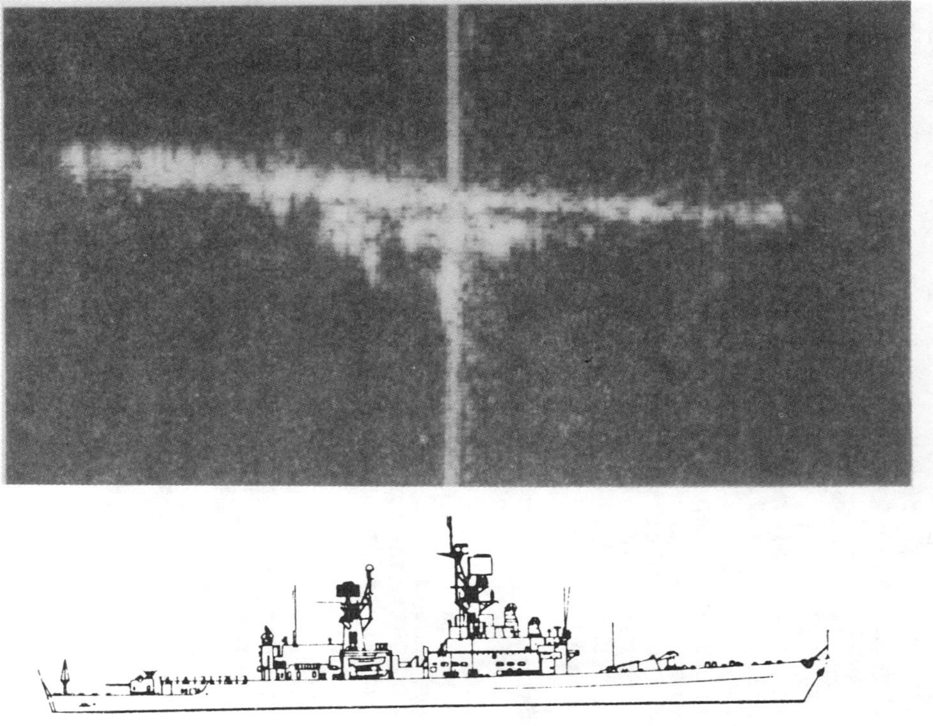

Figure 21. ISAR image of U.S. Belknap Class cruiser. (From R.J. Drazovich and R.X. Lanzinger, IEEE 1981 Pattern Recognition and Imaging Processing Symposium Proceedings, Aug. 3-5, 1981. Copyright 1981 by IEEE. Used with permission.)

the bottom of the ship had a positive doppler shift. One could of course easily turn the image upside down. Figures 22 and 23 show ISAR images obtained of B-52 and Space Shuttle models using microwave tomography imaging techniques which are basically equivalent to the ISAR techniques described above [1,27-29].

2.3 OVER-THE-HORIZON (OTH) RADARS

Over-the-horizon backscatter (OTH/B) ground-based radars are being deployed to detect an oncoming attack by U.S.S.R. bombers. These radars, one on the East Coast and one on the West Coast of the United States, are being built by General Electric Company for the U.S. Air Force. Two additional OTH/B sites are planned, one in Alaska and one in central U.S.A. looking south and possibly east and west. The Raytheon Company is building a relocatable OTH radar for defense of the U.S. Fleet against an attack by U.S.S.R. bombers. Recently there has been claimed to be an interest in HF radars because of their effectiveness against future stealth targets [30]. It is said to be impractical to use radar absorbant material (RAM) at these frequencies [30]. As a result, OTH HF radars could potentially be used for monitoring the development of U.S.S.R. stealth bombers. It is claimed that a forward-scattering OTH radar is used to monitor U.S.S.R. and Chinese missile traffic [31]. OTH systems operate in the frequency range from 5 to 28 MHz and typically scan only in the azimuth direction. At these frequencies the ionosphere acts like a mirror and as a result the transmitted signal reflects downward from the ionosphere and allows the detection of aircraft beyond the horizon, to about 2000 miles (3700 km) (see Fig. 24). The OTH/B radar being built by the General Electric Co. uses a 3630 ft long transmitting antenna located at Moscow/Caratun, Maine; see Fig. 25. The antenna's height ranges from 35 to 135 ft. Twelve 100 kW average-power transmitters provide the power for one 60° sector; three such sectors provide a total coverage at Maine of 180°. The receiving antenna consists of a 4980 ft long, 64 ft high array of 246 elements. It is located 110 miles southeast at Columbia Falls, Maine.

The Relocatable Over-the-Horizon Radar (ROTHR) being developed by the Raytheon Co. for the Navy is a tactical land-based, bistatic ionospheric backscatter radar system. Its mission is to provide wide area over-the-horizon surveillance of both aircraft and ships in support of tactical forces in locations of national interest. ROTHR extends the range of horizon-limited microwave radars and provides the Navy with early warning of approaching naval and airborne threats at any altitude. The range is extended from the present capability of 20 nmi (37 km) for low flying aircraft and 200 nmi (370 km) for high-flying aircraft to more than 1000 nmi (1850 km) [39]. Typically it will scan a 60° sector [39].

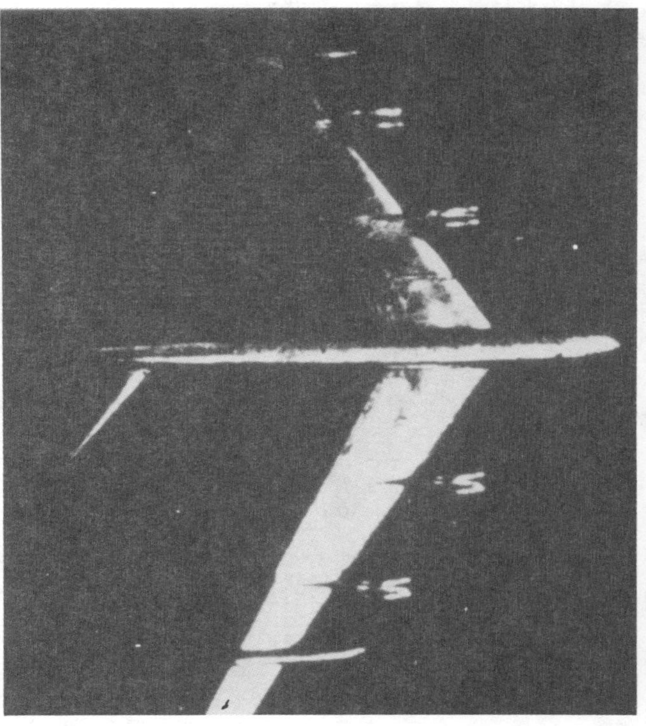

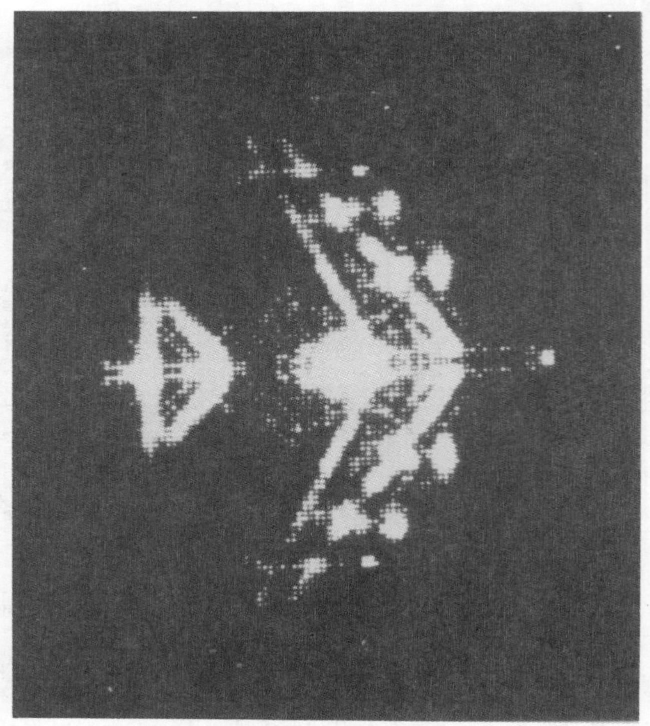

Figure 22. ISAR image obtained of model of B-52 using microwave tomography. Model on left side; image on right side. (From: "High Resolution Wideband Radar Adds New Dimension to Imaging" by B.E. Manz, 1983, Microwaves and RF [28]. Used with permission.)

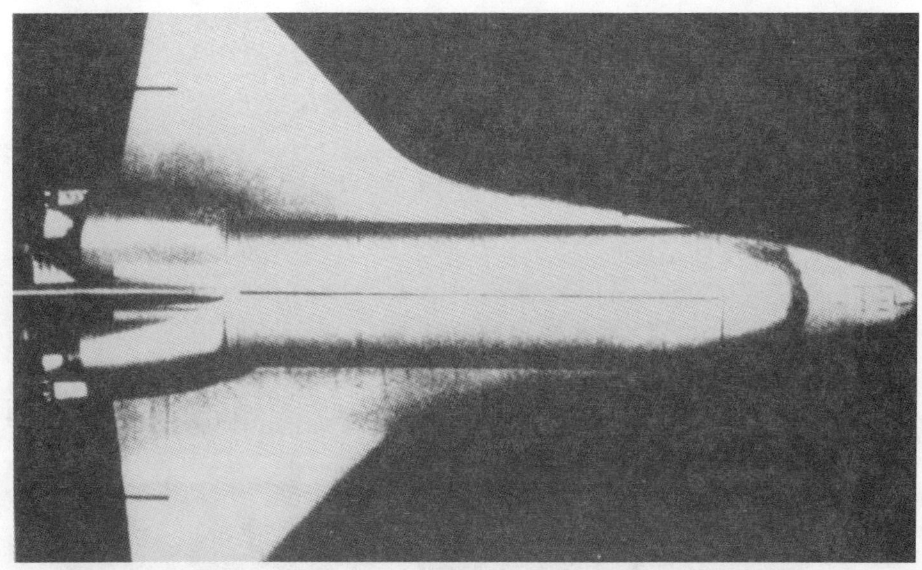

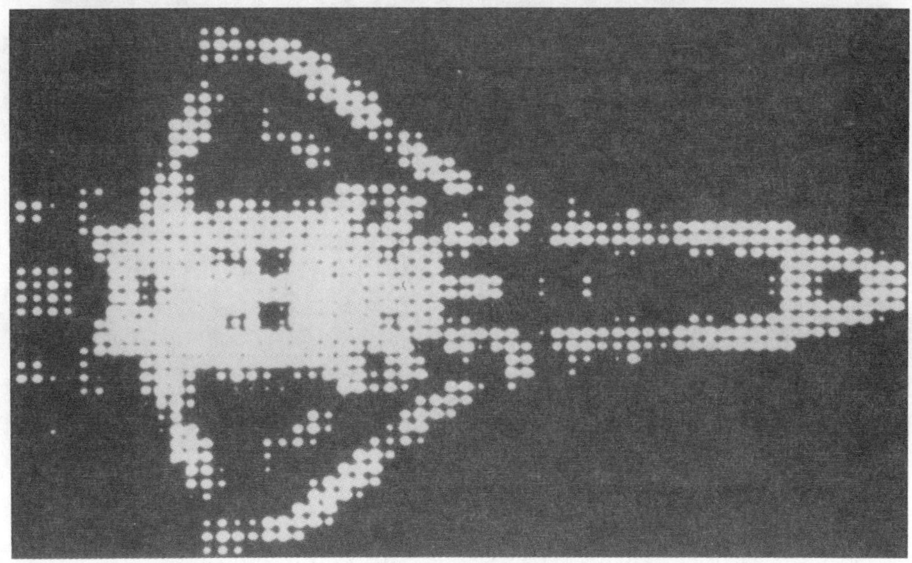

Figure 23. ISAR image obtained of model of Space Shuttle using microwave tomography. Model on top; image on bottom. (From: "High Resolution Wideband Radar Adds New Dimension to Imaging" by B.E. Manz, 1983, Microwaves and RF [28]. Used with permission.)

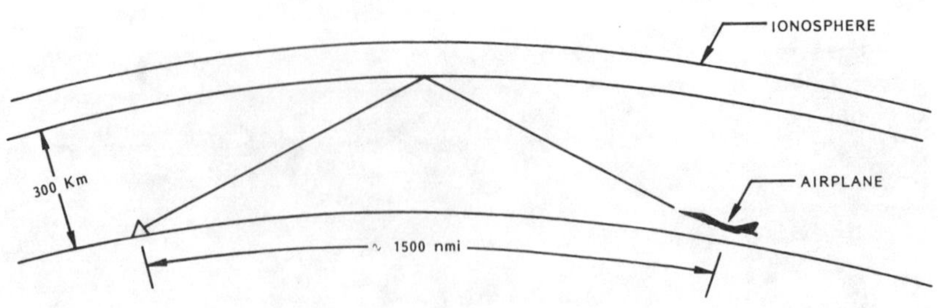

Figure 24. Over-the-Horizon Backscatter (OTH/B) radar.

Figure 25. AN/FPS-118 Over-the-Horizon/Backscatter (OTH/B) transmit antenna. (Courtesy of M. Toner of General Electric Company.)

In contrast to the OTH/B of Fig. 25, the ROTHR transmitter uses an solid-state transmitter for better reliability and ease of maintenance. Figures 26 and 27 show respectively the transmitting and receiving antennas, which are 50 (90km) to 100 nmi (180km) apart [39].

Operationally, ROTHR can automatically detect and track both aircraft and ships. At the operators options, the system can be used for overall surveillance and tracking within the coverage area, spotlighting specific regions to handle targets of special interest, or assessment of the size of an attack raid. All target information is forwarded to the user command center.

ROTHR is designed to be relocatable to prepared sites in support of tactical surveillance missions. This cost-effective feature preserves the high value assets and eliminates the need for on-station personnel at non-operational sites. Because ROTHR is relocatable, it must adapt to changing ionospheric conditions throughout the world. This requirement is met by eliminating the need for an external ionospheric sounder. ROTHR uses a collocated quasi-vertical incidence (QVI) sounder and a radar backscatter sounder for Propagation Management Assessment (PMA). This unique approach automatically adapts to the ionospheric conditions of the operational sites.

The first system, after completing its testing in Virginia, is scheduled to be moved to Amchitka, Alaska in 1989. There it will provide surveillance across the Northwest Pacific into the U.S.S.R. Eleven sites are to be prepared over the next five years. ROTHR is also being considered for use by the Departments of Transportation and Treasury in support of ocean surveillance and drug enforcement, respectively.

3.0 SYNTHETIC APERTURE RADAR (SAR) [1,24]

Space-based radars such as the SEASAT, SAR-A and SIR-B have provided high range-resolution ground maps of the earth [1,24]. Planned future space-based radars will also provide such maps [32,33]. To obtain high resolution in the slant-range direction one simply uses a high range-resolution radar waveform, or equivalently, a wideband radar waveform. To obtain high resolution in the cross-range dimension it is necessary for the radar to effectively have a very long antenna. For example, for a cross-range resolution of 25 m as provided by the L-band SEASAT radar, it is necessary for the space-borne antenna length along the direction of flight to be 7840 m (4.2 nmi) long, the cross-range resolution ΔR_A being equal to $R_S \lambda / L_R$ where L_R is the length of the antenna in the along-track direction of the satellite and R_s is the slant range to the region being mapped, about 450 nmi (830 km) [24]. Obviously it is not possible to deploy such a long antenna. What was done in the SEASAT system was to synthesize such a long antenna by making use of the motion

Figure 26. Relocatable Over-the-Horizon Radar (ROTHR) tramsit site. (Courtesy of Raytheon Co.)

Figure 27. ROTHR receive site. (Courtesy of Raytheon Co.)

of the satellite. Specifically, one processes the echo received during the time that the satellite travels a distance equal to 2.1 nmi (3.9 km). This distance of travel then becomes the length L_S of the synthesized antenna. The synthetic antenna length need only be half of the 4.2 nmi (7.8 km) needed for a real antenna [24]. This synthesized antenna provides the desired 25 m cross-range resolution which is $\Delta R_A = R_S \lambda / 2L_S$. The velocity of the satellite is approximately 4 nmi/sec (7.4km/sec). Thus the time it takes for the satellite to travel 2.1 nmi is 0.53 sec. This 0.53 sec represents the coherent processing time needed to achieve the 25 m cross-range resolution. (For simplicity these calculations assumed a flat earth. It is a straight forward matter to extend the results to the actual curved earth case.)

It is also possible to explain the basis for cross-range resolution obtained with a SAR processor by viewing it as being obtained from doppler processing just as the cross-range resolution was obtained for the ISAR image of Fig. 19. For a SAR radar one typically looks perpendicular to the sensor velocity vector (i.e., abeam) as shown in Fig. 28. Without synthetic-aperture processing the resolution in the along-track direction (or equivalently, what was called the cross-range direction for the ISAR images) is equal to the real antenna beamwidth times the slant range. For the SEASAT radar the antenna length along the velocity track is actually 10.7 m. The carrier frequency of the radar is 1.275 GHz so that its wavelength λ=0.235 m. Hence the antenna beamwidth is about 0.235 m/10.7 m=22 mrad. As a result the doppler spread across the beam in the along-track direction is equal to the antenna beamwidth times the velocity of the satellite or (0.022 mrad)(7400 m/sec)=163 m/sec. This results in a doppler spread across the beam of 1387 Hz. As indicated above, in order to obtain the required synthetic-aperture length of 2.1 nmi, the coherent processing time has to be 0.53 sec. As a result the doppler resolution achieved with the radar for this processing time is one over this time, or 1.9 Hz. Hence the 1387 Hz across the beam is broken up into 1387 Hz/1.9 Hz=730 doppler cells. The width of the beam in the along track direction is 18,300 m. This width is broken up into 730 cells or equivalently into 25 m resolutions cells.

Figure 29 shows a 25 m resolution SAR map made of Los Angeles City. This image was obtained by digital processing in non-real time [1,34]. (Real-time processing is available [1].) It is possible to obtain stereographic 3-dimensional images of the earth by recording SAR maps from two different viewing angles. This can be achieved by recording SAR maps of a given region on two successive orbital passes. Because the earth rotates relative to the satellite orbit plane, the satellite will be viewing the region being mapped on two successive passes at two different angles (see Fig. 30). Such mapping provides topographical maps [7]. Figure 31 shows one such map

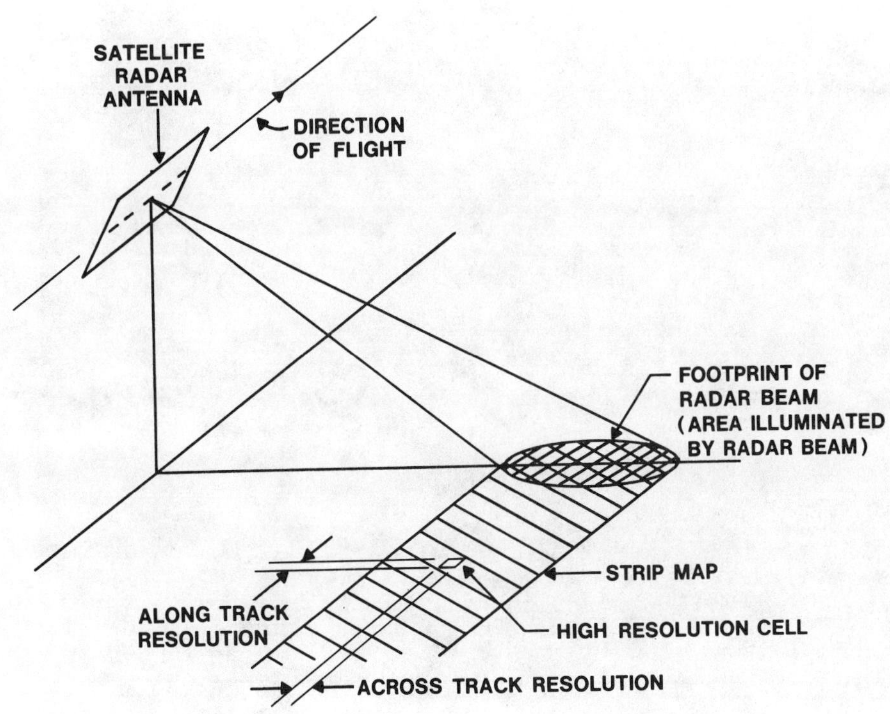

Figure 28. Geometry of high-resolution Strip Mapper. (From Brookner, E., "Radar Imaging for Arms Control," Chapter 11 in _Arms Control Verification, The Technologies That Make It Possible_, Edited by K.T. Tsipis, D.W. Hafemeister and P. Janeway, Pergamon-Brassey's, 1986. Used with permission [24].)

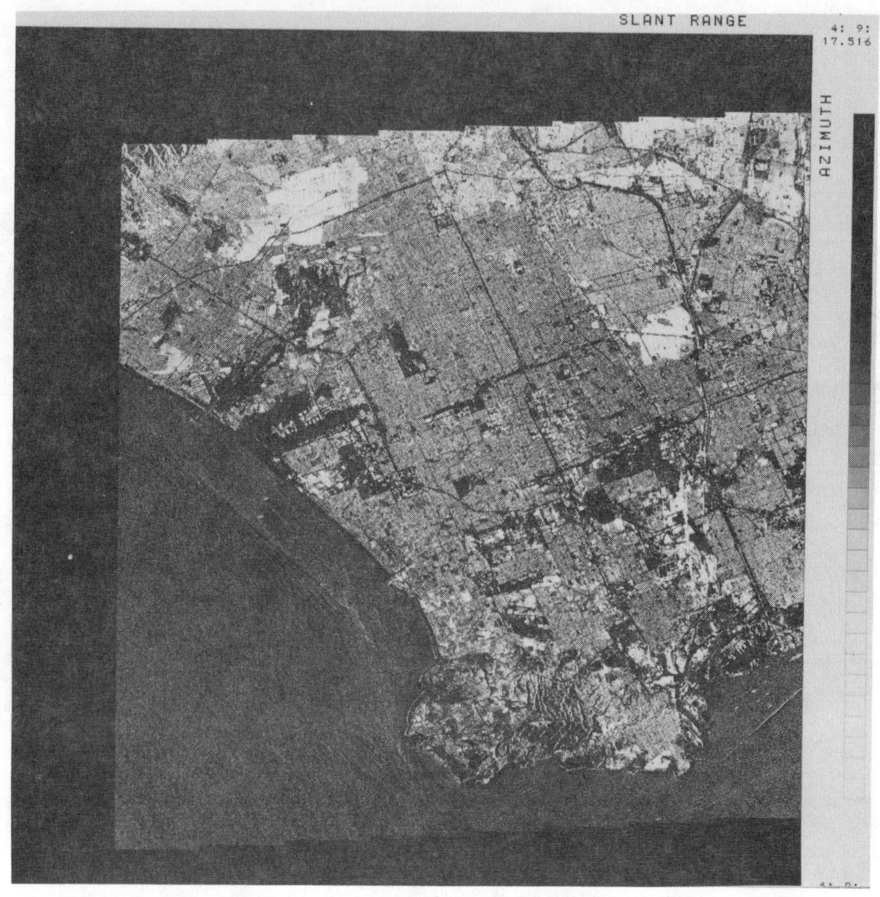

Figure 29. 25m high-resolution SAR Map of Los Angeles, obtained August 12, 1978 with SEASAT. (Courtesy of C. Elachi of Jet Propulsion Laboratory.)

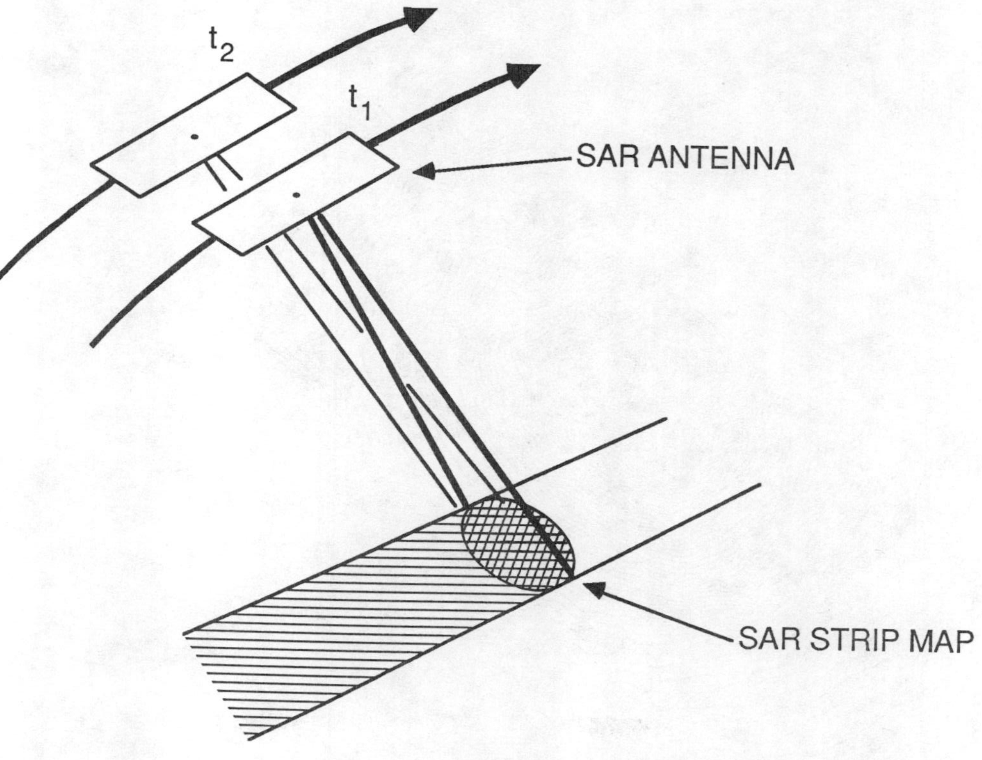

Figure 30. Two-pass Strip Maps of the same region for stereographic 3-dimensional imaging.

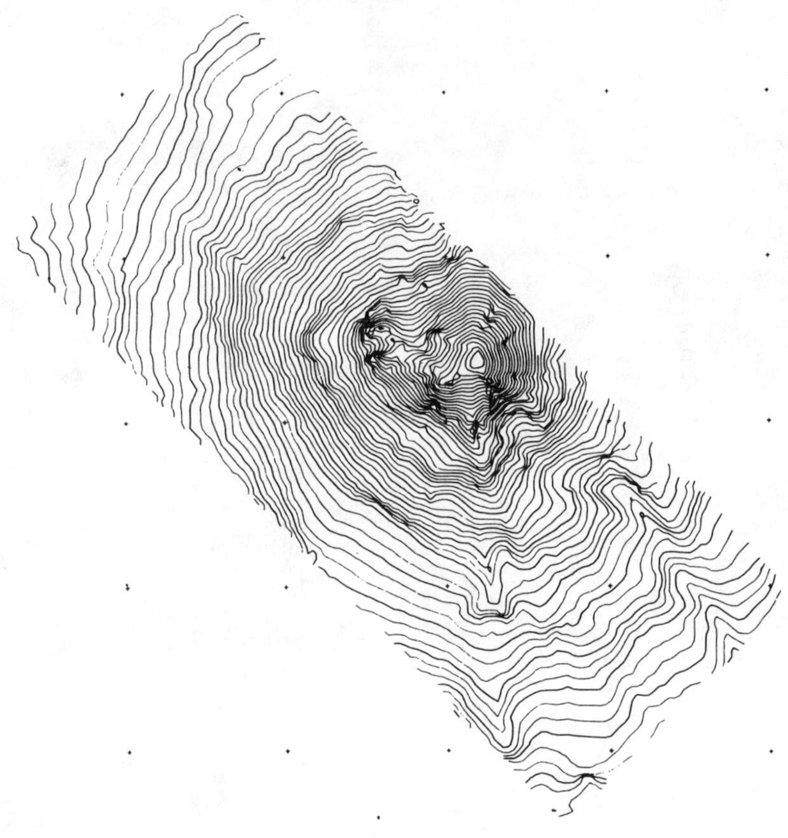

Figure 31. A contour plot of Mt. Shasta at intervals of 200 m produced from SIR-B two stereo SAR images. (From [37].)

Figure 32. Several computer-generated perspective views of Mt. Shasta (California) derived from two stereo SAR images obtained with Shuttle Imaging Radar (SIR-B). (Courtesy of A.H. Richardson of the Jet Propulsion Laboratory [37].)

Figure 33. One blown up perspective of Mt. Shasta. (Courtesy of A.H. Richardson of the Jet Propulsion Laboratory [37].)

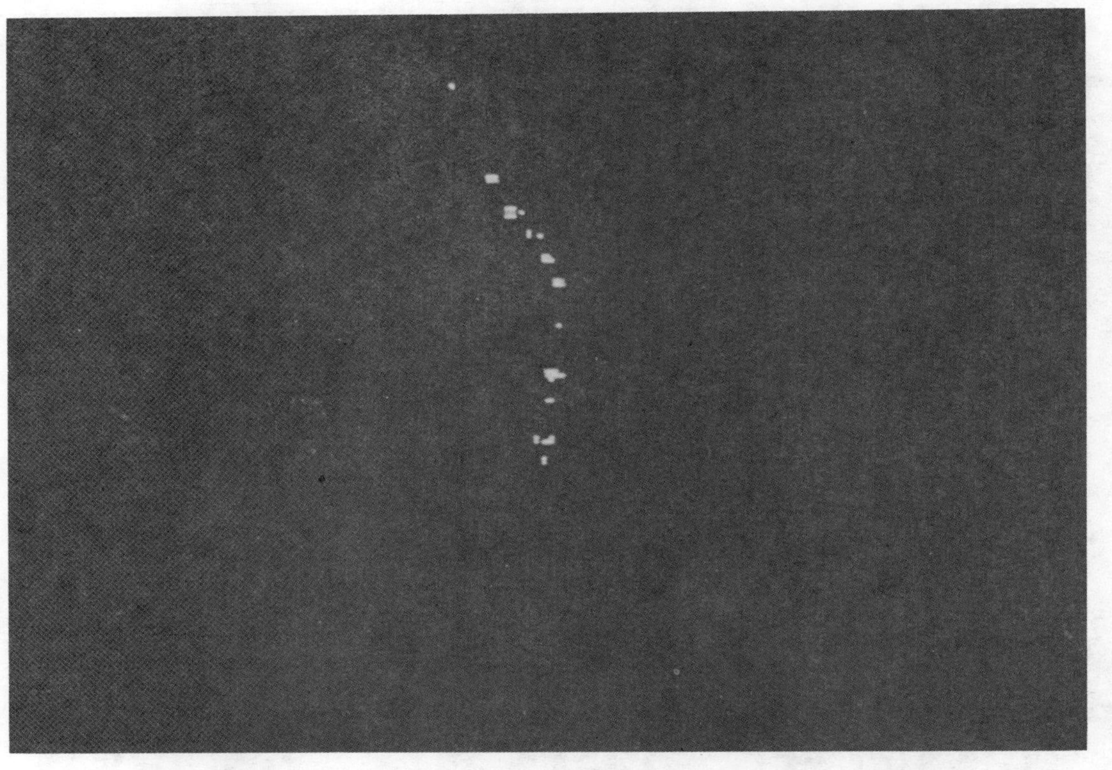

Figure 34. SAR Image of Tank/Truck Column. Image Obtained With General Electric Corp. Airborne Multimode Surveillance Radar (MSR). (From Krass, A.S., Verification, How Much is Enough? Taylor and Francis, 1985.)

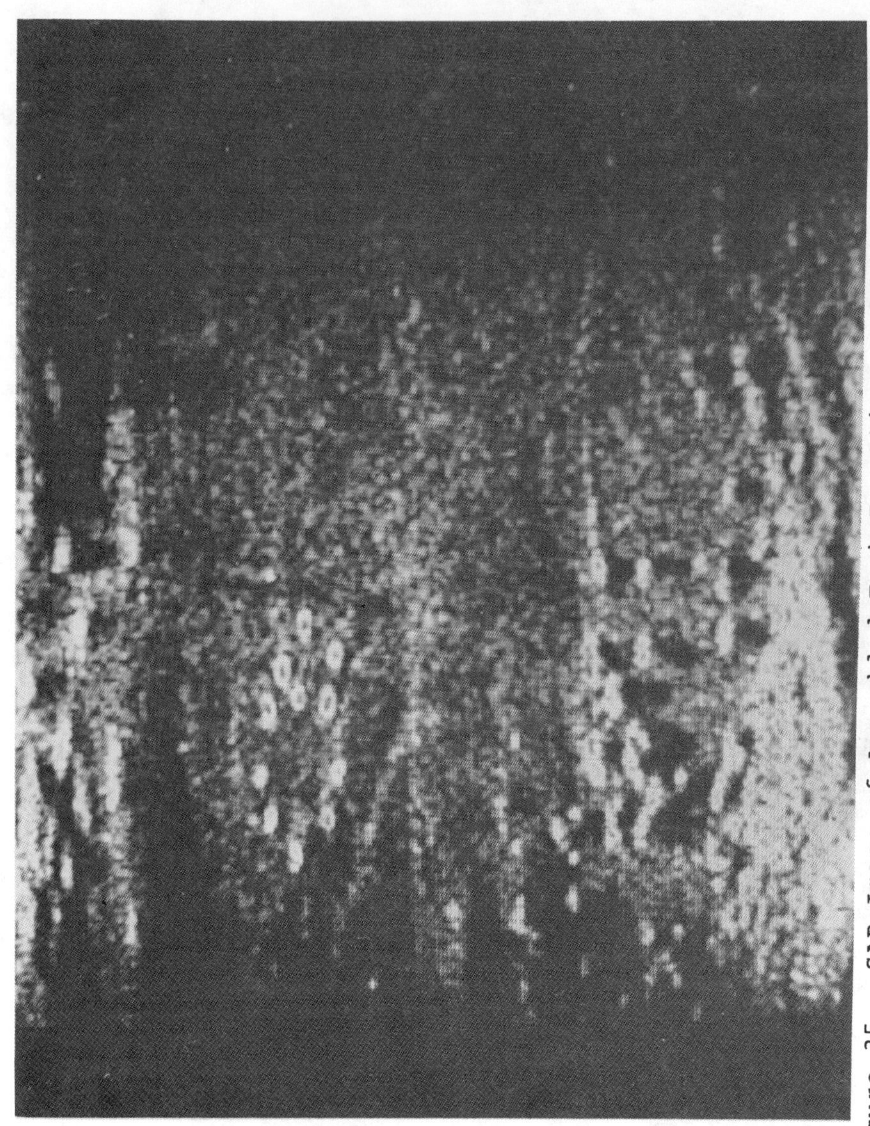

Figure 35. SAR Image of Assembled Tank Formation. Image Obtained With General Electric Corp. Airborne Multimode Surveillance Radar (MSR). (From Krass, A.S., Verification, How Much is Enough? Taylor and Francis, 1985.)

obtained of Mt. Shasta in California. Using such two-look data the Jet Propulsion Laboratory has through computer processing synthesized perspective views of Mt. Shasta (see Fig. 32). A blow up of one of these perspective views is given in Fig. 33.

Figure 34 shows a SAR image of a tank/truck column; Figure 35 shows an assembled tank formation. Both were obtained with an airborne system.

The advantage of SAR processing is that it provides day-night viewing of the earth. Also, it can penetrate clouds and even function when it is raining.

4.0 REFERENCES

1. Brookner, E., Chapter 2, "Trends in Radar Systems and Technology to the Year 2000 and Beyond," in Aspects of Modern Radar, Artech House, 1988.
2. Brookner, Eli, "Array Radars: An Update," Microwave Journal, Part 1, Vol. 30, No. 2, Feb. 1987, pp. 117-138 and Part 2, Vol. 30, No. 3, Mar. 1987, pp. 167-174; or: Brookner, E., " Radar of the '80s and Beyond--An Update," IEEE Electro/86, Session 25, May 13-15, 1986.
3. Brookner, E., "Radar of the '80s and Beyond," IEEE Electro/84, Session 4, Boston, MA, May 15-17, 1984.
4. Brookner, E., "Phased-Array Radars," Scientific American, Vol. 252, No. 2, Feb. 1985, pp. 94-102.
5. Brookner, E., "A Review of Array Radars," Microwave Journal, Vol. 25, No. 10, Oct. 1981, pp. 25-42, 61, 114.
6. Brookner, E., "Radar Trends to the Year 2000," Interavia, May 1987, pp. 481-486.
7. Brookner, E. (Ed.), Radar Technology, Artech House, 1977.
8. Filer, E. and J. Hartt, "COBRA DANE Wideband Pulse Compression System," 1976 IEEE EASCON, Paper No. 61.
9. Coraine, R.T., "COBRA DANE (AN/FPS-108) Radar System," Signal, May/June 1977.
10. Johnson, R.F. and H. Jasik, Antenna Engineering Handbook, McGraw-Hill Book Co., New York, NY, 1984.
11. Brookner, E., "Radar Technology," Boston IEEE Radar Lecture Notes, 1984.
12. Daley, E.J. and F. Steudel, "Modern Electronically-Scanned Array Antennas," Electronic Progress (Raytheon), pp. 10-17, Winter 1974.
13. Brookner, E., "Practical Phased-Array Systems," Microwave Journal Intensive Course Notes, Lecture 2 and 3, Apr. 1975.
14. Skolnik, M.I., in R.E. Collin and F.J. Zucker (eds.), "Antenna Theory," Pt. 1, McGraw-Hill, 1969; or: Skolnik, M.I., J.W. Sherman III, and F.C. Ogg, Jr., "Statistically Designed Density-Tapered Arrays," IEEE Trans. Antennas and Propagation, Vol. AP-12, July 1964, pp. 408-417.
15. Hanfling, J.D. and F. Beltran, "Elements for Array

Antennas," Electronic Progress (Raytheon), pp. 33- 44, Winter 1974.
16. Stein, K.J., "COBRA JUDY Phased-Array Radar Tested," Aviation Week and Space Technology, Aug. 10, 1981, pp. 70-73; see also "USAF Shipborne Radar Detailed," Aviation Week and Space Technology, Aug. 17, 1981, p. 105.
17. Campbell, R.B. and D.J. Hoft, "Solid State Transmitters for Radar Applications," International Conf. on Radar, Dec. 4-8, 1978, pp. 487-496, Paris.
18. Stein, K.J., "PAVE PAWS Phased Array Radar Nears Acceptance," Aviation Week, Apr. 9, 1979.
19. Stein, K.J., "BMEWS Update Program Progresses at Thule Site," Aviation Week and Space Technology, Aug. 20, 1984, pp. 87-91.
20. Foley, T.M., "Raytheon Proposes Rail-Mobile Radar for Midcourse SDI Sensing," Aviation Week and Space Technology, January 11, 1988, pp. 22,23.
21. Mohr, C., "US and Soviets Discuss Whether Moscow Violated Terms of 2 Arms Pacts," The New York Times, Wednesday, Oct. 5, 1983, p. A8.
22. Mann, Paul, "Administration Disputes Findings of U.S. Visit to Soviet Radar," Aviation Week and Space Technology, Sept. 14, 1987, pp. 26-28.
23. Broard, William J., "Soviet Radar on Display," New York Times, September 9, 1987.
24. Brookner, E., "Radar Imaging for Arms Control," Chapter 11 in Arms Control Verification, The Technologies That Make It Possible, Edited by K. Tsipis, D.W. Hafemeister, and P. Janeway, Pergamon-Brassey's 1986.
25. Chen, C. and H.C. Andrews, "Target-Motion-Induced Radar Imaging," IEEE Transections on Aerospace and Electronic Systems AES-16, 1 (January 1980), pp. 2-14.
26. McCune, B.P. and R.J. Drazovich, "Radar with Sight and Knowledge," Defense Electronics, August 1983.
27. Munson, Jr., D.C., J.D. O'Brien and W.K. Jenkins, "A Tomographic Formulation of Spotlight-Mode Synthetic Aperture Radar," Proceedings of the IEEE 71, 8 (August 1983), pp. 917-925.
28. Manz, B.E., "High Resolution Wideband Radar Adds New Dimension to Imaging," Microwaves and RF 22, 4 (April 1983), pp. 35-38.
29. Farhat, N.H., T.H. Chu and C.L. Werner, "Tomographic and Projective Reconstruction of 3-D Image Detail in Inverse Scattering," Proceedings of the IEEE 10th International Optical Computing Conference (1983), pp. 82-88.
30. Pengelley, R., "Ground Air Defense Sensors for the 1990s; Technologies Enter Gradually," International Defense Review, December, 1987, pp. 1619-1625.
31. Tsipis, K., "Arsenal: Understanding Weapons in theNuclear Age," A Touchstone Book, Simon and Schuster, Inc., 1983, p. 248.
32. Brookner, E. and T.F. Mahoney, "Derivation of a

Satellite Radar Architecture for Air Surveillance," IEEE EASCON '83 Conference Record, Sept. 19-21, 1983, pp. 465-475.

33. Brookner, E. and T.F. Mahoney, "Derivation of a Satellite Radar Architecture for Air Surveillance," Microwave Journal, Vol. 29, No. 2, Feb. 1986, pp. 173-191; see also M.I. Skolnik (Ed.), Radar Applications, IEEE Press, N.Y., N.Y., 1988
34. Brookner, E. and B.R. Hunt, "Synthetic Aperture Radar Processing," Trends and Perspectives in Signal Processing, (Prof. Alan V. Oppenheim, Editor, PO Box 157, Waban, MA 02168), Vol. 2, No. 1, Jan. 1982, pp. 2-4
35. Broad, W.J., New York Times, Private Communication.
36. Drazovich, R.J. and F.X. Lanzinger, "Radar Target Classification," Proceedings of IEEE Pattern Recognition and Imaging Processing Symposium, Aug. 3-5, 1981, CH1598-8/81/0000/0498, pp. 496-501.
37. Cimino, J.B., B. Holt and A.H. Richardson, "The Shuttle Imaging Radar B (SIR-B), Experiment Report," Jet Propulsion Laboratory, JPL Publication 88-2, March 15, 1988.
38. Elachi, C., "Spaceborne Radar Remote Sensing: Applications and Techniques," The Institute of Electrical and Electronics Engineers, Inc., New York, 1987.
39. "New Navy Radar Can See Well Beyond Horizon," Raytheon News, Raytheon Company, Lexington, MA, February 1988, pp. 1,2.

CHAPTER 8

INFRARED MONITORING OF NUCLEAR POWER IN SPACE

David W. Hafemeister
Center for International Security and Arms Control, Stanford University
and
Physics Department, California Polytechnic State University

ABSTRACT

Using parameters from unclassified astronomical observatories based on Maui and on the Kuiper Airborne Observatory, we have determined the level of confidence of monitoring a ban on nuclear power in earth orbit. Existing military and astronomical observatories can detect and identify operating nuclear power sources on satellites, such as the Soviet RORSAT and American SP100, with a very high level of confidence to distances beyond geosynchronous orbit. A cold reactor can be detected with a medium level of confidence with visual observations by close-flying reconnaissance satellites with medium confidence, and in the future with very high confidence with the interrogation of neutrons. The smaller thermal sources, RTG and DIPS, could be detected with medium level of confidence under certain conditions. Large pulsed reactors can be detected with a medium confidence level with visual observations from close satellites, and with a very high level of confidence with neutron interrogation.

I. INTRODUCTION

This paper analyzes the detection and identification of nuclear power sources in earth-orbit with infrared monitoring systems. Various proposals have been made for the use of nuclear power in earth orbit. The U.S. Strategic Defense Initiative Organization (SDIO) has estimated the power requirements for space assets as indicated in Table 1. A companion paper analyzes these power requirements in more detail.

Mode of operation	Base	Alert	Burst (battle)
BSTS	4-10	4-10	4-10
SSTS (IR)	5-15	5-15	15-50
Ladar	15-20	15-20	50-100
Ladar imager	15-20	15-20	100-500
Laser illumination	5-10	5-10	50-100
Doppler ladar	15-20	15-20	300-600
SBI carrier	2-30	4-50	10-100
Chemical laser	50-100	100-150	100-200
Fighting mirror	10-50	10-50	20-100
NPB/SBFEL	20-120	1,000-10,000	100,000-500,000
EML (railgun)	20-120	1,000-10,000	200,000-5,000,000

Table 1. Estimated Average Power Requirements for Space Assets (kW). (SDIO in OTA, ref. 1).

The U.S. is initially considering[1] obtaining power from the following three systems: the Radioisotope Thermoelectric Generator (RTG), the Dynamic Isotope Power System (DIPS) and nuclear power (the proposed SP100 system). In addition the

US is considering multi-megawatt systems and burst power systems. The Soviet Union has nuclear power reactors in their Radar Ocean Reconnaissance Satellites (RORSAT).

Table 2:	POWER	EFFICIENCY	HEAT REJECTED	AREA	TEMP(OUT)
RTG	0.3 kWe	7%	4.1 kW	1.2 m^2	500 K
DIPS	6 kWe	24%	19 kW	25-48 m^2	330 K
SP100	100 kWe	4%	2.4 MW	90 m^2	800 K

The rejected heat from the nuclear power sources is considerable. The infrared signature of the present generation of nuclear power plants in earth obit is already easily observable with the unclassified astronomical observatories:

(1) The Air Force Maui Optical Station (AMOS), based on Mt. Haleakala, Maui, Hawaii at an altitude of 10,000 feet,

(2) The Kuiper Airborne Observatory (KAO), based on a C-141 aircraft, stationed at the NASA Ames Research Center, Moffett Field, CA at an altitude of 45,000 feet.

The outline of this paper is as follows:

Sec. II: The IR signature,
Sec. III: IR temperature determinations,
Sec. IV: IR signatures for various nuclear power sources,
Sec. V: IR monitoring systems based on (1) land, (2) in high-altitude aircraft, (3) satellites in low-earth orbit, (4) satellites in geosynchronous orbit, (5) inexpensive sounding rockets for a close-look, and (6) the space-shuttle,
Sec. VI: Countermeasures to IR monitoring,
Sec. VII: Analysis of IR Monitoring of Nuclear Power in Earth-orbit,
Sec. VIII: Conclusions.

II. THE INFRARED SIGNATURE

The generation of electricity, from either thermoelectric effects or from traditional thermodynamic heat engine cycles, requires the rejection of heat by thermal radiators in order to maintain a temperature difference. For these kinds of systems, the rejection of heat can be described by Planck's radiation laws. On the other hand for comparison sake, the exhaust from a missile is not a black-body radiator, but it is dominated by the emission lines from the formation of CO_2 and H_2O. Because of these discrete transitions, it is relatively easy to monitor the plume of the various missiles stages (2-100 MW) against the earth's background from geosynchronous orbit[2]. Figure 1 gives some idea of the spectra from the missile plumes as compared to the almost black-body radiation spectra from the sun at 6000 K and the upper atmosphere of the earth at about 240-270 K.

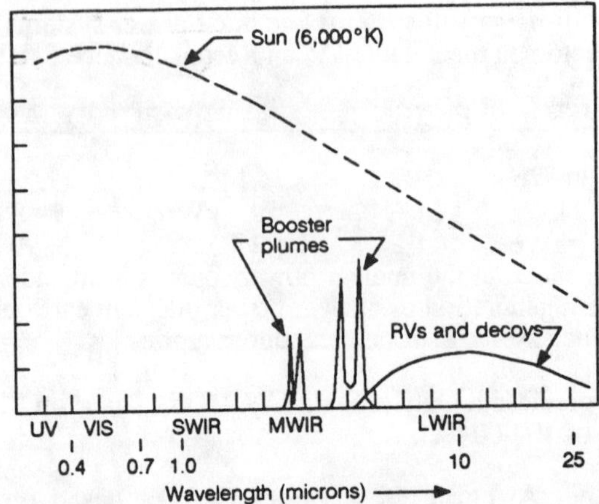

Fig. 1. The sun's radiation spectrum predominates in the visible and short- and mid-wave infrared regions (SWIR/MWIR), while during the night the earth's radiation spectrum dominates in the long-wave infrared (LWIR) region. The hot exhaust from missiles stages are dominated by molecular emission lines in the MWIR region. (OTA, ref.1)

The spectral radiant emittance of a black body is given by[3]

$$W_L = \varepsilon(\lambda)C_1/[\lambda^5(e^{C_2/\lambda T} - 1)] \tag{1}$$

where W_L is in W/cm$^2\mu$, λ is the wavelength in μ (10^{-6} m), T is the temperature in Kelvin (K), C_1 is 3.74×10^4, C_2 is 1.44×10^4 and $\varepsilon(\lambda)$ is the emissivity at the wavelength λ to account for nonideal radiators. The black-body spectrum is similar to a bell-shaped curve, which peaks according to Wien's law at the wavelength (energy)

$$\lambda_{max} = 2897/T, \qquad (E_{max} = 1.24/\lambda_{max}) \tag{2}$$

where the energy E is in eV. Integrating over all wavelengths gives the radiant emittance,

$$W = \varepsilon \sigma T^4 \tag{3}$$

where W is W/cm^2, σ is the Stefan-Boltzmann constant with a value of 5.67×10^{-12} W/cm^{2}K^4 and ε is the averaged emissivity. The radiant flux (F) at the detecting system for a spherical radiator of area A is

$$F = WA/4\pi R^2 \tag{4}$$

where R is the distance to the detector. This value must be reduced by the absorption in the atmosphere. On the other hand, the radiant flux will be enhanced by the optical system, the diameter and focal length of the mirror. The rate of photon absoption can be determined with a new distribution function proportional to λ^{-4}, or it can be approximated by

$$N = F\,Q\,f\,T_r\,A\,/\,E \quad \text{(counts/sec)} \tag{5}$$

where Q is the quantum efficiency of the detector (usually between 20 and 50%), f is the fraction of the radiation within the absorption band, T_r is the transmissivity of the atmosphere, A is the area of the detector, and E is the average energy of the photons in the band. Avalanche diodes can be used to detect individual IR photons, but at higher counting rates, photovoltaics or phototoconductors which integrate the charge are used. Because the background corrections for infrared radiation dominate when looking through the atmosphere, or when looking against the earth's background, we will not attempt to determine signal to noise ratio directly, but rather we will use the measured sensitivity functions for various existing infrared facilities to determine the detectability of power sources in space.

III. IR TEMPERATURE DETERMINATIONS

There are several methods that could be used to determine the temperature of radiators in space. By observing the peak in the blackbody radiation, one can directly use Wein's law to determine the temperature. For example, if the distribution peaked at 10 μ, this would imply a temperature of 290 K, while a peak at 4 μ would imply a temperature of 725 K. Since the usual temperatures of space crafts are about 300 K, it is clear that the thermal radiators for heat rejection must dominate for Wein's law to be effectively used without complications. Alternatively, one could make several measurements of the distribution function with multispectral detectors, and fit the data to the Planck formula to determine the temperature. (If the thermal radiators have a relatively small power, or if their temperature is close to 300 K, then the subtraction of the normal radiant power of the space craft from the data must be accurate.) As an example, consider very narrow band detectors at 4 μ and 10 μ. The ratio of the intensity at 4 μ to that at 10 μ varies greatly as the radiator's temperature is raised from 300 K to 800 K, by a factor of 77, from 0.072 at 300 K to 5.5 at 800 K (Fig. 2).

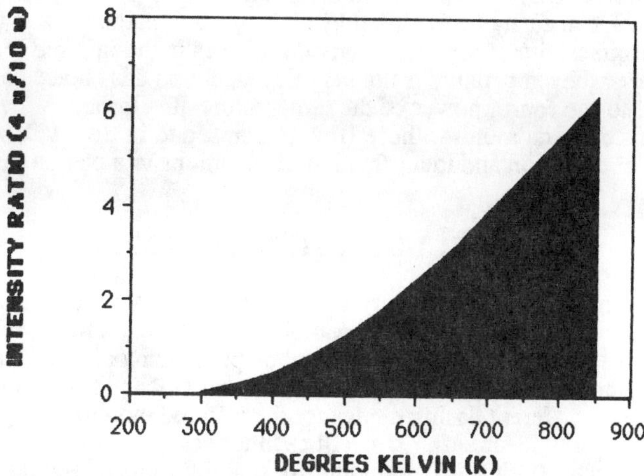

Fig. 2. Intensity Ratios Determine Temperatures. The ratio of the intensity at 4 μ to the intensity at 10 μ is plotted vs. temperature for very narrow band IR detectors.

Because of the great sensitivity of the ratio to the temperature, it is not surprising that infrared imaging technologies are capable of measuring temperatures to an accuracy of 0.1 or 0.01 degrees Kelvin. However, actual measurements are more complicated, because of two reasons. First, the background must be accurately subtracted so that the intensity values are meaningful. Secondly, the finite band width of the detectors tends to reduce the sensitivity of the ratio. Using the universal black-body curves[3], one can easily obtain the fraction of the radiation emitted between the two limits of a filter. For a body at 800 K, 32% of its energy is radiated in the 3-5 μ band, and 17% in the 8-13 μ band. The intensity ratio becomes:

$$I(4\mu)/I(10\mu) = 32\%/17\% = 1.9,$$

or considerably less than the ratio of 5.5 for the very narrow band detectors. Nevertheless, infrared thermography from space can measure temperatures on the earth to about 1 K. A third method to determine the temperature is to determine the total power radiated by the system of a known area, and apply Eq. 3 to determine the temperature. Of these three methods, the use of several narrow band detectors/filters to determine ratios is the most likely to give the most accurate determination of the temperature. By using three measurements, giving two ratios, one can make "color-color" plots to improve the analysis.

IV. NUMERICAL VALUES OF IR SIGNATURES

Starting with the area of the thermal radiators for the SP100, DIPS, and RTG systems and their radiated power, the temperatures of the radiators have been calculated with Eq. 3. We have also considered the case of the SP100 reactor after it has been turned off for about 20 hours, reducing its thermal output to about 27 kW (0.1% of the initial power). At that point the radiators would have a temperature of 270 K, about the temperature of a spacecraft. As a benchmark, we have also calculated the case of a unit area of 1 m^2 at 270 K for calculating the radiation from the space craft. This value is similar to that of a reentry vehicle. Two different detection ranges have been assumed for our calculations of the radiant flux and infrared counting rates at the sensors, 300 km for low-earth orbit (LEO) and 40,000 km for geosynchronous orbit (GEO). The results, shown in Table 3, can easily be modified by chosing other parameters in the spreadsheet program. The emissivity of practical thermal radiators in the infrared region is close to one. The value of the temperature is not very dependent on this choice because radiation is proportional to the fourth power of the temperature, T^4. In Sec. V, we will describe various systems that can monitor these IR signatures, and in Sec. VII, we will discuss the likelihood of detection and identification of violations to a ban on nuclear power in earth-orbit.

V. IR DETECTION SYSTEMS

Since parameters of the military monitoring systems are classified, we will use the parameters from various infrared astronomical observatories for our calculations and conclusions. This assumption is a lower bound on possible monitoring capabilities since we assume that the military monitoring systems are better than those in the public sector. We will consider 6 different basing modes for the infrared monitoring systems based on (1) land, (2) high-altitude aircraft, (3) satellites in low-earth orbit (LEO), (4) satellites in geosynchronous orbit (GEO), (5) sounding rockets for a close-look, and (6) the space shuttle.

Table 3. IR Signatures for the SP100, the SP100 (20 hours after turn off), DIPS, RTG, and 1 square meter of radiator at 270 K. The radiant flux is calculated at LEO and GEO for 3-4u and 8-13u.

REACTOR CASE	SP100 LEO	SP100 GEO	SP100 OFF L	NonNuc Bkg L	DIPS LEO	DIPS GEO	RTG LEO	RTG GEO
ELECTRICAL PWR (kW	100	100	0	0	6	6	0.3	0.3
RADIATED PWR (kW)	2400	2400	27	0.305	18.6	18.6	4.1	4.1
AREA (m2)	90	90	90	1	28	28	1.15	1.15
EMISSIVITY	1	1	1	1	1	1	1	1
TEMP (K)	827	827	269	270	329	329	500	500
LAMBDA (microns)	3.50	3.50	10.76	10.71	8.82	8.82	5.79	5.79
DISTANCE(km)	300	40000	300	300	300	40000	300	40000
TOTAL FLUX (W/cm2)	2.12E-10	1.19E-14	2.39E-12	2.70E-14	1.65E-12	9.26E-17	3.63E-13	2.04E-17
FRACTION IN 8-13μ	0.15	0.15	0.29	0.29	0.36	0.36	0.3	0.3
FLUX IN 8-13μ	3.1847E-11	1.7914E-15	6.9268E-13	7.8247E-15	5.9236E-13	3.332E-17	1.0881E-13	6.1206E-18
QUANTUM EFFICIENCY	0.3	0.3	0.3	0.3	0.3	0.3	0.3	0.3
ENERGY MAX (ev)	0.35	0.35	0.12	0.12	0.14	0.14	0.21	0.21
MIRROR DIAMETER (m	1.6	1.6	1.6	1.6	1.6	1.6	1.6	1.6
PHOTON COUNT/SEC	2.26E+13	1.27E+09	7.81E+11	8.78E+09	4.41E+11	2.48E+07	6.38E+10	3.59E+06
lamda picked, λ	4	4	4	4	4	4	4	4
w(λ)	0.47616239	0.47616239	5.7271E-05	6.0516E-05	0.00063696	0.00063696	0.02732576	0.02732576
w(λ)*A	428546.153	428546.153	51.5437341	0.6051605	178.350012	178.350012	314.246191	314.246191
Diff Flux (W/cm2-μ)	3.7911E-11	2.1325E-15	4.5598E-15	5.3535E-17	1.5778E-14	8.8749E-19	2.78E-14	1.5637E-18
Mirror Diameter (m)	1.6	1.6	1.6	1.6	1.6	1.6	1.6	1.6
AREA (cm^2)	20096	20096	20096	20096	20096	20096	20096	20096
W mirror/micron	7.6186E-07	4.2855E-11	9.1633E-11	1.0758E-12	3.1707E-10	1.7835E-14	5.5866E-10	3.1425E-14

V.1. IR MONITORING FROM LAND-BASED TELESCOPES. The Air Force Maui Optical Station (AMOS) operates telescopes[4] in the visible and the IR for observing satellites and reentry vehicles, as well as astronomical objects. The observatory is located at an altitude of 10,000 feet (3 km), greatly reducing the absorption from water vapor and clouds. The relatively stable climate of dry air of the AMOS site gives a good astronomical seeing of about 1 arcsecond (arcs), or 5 microradians (μrad). Since the AMOS observatory is located at 20 degrees north latitude, it observes all the Soviet Satellites which are launched from the Soviet Union. If the Soviets were to ever launch satellites in an equatorial plane from a launch site near the equator, these satellites would be readily observable as they passed 20 degrees south of the zenith. The US government operates[5] the much more extensive, but classified Ground-Based Electro-Optical Deep-space Surveillance System (GEODSS). The location of the radar and electro-optical sensors are shown in Fig. 3.

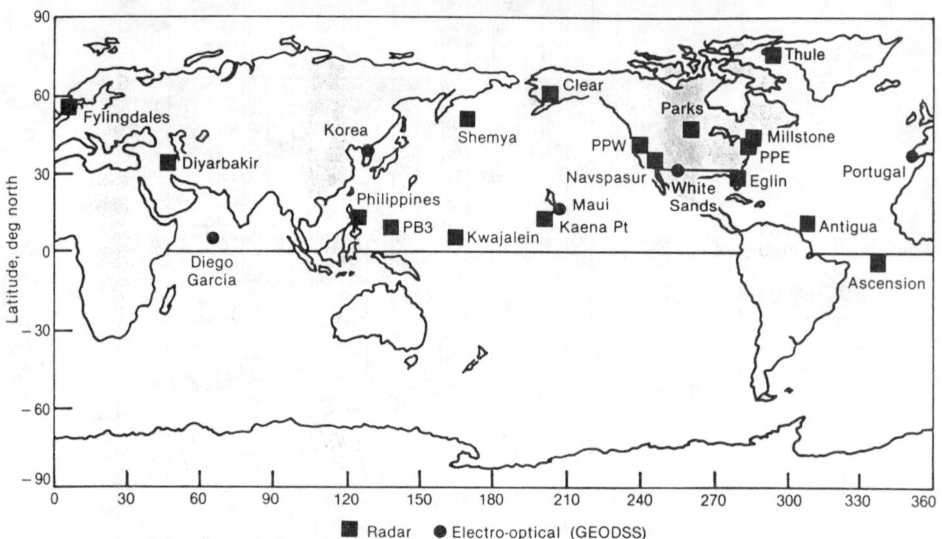

Fig. 3. The GEODSS radar and electro-optical sensor locations of the GEODSS system. (Aerospace Corporation in OTA, ref. 5)

The AMOS facility has one 1.6 meter and two 1.2 meter diameter Cassegrain telescopes that can be used for measurements in the visible and the infrared. The AMOS telescopes are mounted on high performance, three-axis mounts for fast tracking. The observatory has a Beam Director/Tracker (BD/T) system which expands a laser beam to 0.8 meters in diameter, directs the beam onto a satellite, and tracks the reflected light with a 0.8 meter observing telescope. This BD/T enables the AMOS facility to track satellites with velocities up to 5 degrees/sec, and accelerations of 4 degrees/sec^2. The BD/T system can easily track satellites in LEO with velocities of about 0.5 degrees/sec. In addition, satellites can be tracked by using radar, or by following the diffuse scattered light from the sun or earth, or from its IR emissions. In addition, there are several similar facilities operated by DoD, as well as the 2.3 meter telescope at Kitt Peak, to track satellites.

VISUAL IDENTIFICATION: The AMOS facility is equipped with adaptive optics to remove the blurring effects of atmospheric turbulance. The Compensated Imaging System (CIS) adaptive optics system is used with the 1.6 meter telescope, reducing the atmospheric seeing in the visible region from 1 arcsecond to its diffraction limited value of 0.07 arcseconds (0.4 μrad). This improved capacity gives a resolution of 0.2 meters for a satellite at an altitude of 500 km, and 0.1 m (4 inches) for a satellite at an altitude of 250 km. The high quality of the photographs available from this facility would allow some visual determinations of the properties of a satellite. The advantages of CIS are greatly diminished as one considers the diffraction limits for the longer wavelength IR radiation.

Table 4: L	Angular Resolution	Linear (500/250 km)
0.5 μ	0.07 arcs (0.4 μrad)	0.2m/0.1m
4 μ	0.6 arcs (3 μrad)	1.6 m/0.8m
10 μ	1.5 arcs (7.5 μrad)	4m/2m

Thus, it appears that the adaptive optics CIS will not be necessary for IR monitoring of power sources in space. The Cassegrain telescopes at AMOS have very long focal lengths of about 25 meters, giving a plate scale of 8 arcseconds/mm, or 40 μrad/mm. In the infrared region, pixels are about 100 μ in size, corresponding to an angular spread of 0.8 arcs (4 μrad). In the visible region, pixels are about 10 μ in size, corresponding to 0.08 arcs (0.4 μrad); therefore, smaller sized pixels won't improve the resolution. A large variety of arrays of IR detectors[1] are available, and very large arrays, perhaps with as many as 40,000 pixels, are in the planning stages. When combined with a diffraction grating to spread out the spectrum, one can obtain a measurement of spectra as a function of distance along an extended body if it is large enough.

SPECTRAL MEASUREMENTS: Background can be subtracted in two ways. One approach is to shake or wobble the secondary mirror of the Cassegrain telescopes, spending almost 50% of the time on the object, and almost 50% of the time on the neighboring IR background. These background corrections can be done with great accuracy; it is possible to observe a signal which is about one part in 10^6 of the background. Alternatively, by using a large IR array, one can simultaneously collect the background and the signal in different parts of the array, and then subtract the background directly. The AMOS facility can measure[6] the following limits, or lower bounds, with a signal to noise ratio of one (one standard deviation) after background subtraction with a one second integration time for the following bands:

3 to 4 μ, limit of 10^{-17} W/cm^2
8 to 13 μ, limit of 5×10^{-18} W/cm^2

Longer integration times would enhance the data. For example, a 4 second integration time would give a signal to noise ratio of 2. A satellite in LEO can be measured for about 4 minutes (240 seconds), but the improvement is not as large as the square root of 240 because of the variations in the atmospheric background. These two windows have been chosen because the sky radiance is very small in these regions (about 50 microwatts/cm^2). IR spectra have been taken of satellites[7] at AMOS, the ultimate sensitivity can be obtained in a more extended time, about 30 minutes for 24 pixels, at about 5×10^{-18} W/cm^2 with a S/N ratio of 1, and a resolving power of $d\lambda/\lambda = 0.005$ at 10 μ.

V.2. IR MONITORING FROM HIGH ALTITUDE AIRPLANES. For the past two decades, airplanes at an altitude of about 50,000 feet (15 km) have used cooled IR detectors to monitor astronomical objects. At these high altitudes, the absorption from water vapor and clouds has been essentially eliminated, and the absorption due to carbon dioxide has been substantially reduced. The NASA Ames Research Center, Moffett Field, CA operates the Kuiper Airborne Observatory (KAO). The flying KAO observatory carries a 0.9 meter diameter telescope at altitudes up to 45,000. Observatory time on the KAO is relatively inexpensive, about $100,000 per flight. Since the KAO can fly at any latitude, it can observe satellites in all orbits.

To first order, the 1 meter KAO observatory has the about the same ultimate limit in sensitivity as the AMOS facility, $5x10^{-18}$ $W/cm^2\mu$ at 10 μ (S/N = 1). However, the KAO is much better at wavelengths longer than 20 μ because of its higher altitude. The recent measurements[8] by the KAO of the IR spectrum of the 1987 supernova have been impressive. The standard deviations of the data points in the IR spectra were 5×10^{-18} W/cm^2-μ between 3-13 μ. The KAO has used a 10 x 50 array, each pixel having a size of 75 x 75 μ for measuring the IR spectrum between 7 and 25 μ. There are future plans to build an airborne astronomical observatory with a 3 meter mirror, the Stratospheric Observatory for IR Astronomy (SOFIA), improving the light gathering ability by a factor of 10, and the resolution by a factor of 3.

The KAO is not designed to track satellites, but it has observed them as they pass through the stellar field. Rather than modify the KAO to observe satellites, it would be far easier to use existing DoD airborne observatories that already track satellites. Narrow spectra lines, of the order of dL/L = 0.05% are obtainable[9]. In addition, the Department of Defense[1] is constructing the Airborne Optical Adjunct (AOA) for tracking reentry vehicles against the cold space background. Since the radiated power of an RV is only 0.01% of the SP100's radiated power and 2-8% that of the RTGs and DIPS, it is clear that tracking nuclear power in space is much easier than tracking RVs.

V.3. IR MONITORING FROM SATELLITES IN LEO. Excellent visual images can be obtained from close proximity with U.S. reconnaissance satellites, Landsat, the Space Telescope, or the French SPOT satellite. In addition, the Landsat, and SPOT satellites have multispectral capabilities that can be used to monitor the IR signatures of satellites. Against a cold, space background, these satellites can detect very small thermal signatures. For example, the Thematic Mapper[10] uses Si, InSb and HgCdTe detectors, cooled to 95 K, to detect between 0.45 and 12.5 μ. It was able to resolve angles of 9 arcs (43 μrad) in the visible and near-IR, and 35 arcs (170 μrads) in the 10.4 to 12.5 thermal band, which is sufficient for easily observing power sources against the cold background of space and against the warm background of the earth under certain conditions. Long-wave IR observations have been carried out from LEO with IRAS satellite[11], obtaining an angular resolution of 20 arcs (100 μrad) at longer wavelengths up to 100 μ from colder sources.

It would also be possible to launch less expensive rockets from airplanes, such as the B52, at higher altitudes of about 12 km, to insert IR and gamma detectors in LEO. By launching a rocket[12] from an airplane, payloads can be doubled. This approach would reduce the cost per kg of payload; it is projected to cost $6 million per launch. There are a variety of DoD satellites that could perform this monitoring task from LEO, or from somewhat higher orbits. The DoD is presently in the process of constructing the Teal Ruby sensor[1] to be placed in a satellite in LEO to track moving aircraft against the earth's background. Since an airplane has a considerably smaller thermal image than a missile, this will require enhanced capabilities. If the SDIO Program progresses to its second phase, it will be necessary to deploy the Space Surveillance and Tracking System[1] (SSTS) at altitudes above the trajectory of an RV. The SSTS satellites would carry long-wave length infrared (LWIR) sensors in order to track RVs and post-boost vehicles (PBV) against the cold temperature of space. The SSTS would detect the RVs and PBVs in a conical pattern above the horizon of the earth. Since the RVs have a thermal image considerably smaller than the power stations, this approach is clearly viable for the detection of power sources, but rather expensive.

V.4. INFRARED MONITORING FROM GEO. The earth subtends an angle of about 20 degrees from GEO. Detection of power sources above the earth is a difficult task because of the infrared emissions of the earth, and the reflection of sunlight during the daytime. Nevertheless, present DoD satellites are able to detect and localize the 100 MW plumes from the primary boosters of missiles. The American Physical Society[2] has estimated that the third stage of 2 MW would be detectable from GEO. However, these observations of missile plumes are greatly assisted by the discrete molecular transitions between 2.5 and 4.5 μ from the burning process. The blackbody spectrum from a 2.5 MW reactor could be observable from GEO, but it would not warrant the effort because the other options are much easier.

V.5. INFRARED MONITORING FROM POP-UP MISSILES. If one side were to become very suspicious and wished to examine the potential nuclear source at a closer distance, they could launch direct-ascent, sounding rockets to get a close look. The counting rates for IR and gamma rays would be enhanced by a factor of one million if the distance of closest approach was reduced from 1000 km to 1 km. If a system measured a counting rate (N) at its closest approach (b), the total counts for the fly-by is $\pi bN/v$, where v is the relative velocity at closest approach. Such systems have been proposed[1] to measure IR signals from RVs, to supplement the SSTS for the SDI program. Since the mission requirements for the rocket would not be as great as for LEO missions, the costs for this type of sounding rocket have been projected to be in the neighborhood of $1 million.

V.6 INFRARED MONITORING FROM THE SHUTTLE. The shuttle has experienced problems and cost over-runs. The very large cost of a dedicated Shuttle flight is prohibitive at this time. Nevertheless, the Shuttle could be used for very close examination of objects in the lower orbits. By flying in a close-coorbital position, the Shuttle crew could use an array of passive measuring devices for IR, gammas, and neutrons. In addition, the crew could also also interrogate the satellite with 14 MeV neutrons to determine its contents. In 1982, DoD placed the Cryogenic Infrared Radiance Instrumentation[13] (Cirris) payload into the the Columbia Space Shuttle for measurements in the LWIR. The Cirris telescope is capable of resolutions of 20 arcs (100 μrad) over the 0.5-degree field of view.

VI. COUNTERMEASURES

DIRECTED RADIATORS: The other side always has the option to design power systems that cannot be detected by the known monitoring systems. Of course, this is a dangerous policy to pursue since it would clearly be a material breach of the Treaty or Executive Agreement. If a nation wished to covertly design its power sources to avoid detection, it would have to greatly modify the design of the radiators. It is possible to design a heat exchanger that preferentially radiated into a conical beam directed away from the IR sensors. The idea of a "radiator in a waste basket" has been designed[14] by the General Atomics Corporation; "thermal signature control is possible by providing a thermal barrier around the reactor and limiting radiation to one direction only." A directional radiator could be discovered by the following six approaches: (1) The design of the preferential radiator of the 6 kWe GA reactor would have a conical, full-angle of about 90 degrees; the IR could be observed by satellites in higher LEO orbits such as U.S. reconnaissance satellites, Landsat, or SPOT as they pass through the beam. On the other hand, the reactor could be turned off a day in advance to avoid detection. (2) The directional radiator would greatly increase the cross-sectional area of the satellite by about a factor of ten; this would be easily observable with observations in the visible region with the AMOS telescope, or with closely-passing U.S. reconnaissance satellites. (3) The thermally insulated wastebasket would have to be very efficient because of the great sensitivity of the land-based and airborne-based telescopes are easily capable of measuring heat leakages less than 1 kW (0.05% of the SP100). (4) The gamma ray fluxes cannot be made directional; and thus their fluxes could be measured by gamma-ray telescopes. (5) The intelligence community also has SIGINT and HUMINT capabilities to examine other signatures. (6) New pop-up or shuttle systems can examine the satellite on a close-look basis.

PULSED REACTORS: Pulsed reactors have been suggested for electromagnetic launchers (EML, or rail guns), free-electron lasers (FEL), and neutral particle beam (NPB) weapons and active discriminators, and the Fission Activated Laser Concepts (FALCON and Centauraus) for the Strategic Defense Initiative. These reactors might conceivably be as large as 5 GWe. For our example here we will use 100 MWe. If it operated for 100 seconds, the electrical pulse would have 10^{10} Joules. If the pulsed reactor had an efficiency of about 10%, it would be necessary to reject about 10^{11} Joules. In order to hide this large amount of excess heat, it would be very ineffective to heat a large mass. The most effective countermeasure would be to use a phase transition, to turn a solid into a liquid. If the large latent heat of ice was used to mask the heat, it would take about 300 tonnes of ice to hide the full signal. Of course, the burst reactor could be made in smaller sizes, and operated for shorter times, but ultimately the full-scale deployment model would have to be tested in space without being observed, and that would be hard to do because of the large thermal systems.

VII. ANALYSIS

The numerical values for the fluxes displayed in Table 3 should be compared to the capabilities of the IR telescopes list in Section V. In order to quickly summarize the results, we will compare the fluxes in the 8-13 and 3-4 μ windows with the capabilities of the AMOS facility.

Table 5.	SP100	SP100(off)	DIPS	RTG	$1m^2$/270KBkgd
F(8μ-13μ)	$3.2x10^{-11}$	$6.9x10^{-13}$	$5.9x10^{-13}$	$1.1x10^{-13}$	$7.8x10^{-15}$
F/($5x10^{-18}$)	6,400,000	140,000	120,000	22,000	1600
Range(km)	800,000	100,000	100,000	40,000	10,000
F(3-4μ)	$3.8x10^{-11}$	$4.6x10^{-15}$	$1.6x10^{-14}$	$2.8x10^{-14}$	$5.4x10^{-17}$
F/(10^{-17})	3,800,000	460	1600	2,800	5
Range(km)	600,000	6,400	12,000	16,000	700

Table 5. DETECTION/IDENTIFICATION AT LEO. The flux (F) in W/cm^2 from satellites at LEO, the AMOS limits, and range of detection for the 3-4 μ and 8-13 μ IR windows for SP100 (full power, and turned off for 20 hours), DIPS, RTG, and a 1 m^2 radiator at 270 K.

SP100: It is clear that the IR signal for the SP100 is very large, Table 5 indicates that it could be detected out to a distance of about 700,000 km, or about 20 times the distance of the gyosynchronous orbit (for a 1 second integration time). For the SP100, this distance is also very realistic for identification because the SP100 reactor radiates large amounts of heat at about 800 K (3.5 μ), well separated from the normal thermal radiations of non-nuclear satellites. We conclude that the SP100 can be identified with very high confidence well beyond GEO. For the case of the SP100 after it has been shut down for 20 hours, its thermal power is reduced from 2.5 MW to 27 kW, reducing its rejection temperature to 270 K. Since this is about the normal temperature of unused reactors, the IR signature would not identify the reactor, and it would have to be interrogated by visual means or with 14 MeV neutrons.

RTG: Because of its fainter signal, identification of RTGs in GEO would be very difficult. For the case of the RTG reactor in LEO, the more modest temperature of 500 K, and its much more modest reject power of 4.1 kW, would require a subtraction of its non-nuclear thermal image. A 1 m^2 area at 270 K radiates about 0.3 kW. The areas of spacecraft vary between about 20 and 450 m^2. If we assume an area of 20 and 100 m^2, the nonnuclear thermal signature is 6 or 31 kW. It is possible to remove this background if the areas are known very accurately. We obtain the ratios:

Table 6.		100 m^2	20 m^2
Nuc/NonNPwr		4kW/31kw=0.13	4kw/6kw=0.67
Nuclear/NonN	(3-4μ)	5.2	26
	(8-13μ)	0.14	0.69

By using these additional "color" ratios, and curve fitting, it is easy to develop algorithms to estimate the RTG's temperature and heat output, but the accuracy must be discussed. Clearly the degree of confidence in this approach will depend on having good data, and on knowing the heat transfer characteristics of the entire satellite. Since satellites have predictable orbits, and since there are a number of reconnaissance satellites (KH, Landsat, SPOT) in orbit, a close encounter will certainly take place. Since the Pu^{238} power source in the RTGs results from radioactivity which cannot be turned off, it would be difficult to avoid detection during the close encounter. At that time one could not only measure the areas accurately to improve the calculation, but more importantly, one could use an infrared monitoring system (multispectral, or commercial thermal imager) to make a thermal image of the satellite and look for the 500 K RTG radiator. This type of a measurement is essentially a certainty, and it can be done with existing equipment, but it could be masked by rotating the nuclear satellite's sensitive parts in the away direction. We assign a medium probability of identification of the RTG source by using accurate algorithms on well understood satellites, and we assign a high probability of identification of the RTG radiators with thermal imaging (if they have not been covered) over a long period of time. If the thermal radiators disburse the heat into the broader satellite, the IR detection method would be greatly diminished, but this countermeasure could downrate the capability of the RTG. It would also be possible to cover the RTG radiators with a moveable cover during the close-encounter.

DIPS: As in the case of the RTG, we will only consider the identification of DIPS in LEO. For the case of the DIPS reactor, the observability is further complicated because the radiator's temperature of 330 K is very close to the usual temperatures of 270-300 K for satellites. On the other hand, the output thermal power for the 6 kWe DIPS is 19 kW, or 5 times larger than the 4.1 kW of RTG, making a larger contribution of the total power of the satellite. The ratio of intensities for the 6 kWe DIPS is as follows:

Table 7.	100 m^2	20 m^2
Nuc/NonNPwr	19kW/31kW=0.6	19kW/6kW=3
Nuclear/NonN (3-4μ)	2.9	15
(8-13μ)	0.76	3.8

Our conclusions for DIPS are very similar to those of RTG, except that an RTG would be easier to identify with a close-passing thermal imager because the RTG temperature is higher (500 K) than DIPS (330 K). On the other hand, the higher thermal power of DIPS would be more accessible to a measurement of total emitted power of the satellite.

VIII. CONCLUSIONS

The existing facilities based on the ground, on high-altitude aircraft, and in low-earth orbit can easily monitor and identify the IR signatures from nuclear power plants based on U.S. or Soviet satellites. For the cases of identification of RTGs and DIPS, the detection ranges of 40,000 to 100,000 km would have to be reduced to take into account the subtraction of its normal thermal background. After background subtraction, the detection and identification of RTGs and DIPS in LEO should be possible, but with much less confidence than with SP100. For these more difficult cases, the infrared measurments can be supplemented with measurments of the gamma-ray emissions[15] to determine the use of nuclear power in earth orbit. Our estimate for the degree of confidence in the IR monitoring process to determine compliance to a ban on nuclear power systems in earth orbit is as follows:

DEGREE OF CONFIDENCE IN MONITORING

SP100/RORSAT: VERY HIGH CONFIDENCE beyond GEO with existing systems because of high temperature and large power.

COLD REACTORS: Await a close-passing satellite; visual gives MEDIUM CONFIDENCE; future interrogation with neutrons would be VERY HIGH CONFIDENCE.

DIRECTED RADIATOR. Existing satellites in higher LEO: measure beam in IR and satellite in the visible; HIGH CONFIDENCE, but could be avoided by rotating the directed beam, turning reactor off, or covering radiators for isotopic systems.

RTG/DIPS: MEDIUM CONFIDENCE if have (1) good data, (2) known areas of nuclear radiators, (3) a proper algorithm. HIGH CONFIDENCE if await a close-passing thermal imaging of the uncovered RTG radiators; MEDIUM CONFIDENCE for DIPS. On the basis of total power measurements, MEDIUM CONFIDENCE for DIPS.

LARGE PULSED REACTORS: MEDIUM CONFIDENCE from visual observations from close satellites, VERY HIGH CONFIDENCE from neutron interrogation.

ACKNOWLEDGEMENT

I would like to thank Glenn Ashley, Lisa Ensman, Carl Gillespie, Joel Primack, Ted Postal, Frank von Hippel, and Fred Witteborn for useful comments on this work. This work has been carried out as part of the Joint Disarmament Project of the Federation of American Scientists and the Committee of Soviet Scientists Against the Nuclear Threat.

REFERENCES

1. SDI, Technology, Survivability, and Software, Office of Technology Assessment, Washington, DC, 1988.
2. Report To The American Physical Society of the Study Group on Science and Technology of Directed Energy Weapons, Rev. Mod. Physics 59, No. 3, Part II, July 1987.
3. R. Hudson, Infrared System Engineering, Wiley, New York 1969.
4. AMOS User's Manual, Air Force Maui Optical Station, Mt. Haleakala, Maui, Hawaii, 1988.
5. Anti-Satellite Weapons, Countermeasures, and Arms Control, Office of Technology Assessment, Washington, DC, 1985.
6. Obtained with calibrated black bodies and with the star Vega, Glenn Ashley (AMOS), private communication.
7. F. Witteborn, K. O'Brien, and L. Caroff; "Measurement of the Nighttime Infrared Luminosity of Spacelab 1 in the H- and K-Bands, NASA Technical Memorandum 85972, March 1985. F. Witteborn, L. Caroff, D. Rank, and G. Ashley, Nighttime Spectroscopic and Photometric Observations of Spacelab 2 and Other Satellites, to be published.
8. D. Rank, P. Pinto, S. Woosley, J. Bregman, F. Witteborn, T. Axelrod, and M. Cohen; Letters to Nature 331, 505 (1988).
9. R. Rubin, J. Simpson, E. Erickson and M. Hass, Astrophysical Journal 327, 377 (1988).
10. J. Fraser in Arms Control Verification, ed. by K. Tsipis, D. Hafemeister, P. Janeway, Pergamon-Brassey, Washington DC, 1986.
11. G. Neugebauer, et al; Science 224, 14 (1984). G. Rieke, et al, Science 231, 807 (1986).
12. M. Waldrop, Science 240, 1396 (1988).
13. Aviation Week and Space Technology, pg. 20, June 14, 1982.
14. R. Dahlberg, Star-C: Nuclear Power System for Space Applications, General Atomic, San Diego, CA.
15. P. Pinto and J. Primack, to be published.

CHAPTER 9

THE TRANSITION TO A DETERRENCE POSTURE MORE RELIANT ON STRATEGIC DEFENSES

Paul L. Chrzanowski
Evaluation & Planning Program
Lawrence Livermore National Laboratory
Livermore, CA 94550

Introduction

Strategic nuclear deterrence is currently based on the overwhelming capability of the arsenals of the two superpowers. Massive damage would be inflicted upon the military forces and industrial capacity of both sides should nuclear war occur and escalation of conflict not be controlled. Nuclear deterrence has fostered a condition of peace in central Europe and an absence of direct conflict between the U.S. and the Soviet Union. However, some question whether deterrence will remain effective into the indefinite future, and should deterrence fail the consequences are grave.

If deterrence were based on effective strategic defenses -- rather than effective offensive forces -- or on a combination of offensive and defensive forces, there is the possibility of limiting damage in some circumstances where deterrence fails. This possibility, together with the emergence of new technologies pertinent to the development and deployment of ballistic missile defenses, has motivated the President's Strategic Defense Initiative (SDI). We do not at this time know whether survivable and cost-effective defenses are feasible. The goal of the SDI program to provide a basis for a development decision in the early 1990's.

If effective defenses prove to be possible, we need to understand how the transition can be made safely from the present world where offensive weapons dominate to a future world where defensive weapons provide security. Clearly, this is a complicated issue with many facets, including consideration of the technical capability of defenses (how effective are the defenses, particularly in view of possible countermeasures), maintenance of extended deterrence, and strategic stability during the transition. In this paper we focus on the subject of strategic stability and attempt to quantify the impact of defense deployments on strategic stability through the use of simple mathematical models.

We have attempted to approach the issue of stability through quantitative assessments for several reasons. To be quantitative about issues, one has to be precise in the definition of measures of stability and assumptions, which tends to clarify the discussion. And, with quantitative measures, one is able to notice the relative importance of the various factors that contribute and detract from stability under different sets of assumptions. Finally, these analyses lay the methodological groundwork for more detailed future studies that will be needed to support program decisions about specific system proposals.

Measures of Stability

In our studies[1] we have used two measures of stability (or instability). One relates to crisis stability, and the second relates to force balance (or arms race stability).

For crisis stability, both sides must have no incentive to execute a first strike in a time of international crisis. A lack of incentive would certainly be the case if whoever struck first ends up in a worse situation than the side that responds, but this goal is difficult to achieve. It may suffice if either side can achieve no significant advantage by attacking and is comparatively not much worse off than the aggressor after retaliation, should the other side strike first.

The force balance between the superpowers could be disrupted through a unilateral defense deployment by one side. This deployment could lead to (or be perceived to lead to) a strategic advantage that the other side, which may lose its ability to deter aggression, would deem unacceptable. A consequence could be an uncontrolled arms race -- either in offensive weapons or defensive weapons, depending (in part) on comparative cost effectiveness. Also, any strategic advantage due to a force imbalance that occurs during the transition, however temporary, could be important should a crisis arise while the advantage exists.

To "measure" force balance and crisis stability, we use the results from an exchange model that simulates the net outcome of a nuclear attack by one side and the retaliation by the other. A particularly simple exchange model is described in the next section. The model results are used in a procedure illustrated by Figure 1 to quantify crisis stability and force balance.

Consider Figure 1, taken from a paper by Wilkening and Watman[2], which is a decision tree for one of the two competing sides (labeled "BLUE", and the opposing side is "RED"). The decision for BLUE is to strike first or to wait. Should BLUE decide to strike first, the value to him is V_{BB}, which denotes

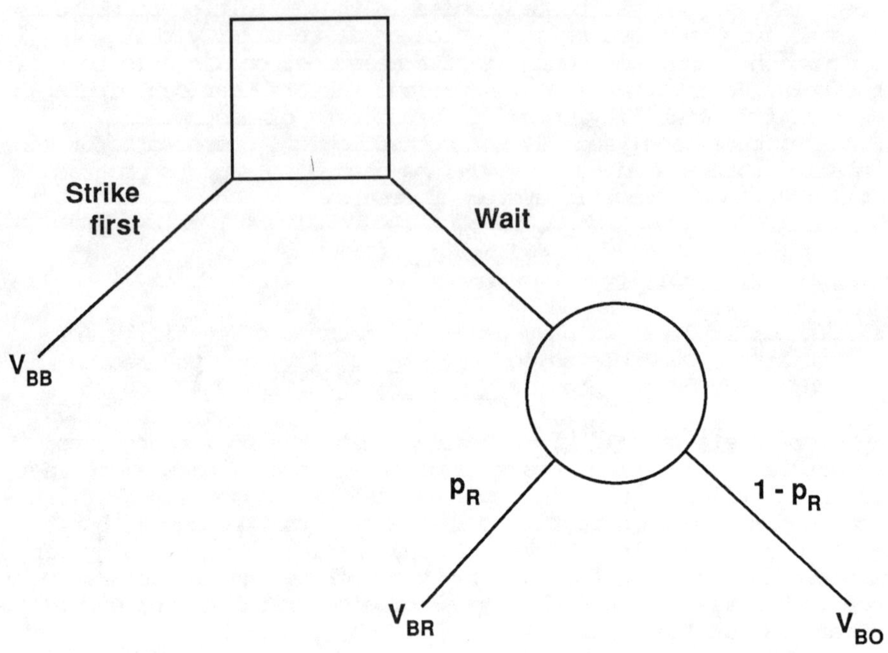

V_{BB} = value to BLUE of a strategic exchange where BLUE strikes first

V_{BR} = value to BLUE of a strategic exchange where RED strikes first

V_{BO} = value to BLUE of the status quo

p_R = probability that RED will strike first if BLUE waits

Decision variable when $p_R \sim 0$: $V_{BB} - V_{BO}$

Decision variable when $p_R \sim 1$: $V_{BB} - V_{BR}$

Figure 1: BLUE decision tree

the value to BLUE of a strategic exchange in which BLUE strikes first. (Our notation is that the first suffix refers to the side for which the value is measured and the second suffix identifies the side that strikes first in the exchange.)

If BLUE decides to wait, there are two possibilities. First, RED may decide to strike first -- the probability is p_R and the value to BLUE of the exchange is V_{BR}. Alternatively, RED may also choose not to strike. The value to BLUE is the status quo value, which we denote V_{BO}. To ascertain whether or not it is to BLUE's advantage to initiate conflict, one must know the values of p_R, V_{BB}, V_{BR}, and V_{BO}. Similarly, p_B and the value of exchanges to RED must be known to determine whether it is to RED's advantage to strike first.

Two simplifications make subsequent analyses more tractable. The first is to consider the case where the value functions are symmetric: the value to RED of any exchange is the negative of the value to BLUE. Such would be the case, for example, if the value functions were

V_{BB} = [damage to RED inflicted in a BLUE first strike]

- [damage to BLUE inflicted in RED retaliation]

and

V_{RB} = [damage to BLUE inflicted in RED retaliation]

- [damage to RED inflicted in a BLUE first strike]

$= - V_{BB}$,

and similarly for the other cases. The status quo values (V_{BO} and V_{RO}) need not be equal in magnitude and opposite in sign.

A second simplification is to consider the two important limits: war is deemed to be highly likely (for whatever set of causes) and war is deemed improbable. In the first case, which would be a crisis situation, BLUE would believe that p_R is near unity and his decision variable whether or not to attack would be $V_{BB} - V_{BR}$ (see Figure 1). Similarly, the decision variable for RED is $V_{RR} - V_{RB}$, which is equal to BLUE's decision variable when the value functions have the symmetry property discussed above.

Hence, we define the <u>crisis stability index</u> (I_{CS}) as

$$I_{CS} = V_{BB} - V_{BR} = V_{RR} - V_{RB},$$

which has been defined so that the greater the value the more unstable the situation is. Basically, I_{CS} is a measure of the regrets one side has for not striking first in a crisis situation. This measure of crisis stability has been used in numerous studies at Lawrence Livermore National Laboratory.[3]

For the case where the probability of war is deemed to be low ($p_R = 0$), Figure 1 illustrates that the decision variable for BLUE is $V_{BB} - V_{B0}$ (and, correspondingly, $V_{RR} - V_{R0}$ for RED). A significant complication is the fact that we do not know the value of V_{B0} or V_{R0} (the status quo) relative to the value functions of a nuclear exchange. Hence, there is, in effect, no obvious, convenient way to "normalize" the value of a nuclear exchange. One might expect both V_{B0} and V_{R0} to have large positive values because there is little incentive for either side to strike first in a time when a strike is deemed unlikely by the opposing side.

We have found it useful to define and use a force balance index (I_{FB}) given by

$$I_{FB} = \text{Max}\,[V_{BB}, V_{RR}].$$

Our force balance index has the property that it is large in cases where one side or the other can, in some sense, "win" by striking first, which is indicative of an imbalance between the strategic forces of the two sides. We use the word "win" advisedly because there is no clear way to compare the status quo to the value of inflicting relatively more damage on the opposition (large V_{BB} or V_{RR}). If I_{FB} is large, there will be strong incentives for the weaker side to redress the imbalance, which can drive an arms race.

Other measures of stability have been used for study purposes. As an example, Wilkening and Watman[2] define a BLUE strike incentive function (I_B) based on the decision tree depicted by Figure 1:

$$I_B = V_{BB} - p_R(I_R)\, V_{BR} - [1 - p_R(I_R)]\, V_{B0},$$

and similarly for RED:

$$I_R = V_{RR} - p_B(I_B)\, V_{RB} - [1 - p_B(I_B)]\, V_{R0}.$$

Notice that this is a coupled set of equations in that the probability that the opposing side strikes first is a function of that side's incentive function. Wilkening and Watman show that for plausible choices of the functional form of $p(I)$, there can be more than one solution to the set of equations -- I_B and I_R are not necessarily unique.

If the value functions have the symmetry property discussed above, I_B and I_R are related to our stability indices by

$$I_{CS} = I_B = I_R$$

when $p_R = p_B = 1$, and

$$I_{FB} = \text{Max}\ [I_B, I_R]$$

when $p_R = p_B = 0$ and $V_{B0} = V_{R0} = 0$.

Strategic Exchange Models

This analysis approach for studying transition stability issues is very general and requires specific choices for the exchange model and the value functions for implementation. We have chosen to use an exchange model detailed enough to illustrate many significant issues associated with defense deployment but simple enough to facilitate an understanding of analysis results.

Figure 2 depicts a simple, two-step exchange model which we have used to study the results of an exchange between RED and BLUE forces. The relevant force structures on both sides are described by six entitities: alert bombers, survivably-based ballistic missile systems (including alert SLBMs and mobile ICBMs, and ICBMs in survivable silos), ICBMs in vulnerable silos, ballistic missile interceptors, high-value targets, and low-value targets.

We have assumed, in this model, that the interceptors are sufficiently survivable that the option chosen by the attacker is to overwhelm the defense rather than to attack it directly. Also, the survivably-based missiles and alert bombers are not targetable in the first strike. The non-alert bombers and SLBM forces would be included in the set of targets.

In the first step of the exchange, one side (as depicted, RED) launches its forces at some combination of the vulnerable ICBMs and the targets, and the second side uses its interceptors in defense of some combination of silos and targets. In specific, the number of surviving BLUE high-value targets (S_{BTH}) would be

$$S_{BTH} = (1 - h_R)\ T_{BH}\ Q(i_H, a_H, (1-h_R)T_{BH}),$$

where h_R is the fraction of the target set attacked (and destroyed) by RED bomber-delivered weapons, i_H is the number of defenders used to protect BLUE high-value targets, and a_H is the number of RVs attacking those targets. Throughout, we use $Q(x,y,z)$ to denote the fraction of a set of z targets that survives when there are x defenders and y attackers ($Q(x,y,z)$ for different types

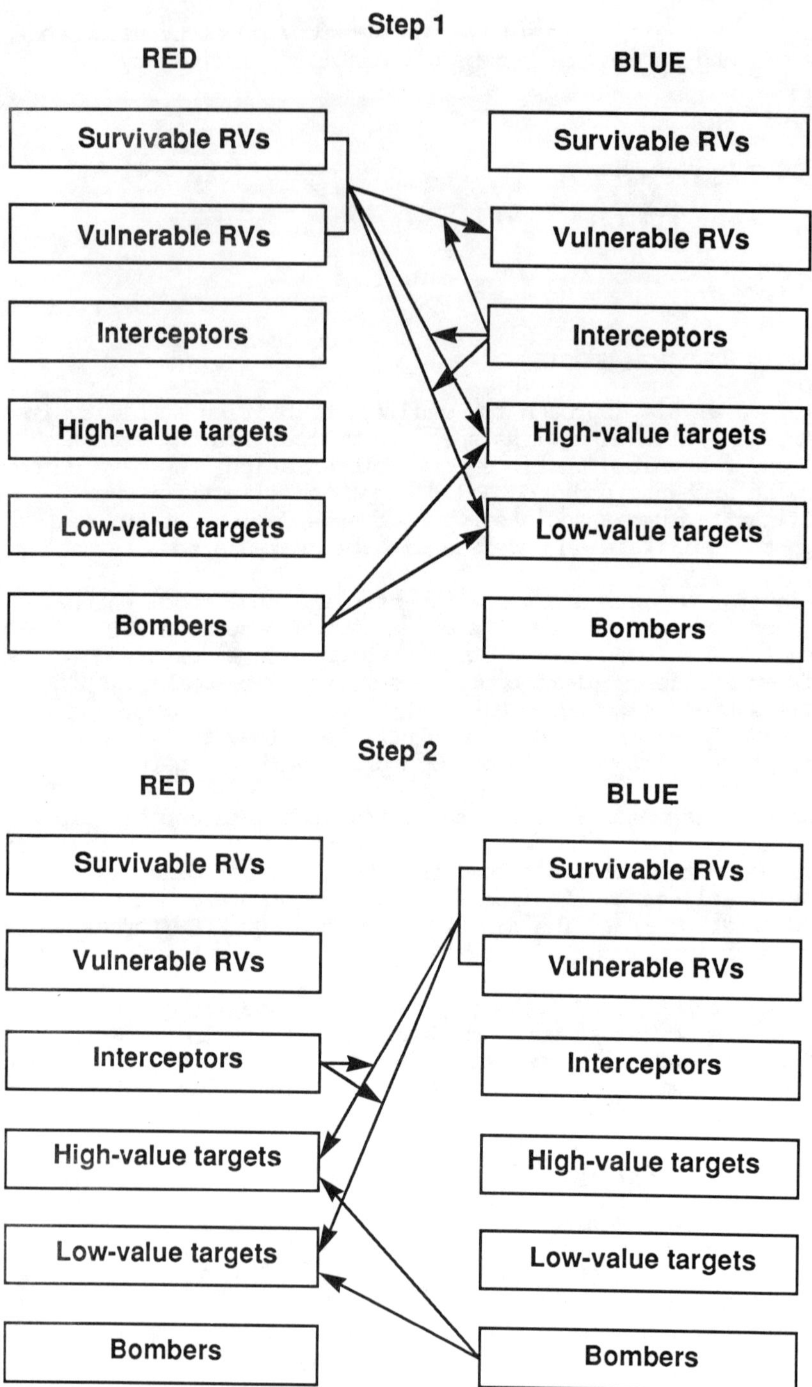

Figure 2: The Exchange Model

of defenses is discussed in the next section). Also, A_B, I_B, and T_{BH} are the number of BLUE RVs, interceptors, and high-value targets (and similarly for RED). A similar formula applies for the survivability of low-value targets (S_{BTL}) with a_L the number of attackers and i_L the number of defenders.

The number of surviving BLUE RVs (S_{BW}) is given by

$$S_{BW} = (1 - f_B)\ g_B\ A_B + f_B\ A_B\ Q(I_B-i_H-i_L, A-a_H-a_L, S_B),$$

where f_B is the fraction of BLUE RVs vulnerably-based, g_B is the alert rate for the survivably-based BLUE RVs, and S_B is the number of vulnerable BLUE ICBM silos. Finally, A is the number of RED RVs on alert:

$$A = (1 - f_R)\ g_R\ A_R + f_R\ A_R.$$

In the return strike, BLUE launches his surviving weapons against RED targets, with the result that

$$S_{RTH} = (1 - h_B)\ T_{RH}\ Q(i, a, (1-h_B)T_{RH})$$

and

$$S_{RTL} = (1 - h_B)\ T_{RL}\ Q(I_R-i, S_{BW}-a, (1-h_B)T_{RL}),$$

where a is the number of attackers against RED high-value targets and i is the number of defenders.

At the end of the exchange all offensive and defensive weapons have been used and what remains are some surviving targets on both sides. The number of surviving targets can be used to construct a value function for the exchange:

$$V_{RR} = V_{RR}(S_{RTH}, S_{RTL}, S_{BTH}, S_{BTL}),$$

which is optimized through the choice of a_H, a_L, a, i_H, i_L, and i. The optimum is obtained by choosing the value of a_H, a_L, and i such that V_{RR} is maximized from RED's point of view, given that BLUE chooses i_H, i_L, and a values that minimize V_{RR}.

The analyst is free to choose the functional form for V, and in some of our studies we have investigated a variety of possibilities.[1] A very simple "representative" game value is the difference in the fraction of the target value that survives (attacker minus responder). That is, V_{RR} satisfies

$$V_{RR} = -V_{BR} = W_R\ S_{RTH}/T_{RH} + (1 - W_R)\ S_{RTL}/T_{RL} - [W_B\ S_{BTH}/T_{BH} + (1 - W_B)\ S_{BTL}/T_{BL}],$$

where V_{RR} and V_{BR} are (respectively) the game value from RED's and BLUE's point of view and W is the fraction of the total target value residing in the high-value targets. With this choice, V_{RR} ranges in value from - 1 to + 1.

The analysis becomes more complicated if different value functions are chosen for the two sides. As an example, each side might put more weight on saving its own targets in comparison to damaging the other side's targets. In such a case there is an ambiguity with regards to the solution of the minmax problem (determination of the values of the i's and the a's). In planning the attack, does the attacker expect the responder to attempt to minimize the attacker's value function or to maximize the responder's value function? With the specific value function defined above, the two choices are equivalent.

Finally, it should be noted that we have made two particularly important simplifications in this model. First of all, the formulas reflect an extremely simple model for the impact of bomber-delivered weapons. In short, the target set is partitioned between targets attacked (and destroyed) by bomber weapons and those attacked by ballistic missile systems. The variable h_R represents the fraction of BLUE targets destroyed by RED bombers (and similarly for h_B). Obviously, more realistic (and complicated) models are possible, including variants with coverage of some targets with both bomber and ballistic missile weapons.

A second significant simplification in the above is that the targets are categorized as being either high-value or low-value. If there is a greater gradation in target values, the formulas presented here quickly become untractable and other approaches need to be considered.[4] If targets all have the same value, the above discussion simplifies somewhat, and there are only two variables (a_H and i_H) to fixed by the minmax problem.

The Defense Model

The functional form of Q(x,y,z) depends on the characteristics of the defenses. We present formulas here for two idealized, bounding defense cases: a randomly subtractive defense and an adaptive preferential defense. In both cases, the formulas we show are expressed in terms of the probability that an interceptor (if launched) destroys an RV (P_I) and the single-shot probability of target damage from an RV that leaks through the defense (P_d). For any particular target the probability of survival (P_s) is

$$P_s(k,m) = \{1 - P_d + P_d\,[1 - (1-P_I)^m]\}^k,$$

where k is the number of attackers per target and m is the number of interceptors (launched in salvo) against each RV.

For a randomly subtractive defense, the defender does not know the intended target for each RV. Hence, the probability that any particular RV is targeted by the defense is I/A for the case where there are fewer interceptors than attacking RVs. A randomly subtractive defense is a reasonable model for a mid-course defense against manuevering RVs, for a boost-phase defense, and for defense against post-boost vehicles. In each of these cases the specific targets of the RVs are not known.[5]

The formula for Q for a randomly subtractive defense is most easily expressed in terms of two integers:

$k = \mathrm{Int}(A/T)$: each target is attacked by k or k+1 RVs

$m = \mathrm{Int}(I/A)$: each RV is intercepted by m or m+1 defenders,

and two constants:

$I^* = I - m\,A$: I^* attackers face m+1 defenders

$A^* = k\,T\,[1 + k - A/T]$: A^* attackers used k-per-target.

In terms of these variables and $P_s(k,m)$ defined above, Q satisfies

$$Q(I,A,T) = (1 - A^*/A)\ [(1 - I^*/A)\ P_s(1,m) + (I^*/A)\ P_s(1,m+1)]^{k+1} + (A^*/A)\ [(I^*/A)\ P_s(1,m+1) + (1 - I^*/A)\ P_s(1,m)]^k.$$

When there are k RVs aimed at every target and fewer interceptors than attackers, this formula reduces to

$$Q(I,A,T) = \{(1 - I/A)\ (1 - P_d) + (I/A)\ [1 - P_d\ (1-P_I)]\}^k,$$

which becomes

$$Q(I,A,T) = [P_I\ I/A]^k$$

in the limit of $P_d = 1$. In this last equation, I/A is the probability that an interceptor is targeted at any particular RV and P_I is the probability that the intercept is successful so that $[P_I\ I/A]^k$ is the probability that all k RVs aimed at any single target are defeated.

An adaptive preferential defense is the other bounding case. A preferential defense allows for optimal use of interceptors and protection of the maximum number of targets. However, such a defense must have either a large defended area (i.e., an interceptor is able to defend any one of many targets) or must rely on deceptive basing of interceptors so that the attacker does not know which targets are protected. When there are not enough interceptors to defend all targets, some must be sacrificed and the rest of the targets are protected as well as possible. Roughly, the fraction of the target set that can be protected is I/A.

It is essential to the success of the preferential defense that the choice of defended targets not be known to the attacker. Hence specification of what the U.S. would defend with a partially effective (preferential) defense would impair performance. However, some types of targets are not suitable for preferential defense because of their fixed location and value. Preferential defense fails if the attacker is afforded several opportunities to shoot, assess target damage and defense tactics, and shoot again. Military targets that might be moved or used before they are re-attacked are more defendable than fixed industrial facilities.

For a preferential defense, Q is given by

$$Q(I,A,T) = [I^* P_s(k,m+1) + (A^* - I^*) P_s(k,m)]/(k T) + (A/T - k) P_s(k+1,m)$$

when $I^* < A^*$; and when $I^* > A^*$, one has

$$Q(I,A,T) = (1 + k - A/T) P_s(k,m+1) + [(A - I^*) P_s(k+1,m) + (I^* - A^*) P_s(k+1,m+1)]/[(k+1) T].$$

These formulas reduce to

$$Q(I,A,T) = (1 - I/A) (1 - P_d)^k + (I/A) [1 - P_d (1-P_I)]^k,$$

when there are k attackers per target for all targets and fewer interceptors than attackers. Furthermore, one has

$$Q(I,A,T) = P_I^k I/A$$

when $P_d = 1$, which is easily interpretted. I/A is the fraction of the targets that can be defended and P_I^k is the probability that all k intercepts in defense of a target are successful.

The minmax problem to determine V_{RR} is simplified for the case where the ratio of interceptors to attackers is the same for each class of targets (e.g., "high-value targets" or "silos"):

$$I_B/A = i_H/a_H = i_L/a_L$$

and

$$I_R/S_{BW} = i/a.$$

These relationships hold exactly for a random subtractive defense (because the intended targets are not known) and are nearly the optimal defensive tactics for a preferential defense. We use these formulas in our simple models to reduce the complexity of the minmax problem that must be solved to determine optimal strategies for both sides.

For all examples shown below, we constrain forces to roughly the levels proposed for START and consider the case where there are 500 high-value and 1500 low-value targets on each side with half the target value ascribed to each subset ($W_R = W_B = 0.5$). The P_d for attacking targets is taken to be 0.90 and 0.80 for attacking silos. Also, we use $P_I = 0.90$. The number of RVs on each side is 5000 and there are 3 RVs per missile. We will vary the number of RVs vulnerably based, the effectiveness of the bomber force, and the alert rate for the non-vulnerable RVs.

Figure 3 illustrates a significant difference between the performance of a randomly subtractive defense and a preferential defense. The graphs are in the format used by Kent[5] and Wilkening and Watman[2] in which the ordinates are the size of defense deployments by RED and BLUE scaled by the size of the opponent's offensive forces (I_B/A_R or I_R/A_B). Hence, in these graphs, a transition to defense dominance is described by a path from essentially no defenses (the lower, left-hand corner) to highly effective defenses on both sides (the upper, right-hand corner). Our interest is in finding a crisis-stable transition path which roughly preserves force balance.

The graphs depict the force balance index for both types of defense for the case where there are no bombers and all missiles are survivably based and are on full alert. Because the responder's weapon systems are survivable, all attackers are used against the targets. Thus, there is no advantage to striking first and the crisis stability index is zero. Force balance is the significant stability issue if forces are survivable.

The most important feature in Figure 3 is the varying width of the region of approximate force balance for the randomly subtractive case. The "channel of stability" is very wide for small deployments of interceptors on both sides because the defenses are largely ineffective and offenses continue to dominate. When deployments are such that almost all of the RVs can be negated (I/A near unity), a small difference in the relative size (or relative P_I) of defenses can result in a large imbalance

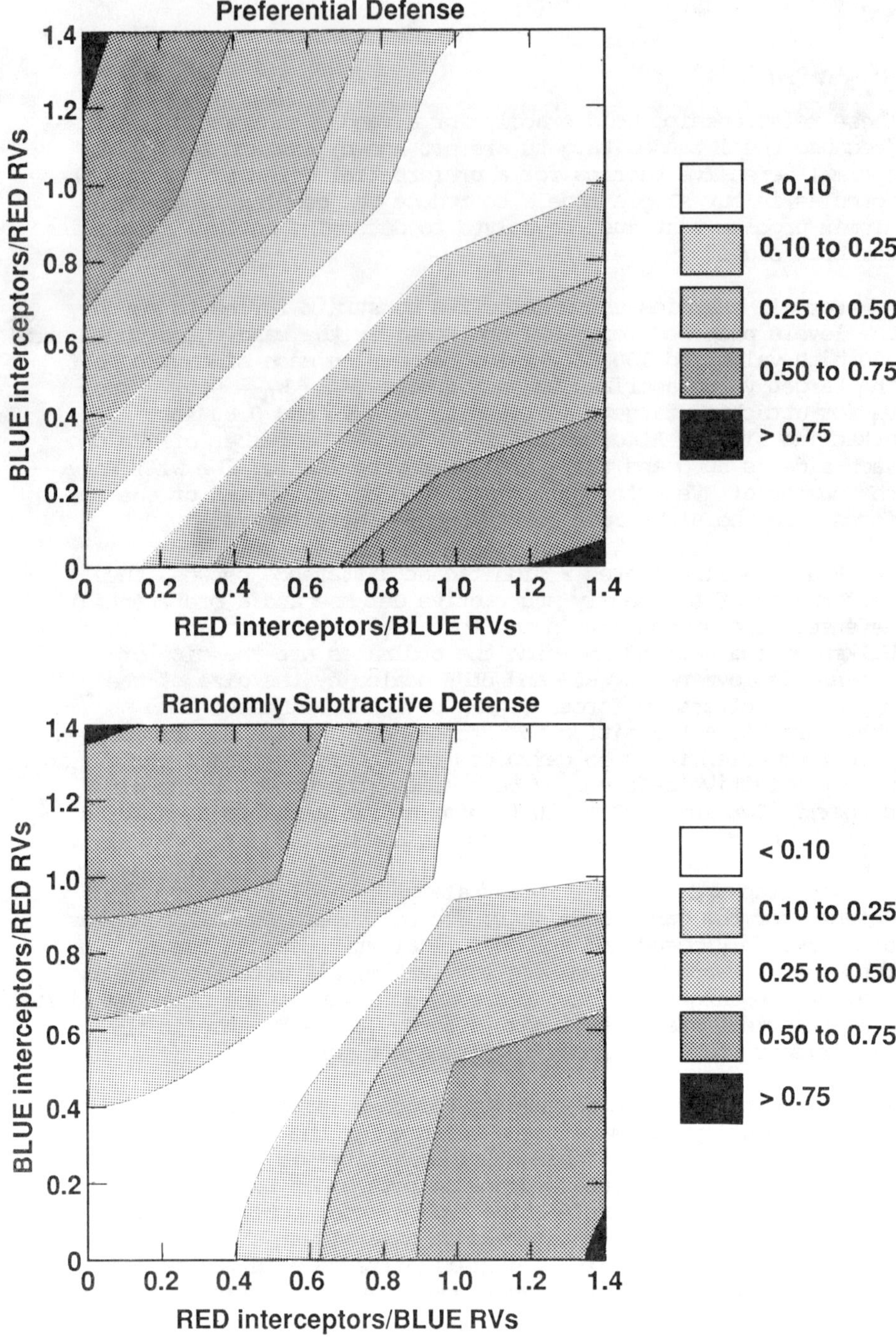

Figure 3: Force Balance Index for Highly Survivable Forces

in force effectiveness. By comparison, a preferential defense deployment has a more uniform impact and the channel of stability has rather constant width through the course of defense deployments. In both cases, when there are more interceptors than attackers (I > A) the channel of force balance opens up.

Silo-Based ICBM Survivability

The intrinsic survivability of each side's forces (survivability in the absence of deployment of ballistic missile defenses) is a critical factor affecting the stability of the transition. Survivable forces are important. Here, we confine our discussion to offensive forces and assume that the defenses are sufficiently survivable that direct attack to suppress defenses is not the tactic of choice.

The survivability issues for the vulnerable forces (such as ICBMs in silos) are very different than those for the invulnerable forces (such as SLBMs). Silo-based ICBMs, of course, can be attacked. If the silos are defended, there are fewer interceptors available for the defense of other targets. If the silos are not defended, the responder's strike will be reduced in size with the probable consequence that the attacker's defense would be comparatively more effective. For SLBMs, the issue of concern is the impact of variations in the alert rate of the two sides' forces. Alert-rate issues are discussed in the next section.

To highlight the consequences of ICBM vulnerability during the transition, consider the case where the SLBMs are on 65% alert ($g_R = g_B = 0.65$) and bomber forces are able to damage 20% of the target base ($h_R = h_B = 0.20$). Figure 4 shows the results when 25% of the forces are in vulnerable silos ($f_R = f_B = 0.25$), and for Figure 5, 75% of the forces are in vulnerable silos.

Figure 4 depicts a situation where the crisis stability index is small, independent of the path of mutual defensive buildup. (It is, however, somewhat greater in the randomly subtractive defense case.) Rather, the issue is strategic balance, which would be upset if one side deployed defenses at a much faster rate. In contrast, Figure 5 shows that if the forces are vulnerable, all buildup paths are relatively crisis-unstable. In addition, the channel of acceptable strategic balance can be very narrow (or nonexistent) when offensive systems are vulnerable.

Although Figure 5 (with 75% of the forces in vulnerable silos) appears to be very crisis unstable, the possibility that the responder might launch his ICBMs on warning of attack or while the attack is underway provides a strong deterrent. How should the attacker optimize his attack if he thinks launch on warning is a

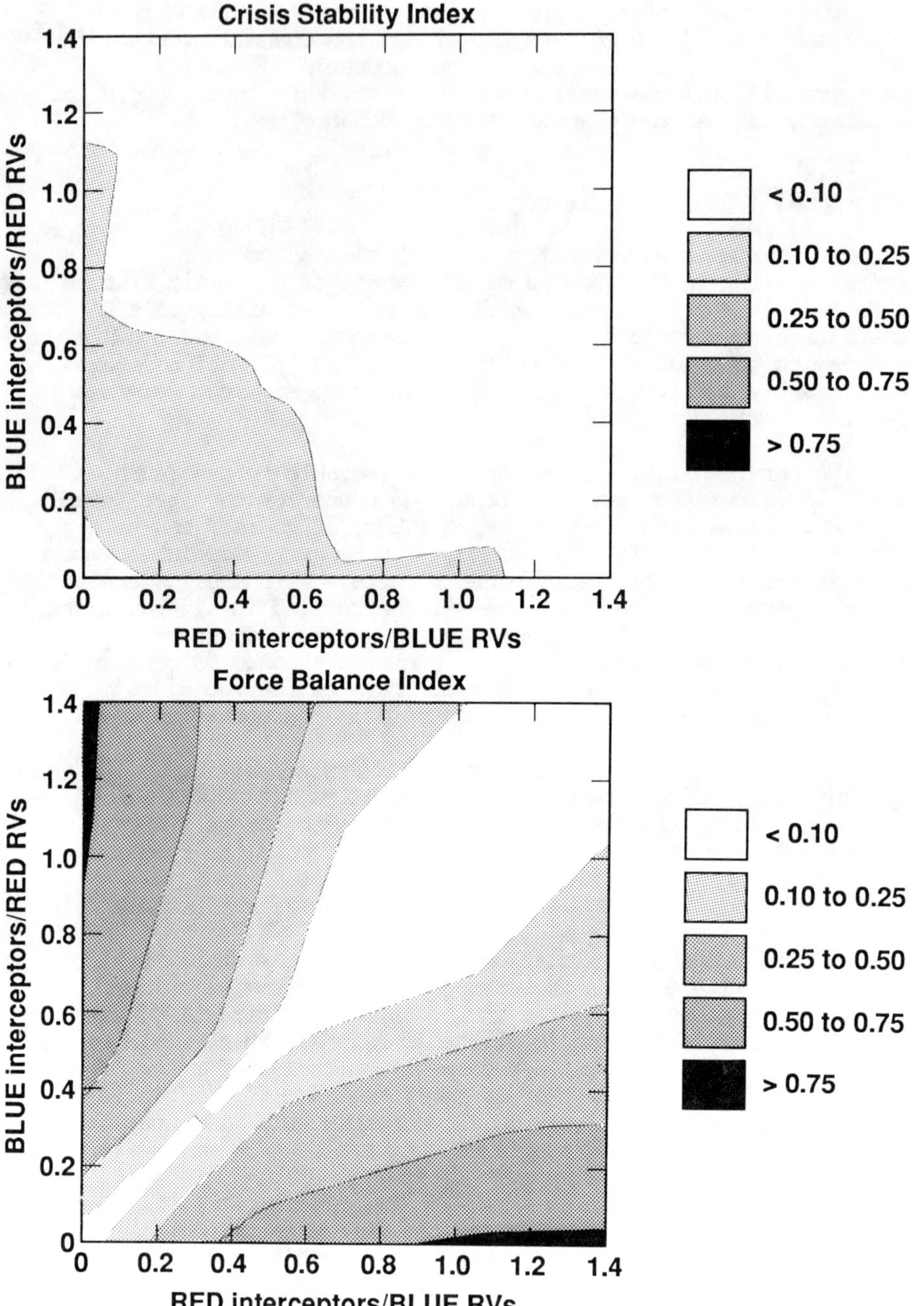

Figure 4a: Crisis Stability and Force Balance for 25% Vulnerable Forces, Preferential Defenses

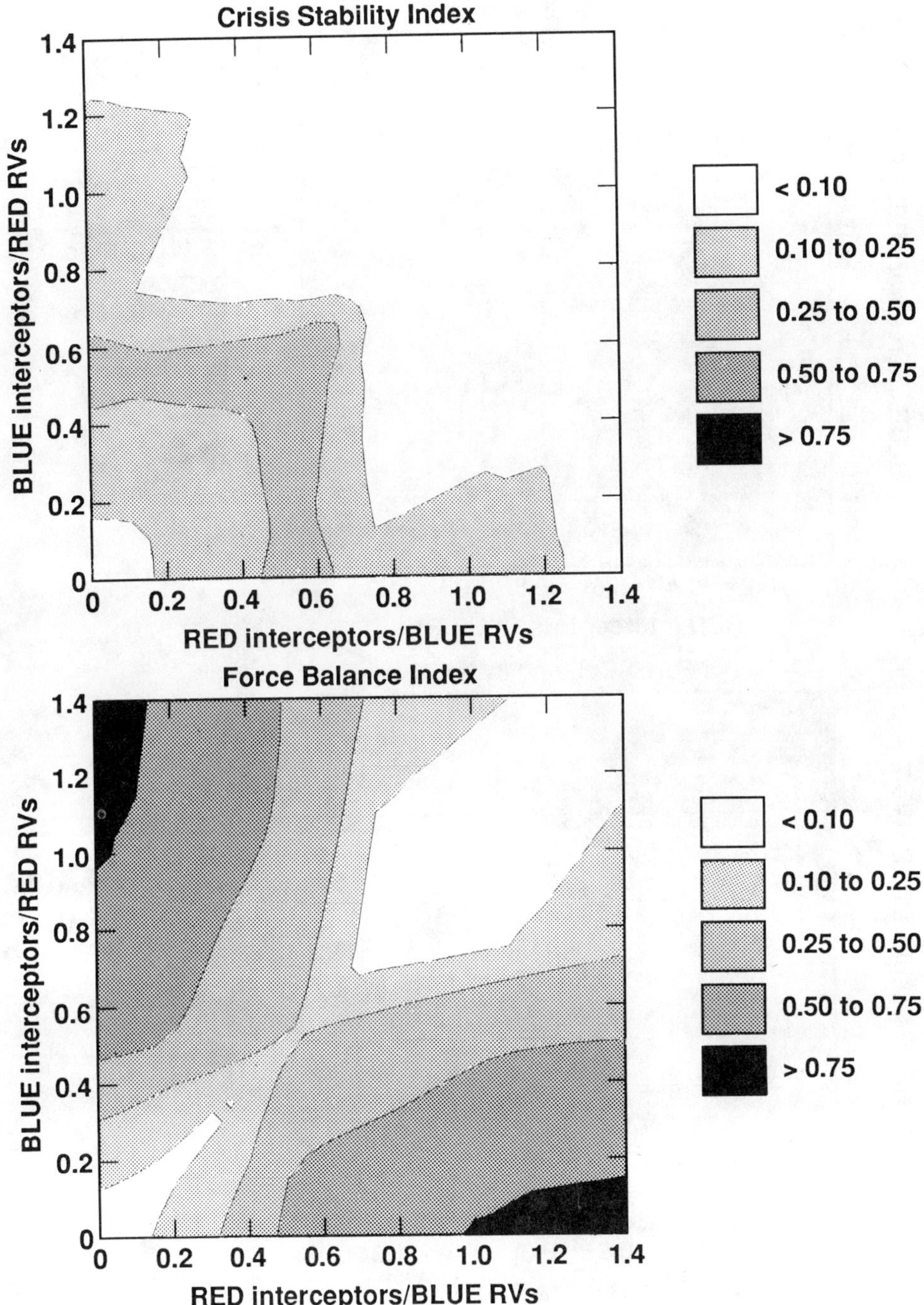

Figure 4b: Crisis Stability and Force Balance for 25% Vulnerable Forces, Randomly Subtractive Defenses

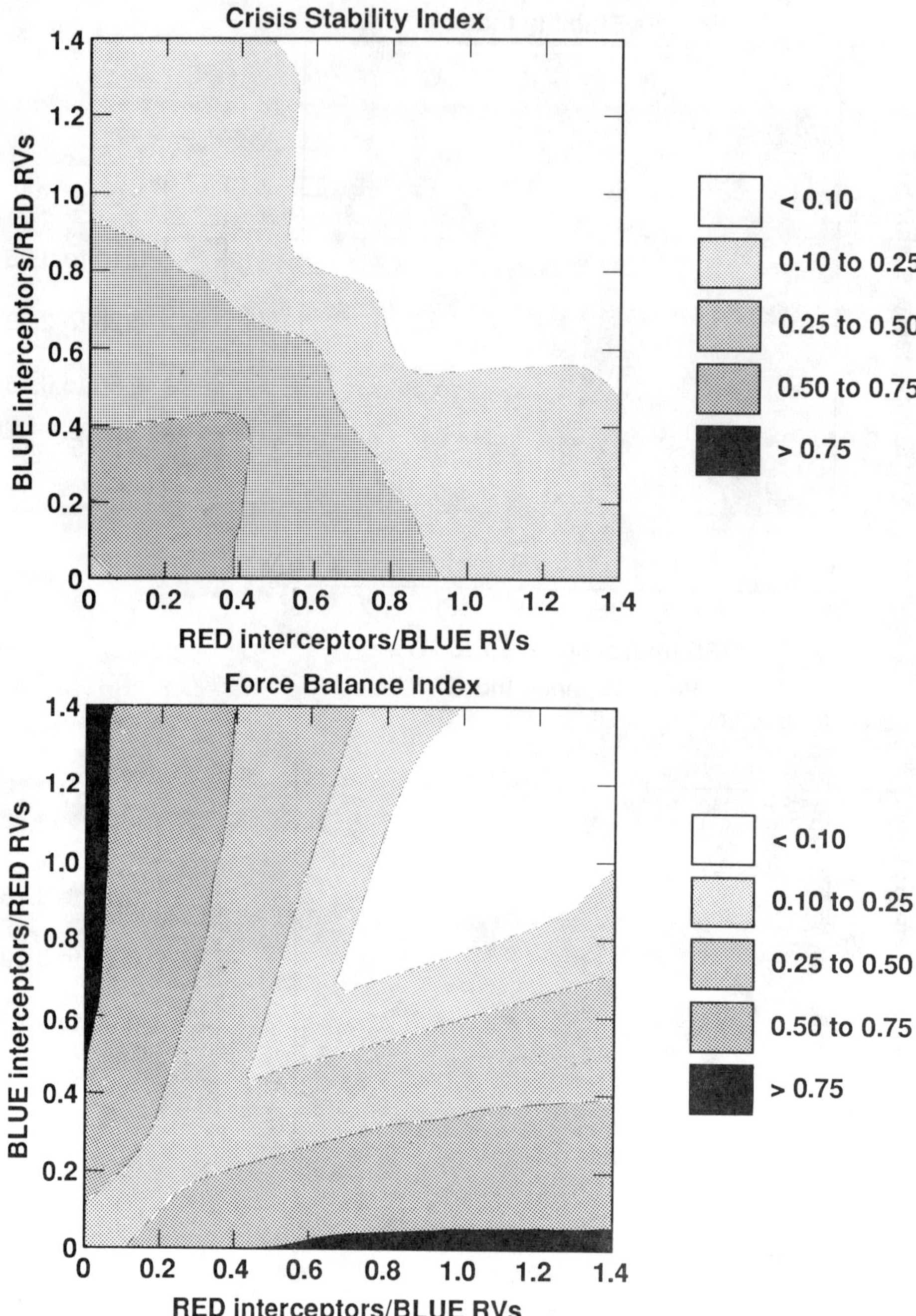

Figure 5a: Crisis Stability and Force Balance for 75% vulnerable Forces, Preferential Defenses

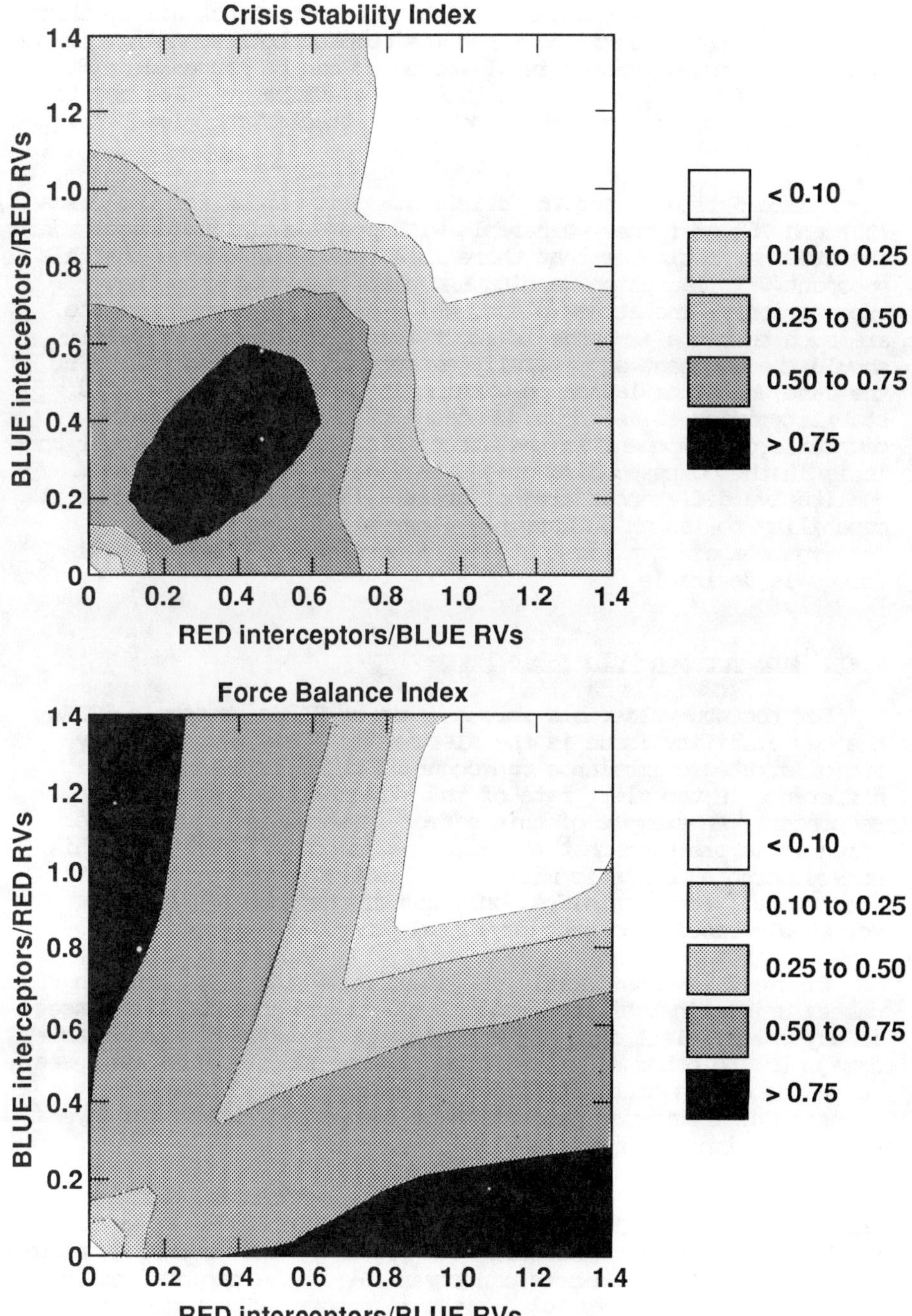

Figure 5b: Crisis Stability and Force Balance for 75% Vulnerable Forces, Randomly Subtractive Defenses

possible or probable response? If he does not attack silos, there is no advantage in striking first (as opposed to responding) if the attacker's forces are survivably based or can be launched on warning. If he emphasizes a counterforce strike at silos and the responder executes a prompt launch, the attacker can "lose" the exchange.

Figure 6 illustrates the crisis stability index for two cases (25% and 75% of forces vulnerable with preferential defenses) where the attacker estimates that there is a reasonable chance that his opponent will launch on warning and plans his attack on that expectation.[6] The attack plans, which entail less counterforce, are such that the exchange is about even ($- 0.10 < I_{CS} < 0.10$) should the opponent successfully execute a prompt response. For the case where the launch on warning is unsuccessful, the crisis stability index is as shown. Because of the reduction in the counterforce component in the attack plans, the index is lower than it is in the corresponding graphs in Figures 4 and 5, but the qualitative differences are not great. Thus, while the technical capability to launch on warning vulnerable forces enhances deterrence against a first strike, a more secure means of basing forces is desirable.

Alert Rate for Non-Vulnerable Forces

For the non-vulnerable forces (such as SLBMs or mobile ICBMs) the key stability issue is the alert rate. Crisis instability and/or strategic imbalance can occur if there is a substantial difference in the alert rate of the attacker and that of the responder. An example of this effect is shown in Figure 7, for which it is presumed that the attacker is on full alert with his non-vulnerable ballistic missiles while the responder has a 65% alert rate. For both sides, only 25% of the forces are vulnerably-based, and preferential defenses are assumed.

Figure 7 is considerably different from Figure 4. In both cases, crisis instability is not large in the absence of defenses if day-to-day alert forces are sufficient to extract considerable damage to the other side. With very large and effective defenses, again the crisis stability index is small. But, during the transition, a surprise generation of forces and attack can have a pronounced impact on crisis stability.

A similar effect occurs if there is a substantial difference between the alert rates for bomber forces. In Figure 8, we have used the assumptions discussed with Figure 7, and we further assume that the attacker's bombers can cover 40% of the targets and the corresponding figure is 20% for the responder. The region of instability persists in Figure 8 beyond full deployment of

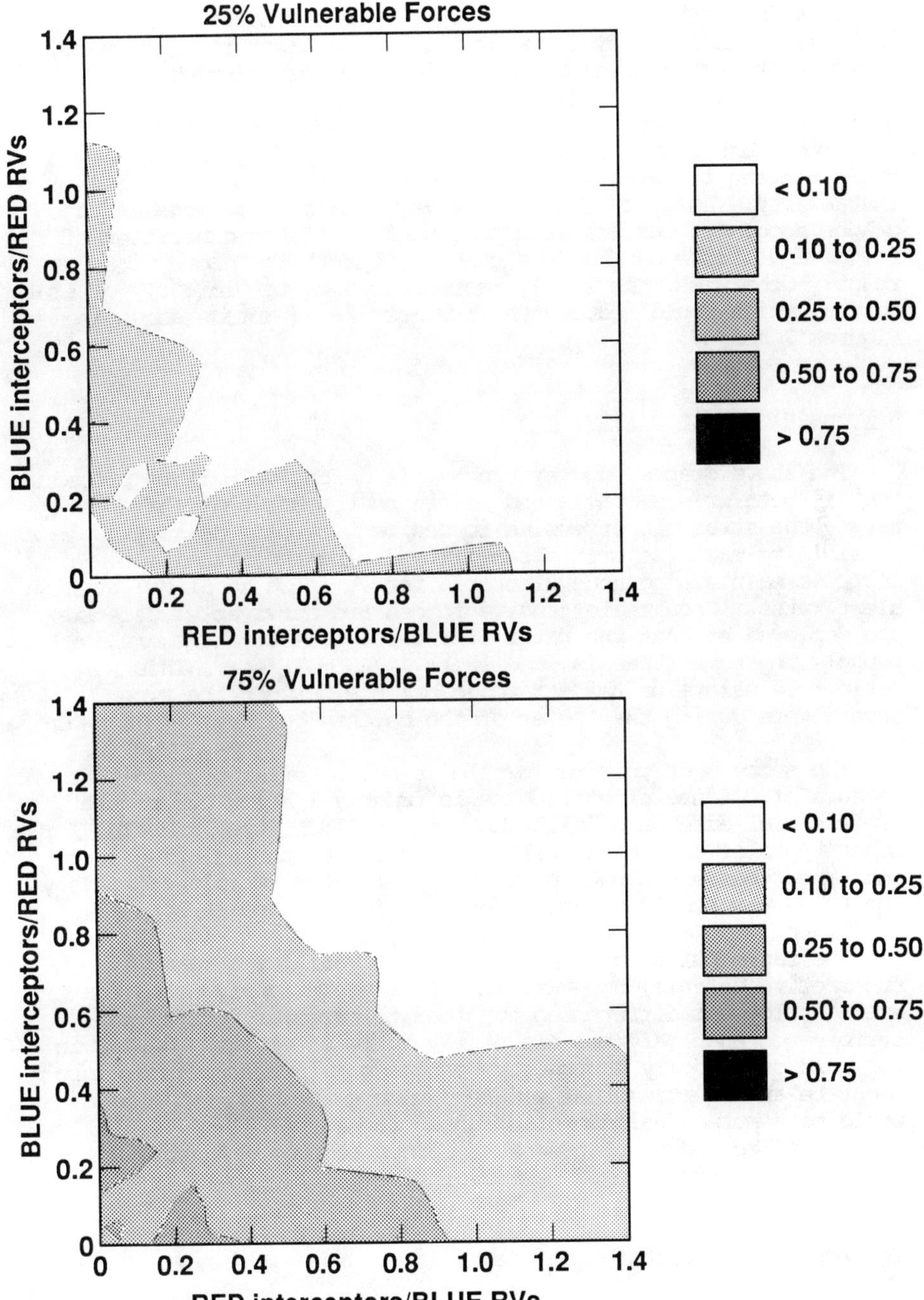

Figure 6: **Crisis Stability Index — Attack Plans Consider Possibility of Launch On Warning — No Launch On Warning, Preferential Defenses**

effective ballistic missile defenses. The source of the instability is the difference in the bomber alert rates because, after the transition, bombers are the only effective weapons unless, of course, effective air defenses are also deployed.

From these examples, it is clear that a difference in alert rates between the attacker and responder contributes to crisis instability. However, the problem may not be as serious as the graphs suggest. Our analysis only applies if the generation of forces and strike can be done quickly, before the other side can react. Otherwise, the likely response may be to similarly generate forces, which would reduce the first-strike advantage shown in Figures 7 and 8.

A Transition Possibility

The above graphs suggest a possible defensive buildup procedure that is both crisis-stable and maintains the strategic balance. First, the strategic offensive forces must be made as survivable as possible to maximize crisis stability during the transition. This includes maintaining non-vulnerable forces (such as SLBMs) at high alert rates. Once the offensive forces are survivable, defenses are deployed so that the ratio of interceptors to attackers is roughly the same (that is, one seeks $I_R/A_B = I_B/A_R$). This ratio rule maintains the strategic balance between the two superpowers during the course of the defense buildup.

The success of this ratio rule is not surprising given that the measure of defense effectiveness is roughly I/A for our simple preferential defense model. The ratio buildup rule is useful for other types of defense as well, provided that the relative force and target bases are not too different for the two sides.[7] Also, the P_I values should be about the same for RED and BLUE.

The ratio build-up rule works fairly well for a randomly subtractive defense; however, it has been previously noted that small deployment differences can lead to large asymetries in performance when defenses are nearly fully effective. Hence, in the case of randomly subtractive defenses, a preferred procedure might be an abrupt, mutually agreed-upon transition. The idea would be a mutual buildup of defenses to, say, I/A = 0.40, followed by a rapid builddown of offensive forces to push I/A above unity in a very short, prearranged period of time.

The Impact of Uncertainties

It is clear from our discussion that the transition will be attended by a large number of uncertainties. Some uncertainties

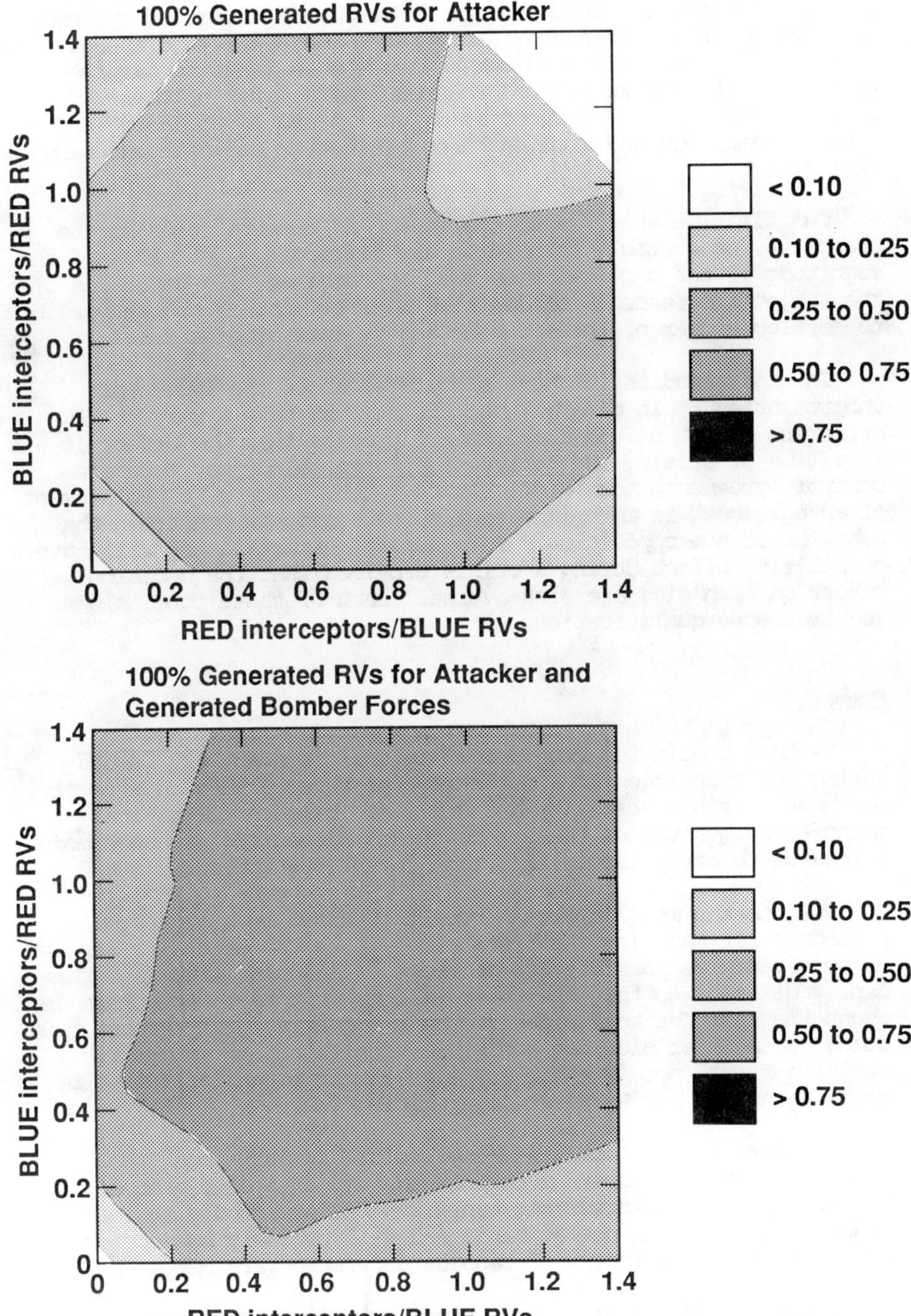

Figures 7-8: Crisis Stability Index for Generated Attacker Forces, 25% Vulnerable Forces, Preferential Defenses

are of a technical nature (the performance and number of systems), some are operational (system alert rates and ability to manage the battle), and some relate to subjective values (what to target, what to defend, and how to "score" the results). A few of these uncertainties will be reduced as defensive system technologies are better understood and architectures are better defined, but many will not.

Quantitative analyses of the impact of uncertainties on the prospects for a stable transition are difficult. Estimates of the magnitude of the uncertainties must be combined with estimates of the subjective response of decision makers. It may not be possible to develop either of these two necessary sets of data.

In other studies, we have speculated about the impact of uncertainties.[7] In general, uncertainty may have a very different effect on the arms control process than the effect it has at a time of crisis. We believe that uncertainty may make arms control agreements more difficult to reach; however, establishment of an agreement is more important to each side to constrain the behavior of the opposition. In contrast, uncertainties will have a stabilizing effect during a crisis provided that the decision makers on each side are risk-averse. Each of these conclusions requires some qualification.

Summary

We have considered two factors to explore stability issues during the transition to a deterrence posture reliant on effective defenses: crisis stability and strategic balance. A general methodology quantifies these factors and a specific exchange model allows us to study issues important to the transition.

Perhaps the most important result obtained is a suggested procedure for managing the transition: make the offensive forces as survivable as possible (to increase crisis stability) and then deploy defenses so that the ratio of interceptors to attackers is about the same for both sides (to maintain strategic balance). By doing so, neither side can achieve a significant advantage by striking first, nor is there any significant potential regret by choosing not to strike first in a crisis.

In spite of the fact that many useful insights about the transition can be discerned through the use of a simple model, a thorough understanding of the many issues associated with the transition period is not at hand, and will probably not emerge until the characteristics of candidate defenses are better defined.

There is much to be gained in understanding through the development and use of more sophisticated models. Nonetheless, it may be of continuing importance to relate results from more detailed studies to simple models, such as this one, in order to convey to a wide audience our belief that, with proper management after a successful developmental program, a stable transition to defense dominance may indeed be possible.

References

[1]Chrzanowski, Paul L., "Transition to Deterrence Based on Strategic Defense," Energy and Technology Review, UCRL-52000-87-1-2 (Lawrence Livermore National Laboratory, Livermore, CA, 1987), p. 31; and references cited therein. Portions of this paper are taken directly from reference 1.

[2]Wilkening, D. and Watman, K., "Strategic Defenses and First-Strike Stability," R-3412-FF/RC (RAND Corporation, Santa Monica, CA, 1986).

[3]Early work using exchange models and differences between exchange results to measure crisis stability was reported in Reinhardt, G. C. and Squire, R. K., "Using Game Theory to Evaluate the Stability of Strategic Forces," UCID-19297 (Lawrence Livermore National Laboratory, Livermore, CA, 1984). More recent work by one of these authors on the impact of defenses is reported in Reinhardt, G. C., "Defense and Stability," UCRL-53743 (Lawrence Livermore National Laboratory, Livermore, CA, 1986).

[4]At Lawrence Livermore National Laboratory, both sequential optimization schemes (see reference 3) and linear programming schemes (Chrzanowski, Paul L., to be published) have been developed to deal with exchange optimizations with non-uniform target values.

[5]Boost-phase defenses can be used preferentially against certain classes of boosters (e.g., boosters with RVs of sufficient accuracy to be effective against hard targets such as silos). This case was analyzed by Kent, G. A. and DeValk, R. J., "Strategic Defenses and the Transition to Assured Survival," R-3369-AF (RAND Corporation, Santa Monica, CA, 1986).

[6]To plan the attacks whose results are shown in Figure 6, we optimized an average value for V_{RR} (with and without a launch on warning response). When there is no prompt response, the silos are defended. If there is a launch, the silos are not defended. Other methods could have been used to develop attack plans that hedge against the possibility of launch on warning.

[7]For additional discussion, see reference 1.

CHAPTER 10

OBSERVATIONS ON THE FEASIBILITY AND SURVIVABILITY OF NEAR-TERM STRATEGIC DEFENSES

Anthony Fainberg
Office of Technology Assessment, U.S. Congress, Washington, D.C. 20510[1]

ABSTRACT

This paper reviews a possible early (mid-1990s) missile defense system utilizing space-based components and employs some elementary considerations to evaluate its usefulness. Topics covered include timelines for intercepts, system mass in orbit, U.S. launch capability, survivability, and some countermeasures. This review concludes that a degree of skepticism is warranted.

INTRODUCTION

The purpose of this paper is to present a discussion of some elementary considerations regarding widely-discussed concepts for space-based ballistic missile defense systems. The topics outlined below will be analyzed in the light of first principles and of information gleaned from open sources.

This paper will discuss the prospects of a possible space-based early missile defense system from the perspectives of feasibility, resilience against obvious countermeasures, and survivability. Where necessary, open sources will be cited to justify the choice of values for various physical parameters used in the analysis.

Strategic defense is a vast enough technological field that a semester, rather than an hour, would be required to give a comprehensive analysis; therefore, the topics will be selected and abbreviated. For a few of them, this paper presents some sample calculations. For others, only a cursory mention is given. This does not mean that in the latter cases problems are non-existent; rather it means that there is not enough room for a satisfying quantitative discussion of those matters.

THE "PHASE ONE" SYSTEM

A proposed system for space-based missile defense, intended for deployment about ten years from now, would include a number of space-based interceptors (for boost-phase and post-boost phase intercept), and ground-launched missiles (for late midcourse intercept). There would be on the order of 300 carrier vehicles circling the Earth in relatively low orbits. Each carrier vehicle would hold on the order of ten interceptors.[2] These weapons might be guided by several sets of space-based sensors: high altitude launch detectors, low altitude trackers, and homing sensors on the interceptors' kill vehicles. They would operate in the infrared and microwave regions of the electromagnetic spectrum: in the boost phase, the short-wave infrared (SWIR) region of a few microns (where the boosters radiate most of their energy) would be used. Longer wavelengths would be employed in later phases, where targets (post-boost vehicles (PBVs), reentry vehicles (RVs)) would be cooler. In the late midcourse phase, radars might also be used for tracking. The system might attempt discrimination between warheads and decoys both by passive infrared sensing and by radar imaging.[3]

EFFECTIVENESS

The chief systemic problem with the types of space-based interceptors being suggested is the time available for intercept. Given the velocities these rockets are likely to achieve, and given the altitudes at which they orbit, only the slowest long-range ballistic missiles will be threatened in the boost phase (before burnout). The defense would like to attack in the boost phase so that all RVs and decoys aboard one missile would be destroyed by one hit before dispersal. Also, boosters are easily visible and vulnerable then. An attack early (within ten or twenty seconds) in the post-boost phase would be nearly as good from the point of view of leverage, but the post-boost vehicle would be more difficult to find. Further delays would permit at least some of the decoys and RVs to escape the interceptor. Interception is progressively less effective as the post-boost phase progresses.

I would like to point out a few simple and rough (but correct) observations based on first principles. If the satellites carrying the interceptors are at 500 km (taken from the lowest--most favorable from the timeline point of view--altitude given in the Marshall Institute Report,[4] a publication advocating an early deployment of this sort), and a missile burns out at 200 km, the best that a (again, this is a Marshall Institute Report specification) 6 km/sec interceptor can do is to descend in 50 seconds to the altitude of burnout. Actually, if the interceptor thrusts straight down, it will wind up traveling a slightly longer distance than a straight line, coasting along an ellipse, which means that it would take somewhat more than 50 seconds. After all,

$$v\,T = d, \qquad (1)$$

where v is velocity, T time elapsed, and d distance. This, however, assumes instant acceleration, which is clearly not the case. For realistic, finite acceleration,

$$d = 1/2\ a\ t_a^2 + vt \qquad (2)$$

where a is acceleration, t_a is time of acceleration, and t is the coasting time. The total time flown is

$$T = t_a + t. \quad (3)$$

The first term, representing the distance traveled before attaining full velocity, adds some 10 seconds time, given reasonably optimistic assumptions: a ballpark number, 30 g acceleration (a relatively fast-accelerating rocket), is used:

$$t_a = v/a = \frac{6 \text{ km/s}}{0.3 \text{ km/s}^2} = 20 \text{ s} \quad (4)$$

Then, for the first 20 seconds of flight, the interceptor travels, on the average, at half its final velocity, adding 10 seconds (over the mythical instant acceleration case) to its total time of flight.[5] If the acceleration were slower, say 15 g, the added time would be doubled to 20 seconds.

In addition, note that Gen. Abrahamson, in testimony before the Senate Defense Appropriations Subcommittee in March 1987, remarked that it would probably take the system about 70 seconds to respond (including time for the ICBM to clear clouds). That would leave about 130 seconds, in the best of cases, from ballistic missile launch to intercept. For these parameters, then, ballistic missiles with a burnout time of 130 seconds (or less) would be immune to boost phase attack from such a system. Note that the situation does not improve if one increases the number of interceptors.

Actually, of course, the situation is worse: it will be very rare that the interceptor will be located so that it will only have to accelerate straight down in order to reach its target. It will, in general, have to accelerate to the side and either forwards or backwards as well.

To discover the size of this effect, we need to know something about the disposition of the "constellation" of space-based interceptors in the defensive system. In his March 1987 testimony, Gen. Abrahamson referred to a possible architecture for the "Phase One" system that contained 3000 interceptors on 300 carrier vehicles. Let us assume these numbers and take 12 rings of 25 vehicles each in near-polar orbit, providing relatively even north-south v. east-west spacing along the equator.[6] An interceptor will then, on the average, have to travel 300 km down, but also 400 km ahead or backwards and 200 km to the side, resulting in 540 km travel distance. That means, for example, that on the average, the 6 km/sec vehicle would take around 100 seconds to reach its target or about 170 seconds after ICBM launch: about half the vehicles that were fired would then arrive after this period of time.

The SDIO report to Congress in 1987[7] noted that current ICBMs have 150-300 second burn times, so one deduces that some current ICBMs would be highly resistant to such a system. Future ICBMs, which the defense system would have to face some ten years from now, could be made to burn out somewhat faster, stressing the defense even more.

One can argue, however, that the system would then be effective in the post-boost phase. On the other hand, though, the post-boost phase for single warhead missiles, such as the solid-fueled SS-25, is rather short, so serious problems would remain for the defense. It would then be interesting to compare the costs of, say, 3000 SS-25s with the cost of erecting the proposed Phase One system, now estimated at up to $150 billion.

ADAPTIVE PREFERENTIAL DEFENSE

Even assuming that many Soviet missiles would be slower than the fast end of the distribution cited above (150 s), and thus susceptible to attack by the system, the effectiveness of a 3000 interceptor system would not be high. The absentee ratio (ratio of interceptors to those actually in position to attack the missiles) might be on the order of 10-30. meaning that, at most, 300 interceptors would attack up to 1400 incoming ICBMs (assuming that the missile inventory is about what it is today and that all are launched in a single salvo) and the total effectiveness for the boost and post-boost phase intercepts could not be more than about 20%. Gen. Abrahamson indicated in his testimony that the calculated effectiveness of the whole system (including the late midcourse layer) was classified, but greater than 10%, so that gives us some idea.[8]

Nevertheless, a strategic concept has been advanced to show that even a system that is not highly effective could be strategically useful. It is called "adaptive preferential defense." The idea is to arrange the targeting of incoming RVs so that at least a fraction of previously-defined highly-valued targets in the U.S. can be guaranteed survival. One doesn't know in advance which ones will surive, just that a certain fraction will.

Remember that there are two layers to the Phase One SDI system: the space-based interceptors (SBIs) that attack in the boost and post-boost phase and the ground-based ERIS interceptors that attack surviving RVs in late mid-course.

The idea is the following. After the first layer has operated, one looks to see where the remaining RVs are going (by the way, one needs an excellent tracking system to do this), and attacks only those that are headed towards the targets one wishes to defend. Those targets with only one RV headed their way are defended first, then those with 2 RVs, etc., until the defense runs out of interceptors.

There are, however, some difficulties with such a system besides the stringent requirement for extremely capable battle management software. One is that since the number of ERIS interceptors is limited, the problem of discriminating decoys from RVs would have be resolved (more on this later). Second, if the offense used MaRVs (maneuverable reentry vehicles), it couldn't know in advance which U.S. targets were under attack (the MaRVs change course radically upon reentry), so the system wouldn't work. Finally, if the probability of kill of the ground-based interceptors is not very high (well above 0.8), it won't work either.

WEIGHT-TO-ORBIT

The following discussion is intended to provide a framework for understanding the weight-to-orbit requirements of a Phase One system, and to provide the various uncertainties that are presumably the subject of research and development. Assume an optimistic specific impulse, I_{sp} of 300 sec for the interceptor, a mass fraction (ratio of fuel to total weight of a stage) of 0.85, and three stages with identical mass fractions. How heavy would the interceptor have to be? The rocket equation says that

$$v = I_{sp} \quad g \ln(M_i/M_f) \tag{5}$$

for each stage, where I_{sp} is specific impulse, g is gravitational acceleration, and the M's are initial and final stage masses, respectively.

We define the mass fraction of a stage as:

$$f = \frac{M_{shell}}{M_{shell} + M_{fuel}} \tag{6}$$

Then, for a given stage,

$$M_i = M_{rocket} + M_{payload} = M_f \, e^{(v/I_{sp} \, g)} =$$

$$[(1-f)M_{rocket} + M_{payload}] \, e^{(v/I_{sp} \, g)} \tag{7}$$

For convenience, we define Z to be $e^{(v/I_{sp} \, g)}$, the mass ratio. Now,

$$M_{rocket} = \frac{M_{payload}\,(Z-1)}{[1 - (1-f)Z]} \tag{8}$$

and

$$M_i = [(1-f)\,M_{rocket} + M_{payload}]\;Z = \frac{M_{payload}\;Z\;f}{[1 - (1-f)Z]}\,. \tag{9}$$

Here, the $M_{payload}$ of a stage is the total mass of the following stages, including the final payload of the vehicle. Then, assuming identical mass fractions, mass ratios, and I_{sp}'s for each stage, the total mass of the rocket is, by recursion,

$$M = \frac{(Z\;f)^N\;\;M_{payload}}{[(1-Z(1-f)]^N} \tag{10}$$

where N is the number of stages and $M_{payload}$ is now the payload of the final stage, that is, the final payload of the vehicle.

For a 6 km/sec rocket, one might take 2 km/sec for each stage, and calculate the mass of the interceptor, given assumed numbers for the mass of the kill vehicle, which is the payload of the last stage. Using the assumed numbers, Z = 1.97. Then,

$$M = M_{payload} * 13.5 \tag{11}$$

For a one kg payload, the rocket needs a mass of only 13.5 kg; a 5 kg payload would have a mass of 68 kg.

The total mass will be very sensitive to mass fractions and specific impulses, as the reader can verify for him/herself. If optimistic goals are achieved, an interceptor below 100 kg might be possible, which would be quite impressive. Three thousand interceptors would have a mass of some 200,000 kg; multiply by 2 (as an estimate to account for the carrier vehicle mass), and one gets 400,000 kg in orbit--nearly a million pounds, and about 30 shuttle loads, assuming that one could launch the shuttle from Vandenberg (to provide coverage against submarine-launched missiles, the interceptors must be in near-polar orbit).

Note that the kill vehicle would have to have communications equipment, homing sensors, and a divert rocket, so it is a real challenge to package all of this into a few kg. Note also that the mass of the whole system is roughly proportional to the mass of the kill vehicle.

As an excursion, note that if mass fractions were only 0.75, equation (11) would be

$$M = M_{payload} * 25.0 \tag{12}$$

and the total weight would nearly double. If, in addition, I_{sp} were only 270 sec, one would get

$$M = M_{payload} * 39.8. \tag{13}$$

Therefore, the weight of the entire system is heavily dependent, not only on the achievable weight for the kill vehicle, or warhead, but even more so on the characteristics of the small rocket on which the vehicle will be carried. Probably on the basis of these considerations, the SDIO apparently is allowing for 3200 kg/carrier vehicle, or 960,000 kg to orbit.[9] For comparison, before the shuttle disaster, the U.S. launched about 250,000 kg/year into low Earth orbit.

SPACE TRANSPORTATION

Without the shuttle or a new launch system (the planned one is called the Advanced Launch System with a proposed payload hovering around 50,000 kg to polar orbit),it would take inordinately long to deploy the system in space. Table I shows the current inventory (and ordered units) of Atlas's, Deltas, Delta 2s, Titan 34Ds, future Titan 4s, and the old decommissioned Titan 2 ICBMs.

Table I

Current and Planned U.S. Launcher Inventories

Missile Type	Number in Inventory or on Order	Thousands of Kg to Near-Polar Low-Earth Orbit per Launcher	Total Thousands of Kg to Low-Earth Orbit
Atlas	13	5.5	67.5
Delta	8	2.7	21.6
Delta 2	7	3-3.8	21-26.6
Titan 2	60-70	1.9	114-133
Titan 34D	6	15	90
Titan 4	23	18	414
		Total	728-753

The total would fall at least 150,000 kg short of the the required payload capability; further, it would take about 100 launches--one per week over two years--to launch the entire inventory. This is theoretically possible, but unrealistic. For one thing, except for the insignificant Titan 2s, the other rockets are already committed to other payloads, mostly for national security missions. Moreover, for major rockets, it often takes around four months to refurbish launch pads. Also, all would have to be launched from Vandenberg, otherwise the payloads would be greatly reduced if a polar orbit had to be achieved by planar changes from the orbits attained at Cape Canaveral. Only two current pads could be used for the heavier vehicles. One could build more pads at Vandenberg, but there is hardly any more room--further, a pad takes over 6 years to build. Realistically, the 100 launches would take many years.

However, about 30 shuttle launches or 18 ALS launches would suffice. A realistic estimate of the time needed for launch using the ALS would be about 3-4 years. If the ALS were not available, the system could not be deployed in a reasonable amount of time.

SENSOR REQUIREMENTS

The launch of ballistic missiles would be detected by SWIR focal plane arrays, located aboard the BSTS satellites. These would be in high orbits and would have quite large numbers of elements in their arrays. Sensors at lower altitudes would generate more precise track information to handoff to the kill vehicles of the SBIs and of ERIS. These might be located on ground-launched probes, carrier vehicles or elsewhere. Manufacturing and deploying sensors capable of performing the necessary tracking and pointing appears challenging but feasible. Some discrimination capability in the midcourse phase is hoped for from passive infrared sensing and, during the terminal phase, from radar. It is not clear how successful the discrimination would be in the face of plausible decoys, not to mention chaff and jamming possibilities.

DISCRIMINATION

In the midcourse phase, there will be a considerable amount of debris and, probably, decoys that the defense will have to separate from the real targets--the RVs. The hope apparently is to use infrared emissions to detect differences in temperature and emissivity in order to separate the more massive (actually, thermally massive) RVs from the necessarily lighter decoys. Also, if imaging radars could observe mass-dependent differences in the motions of the bodies, this might serve to aid discrimination. It is currently highly questionable whether such methods would be useful in the time frame of the mid-1990s, given the possibilities for offensive countermeasures, such as decoys or radiofrequency jammers. We discuss some of these point further below.

COUNTERMEASURES

FASTER-BURNING BOOSTERS; FASTER DISPENSING POST-BOOST VEHICLES.

One does not need a super 50-80 second booster to cause the system to deteriorate significantly. As noted, boosters only slightly faster than current ones would remain relatively immune in the boost phase.

The Soviets have replaced their missile fleet over the last 15 years at the cost of about $120 million. They might be expected to continue modernization, including the introduction of boosters with shorter burn times, if the U.S. were to head in the direction of a missile defense. We mentioned above that the solid-fueled SS-25 would be relatively immune to the proposed system, even in the post-boost phase, since it has only one warhead. A slightly faster follow-on could be totally immune. How much would it cost to deploy a few thousand of these objects? Estimates vary from $40 million to $90 million per missile. The latter numbers come from the assumption that they would all be deployed in a mobile mode, in which case the carry vehicles, and operation and maintenance would be very expensive.

The SDI system is now estimated to cost up to $150 billion--the estimate more than doubled over the past year. For this sum, the Soviets could produce thousands of new SS-25s, each of which might not be harmed by the boost-phase defense. It is worth remarking that this tactic would use currently available technology, whereas the defense system would use projected near-future capabilities.

More complicated would be the option of faster-dispensing post-boost vehicles. These would probably exact a penalty in payload, but multiple warheads could still be placed on a single missile, resulting in some economy relative to single-warhead missiles. This technology is somewhat further away than that for faster-burning boosters, and it is an open question whether such equipment would be ready in time to face a Phase One SDI system.

DECOYS, CHAFF, JAMMING

Many anti-sensor countermeasures have been proposed. Prominently mentioned are ways of deploying decoys to look like RVs. In general, not only would one want to make decoys look like RVs but also to make RVs look like decoys. One openly-discussed suggestion has been to place aluminized mylar balloons around RVs and to deploy them with many empty balloons. It would be difficult to tell from infrared emissions which were which, particularly if some care had been taken to isolate the RV from the balloon. Many other suggestions along these and similar lines have been made. The offense's response could be to try to detect differences in motion of the empty from the full balloons.

One could try to use aerosols to surround RVs and decoys, concealing them from infrared observation and making the discrimination task harder. One could also use chaff similarly to confuse radars; and one could, in addition, try to jam radars by means of various well-known (and, perhaps, some lesser known) techniques.

It is no means certain whether the discrimination techniques that have been proposed in the short term, i.e., passive infrared emissions and doppler imaging radar, would be sufficient to detect simple and plausible anti-sensor countermeasures.

If radar were to be used for late midcourse sensing and discrimination, chaff and jamming would be obvious potential countermeasures. Chaff has been used for nearly half a century--a cloud of very light conductors might conceal an RV, a decoy, or nothing. Jamming has also been much studied. At distances of hundreds of km, a very small, weak jammer could confuse radar information. There are countermeasures to both chaff and jamming, of course. Both involve large efforts in signal processing. Unfortunately, there is no time to go into details here, and I will leave matters at that.

SURVIVABILITY

Survivability means the survivability of a <u>system</u> so that it can accomplish its mission--not the survivability of <u>one</u> particular asset. Virtually any space asset can be destroyed if enough resources are devoted to that end. What counts is whether the entire architecture can function through the loss of some of its elements.

The most discussed method in the short term for attacking defense assets in space is the direct ascent anti-satellite weapon. This is simply a rocket, such as the 15-year-old Soviet Galosh, that can rapidly reach the altitude of an orbiting satellite. It may be nuclear-tipped (DANASAT) or have a non-nuclear hit-to-kill weapon (DANNASAT). If the ASAT can reach the satellite before it can maneuver beyond a lethal radius, it does not have to be a homing device, but can simply operate under inertial guidance.

The defense can use various means to protect itself against such an attack. It can try to move out of the way, it can try to shield itself, it can try to destroy the attacking vehicle before it comes within deadly range (shootback), or it can try to deploy decoys of its own or try to confuse the offense's sensors in other ways. Many of these techniques can act in synergy--strengthen each other--when used in combination. Thus, maneuvering plus shielding are mutually reinforcing techniques. More maneuver capability means less shielding needed; more shielding means less maneuver required; double the maneuver capability and (due to the inverse square law) the required shielding against x-rays from a nuclear blast will be reduced by a factor of four.

The reason that a DANASAT is so serious a threat is that it inherently counters many of the techniques contemplated for self-defense, eliminating the anticipated synergy of several of them. A nuclear warhead may have a lethality radius large enough that maneuvering may no longer be a reasonable option. It may not need to home, so it won't have sensors to be confused. Further, it can deploy decoy warheads that may force the defense to shoot back at dummy warheads, thus exhausting this option. Decoy platforms must either be deployed in advance, giving the offense a chance to study them at leisure, or deployed on attack, in which case they may not get far enough from the original satellite in order to be relevant.

To illustrate the situation, take the following case. Assume that one has a 1-megaton warhead on a rocket that can reach 500 km in 250 seconds. Since some ICBMs can burn out in 150 seconds and are then traveling at some 7-8 km/sec, we can calculate that one can build a rocket that has an average acceleration of $v/t = 53$ m/sec^2 or 5.4 g at an angle of 20-30 degrees to the horizontal. (In fact, some missiles, for example, anti-ballistic missiles, can accelerate at much higher rates.) If one shoots straight up instead, the force of gravity will act directly against the applied acceleration, so we'll assume that one can accelerate vertically at 4.4 g (neglecting the reduction in the Earth's field at these altitudes). To reach 500 km, note that

$$t = (2s/a)^{1/2} \tag{14}$$

and then,

$$t = (2 \times 5 \times 10^5/43)^{1/2} = 153 \text{ s.} \tag{15}$$

Actually, the ASAT will not only rise but will travel down range as well. Some 250 seconds would be needed for this direct ascent rocket to achieve a reasonable range of 700 or more km;[9] for less range, less time is needed.

In space, a megaton warhead puts out about 10^{15} calories, mostly as soft x-rays; at 100 km, the fluence reaches about 0.8 cal/cm^2. The interplay between shielding and maneuverability is clear. We can ask the following question: if a defense platform can maneuver at, say, 0.5 km/sec and can accelerate at 1 g, how far will it travel in 250 sec?

Note that,

$$v/a = t = 51 \text{ seconds acceleration time;} \qquad (16)$$

$$s = 1/2 * a * t^2 = 1/2 * 9.8 * 51^2 = 13 \text{ km;} \qquad (17)$$

so the platform moves 13 km during acceleration; then, it coasts for 199 sec at 0.5 km/sec = 100 km, or 113 km total. Therefore, against this sort of threat, shielding of about 0.6 cal/cm^2 would be sufficient.

But wait, I forgot an interesting point: the target will not maneuver right away: it needs time to respond. If this time is 70 seconds, as for the SBIs, it only has 129 s, not 199 s, to coast away; the distance from the detonation is only 77.5 km, and the fluence is $(112.5/77.5)^2 = 2.1$ times greater; so the shielding needs to be over twice as strong.

Figure 1 shows a plot of the energy fluence at the target as a function of the distance of the target from the detonation, and figure 2 shows the fluence as a function of the velocity change attained during the 250 s evasive maneuver, assuming a 1 g acceleration, for three nuclear explosive yields.

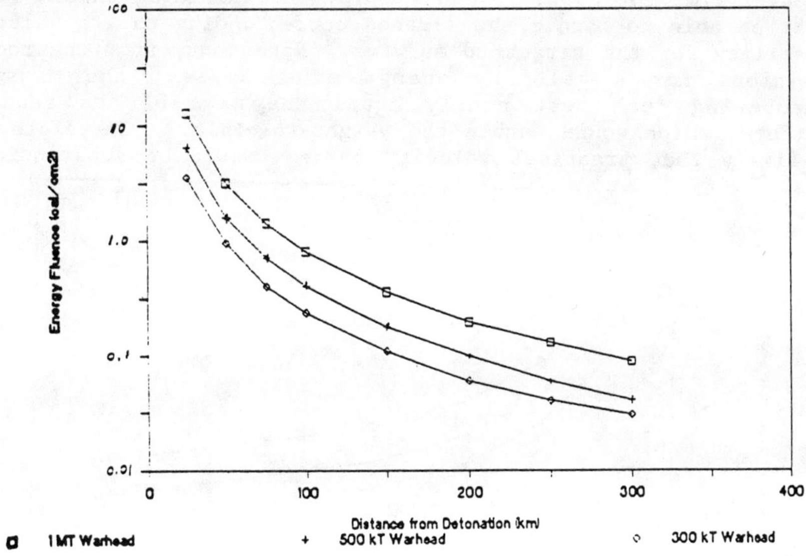

Figure 1. Energy Fluence v. Distance from Detonation for Three Nuclear Yields.

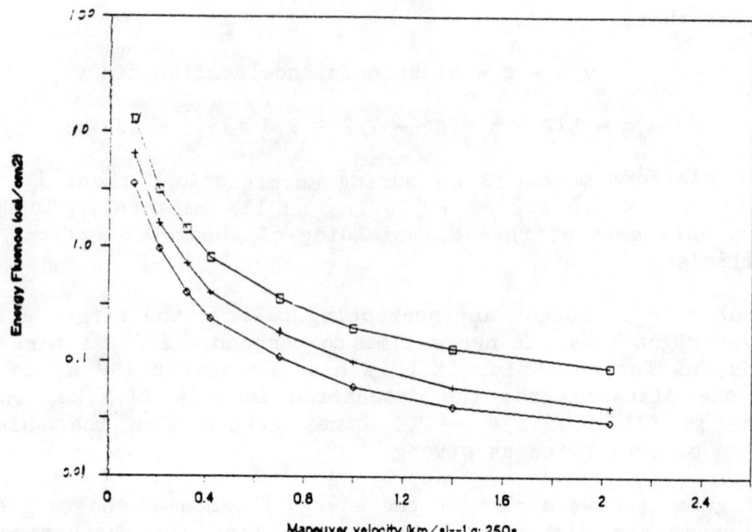

Figure 2. Fluence v. Maneuver Velocity: 250 s Maneuver Time

Figure 3 similarly shows the fluence as a function of velocity, but assuming a 70 s reaction time, which reduces the maneuver time to 180 s. In principle, the shielding should be at least be able to handle the fluence corresponding to its velocity capability for the target to survive. Note that, from the rocket equation, for a velocity change of 2 km/sec, the mass of maneuvering fuel must nearly equal the mass of the rest of platform, which would double the weight-to-orbit. Therefore, it is likely that practical velocity changes would be substantially less.

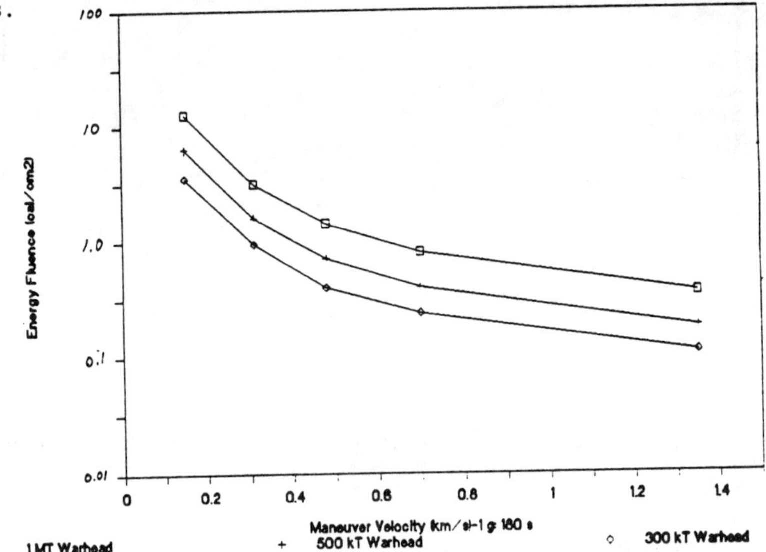

Figure 3. Fluence v. Maneuver Velocity: 250 s Maneuver Time

All one needs to do now to find the system survivability against this sort of threat is to put in the real numbers, run a simulation program, and make assumptions as to the robustness of practical shielding. I think you can see that there are enough parameters to adjust that one can frequently achieve the desired results, if one were so inclined.

Non-nuclear ASATs would have much smaller lethality radii and would have to have homing devices. Their success would depend on the abilities of sensors to resist electronic and electro-optic countermeasures.

OTHER THREATS

SPACE MINES

Space mines have frequently been mentioned as a possible threat to assets in space. A small, simple device, a space mine has an explosive (nuclear or non-nuclear) charge, and approaches a target to within a lethal radius, at which point it may destroy the target on command, by pre-programmed instruction, or upon an attack on itself.

The defense will presumably try to destroy the mine before it gets to within this range. Thus, "keep-out" zones might be declared and, in a simplified scenario, if any object gets within a the volume of space around the defended satellite, it is destroyed. Of course, the question is, who destroys whom first? It is not clear which side would win...tactics and levels of sensor of technology would be determinant.

A mine might be predeployed and disguised, just waiting for the other side to deploy a space defense system, or it may be popped up upon need; in the first case, it may be harder for the defense to deal with, in the second, the offense does not have to worry as much about long-term reliability.

SPACE-BASED INTERCEPTORS

Of course, one could imagine space-based interceptors that are not quick enough to catch ballistic missiles before burnout, but are quite quick enough to catch lumbering defense satellites with limited maneuvering capability.

Here, we reach a conclusion that can be extended to many different technologies, a conclusion that is obvious and occasionally stated but not emphasized often enough: **anti-satellite technology will precede anti-missile technology.** There are some caveats and exceptions, but, generally, this makes sense: more time is available to attack satellites than boosters, particularly since satellites are in generally predictable orbits and cannot maneuver indefinitely.

The major objection to this observation is that satellites can be shielded better than boosters if one has enough time and launch capacity to place heavy shielding in orbit. This is more relevant in the case of defense against radiation from explosions and directed-energy weapons than in the case of hit-to-kill interceptors, however. The required weight-to-orbit would be significantly increased.

If it is true in general that anti-satellite defenses are easier and will become available earlier than anti-missile defenses, one could argue that whichever side wished to deploy a missile defense system in space would need to control its adversary's access to space first, else the adversary might, even with inferior technology, deny the ability to deploy the system.

SURVIVABILITY DURING DEPLOYMENT

Obviously, while a defense system is being deployed, it is less able to defend itself than when it is fully operational. Sensing equipment, shoot-back capability, and mutual defense possibilities are necessarily limited before full deployment. Particularly, if one were to assume that policy constraints would not prevent the use of nuclear explosives in time of peace (say, over the one's own territory), one side might be able to deny the other the possibility of deploying defensive space assets.

CONCLUSIONS

There are several problems with an early missile defense system, based on space-based interceptors. The first is that the space-based rockets will not be able to attack moderately fast-burning boosters in the boost phase, nor would they be very effective against current single warhead ICBMs. The second is that the needed weight-to-orbit requirements would exclude deployment for at least 8-10 years and probably significantly longer. The third is that the survivability of the system against straight-forward ground-based attacks might be a problem, depending on achievable shielding. And the fourth is that survivability is even more difficult to assure during the period of deployment.

The above problems concern the boost and post-boost phase defenses. In the late midcourse phase, there are also other problems, such as discriminating warheads from decoys and assuring that radars can function in the presence of jamming.

In general, anti-satellite capability will precede ballistic missile defense capability. This has interesting implications for the survivability of space-based elements of a missile defense system.

Overall, there are enough serious concerns to dictate a healthy skepticism regarding the capabilty of an early space-based missile defense system. This does not, of course, rule out the usefulness of more advanced missile defense systems, for example those based on directed-energy weapons. Those would have to be analyzed on their own merits.

REFERENCES AND FOOTNOTES

1 The opinions expressed in this article are those of the author alone and do not necessarily reflect the opinions of the Office of Technology Assessment.

2 These numbers were cited in testimony given by Lt. Gen. Abrahamson, Director of the Strategic Defense Initiative Organization, to the Senate Subcommittee on Defense Appropriations on March 19, 1987.

3 Six essential components of the proposed system were approved by the Defense Acquisition Board in September 1987 for demonstration and validation of concepts.

In current Department of Defense parlance, this was a "Milestone Ones" decision. If and when DoD feels that development is sufficiently advanced for full-scale engineering development, that decision would be called a "Milestone Two" decision. The "Milestone Three" decision would be a go-ahead for construction and deployment.

The six approved elements were: a space-based interceptor system, the Boost-Phase Surveillance and Tracking System (BSTS--sensors on high altitude satellites), Space Surveillance and Tracking System (SSTS--sensors on lower altitude satellites), Ground-Based Surveillance and Tracking System (GSTS--sensors on a "pop-up" rocket to be launched when needed), Exo-atmospheric Reentry vehicle Interceptor System (ERIS--ground-launched interceptor rockets), and a battle management system.

4 Report of the Technical Panel on Missile Defense in the 1990s, (George C. Marshall Institute, Washington, D.C., 1987). The members of the panel include the Chairman of the Science Advisory Panel to the Strategic Defense Initiative Organization (SDIO), (F. Seitz), another member of that Panel (W. Nierenberg), and a former high-ranking official in SDIO (J. Gardner), as well as a contractor for SDIO (E. Gerry). Although this work is not an official DoD publication or proposal, the positions of the authors may indicate a certain familiarity with the thinking of at least some players in SDIO.

5 Since $d = 300 \text{ km} = 1/2 * (0.3) * 20^2 + 6 * t$, $t = 50 - 10$, giving $T = 40 + 20 = 60$ s instead of $300/6 = 50$ s, which is the case for instantaneous acceleration.

6 In fact, the spacing should be made even around latitudes of 50° to 60° since that is where the Soviet missile fields are concentrated. In that case, a 10 x 30 or 9 x 33 constellation might be appropriate. The conclusions would not be significantly different, however.

7 Report to the Congress on the Strategic Defense Initiative, (Strategic Defense Initiative Organization, Washington, DC, 1987), p. A-5.

8 As reported in The Hartford Courant, March 20, 1987, p.1.

9 The Hartford Courant, ibid.

10 Assume about 200 seconds at an average velocity of 3 km/s along the surface of the Earth.

CHAPTER 11

THE SPACE-BASED INTERCEPTOR

CHRISTOPHER T. CUNNINGHAM
LAWRENCE LIVERMORE NATIONAL LABORATORY

ABSTRACT

We calculate the effectiveness and marginal cost-effectiveness of SBI defenses against a variety of possible Soviet threats. Defense effectiveness is expressed in terms of participation fraction, which depends on SBI flyout distance and threat basing, and in terms of the number of interceptors needed to accomplish a given mission. Optimal interceptor sizing favors large high velocity rockets. With such optimal design, the defense appears marginally cost-effective with a 50% kill criterion against current or responsively modified large liquid boosters, although not against more stressing threats. Survivability of direct attack is not assessed, although survivability enhancement options are briefly discussed.

I. Introduction

The concept of using space-based interceptors (SBI) in a ballistic missile defense is not new. Studies done in the late 1960's postulated very large interceptors due to the mass needed for sensors, guidance and control of the homing front end. Now it appears that technology available in the next decade, or perhaps today, will allow front-end masses of at most a few kilograms, making the concept much more viable. As a result, the SBI has become the centerpiece of recent work sponsored by the Strategic Defense Initiative Organization (SDIO). In June to September of 1987, a system concept relying on SBIs as the primary defense mechanism was reviewed by the Defense Acquisition Board (DAB), ultimately receiving approval to proceed with demonstration and validation experiments, the first formal step toward developing a military system. (While not precisely the system presented to the DAB, the defense in Ref. 1 is fairly representative of current concepts.) In April 1988, the Defense Science Board's SDI Milestone Board published a report[2] outlining the many preliminary steps which must be taken before such a defense could be put in place and emphasizing the early steps, such as the development of sensors and ground-based interceptors rather than SBIs. This view was reinforced in a June 1988 meeting of the Defense Acquisition Board. Therefore, at this writing, the future of the space-based interceptor is uncertain, and the program can be expected to have changed radically by the time this is read.

In order to be a viable concept, it is generally conceded (both inside and outside of SDIO) that the SBI, or any other defense element, must be 1) effective against the probable Soviet threat against which it is is deployed, 2) sufficiently survivable against direct attack to perform its primary mission, and 3) cost-effective at the margin with regard to incremental responses by the offense and counter-responses by the defense.

Accordingly in Secs. II-V we discuss the components of an SBI defense and its effectiveness. Sec. II presents several possibilities for a defense architecture. The architecture which was presented to the DAB relies on fire control assets associated with the carrier vehicle (CV) satellites, which contain the SBIs, and these satellites become very expensive as a result. Four alternative architectures, currently under study, are discussed; each presents a different approach to alleviating this problem.

In Sec. III we consider an important performance measure, the participation fraction or fraction of deployed SBIs which can enter the battlespace. The distance flown by an SBI in an engagement, relative to its carrier, is seen to be the primary determinant of this measure, although threat basing is also an important consideration.

In Sec. IV the inventory levels of SBIs required in specific missions against specific threats are presented as a function of the axial velocity with which an SBI leaves its carrier. A roughly linear dependence of inventory on axial velocity is noted.

Finally, in Sec. V the velocity required for midcourse diverts is assessed. Booster maneuvers, employed as a countermeasure, may require a divert capability of 2-3 km/s.

In Sec. VI we consider the survival of the defense to direct attack. The primary antisatellite (ASAT) threat we consider is a large nuclear warhead detonated at a pre-determined point in space (PINS), although a maneuvering ASAT is also considered. Against these threats, the defense has a variety of options: nine are discussed in the text. The primary options are to harden to reduce the lethal radius of the ASAT, to maneuver to prevent the ASAT from closing within the lethal radius, and to deploy redundant assets to compensate for loss. Survivability features of specific architectures are also discussed in Sec. II.

Finally, the issue of cost and cost effectiveness is addressed in Sec. VII. We begin by considering the trade between axial velocity and numbers of interceptors. For reasonable estimates of SBI production costs and launch costs, assuming a heavy lift capability, SBIs with axial velocities of around 10 km/s (and masses of about 550 kg) appear the more cost-effective. For such a design, launch costs associated with the SBIs are significant compared with production costs. Total production and launch costs of the SBI inventory required against specific threats begin at about $2B for 50% kill of 3000 RVs from current liquid boosters and increase dramatically against more stressing threats. Total life cycle cost of the SBI tier of the defense is about a factor of 10 greater. Assuming a $100M life-cycle booster cost, our estimate of marginal cost exchange is favorable for the defense for current or responsively modified large liquid boosters (about $7M per RV for 50% kill or for deployment), but becomes unfavorable for more stressing threats and missions (such as solid-fuel, medium MIRV missiles or a 90% kill criterion).

II. Defense Architectures

The baseline defense system presented to the Defense Acquisition Board in June 1987 is illustrated in Fig. 1. It consists of four space-based elements:

1) Boost Surveillance and Tracking System (BSTS)
   These are large satellites in near-geosynchronous or higher orbits responsible for launch detection and booster tracking.

2) Space Surveillance and Tracking System (SSTS)
   These large satellites in medium earth orbit track all threat objects and perform midcourse discrimination using passive radiometry.

3) Carrier Vehicles (CV)
   These satellites house the interceptors (typically about 10 SBIs per CV). In the baseline architecture, they process much of the BSTS and SSTS data, determine weapon assignments, track targets, and issue fire control commands. An exemplary carrier vehicle is illustrated in Fig. 2.

4) Space-Based Interceptors (SBI)
   A typical three-stage SBI is shown in Fig. 3. The first stage propels the rocket away from its carrier toward its target at a large velocity. A solid propellant is used to give large specific impulse and mass fraction. The second stage must make midcourse corrections following fire control commands from the CV. It is therefore a liquid stage. The kill vehicle itself tracks the target using an onboard sensor and diverts to intercept. The sensor must be capable of seeing the targets, booster and PBV hard-bodies or RVs: passive, semi-active, and active schemes have all been proposed.

In addition, a ground-launched interceptor, the Exoatmospheric Reentry-Vehicle Intercept System (ERIS), and a ground-launched sensor similar to SSTS, the Ground Surveillance and Tracking System (GSTS) are part of the architecture. GSTS may be augmented by ground radars.

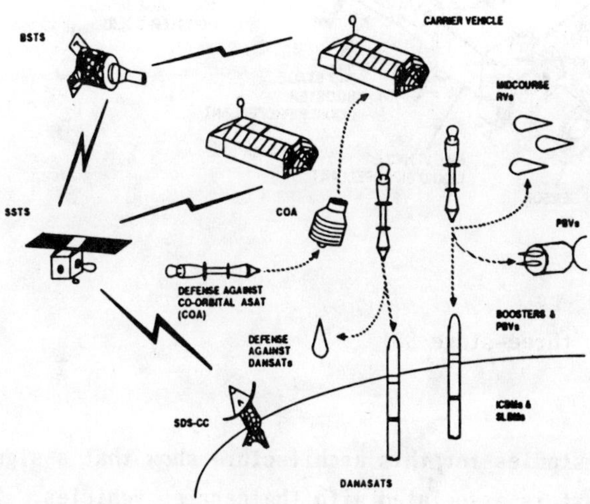

Figure 1. The baseline defense system presented to the Defense Acquisition Board

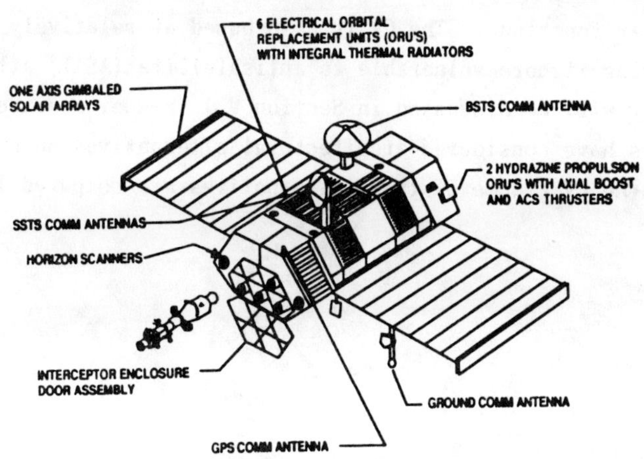

Figure 2. An exemplary carrier vehicle

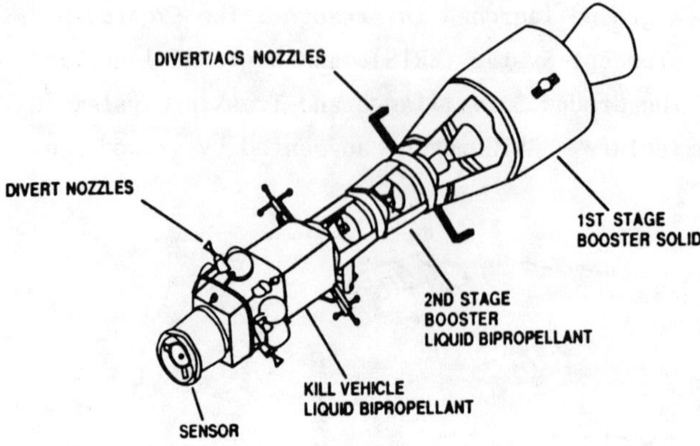

Figure 3. A typical three-stage SBI

Detailed design studies for this architecture show that a significant proportion of its cost is associated with the carrier vehicles. In order to perform their fire control functions against cold targets at thousands of kilometers, they must have capable and expensive sensor, attitude control, communications and power subsystems. Furthermore, there are a few hundred of them, many more than the SSTS and BSTS satellites which perform similar functions. The CV is also based at relatively low altitudes making it more vulnerable to antisatellite (ASAT) attack. (Survivability will be discussed in Section V.) Hence, a number of recent studies have considered architectural alternatives which simplify or eliminate the CV. Some of these alternatives are compared in Table I.

Table I. Alternative Architectures

Function	Baseline	Option 1	Option 2	Option 3	Option 4
Launch detection	BSTS	BSTS	BSTS	WCP	SBI
Booster track	BSTS, SSTS	BSTS, SSTS	BSTS	WCP	SBI
PBV track	SSTS	SSTS	(SSTS, SBI)	WCP	SBI
Midcourse track & discrimination	SSTS	SSTS	(SSTS)		
ASAT track	SSTS	SSTS	(SSTS)		
Battle management	CV	SSTS	BSTS	WCP	SBI
Fire control	CV, SBI	SSTS, SBI	BSTS, SBI	WCP	SBI

In Option 1, the CV's fire-control function is taken over by SSTS. This approach preserves all the capability of the defense and should lead to a significant cost reduction. However, it is now completely dependent upon SSTS; loss of a few platforms to ASAT attack would lead to catastrophic failure.

Option 2 reduces concern about SSTS arrival by reducing reliance on SSTS. Here BSTS performs battle management and fire control in lieu of the CV and communicates directly with the interceptors. The survival of BSTS is critical, as in any architecture which relies on BSTS for attack warning. (However, BSTS has considerable inherent survivability by virtue of its high orbit, making it hard to find and requiring a long time for a ground-launched intercept.) If SSTS fails, the capabilities to intercept RVs and certain types of ASATs are lost (although individual SBIs can probably be protected from attack through dispersal (see Sec. VI.)) In this event, PBVs are not tracked per se, but are reacquired by the SBI during the end-game.

Option 3 is an architecture in which both CVs and SBIs are greatly simplified. The SBIs are here command-guided all the way to intercept by sensors on weapon control platforms (WCP). Thus, the kill vehicles exchange their sensors for much cheaper and lighter radio or optical receivers. The stressing elements for this architecture are the sensors on the WCPs which must track targets accurately enough to perform total command guidance. As very many WCPs are proposed for survivability, it would seem that this approach only exacerbates the problem of expensive fire control.

A completely different strategy ("Brilliant Pebbles") is proposed by Lowell Wood at the Lawrence Livermore National Laboratory (Option 4). Here, it is the SBIs themselves, as independent satellites, which acquire the fire control function, with launch detection, track and battle management as well. Thus, the interceptors become the entire architecture, except for the communications required for status checking and wartime enablement. Such a high degree of proliferation and autonomy gives a survivability advantage, and reliance on either BSTS or SSTS is eliminated. However, as this architecture places greater demands on the highly proliferated SBI sensor(s) to perform surveillance and fire control, it could tend to drive up the cost of the SBIs.

III. Defense Participation

Provided that the defense architecture fulfills the functions given in the previous section (launch detection, battle management, fire control, etc.), its performance will depend primarily on characteristics of the SBIs themselves: numbers and velocities; their basing, altitude and inclination; the time delay for warning, battle management, and SBI launch; and the threat, its geographic distribution and temporal characteristics of launch, boost and PBV operation.

The major determinant of performance is the availability of SBIs in the battle. This depends on their basing relative to the threat and on the kinematics of their flight. During flyout to boosters and PBVs (as opposed to RVs), gravitational accelerations on the SBI and its parent CV are roughly equivalent. This implies that, as seen by the CV, the SBI flies on a straight line with constant velocity. Its distance from the CV is thus, simply

$$D = V(t - V/2a) \qquad (1)$$

with V the relative velocity between CV and SBI, t the flight time, and a the acceleration during SBI boost. A typical distance might be 1200 km at 6 km/s for a 200 s flight.

To approximate gravitational effects, assume that the SBI remains at roughly constant altitude during its flight, so that the gravitational acceleration is of constant magnitude. Then trajectories are analytically quite simple:

$$\vec{X}(t) = \vec{a} \sin \omega t + \vec{b} \cos \omega t \qquad (2)$$

where $\vec{a}$ and $\vec{b}$ are constants and ω is the angular frequency of a circular orbit at the reference altitude. Thus the corrected distance flown from the platform is

$$D = (V/\omega) \sin \omega(t - V/2a) \qquad (3)$$

The correction relative to (1) is small for $\omega t \lesssim 1$ or $t \lesssim 800$ s. Thus, it need be applied only for midcourse intercepts.

As well as flying the required distance to its target, the SBI must remain above some minimum altitude, probably about 100 km, to avoid heating its sensor. Assume that the CV and the target are at roughly comparable altitudes, then the point of extremal altitude will occur halfway through the flight. In the limit of short flyout distance D, the change in altitude from the CV to the extremal point is, from (2),

$$\Delta h = - (1 + 2 \cos\theta V_o/V)D^2/8R_o \qquad (4)$$

where θ is the angle of the SBI relative velocity $\vec{V}$ with the CV orbital velocity $\vec{V}_o$ and R_o is the radius of the CV orbit. Assume CV and target altitudes of 500 km, a minimum altitude of 100 km, an SBI axial velocity of 6 km/s, and the worst-case $\theta = 0$, then D<2400 km or t<400 s. Thus, the minimum altitude constraint may be important for PBV intercepts.

Sensor line-of-sight constraints may also be important for PBV and RV intercepts. The line-of-sight from A to B will remain above some minimum radius R_{min} from the center of the earth provided that the distance between them is less than

$$D_{los} = (R_A^2 - R_{min}^2)^{1/2} + (R_B^2 - R_{min}^2)^{1/2} \qquad (5)$$

where R_A and R_B are distances from the center of the earth. For cold objects viewed in the LWIR, a 50 km limb altitude (R_{min} = 6421 km) is appropriate. The maximum l.o.s. distance to a PBV at 500 km altitude is about 5000 km. For architectures like the baseline in which the CV must track the target throughout the engagement, this limits the maximum distance of the SBI from the carrier. A more serious limitation arises if the CV must have line-of-sight at the moment of SBI release, as in the baseline and Option 4 architectures. Suppose that release is desired 60 sec after an SS-18 booster breaks clouds at 10 km. At this time it is at about 50 km altitude and the maximum distance to a 500 km CV is about 3400 km for line-of-sight above ground. Since a south-moving CV with a 6 km/s SBI could attack the booster itself from a distance of about 3500 km at SBI launch, the effect of the constraint for PBV intercept will be significant.

Figure 4 plots the positions at the moment of threat launch of carrier vehicles in a polar constellation at 500 km altitude, which can deliver interceptors with 6 km/s axial velocity and a 60 s launch delay to an SS-18-like target launched from 45°N, 90°E. The plot shows CV positions in the Northern Hemisphere only.

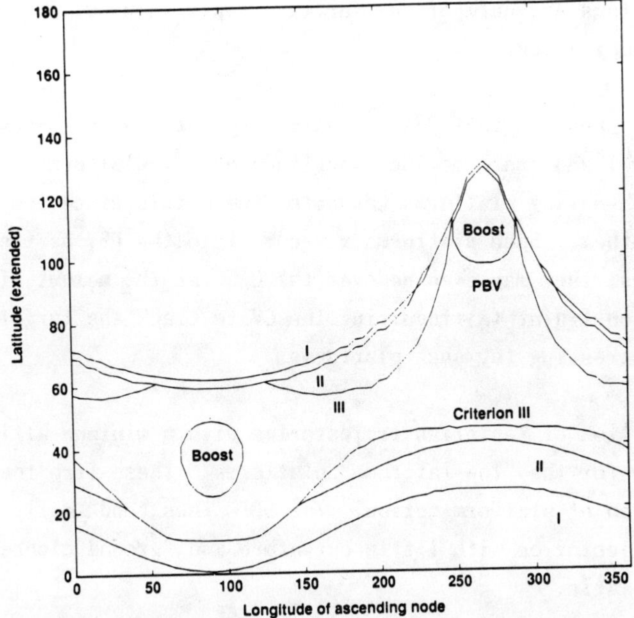

Figure 4. Carrier vehicle position at threat launch for intercept of an SS-18 (see text).

Platforms with longitudes of the ascending node less than 180° are north-moving in the Eastern Hemisphere and south-moving in the Western Hemisphere. The latitude scale is extended to measure angle from the ascending node, so that 180° "latitude" denotes a south-moving platform at the equator. Contours show the position at the moment of booster launch of platforms which can intercept boosters (up to 300 s after launch) and PBVs with at least one RV remaining (up to 705 s after launch). Separate contours are plotted (for PBV intercept) according to whether the intercept capability is calculated using the criteria

1) simple distance from CV to target at intercept, by eq. 1.
2) the assumption of a Keplerian trajectory for the SBI with a minimum altitude above 100 km.
3) Keplerian trajectories with the requirement that the CV be able to track the target during the SBI flight.

These distinctions are unimportant for intercepts in boost phase; eq. 1 is a good approximation.

The figure shows a great deal of asymmetry between north-moving (longitude ~90^0) and south-moving (longitude ~270^0) platforms: many more south-moving platforms can enter the battle since the threat is flying toward them. Such platforms may come into the PBV battle from large distances; they may even be over the U.S. at the moment of threat launch. The constraint (3) requiring the CV to track the target is particularly stressing for such platforms.

The imposition of Keplerian trajectories with a minimum altitude is most stressing for the "low-latitude" platforms. These fire their SBIs in the direction of platform motion. The SBIs thus tend to fly hyperbolic trajectories with little curvature and "ground clobber" becomes problematic.

Platforms in a polar constellation will be uniformly distributed in the coordinate system of Fig. 4. Thus, the area enclosed by the curves will be proportional to the fraction of the constellation which participates in the boost or PBV battle. From this we determine a participation of 1.1% in boost phase and 18.1%, 14.1% and 10.8% in PBV phase, according to restrictions 1), 2), 3), respectively, for a 6 km/s SBI against a single SS-18.

As a simple approximation for these results, we note that at the moment of intercept, participating CVs may be no farther from the target than the maximum interceptor flyout distance, D. This gives for the participation

$$f = (D/R_0)^2/4 \tag{6}$$

assuming a uniform (rather than polar) distribution of CVs on a sphere of radius R_0. This simple formula predicts 1.1% boost and 7.9% PBV participation for the above example. This estimate for PBV participation is lower than observed in Fig. 4 due primarily to the assumption of a uniform, rather than polar, constellation (the polar constellation is concentrated in the vicinity of the late PBV intercepts); and since it estimates late PBV intercepts only (excluding some CVs which can participate, but only earlier).

The above example considers intercept of a single booster only. For the attack of many boosters, it is valid only if the launch region is small compared to the flyout distance. A more extended distribution of launches allows more CVs to participate. Fig. 5 shows the current basing of Soviet missiles. While the SS-11's, which are being retired, are based across the Soviet Union, the more modern missiles, SS-18s and SS-19s, are rather concentrated in regions about 1000 km in extent.

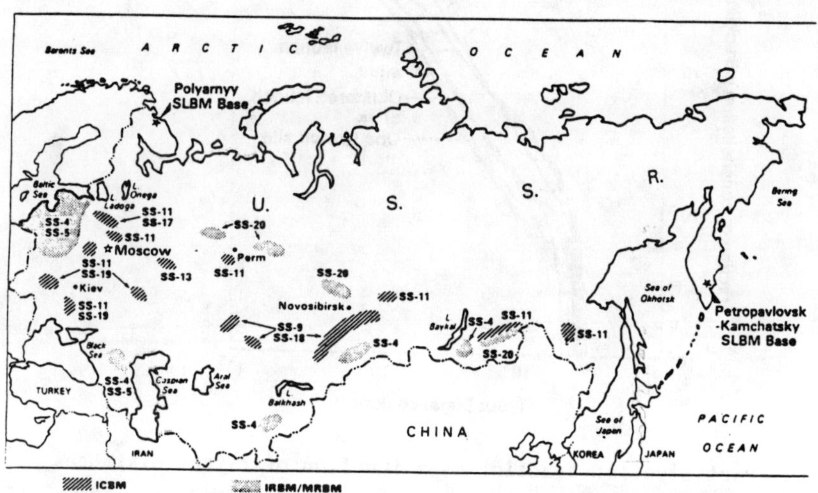

Figure 5. Current basing of Soviet missiles

Thus, three basing assumptions are of interest:

1) Launch complexes spread across the Soviet Union.
2) Complexes concentrated within 1000 km (near SS-18s).
3) Complexes covering a small region (near SS-18s) compared to the flyout distance (~1000 km).

A number of kinematic simulations for each of the above basing assumptions were performed to determine the fraction of the constellation participating as a function of time in the attack as shown in Fig. 6. In these simulations, SBI trajectories were constrained to lie above 100 km, but CVs were not constrained to track targets. The CV constellation was inclined at 60° to maximize participation early in the engagement (at the expense of late PBV intercepts). The axial velocity was varied in 1 km/s intervals between 5 and 10 km/s, demonstrating that it is the flyout distance which largely determines participation, rather than velocity or time-of-flight individually.

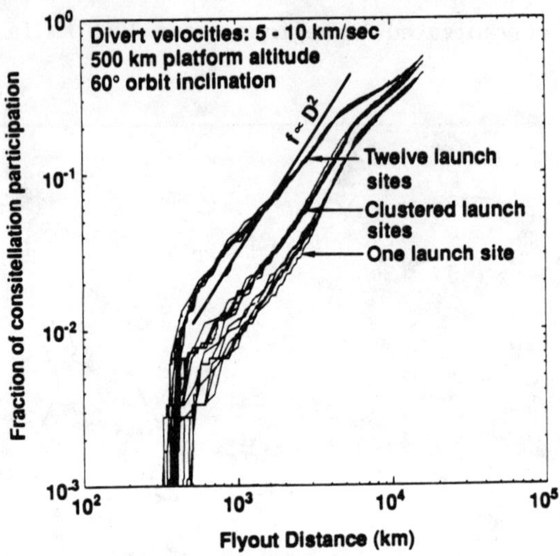

Figure 6. Participation fraction as a function of flyout distance

While eq. 6 predicts participation proportional to the square of the flyout distance, Fig. 6 shows a more nearly linear behavior for flyout distances less than about 2000 km: an empirical fit of its curves gives

$$f \sim \begin{cases} 0.32\ (D/R_0)^{1.1} & : \text{sites across Soviet Union} \\ 0.16\ (D/R_0)^{1.2} & : \text{1000 km region} \\ 0.14\ (D/R_0)^{1.4} & : \text{small region} \end{cases} \qquad (7)$$

The main reason for the difference from eq. 6 is that here the constellation is inclined at 60°, augmenting performance for short flyout distances; in addition, the greater the extent of the launch region, the weaker is the dependence on flyout distance.

IV. Performance against Representative Threats

The analysis of the previous section considered boost and deployment times typical of current heavy liquid boosters (like the SS-18). However, current solid boosters like the MX or SS-24 with fewer RVs on board could have shorter boost and deployment times. Modest near-term responsive modifications could cut these times further. Table II presents boost and deployment times for a representative collection of near-term threat options. (Most threat times are from Ref. 3, however, "modifications" are our own; see also Ref. 4.)

Table II. Threat Options and Participation*

		Boost		Deployment	
System	MIRV	Time(s)	f (%)	Time(s)	f (%)
Current liquid	15**	300	2.5	450	12.0
Modified liquid	15**	240	1.7	300	5.6
Current solid	10	180	1.1	300	4.8
Modified solid	10	150	0.8	240	3.6
Fast-bus solid	10	150	0.8	30	1.1
Fast-bus, fast burn	10	100	(0.3)***	30	(0.6)***

* 6 km/s interceptor, 60 s start delay, 1000 km launch region, 60° constellation

** A possible loading for a follow-on to the SS-18

*** Basing below 500 km required

Most stressing of these options is the fast bus system, which deploys all its warheads in 30 s. This capability requires that each warhead be deployed individually by its own deployment module (mini-PBV). Since each module must carry its own guidance and propulsion systems, either considerable additional cost must be incurred or new, cheaper guidance technology must be developed. Thus, this threat might be viewed as farther term than the "modifications".

A solid fuel booster with a boost time of less than 100 s is also a farther-term option. While not technologically stressing, it would require a completely new design and could entail a significant weight penalty. Defense performance was not evaluated against this very stressing option. Other types of defenses must be postulated.

Suppose 200 follow-on SS-18 boosters (3000 warheads) are to be intercepted in boost phase: then the participation fraction indicates that at least 8,000 interceptors would be needed, while only 1700 would be needed if they were intercepted in PBV phase.

For the modified solid, about 4.5 times as many SBIs would be required. (Participation is lower by a factor of about 3 and there are 1.5 times as many boosters to deploy the same number of RVs.) Of course, intercepts in boost phase kill all the RVs aboard while PBV intercept kill only partial loads. Therefore, a mission requirement that a high fraction of the threat be killed would not allow PBV intercepts.

We consider a START-constrained threat of 6000 warheads total; of these, about half might be ICBM warheads in a strategic attack. The SBI defense might be required to kill 50-90% of these. We have performed engagement simulations to determine the number of interceptors required to meet these goals against each threat type separately (with perfect interceptors, SSPk = 1). The 200 liquid boosters were based in current SS-18 silos; the 300 solid boosters were based in the western-most SS-11 and similar silos, a somewhat more advantageous basing for the defense than SS-18 silos. Results are given in Fig. 7 as a function of axial velocity.

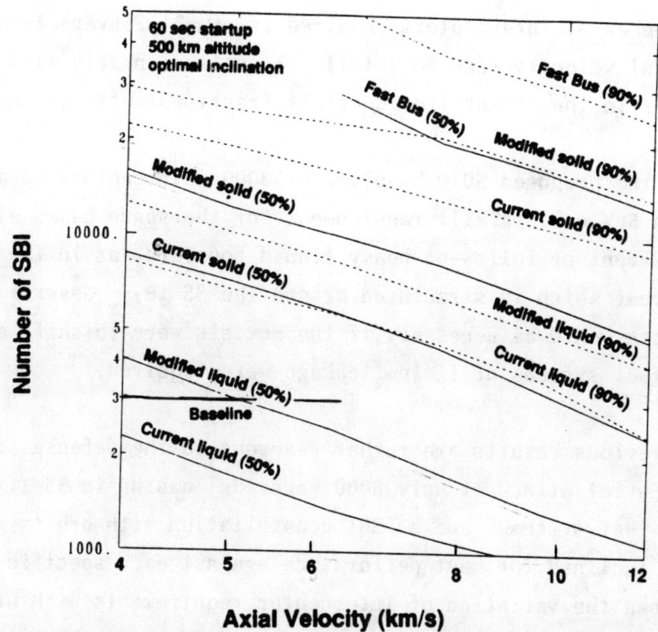

Figure 7. Number of SBIs required as a function of axial velocity against various threats for 50% and 90% kill criteria

For a 6 km/s velocity, the number of SBIs required is also tabulated below.

Table III. Interceptor Requirements (thousands) vs 3000 RV raid*

	50% Kill	90% Kill
Current liquid	1.5	6
Modified liquid	3	9
Current solid	6	16
Modified solid	10	23
Fast-bus solid	30	60

*** SSPk = 1, 6 km/s axial velocity, optimized orbits**

The number of interceptors required is roughly inversely proportional to the axial velocity (see Sec. III); it scales linearly with the number of boosters in the threat (for constant fractional effectiveness).

Thus, the proposed SDIO baseline of 3000 interceptors is appropriate only for a 50% or less kill requirement for the space based elements against current or follow-on heavy liquid boosters (as in the current Soviet threat which is structured around the SS-18). Several times more interceptors would be necessary if the Soviets were to shift decisively to solid-fuel systems or if low leakage were required.

The previous results are rather generous to the defense since they assume a Soviet attack of only 3000 warheads, basing in SS-11 silos for the solid-fuel systems, and an SBI constellation with orbits which are optimally inclined for best performance against each specific threat. Fig. 8 shows the variation of interceptor requirements with orbit inclination. It indicates that these requirements may be cut by more than half if the inclination is optimal against the threat.

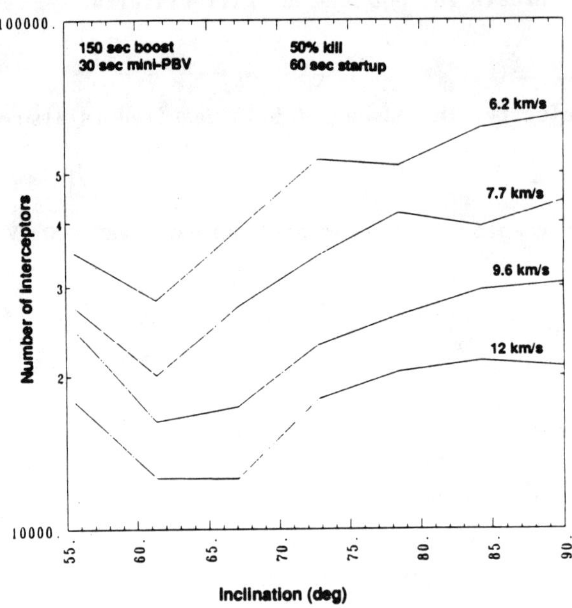

Figure 8. Variation of interceptor requirements with orbit inclination

The problem with such an assumption is that the inclination may not be changed after the constellation is once deployed, whereas a responsive threat may well be mobile. Thus, while some benefit from proper inclination is likely, it will probably not be as great as the factors of 2 shown here (and included in the estimates of Table III).

Basing at lower altitudes than our 500 km baseline or orbit modification to give a low perigee during periods of likely attack is often invoked as a means of dealing with stressing threats. A detailed study of this option[5] concludes that if the SBIs can fly at least 500 km or so during the engagement (and this is true for all the threats considered here), then basing between 400 and 500 km altitude is desired. Basing at lower altitudes is not cost-effective due to more stressing stationkeeping requirements.

V. Divert Requirements

After the SBI has burned its first stage thrusters and is flying toward its impact point, it will be necessary to periodically correct its heading based on subsequent information from external trackers or its own sensors. The magnitude of these corrections may be quite significant, especially to compensate for booster yaw maneuvers.

The amount of additional booster velocity Δv required for such a maneuver is only

$$\Delta v \sim (1-\cos\Delta\theta)/(1/v - 1/v_F) \tag{8}$$

where $\Delta\theta$ is the magnitude of the yaw, v is the booster velocity at the moment of yaw and v_F is the final booster velocity, about 7.2 km/s. For a large yaw of an SS-18, shifting the missile's target point across the U.S. (23^0, about the largest single yaw consistent with targeting), made 60 s after launch (when the instantaneous velocity is about 1.4 km/s), this velocity penalty is only 0.1 km/s, imposing a 1.4% payload penalty. On the other hand, since the booster is accelerating, the SBI would have to counter the yaw with a much greater velocity, up to about $v_F\Delta\theta$ or 2.8 km/s.

Thus, although the booster may yaw many times for little penalty, the SBI cannot respond to each motion of the projected intercept point without using impossibly large amounts of fuel. A clever strategy is needed. Even so, midcourse divert velocities of at least 2-3 km/s would appear to be required. This consideration might be used to size the second stage of a three-stage interceptor. Faster burning boosters are likely to be less problematic in this regard since the yaw (which must occur after the SBI's axial burn) will take place relatively later in boost phase, at a higher instantaneous velocity, and, hence, impose a greater penalty on the booster.

Additional divert capability is needed in the terminal phase of the intercept, during which the SBI navigates solely on its own. Studies at the Air Force's Rocket Propulsion Laboratory[6] show that this divert requirement can be made relatively small, even against boosters, varying between 0.2 km/s for 2 km initial error in impact point prediction to 0.4 km/s for 5 km error.

VI. Survivability

The defense system must be capable of functioning in a hostile environment against nuclear enhanced backgrounds and direct attack by nuclear weapons. Perhaps the simplest and most effective antisatellite (ASAT) concept is a large (say 1 Mt) warhead, detonated at some predetermined point in space (PINS) near its target, although homing concepts should also be considered. Against such threats the defense has a number of options:

i) Hardening - The defense must ensure that the lethal radius of a nuclear ASAT warhead is less than a few hundred kilometers, otherwise escape by maneuver is impossible and multiple kills per weapon become likely. For a 1 Mt burst this implies design against about 1 cal/cm^2 total fluence of bomb x-rays. Protection of sensitive surfaces, suppression and shielding of system-generated electromagnetic pulse, and hardening and shielding of electronics against penetrating ionizing radiation are required at this level of irradiation.

ii) Maneuver - Upon warning of ASAT launch, the asset under attack can try to run away to force a volume barrage by a PINS threat (a winning game for the defense). For a lethal radius of about 100 km, and 100-200 s for maneuver (from positions over the Soviet Union) the asset must have a maneuver velocity of 0.5-1 km/s, a significant but not overwhelming penalty.

iii) Proliferation - The more targets the more costly is an ASAT attack. This favors more CVs, each carrying fewer SBIs. The ultimate in proliferation is the singlet concept discussed in Section II.

iv) Dispersal - This tactic combines the above; upon warning, platforms under attack release their SBIs, which fly away to escape the ASAT threat. They are made capable of surviving for at least a few minutes on their own and of receiving targeting commands from an outside authority (not the CV, which is expendable).

v) Decoys - Upon attack warning, the CV can maneuver and deploy replicas of itself. This tactic is more appropriate against a low-yield maneuvering ASAT than against a big PINS. Decoys might also be deployed in periods of crisis, providing that one is confident of their credibility over extended periods of observation, in which case they could be effective against the PINS threat as well.

vi) Shoot-back - Some of the SBIs can be used to protect CVs or other assets. This tactic might be countered by a PINS threat by its deployment of decoys. (Decoy deployment is more difficult to perform credibly for a maneuvering ASAT.) Decoys or not, shootback may be necessary to protect high-value assets, like SSTS.

vii) Orbit maneuvers - In peacetime or periods of crisis, the CV can make random orbit changes to confound Soviet tracking. Advocates of such tactics generally assume that the Soviet Union will not significantly upgrade its current spacetrack capability.

viii) High altitude-low observables - These tactics are particularly appropriate to BSTS which might be near or above geosynchronous altitudes. More generally, the higher the asset, the longer it has to maneuver away from an upcoming ASAT threat. (Above about 700 km, however, a CVs capability to participate in the battle becomes impaired.)

ix) Redundancy - Regardless of what other survivability options are employed, the defense must be redundant enough to maintain its effectiveness after having suffered the attrition inevitable from a serious ASAT attack. As the threat becomes progressively more stressing, this becomes more difficult: The extra number of platforms needed in the defense to provide redundancy for each platform lost in the ASAT attack is the inverse of the participation factor. Furthermore, as the threat timeline declines, the ASAT battlespace becomes smaller and its locus moves over the Soviet Union, where the attack will be more successful. (It is a shortcoming of this report that we have not quantified these issues.)

Of course, this list of survivability options is by no means exhaustive nor are these tactics mutually exclusive. Detailed studies show that having most or all of these options is more cost-effective than having only a few. (Note the synergism among hardening, maneuver, and dispersal, for example.) Current studies also indicate that building in adequate survivability is likely to be a cost penalty, rather than a showstopper, even against quite robust threats.

VII. Cost

Cost studies of the type of "baseline" defense described in Section II show the major cost driver to be the production of fire control and associated systems on the CV. However, also in that section, we presented several ways, currently under study, which may well mitigate this problem, yielding defenses whose costs are dominated by the cost of the SBIs themselves.

The cost of an SBI might be divided into two parts--the production cost of its payload and its deployment cost. The payload cost will be dominated by that of its sensor and its guidance navigation and control. Current IR focal plane arrays appropriate to this application are available, but at several million dollars each. However, it is expected that maturation of production techniques and economies of scale can drive this cost down substantially. A reasonable estimate of the average production cost of an SBI is about \$1.5M[1] although some serious cost studies have yielded lower estimates.

Let us assume that the deployment cost per SBI is proportional to its mass, which we shall relate to its velocity by the single-stage rocket equation. Thus, for the total C_T of production C_p and deployment C_D costs

$$C_T = C_p + C_D \quad (9a)$$

$$C_D = (C_p/f) \exp (\Delta V_A/gI_{sp}^*) \quad (9b)$$

$$f = C_p/(kM_p \exp (\Delta V_D/gI_{sp}^*)) \quad (9c)$$

where ΔV_A is the axial velocity capability, which we wish to determine; gI_{sp}^* is a scale velocity related to the specific impulse of the motor (with a correction for the mass penalty of the motor case-2.5 km/s is reasonable); k is the deployment cost per unit mass (currently about \$5/g, but development of the Advanced Launch System should reduce this significantly; we assume \$1/g); M_p is the interceptor payload mass, exclusive of divert capability (3 kg is assumed); and ΔV_D is the total divert capability needed (we assume 0.5 km/s disperse, 2 km/s midcourse, 0.5 km/s homing; for 3 km/s total).

The combination of system parameters f of eq. (9c) determines the functional relation between C_T and $\Delta V_A/gI_{sp}^*$. For the baseline parameter values described above

$$f \sim 150$$

In Section III, we saw that the participation fraction, hence, the number of interceptors required to perform a given mission, is well approximated by a power law in the axial velocity

$$N_{SBI} \propto (\Delta V_A)^{-\alpha} \tag{10}$$

$$1 < \alpha < 2$$

Hence, the total cost of the inventory may be expressed as

$$N_{SBI} C_T \propto [f + \exp(\Delta V_A / gI_{sp}^*)] (\Delta V_A)^{-\alpha} \tag{11}$$

This function of ΔV_A is plotted in Fig. 9, for $100 < f < 1000$. (Large values of f correspond to interceptors which are more expensive to produce, cheaper to deploy, less massive, or require less divert capability than our baseline values.) In addition, the case $f = 0$ is shown, corresponding to mass in orbit as the sole figure of merit (system "cost").

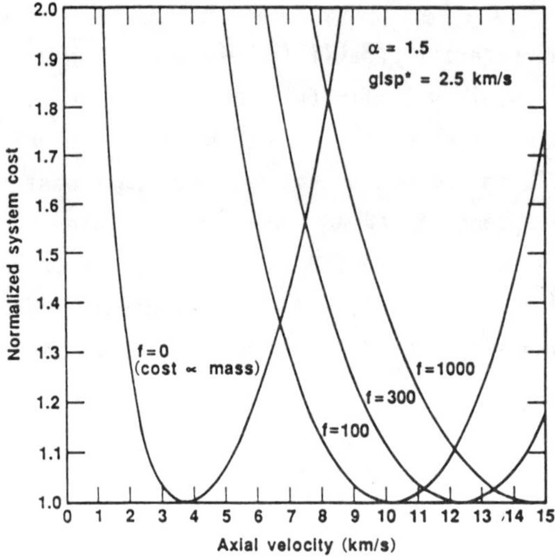

Figure 9. Normalized system cosst as a function of axial velocity (see text).

The figure shows that for minimum mass (f = 0) the interceptors should have axial velocities of 3-4 km/s, but for minimum cost, the axial velocity should be much higher, around 10 km/s. The major differences between a 3 km/s and a 10 km/s system are illustrated below:

Table IV. Axial Capability and System Characteristics

Axial	Interceptor*		Total System**		
Capability	Mass	Cost+	Number	Mass	Cost+
3 km/s	33 kg	$1.53M	6.0	0.36	4.0
6 km/s	110 kg	$1.61M	2.2	0.44	1.6
10 km/s	550 kg	$2.05M	1.0	1.00	1.0

* All payloads cost $1.5M for 3 kg mass

** Normalized values

\+ Production and launch of interceptors only

Overall there is a factor of three difference in the total mass of these systems and a factor of four difference in their costs. An intermediate 6 km/s axial system is presented as a compromise: it has very nearly the minimum total mass (compared to 3 km/s) for a 60% penalty in total cost (compared to 10 km/s).

For a cost minimum (11) predicts that C_p and C_D should obey

$$C_p/C_D = f/\exp(\Delta V_A/gI_{sp}{}^*) \tag{12a}$$

$$= [\ln f - \alpha - \ln(C_p/C_D)]/\alpha \tag{12b}$$

$$\sim 2$$

Thus, production costs should dominate deployment costs, but the two should be comparable. Comparable costs, expensive interceptors, and an efficient deployment system (large f) imply high velocity interceptors (large $\Delta V_A/gI_{sp}{}^*$), as we have seen.

Normalized values only for total numbers, masses and costs were presented in Fig. 9 and Table IV. In Fig. 10 we show the total cost required to fulfill 50%-90% missions against 3000 RVs from the different threats discussed in Sec. IV. These curves show a broad cost minimum above about 6 km/s axial velocity except for the modified-solid fast-bus threat for which velocities above 7 km/s are desired.

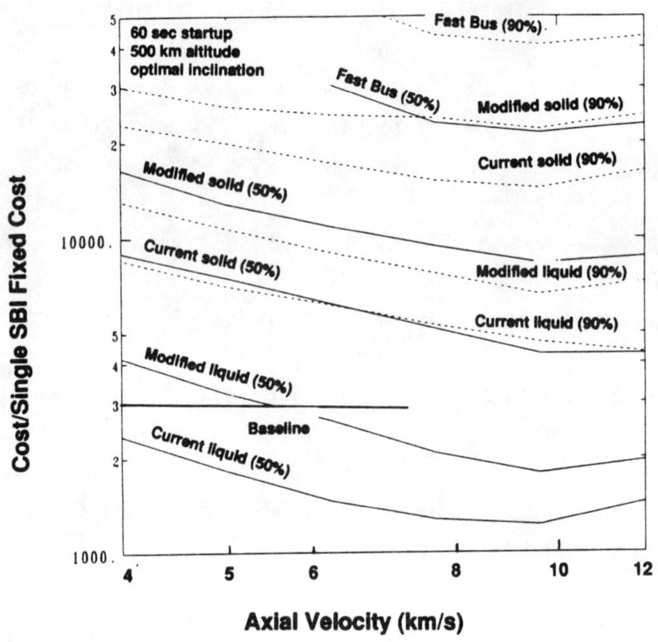

Figure 10. Total cost required to fulfill 50%-90% missions against 3000 RVs from the different threats discussed in Sec. IV. These curves show a broad cost minimum above about 6 km/s axial velocity.

The system cost at optimum axial velocity for these missions, obtained from Fig. 10 is presented in Table V.

Table V. Interceptor Production and Launch Costs
(Unit = 1000 x single payload cost)

Kill Criterion:	50%	90%
Current liquid	1.2	4.5
Modified liquid	1.8	6.6
Current solid	4.4	14.1
Modified solid	8.5	21.5
Fast bus	22.2	40.5

A system designed against fast busses is seen to be 10-20 times as expensive as one designed against current liquid boosters with conventional busses. Increasing the mission from 50% to 90% kill increases the expense by a factor of 2-4.

Above, we advocated a unit interceptor production cost of about $1.5M, indicating production and launch costs for the interceptors of $2-60B against the range of threats and missions considered. For the baseline SDIO defense architecture (fire control on the CVs, see Sec. II) interceptor production costs are only about 7% of the total estimated life-cycle cost of the SBI tier of the defense (which excludes launch costs, BSTS, SSTS, ERIS, etc.).[7] In the Option 2 architecture (see Sec. II) a serious attempt was made to cut non-interceptor costs, yet interceptor production still accounted for only about 13% of life-cycle cost (which again did not include launch).[7] Thus, we should also estimate life-cycle costs to be at least a factor of eight higher than interceptor production and launch, i.e. $14-500B.

Noting that the defenses are designed against 3000 RVs in all cases and that total inventory and, roughly, total cost will scale linearly with the size of the threat, we present the life-cycle cost of the SBI tier of the defense per threat RV:

Table VI. Total Cost of SBI Tier per RV ($M)

Kill Criterion:	50%	90%
Current liquid	5	18
Modified liquid	7	26
Current solid	18	56
Modified solid	34	86
Fast bus	89	162

The life-cycle cost of a modern MIRVed booster is about $100M, based on the MX,[8] for a cost of $7M to $10M per RV deployed. Thus, at the 50% kill level, SBIs would appear to be marginally cost-effective against the current or modified large liquid boosters ($5M or $7M per RV for 50% kill vs $7M per RV for deployment) and not too bad against the current solid boosters ($18M vs $10M per RV). The more stressing threats, modified solids and fast busses, do not appear to allow cost-effective attack, nor does a mission criterion of 90% kill.

Of course, we must qualify these results as based on rough estimates, particularly the giant extrapolation from production to life cycle costs. Nevertheless, our overall costs agree fairly well with more detailed studies.

One of our more interesting results is the cost-effectiveness of very large, very high velocity interceptors. (Mass on orbit is almost universally taken as the cost surrogate in system studies and leads to much different results.) Central to this finding was the assumption of a fixed payload cost independent of the axial velocity. However, faster interceptors imply longer flyout ranges and, thus, better sensors, which probably drives up the payload cost. Higher velocity also implies fewer interceptors and, hence, a less survivable and more redundant defense (survivability concerns were not treated explicitly). Our parameter baseline considered interceptors with light payloads (3 kg) relatively low total divert and dispersal capability (3 km/s) and a low launch cost ($1/g), and all these choices favor massive interceptors. We also did not include the launch cost associated with the CV, or other life support, or with survivability enhancements, all of which probably scale with the mass of the interceptor.

REFERENCES

1. J. Gardner et al., Report of the Technical Panel on Missile Defense in the 1990s, George C. Marshall Institute, Washington, D.C., 1987.

2. R. Everett et al., Report of the Defense Science Board Task Force Subgroup on Strategic Air Defense, Office of the Under Secretary of Defense for Acquisition, Washington, D.C., May 1988.

3. A. Carter, Directed Energy Missile Defense in Space, Congress of the United States, Office of Technology Assessment, Washington, D.C., April 1984.

4. Soviet Military Power 1988, Department of Defense, U.S. Government Printing Office, Washington, D.C., March 1988.

5. P. Duffy and C. Cunningham, Low-Altitude Orbits for Space-Based Interceptors, Lawrence Livermore National Laboratory, UCID-21423, July 1988.

6. A. Weston, D. Ductor, Guidance Law Impacts on Kinetic Energy Weapon Endgame Propulsion Requirements, Air Force Rocket Propulsion Laboratory, TR-85-086, January 1986.

7. S. Gibson, Titan Systems, Inc., private communication.

8. B. Moore, Congressional Budget Office, private communication.

CHAPTER 12

FREE ELECTRON LASERS AS GROUND BASED SPACE WEAPONS

Gary R. Goldstein

Program in Science and Technology
for International Security
Massachusetts Institute of Technology
Cambridge, Massachusetts 02139
and
Department of Physics, Tufts University
Medford, Massachusetts 02155

ABSTRACT

The free electron laser (FEL) is the most promising directed energy weapon in the SDI program. Its theoretical underpinnings, present achievements and future prospects are reviewed. The general requirements of a ground based laser system are derived and are seen to be quite expensive to implement as well as being far beyond current technical capabilities. Atmospheric propagation effects, particularly Stimulated Raman Scattering, will make the transmission of adequate powers dubious. A summary of existing and proposed FEL parameters shows that, at best, future facilities will be many orders of magnitude away from the required GigaWatt average output powers in the visible or near infrared region. Prospects for FEL midcourse or terminal phase weapons are equally problematic, given the simple countermeasures available to the offense. Use as an ASAT weapon is less technically demanding, but of limited applicability given the vulnerablity of an extensive space based targeting system.

I. INTRODUCTION

In 1983 President Reagan initiated a program to research and develop a space based system capable of defending the United States and its allies against a nuclear missile attack. That program, the Strategic Defense Initiative (SDI), has, as a cornerstone, the notion that weapons can be developed to intercept incoming missiles and their warheads. This interception is to be done in several stages along incoming warheads' trajectories by a set of defensive layers. Critical to the mission of the first tier is the ability to destroy or damage ICBM's in the boost phase, *i.e.* immediately after launch while the initial booster rocket is burning and before the release of independent reentry vehicles or MIRV's.

Envisioning weapons that could intercept offensive missiles in the boost phase, which will involve time scales of only minutes, requires a system that can respond very rapidly to commands, on the order of seconds, and can repeatedly deliver lethal amounts of destructive energy to the missiles in a comparable time span. Weapons that use intense beams of particles traveling at or near the speed of light conceivably could fulfill such a mission. Such weapons, generically called directed energy weapons, have been a prime focus of the SDI endeavor.

The technically most advanced directed energy weapons candidates are lasers and elementary particle accelerators. Imagining the emplacement of high power lasers or accelerators into orbit leads to many possible problems. The projected satellites will be extremely massive, expensive to launch and maintain, and will be vulnerable to damage and attack. A more favorable alternative is to base the directed energy weapons on the ground providing their energy can be directed toward the ICBM targets. Only lasers could possibly satisfy these criteria. This is the case because electromagnetic radiation can be aimed upward through the atmosphere above which space based mirrors or lenses could redirect the beam to boost phase targets. Charged or neutral particle beams would neither survive the propagation through the atmosphere nor be steerable from orbit.[1]

Given this unique and essential role for ground based lasers in the SDI "architecture," can lasers be developed that will meet all the requirements that accompany this mission? Before reviewing those extensive and formidable requirements, however, it can be stated that at this juncture in the SDI program, the "Free Electron Laser" (FEL) appears to be the most promising emerging technology for a directed energy space weapon.

The FEL is *theoretically* capable of high power and intensity, high efficiency and, most importantly, the wavelength of its electromagnetic radiation is completely tunable. Recent budgets of the SDIO reflect an optimism about this direction.[2] Although delayed by changing budget priorities, plans to build a very large (roughly ten miles in length) and expensive (about $3 billion) FEL facility in New Mexico have passed the site selection stage.[3] The FEL certainly has had phenomenal success recently in developing high efficiencies and peak powers;[4] its progress is often cited as evidence for the feasibility of the boost phase interception. [5]

Will the FEL be able to fulfill the mission of a space weapon? What are its theoretical capabilities? At what stage of development is the FEL now and what technological advances would be necessary to bring it to an operational level? In the following, these and related questions will be answered.

In the next section, II, the theoretical principles behind the FEL will be reviewed. After considering the free electron laser itself, in III the structure and requirements of an overall ground based laser space weapon system will be presented. Crucial constraints for lethality on the beam and mirror components are explored in IV. The problems of atmospheric propagation will be reviewed in V. Several recent successful FEL experiments will be analyzed and compared to the critical requirements for lethality in VI. The possibility of scaling up the different designs will be discussed as well.

The overall conclusion of this investigation, presented in Section VII, is that a ground based FEL weapon capable of fulfilling the SDI mission will require so many orders of magnitude improvement over the existing complex of performance capabilities that it is not a sensible direction of intense effort and expense. In the analysis it will be seen that while the FEL weapon prospects are poor, it is an especially clever instrument of applied physics with many scientific and commercial uses.

II. THE FREE ELECTRON LASER — THEORY

What is the free electron laser (FEL)? Stated succinctly here and in detail below, it is a device first conceived over 35 years ago and reintroduced by J. M. J. Madey[6] that uses an electron accelerator to convert the kinetic energy in a beam of electrons to coherent electromagnetic energy. That conversion is accomplished by causing the relativistic electrons to oscillate transversely as they move down a free section of beam line. The oscillation, like the pumping of a conventional laser, results in emission of radiation which then interacts with the electron beam to stimulate further coherent emission. The electron beam itself, when properly 'tuned", is the lasing material. Because the oscillation and the beam energy can be varied continuously, in principle the FEL can be tuned to any desired frequency of electromagnetic radiation. In practice, however, there are many limitations on the frequency as well as the power, as the more detailed discussion below will demonstrate.

In the following a more precise quantitative description of the principles and limitations of FEL's will be presented. Much of the theoretical discussion is presented in greater detail in several recent reviews. The purpose here is to develop a sufficient understanding of the basic principles so that an evaluation of the prospects for weapon applications can be evaluated.

The basic components of the FEL are an electron accelerator, a static magnetic undulator or wiggler that produces the oscillations, and a cavity in which the electron-light interaction takes place. To confuse matters somewhat, there are two different general types of FEL's — oscillators and amplifiers. These distinctions will be discussed more thoroughly in Section VI. Briefly, the amplifiers take an external electromagnetic beam (or EM noise) and use a long undulating electron beam to amplify the EM beam power in one pass through the undulator. The oscillator is more like a conventional laser in that the EM wave (which can be produced spontaneously or by stimulation from a lower power input "seed" laser) is reflected in a cavity to stimulate further coherent emissions by the electron beam bunches, thereby filling the cavity with EM radiation. The electrons are the lasing material for which the "level spacing" is continuous. The spacing is controlled by the accelerator energy and the undulator wavelength as the following will elucidate.

The fundamental process involved in converting the electron's energy into coherent radiation can be easily described.[7] Consider a single electron which has been accelerated to a velocity near c. This free electron enters the region of the undulator (see Fig. 1). The undulator or wiggler is a series of nearly identical permanent magnets or electromagnets with their poles arranged in an alternating sequence so that the resulting magnetic field is nearly spatially periodic and transverse to the beam direction. The incoming electron thereby passes through a transversely varying magnetic field, which can be characterized by a wavelength λ_w and a field strength B_w. In a standard classical description, this field, through the Lorentz force, sets the rapidly moving electron into small transverse oscillations about its linear path. Those oscillations are equivalent to an oscillating charge current. The varying current produces electromagnetic radiation of appropriate wavelength λ_s.

To obtain the relation between the wiggler wavelength and the radiation wavelength it is easiest to use the quantum mechanical description. To begin, assume we are in the incoming electron's rest frame. The electron sees the wiggler field as a periodic EM wave approaching. That wave is equivalent to a set of photons whose wavelength will be λ_w/γ, the Lorentz contracted spatial period of the wiggler. Note that γ is the usual Lorentz

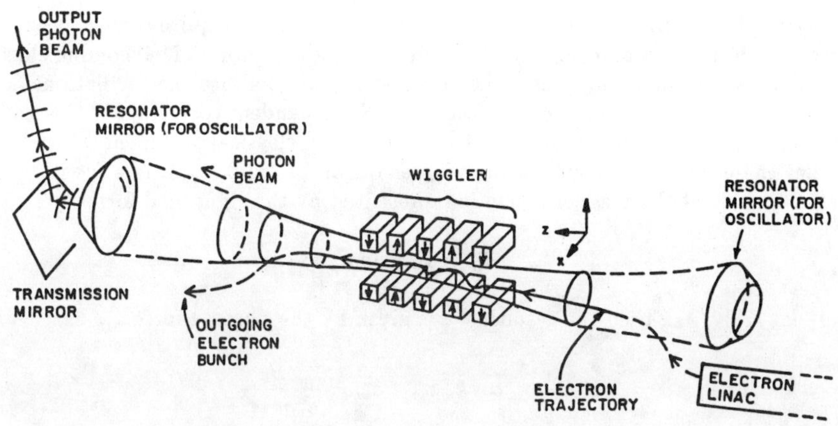

Figure I

factor $1/\sqrt{1-v^2/c^2}$, which is just the total electron energy divided by the rest energy, E_e/mc^2. Viewed in this way the electron and photon can scatter; Compton scattering takes place. The outgoing photon can be *backscattered* to have an emerging wavelength λ_s' in the incoming electron's rest frame. The Compton scattering kinematics relates this backscattered photon wavelength to the "wiggler photon" wavelength via

$$\lambda_s' = \left(\frac{\lambda_w}{\gamma}\right) + 2\frac{h}{m_e c}$$

Lorentz transforming back to the laboratory frame of reference, the outgoing EM wave is Doppler shifted so that

$$\lambda_s = \frac{\lambda_s'}{\gamma(1+v/c)} = \frac{\lambda_w}{\gamma^2(1+v/c)} + \frac{2h/m_e c}{\gamma(1+v/c)} \ ,$$

or

$$\lambda_s \simeq \frac{\lambda_w}{2\gamma^2} \ , \tag{1}$$

where v is approximately c and the Compton wavelength $h/m_e c$ (2×10^{-10} cm) is very small compared to the Lorentz contracted wiggler wavelength. With typical wiggler spatial periods of centimeters, to produce EM radiation in the visible region requires γ's on the order of 100 or electron beam energies of ≥ 50 MeV.

In order to produce coherent EM radiation, the emitted waves from each of the scattering electrons must stimulate subsequent emissions, which will then be in phase with the stimulating waves. That is the basic idea behind the FEL. Its implementation will be examined in detail below.

Before proceeding, however, note that relation (1) is not quite correct because the electron undergoes transverse oscillations in the wiggler region. The incoming electron's rest frame is not the interacting electron's rest frame. Because the oscillations depend on the strength of the magnetic field, there is a B_w dependent correction to the scattered wavelength which can be obtained with the following classical argument.

Let the electron's velocity in the wiggler region be $(\dot{x}, 0, \dot{z})$, where $\dot{x}$ will be much smaller than $\dot{z}$. If the magnetic field is represented by the sinusoidal form

$$\vec{B}_w = B_0 \hat{y} \sin k_w z$$

where $k_w = 2\pi/\lambda_w$, then the resulting $\dot{x}$ is given by the Lorentz force

$$\dot{x} \simeq \frac{c}{\gamma} a_w \cos k_w z \ , \quad a_w = \frac{eB_0}{k_w m_e c^2} \ . \tag{2}$$

Now as the electron oscillates it emits the "scattered" EM wave with electric field component

$$E_x = E_s \sin (k_s z - w_s t) \tag{3}$$

where $w_s = ck_s$, and E_s, the amplitude of the EM wave, is determined by the electron current through Maxwell's equation. In order for the electrons to continue this emission process coherently, the phase change of the transverse motion must coincide with the rate at which the phase of the emitted wave changes. That is, in one period of transverse electron motion, $T_w = \lambda_w/\dot{z}$, the electron has advanced one wiggler wavelength, λ_w, while the EM wave has progressed a distance cT_w. For coherence that latter distance should be precisely one EM wavelength, λ_s, beyond the electron's position, λ_w, *viz*

$$cT_w = \lambda_w + \lambda_s$$

But $cT_w = c\lambda_w/\dot{z}$, so

$$\lambda_s = \left(\frac{c}{\dot{z}} - 1\right) \lambda_w \tag{4}$$

Next note that $\dot{x} << \dot{z} < v < c$. Expanding to second order in $\dot{x}/v$ and $1/\gamma$,

$$\dot{z} \simeq v \left(1 - \frac{\dot{x}^2}{2v^2}\right)$$

and

$$v \simeq c \left(1 - \frac{1}{2\gamma^2}\right)$$

Then using (2) to obtain the RMS value of $\dot{x}$

$$\bar{\dot{x}} \simeq \frac{c}{\sqrt{2}\gamma} a_w$$

we finally obtain the magnetic field correction to the basic equation for the free electron laser,

$$\lambda_s \simeq \frac{\lambda_w}{2\gamma^2} \left(1 + \frac{1}{2} a_w^2\right) \tag{5}$$

In deriving this expression the energy loss of the electrons was ignored. Quantum mechanically each photon emission will reduce the electron's energy by hc/λ_s. For optical photons this is only a few electron Volts. But the lasing condition (5) depends on the square of the electron energy, so the system will "detune" as the beam traverses the wiggler length. This leads to the simple, but elegant idea that to compensate for the energy loss, which would increase the requisite outgoing EM wavelength, the wiggler field should be "tapered". That is, in Eq. (5), as γ decreases along the beam line, a_w or λ_w should decrease correspondingly, to keep λ_s constant. The tapered wiggler will be considered subsequently.

It is also worth noting that in deriving Eqs. (2) and (4) the reaction of the emitted EM fields on the electrons has been ignored. This is appropriate since the force on the electrons due to these transverse electric and magnetic fields nearly cancel. More careful analysis reveals that the combined fields of the EM radiation and the wiggler give rise to a (pondermotive) force on the electrons that acts to keep their longitudinal motion in phase with the EM wave.[8] This is similar to the phase locking mechanism in synchrotrons and results in an important stability with respect to energy variations in the beam.

Now that the condition for lasing has been obtained, it is natural to ask how the gain and efficiency of the laser system depend on the various parameters of the accelerator, wiggler and optics.[9] This is most readily accomplished by noting that the energy lost by an electron traversing the length of the region in which the electron and the EM radiation interact is just the energy gained by the radiation field itself. For a single electron the energy exchange is given by the relation

$$\frac{dE_e}{dt} = e\vec{v} \cdot \vec{E} = e\dot{x}E_x \ ,$$

where $\dot{x}$ and E_x were written above, Eqs. (2) and (3). Hence,

$$\frac{d\gamma}{dz} \simeq \frac{eE_s a_w}{2\gamma m_e c^2} \sin(k_w z + k_s z - w_s t)$$

noting that $\dot{z}$ is near c and $dt = dz/c$. The fractional energy exchanged over electron longitudinal distance from z to $z + dz$ is then

$$\frac{\Delta E_e(z)}{E_e} \simeq \frac{1}{\gamma}\frac{d\gamma}{dz}dz \simeq \frac{eE_s a_w}{2\gamma^2 m_e c^2} \sin\psi \, dz \tag{6}$$

This expression oscillates with the phase

$$\psi \equiv k_w z + k_s z - w_s t \ , \quad \text{so} \ \ \dot{\psi} = (k_w + k_s)\dot{z} - w_s \ . \tag{7}$$

For stationary phase, $\dot{\psi} = 0$, so the coherence condition, Eq. (4), results and ψ is a constant, say ψ_R.

For electrons that do not quite match the condition for coherence there is a "synchrotron" oscillation about the synchronous energy $\gamma_R m_e c^2$. This is the "phase locking" mechanism to which we alluded above; the electron is said to be trapped in the pondermotive potential of the combined wiggler and radiation EM fields. Generally the fraction of beam electrons trapped, δ, depends on the energy spread and angular spread of the beam as well as the beam current and the wiggler parameters. In order that a positive energy transfer occurs, the electrons must be injected somewhat out of synchrony, with energies higher than $\gamma_R m_e c^2$, and, ideally they should undergo half a synchrotron oscillation as they traverse the interaction length L. Hence, setting $\dot{\psi}L/\dot{z} \simeq \pi$ will provide an estimate of the ideal incremental energy δE. The maximum energy extraction for a single pulse is limited to $2\delta E$. Letting $\dot{z} = \dot{z}_R + \delta\dot{z}$ and observing that $\delta\gamma \simeq \gamma_R^3 \delta\dot{z}/c$ gives $\delta\gamma/\gamma_R \simeq \gamma_R^2 \lambda_s/2L$. Then the ideal single pulse efficiency is just

$$2\frac{\delta\gamma}{\gamma_R}\bigg|_{\max} \simeq \frac{1+\frac{1}{2}a_w^2}{2N} \tag{8}$$

where $N = L/\lambda_w$ is the number of wiggler cycles. Note that while the efficiency falls as $1/N$ the gain increases as N, leaving the discouraging result that the gain-efficiency product is insensitive to wiggler length. However, this result applies in the linear or small gain regime, i.e. when the energy loss by the electrons is small enough that the corresponding change in E_s of Eq. (6) can be ignored. For large or exponential gain, or for tapered wigglers, the non-linear system must be considered. Then FEL designers rely on simulations to predict gain and efficiency.

It would seem that injecting the electrons into the wiggler region at energies much higher than the synchronous value would produce the largest energy EM field. However, the electrons' energies cannot exceed a certain amount before being "untrapped". This trapping requirement also limits the fractional gain of the system. For the untapered wiggler this limit is

$$\frac{d\gamma}{\gamma}\bigg|_{\max} \simeq \frac{e}{2\pi m_e c^2}\sqrt{\frac{\lambda_w \lambda_s B_w E_s}{1+\frac{1}{2}a_w^2}} \,. \tag{9}$$

The electric field, E_s, depends on the oscillating current through Maxwell's equations, and that current is proportional to the beam current times the amplitude of the transverse velocity, Eq. (2). The latter, in turn, is proportional to the wiggler field and the wavelength divided by γ_R. For fixed untapered wiggler parameters, the maximum single pulse extraction efficiency in the linear gain regime is thereby proportional to $(1/\gamma_R)^{3/2}$. This observation suggests an essential limitation on low gain FEL's; the *efficiency falls as the laser wavelength decreases* (all else being equal). For tapered wigglers the situation can be improved, as will be discussed.

It should be noted also that as the electrons lose energy to the EM radiation field, the electrons' energies are spread out. This gain-spread relationship constitutes a theorem[10] for the FEL. It imposes limits on the operation of a recirculating system like a storage ring, where a large dispersion of energies leads to growing instabilities in the ring. This ultimately constrains the efficiency of such systems, even with tapered wigglers. With tapered wigglers, however, the gain-efficiency product can be improved. In order to obtain the most optimistic estimates for gain and efficiency, the tapered configuration has become standard in designing new FEL's.[11]

The overall conversion or extraction efficiency is obtained by integrating (6) over the length of the interaction region,

$$\eta = \delta \int_0^L dz \frac{1}{\gamma}\frac{d\gamma}{dz} \simeq \delta \frac{e a_w}{2\gamma_R^2 m_e c^2} \sin\psi_R \int_0^L dz E_s(z) \tag{10}$$

where all but the laser electric field is safely assumed to be slowly varying and near the synchronous values, γ_R and ψ_R, even for gradually tapered wigglers.

The variation of $E_s(z)$ depends on the optics of the interaction region. The approximate magnitude of $E_s(z)$, ignoring the optics, can be obtained from the wave equation for the EM vector potential, $\Box A = j/c$. So the total E_s field is proportional to the electron current of the accelerator. Of course the larger that current, the more energy available for conversion to EM radiation. Note also that the gain will improve for longer wigglers, while the maximum extraction efficiency for untapered wigglers, Eq. (8), declines with length. hence, an appropriate "figure of merit" is the gain-efficiency product.[11]

For a finite beam of electrons, the spatial overlap with the EM field must be taken into account as well. This is determined by the electron beam diameter or radius, r, and the angular divergence, $\langle\theta\rangle$, of the beam. An appropriate invariant specification of the overlap is the normalized emittance, ϵ_N, which is given by

$$\epsilon_N = \beta\gamma r\langle\theta\rangle$$

and is one measure of the beam quality of the accelerator. Another measure of the beam quality is the brightness, $B_N \simeq J/\pi^2\epsilon_N^2$, where J is the beam current. Clearly, to maximize the efficiency the emittance should be minimized so that the overlap region is maximized. Furthermore, the higher the current, the greater will be the gain for the laser, so brightness should be maximized to maximize gain.

An expression for the gain-efficiency product is useful for evaluating the prospects for future near IR or visible FEL's with very high power outputs. Such expressions have been used for simulations in new designs. A typical example of such a formula for tapered wigglers, in which the geometry of the optical cavity, the tapering, and a_W have been optimized, has the form,[11]

$$\text{gain} \times \text{efficiency} = \frac{128\pi^2 e}{mc^2}[\delta \sin\psi_r]^2 \gamma J f_{\Delta E} f_{\epsilon N} \chi^2$$

where χ^2 is a function of the cavity geometry, $f_{\Delta E}$ is a function of the energy spread, roughly inversely proportional to the spread of the beam, and f_{ϵ_N} is a function of the emittance, varying nearly inversely with emittance. Although this expression indicates an increase in the gain -efficiency product with γ, the other factors are also functions of γ. In practice the current density, emittance and trapping fraction will be limited by the beam energy as well. The complexity of the design simulations makes generalization difficult, but experience with actual experiments does support a decrease of the optimized gain-efficiency product with beam energy.

To achieve short wavelength FELs requires electron accelerators capable of energies in the 100 MeV range while high power output necessitates not only high currents, in the 100's to 1000's of Amps, but superb beam quality and high repitition rates as well. Candidate accelerators are the RF linac, for which the beam energy technology is quite mature at moderate currents, and the very high current induction linac, whose scaling to high enough electron energy is under development. The RF linac, because of its rapid repitition rates but peak currents only in the 100's of Amps, can achieve sufficient gain only as an oscillator type FEL, for which it is suitable. However, maintaining beam quality (at the high currents accomplished by "beam bunching"[11]) and properly tuning the resonant optical cavity are quite difficult problems for scaling up the gain of such an oscillator without losing efficiency.

The induction accelerator, with its low repitition rate, operates in a single pass amplifier mode as an FEL. In scaling up the energy, by increasing the length of the induction linac, while keeping very high currents, beam stability problems become serious. Although "ion channeling" has been demonstrated to circumvent beam breakup instabilities[12], beam quality is degraded concomitantly[13].

Then what can be expected? To get a sense of the future prospects for very high power operation of visible or near infrared FELs (more detailed discussion will be pursued in Section VI) consider some of the most optimistic projections for rf linac–oscillator systems. Assuming a series of proposed technological developments are achieved,[13] an rf linac operating at 300 MeV will deliver peak currents of 300 Amps[14] with extraction efficiencies of 20%[15] to produce *peak* powers of $0.20 \times 300 \times 300 \times 10^6 = 18$ GigaWatts. The *average* power in one *train of pulses* or *macropulse*, is obtained by taking as a factor the ratio of the length of the short pulse, or *micropulse*, to the separation in time between micropulses. Researchers will attempt to shorten pulse separation[16] in order to yield 1/500 for this factor to give about 40 MegaWatts in one macropulse. Macropulses can be produced with durations of about $100\,\mu$sec (so the energy in one macropulse would be 4 KiloJoules) and at proposed repetition rates[17] of 50 Hz the duty cycle factor of 1/200 reduces the average power further to 200 KiloWatts. This highly optimistic projection is three orders of magnitude from requirements for lethality. As considered below, average power in the 1 GigaWatt regime (or total energy of about 1 GigaJoule delivered in 1 second or less) will be needed for ground based FELs. These above projections will be compared with typical achievements to date in section VI,[18] wherein it will be seen that macropulse powers are some three orders of magnitude below the proposed 40 MW and six orders below lethal powers.

III. GROUND BASED LASER SYSTEMS

A directed energy weapon, like the free electron laser itself, is but one component in a complex system. To understand the constraints and requirements of the laser, the entire system must be kept in mind. The laser beam must get to its targets. The most direct way for this to occur would be to place the laser in orbit and provide it with one mirror, fixed on the same platform, to aim the beam at a target. Of course, the laser must be in sight of the target. Since low orbit satellites are in motion with periods of 90 minutes, depending on the details of the basing and range assumptions, anywhere from the order of 100 to 400 such lasers would have to be in orbit.[19,20,21,22]

Having so many orbiting lasers has several serious disadvantages. Each station is very vulnerable to attack. To be adequately defended with hardening or interceptors of some kind, the weight of each station becomes considerable.[23] Without such additional defense, the weight of the lasers and their fuel supply makes transport of the satellites into space prohibitive anyway.[19]

An alternative is to place the lasers in high, geosynchronous orbits at 36,000 Km. With this basing far fewer stations are needed; two suffice for covering both hemispheres. However, the fixed high orbits still leave the lasers vulnerable to countermeasures, just as the low orbit basing. Space mines, ASAT weapons in accompanying orbits, and direct attack are all possible problems that threaten the safety of the lasers.

As a result of such considerations, the preferred mode of basing is on the ground.[14,23] The drawbacks are then the difficulties with atmospheric propagation of high intensity light pulses and the necessity for many orbiting mirrors. These issues will be addressed below. But the vulnerability of the lasers is considerably lessened. Secondly, the requirements for space transporting novel, large scale fuel supplies become

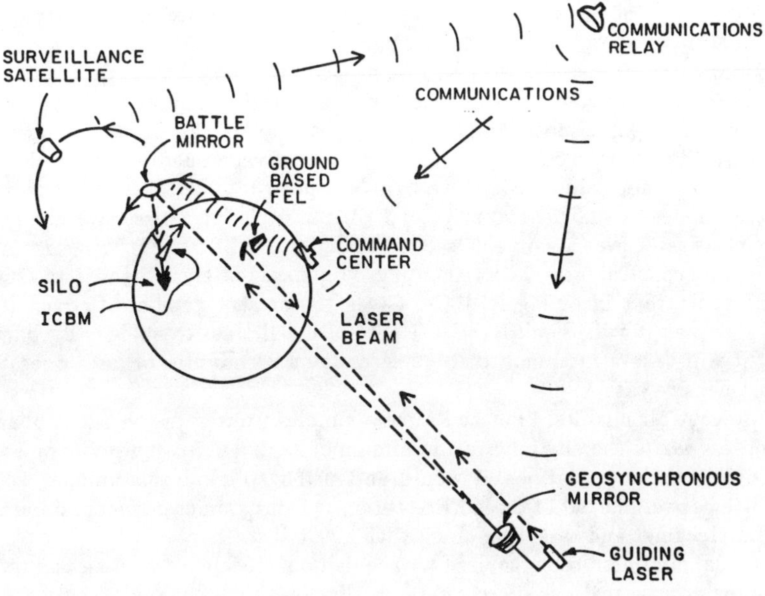

Figure 2

moot with the availability of conventional power sources. (As will be seen, however, the power supply itself is not a trivial matter.) Furthermore, economical, minimum weight design is no longer a priority.

The lasers are the major components of the system, but not the only components. Each laser facility must include a transmitting "telescope" and mirror to aim the beam through the atmosphere to a relay mirror in orbit. For predictability and survivability and to maintain sufficient hardware for aiming and targeting, as well as containing adaptive optics capabilities, these relay mirrors are assumed to be in geosynchronous orbit at 36,000 Km. The geosynchronous mirrors should be as large as possible to minimize beam spreading, as will be explained below. The beam is then directed back to low orbit where another relay or final battle mirror redirects the beam to the target. This configuration is shown in Fig. 2. Of course there are other possible "architectures" that can be envisioned. Other distributions of relay mirrors, in elliptical or high, but not geosynchronous, orbits can be considered, but the accompanying complexity could make the system more vulnerable to coordination and control failures as well as ASAT measures. Given the vagueness of other schemes, the architecture assumed here (and in the APS report[13]) will be taken as a benchmark.

All of these components must be coordinated with surveillance and battle management systems. The overall system must be controlled by a computer hardware and software system that will have unprecedented complexity.[20,21,24] Some critics and proponents alike have emphasized that all of the weapon system hardware will be useless unless the enormity of the software requirements can be addressed successfully in the future.

How many ground based FEL's will be needed? There are two constraints. Even with mountain top basing, local weather is a problem for propagation through the atmosphere. Cloud cover will almost completely absorb a beam in the visible or near infrared range that would otherwise be optimal for the system. Even under clear conditions, only some 70 to 80% of normal intensity blue light will penetrate vertically through the full atmosphere. Hence, a sufficient number of FEL's must be dispersed over the globe to guarantee clear skies over a workable set of them.

The second constraint on the number of ground based lasers is the rate at which any single FEL can attack incoming ICBM's. Given the present number of Soviet ICBM's at 1400, and that it will require about 1 second dwell time to sufficiently damage a missile,[19] it will take a minimum of 1400 seconds or 23 minutes of laser operation to thwart an offensive attack.[25]

To intercept all missiles, launched almost simultaneously, in the boost phase of 3 minutes or less would thereby require a minimum of $23/3 = 7.7 \sim 8$ or more operational FEL's. Given weather variations, it would be sensible to triple this number to assure clear conditions over one out of three FEL stations. This assumes perfect and immediate acquisition, tracking, and targeting along with 100% success rate.

The actual minimum time required for acquisition, tracking, targeting and retargeting, assuming a more realistic success rate, can be roughly estimated. The major time constraints involve the computer processing times for the complex "battle management" system, communications among the various sensors and computers in the network, and the mechanical positioning times of mirrors and sensors.

Computer processing times for the overall system are impossible to estimate without a specific proposed software scheme in mind. While it is anticipated that the software complexity will be enormous, there is little agreement on any design. The computers that could be employed presumably would be the fastest available — supercomputers and array processors. Distributing the next generation of ultra-fast computers throughout the space environment might keep computing times "per kill" down to well below the one second dwell-time scale established for target damage.

Communication and Propagation times would be a significant limit. Consider the steps necessary for destroying an array of missiles.

1. Missile launches are detected and monitored by sensors and surveillance satellites in both low and high orbits. Each missile is discriminated from decoys and deceptions, the trajectories are projected and each missile is tracked.

2. All the surveillance information — numbers of launches, locations and trajectories — is passed on to computers that may be distributed in space and then collected by some well-protected ground based central computer system. That system, which has been continually updated about locations of all relevant hardware and weather conditions, must decide that an attack is truly underway and it must initiate a response. It allocates a target and a set of geosynchronous and low orbit mirrors to each operational ground based laser.

3. The lasers and mirrors are activated and appropriately positioned and aimed.

4. Turbulence conditions of the atmosphere above each laser on the projected upper leg of the path of the beam are monitored and adaptive optics is used to compensate (as will be discussed in Section V).

5. The lasers are fired and the beam directed to the target via the mirrors.
6. Damage is assessed by the surveillance system.
7. If a miss has occurred, steps 1, 3, 4, 5 and 6 are repeated.
8. When a hit has transpired, the full process is repeated — 1 through 6.

Each cycle requires first sending information from sensors to the ground based central system ($1 \rightarrow 2$) via relay stations most effectively in geosynchronous orbit as well. Over a distance of about $40,000\,\text{Km} \times 2$, that initial communication, which occurs at the velocity of light, takes $2 \times 40,000\,\text{Km}/3 \times 10^5\,\text{Km/sec} = 0.27$ sec. Next, the instructions are sent to the relevant components ($2 \rightarrow 3$) requiring another 0.13 sec to reach the geosynchronous mirrors. The turbulence monitoring (4) could have been initiated in the first phase and need not take any extra time. Another 0.27 sec is required for the beam to propagate to the target (5). Then damage assessment information for each target must get back to the central system if a kill has occurred, or to the mirrors and laser otherwise, requiring at least another 0.27 sec in either case. This is now a total of 0.94 sec or about one sec communication time to add to one sec dwell-time for a confirmed kill or a miss in the first pass.

When a miss has occurred the mirrors could have begun their repositioning after the damage assessment information was received. Then retargeting requires only another 0.27 sec, while 0.27 sec elapse in reaching the target from the laser, and damage assessment adds another 0.27 sec. In summary, the time required for light or radio waves to traverse the distances between various components of the system adds about one sec for each initial shot and 0.8 sec for each repeat until a hit is achieved. While it can be envisioned that much of the computing is done in a parallel and distributed manner, with central coordinating computers being consulted less frequently, the time estimate will not be decreased significantly. It will be essential to coordinate and update the extensive incoming data and battle assignments.

The time required to properly *position* the space mirrors will provide a significant constraint on the system as well. After sets of mirrors and lasers have been allocated to targets, each geosynchronous or relay mirror must be aimed to reflect the laser beam onto a battle mirror that will be in the best position to reach the target. The targets could be distributed over longitudes that span half the northern hemisphere. From 36,000 Km the entire earth subtends an angle of 12,000/36,000 radians or about 20°, so the geosynchronous mirrors will need to find their selected battle mirrors over roughly 10°. The Hubble Space Telescope (HST)[22] will be capable of turning through 6° in 20 seconds or 20′ of arc per second. More than an order of magnitude improvement in the steering mechanism would still leave aiming times *on the order of seconds.* While there are alternatives to the "brute force" method of slewing to aim the mirrors, other methods are speculative and their implementation questionable.[26]

Having found the battle mirror, the geosynchronous mirror will have to follow the battle mirror's low orbit motion of about 8 Km/sec which would require tracking through less than 1′ per second, well within the HST capability. The battle mirror itself will be moving laterally with respect to its target at a distance of the order 1000 Km away. At some typical relative speed of 5 Km/sec, the battle mirror could require angular aiming rates of 5/1000 radians per second or about 20 minutes of arc per second — achievable by HST standards. Finally, then, being quite liberal in projecting future developments, the aiming of the geosynchronous mirrors will require at least one second for each firing at the target. Putting these times together, a single shot at a missile will take about one sec dwell-time, one sec communication and propagating time and at least one second aiming time. A further liberal guess that an average of two shots (50% success rate) is required for a kill brings the minimum time per kill to $3 \times 2 = 6$ sec. Hence for boost phase interception requirements the number of ground based lasers should be over an order of magnitude higher than the initial estimate of eight, since a factor of three for weather and six for kill time gives $3 \times 6 \times 7.7$ or approximately 140 ground based FEL's.

With the anticipated development of fast burn solid fuel boosters for the opponent's ICBM's, the boost phase could be shortened to less than two minutes. In that case more than 210 ground based facilities become necessary. Another effective countermeasure is simply the proliferation of ICBM's. Twice as many missiles requires twice as many lasers — up to 420.

It should be pointed out that the cost of such missile proliferation will be drastically less than the cost of the lasers. Taking the conservative estimate of Field and Spergel[22] that an ICBM will cost $50 million and assuuming the $3 billion cost[27] for a prototype FEL drops by an order of magnitude in "mass production", the cost exchange ratio still will be 6 to 1. This does not take account of increases in the number of mirrors, space flights to put mirrors in orbit nor additional surveillance and computer requirements.

Such a large number of ground stations could be reduced by using beam splitters at the outgoing end of the laser so that several independently targetable beams could be produced by each laser. Of course, this splitting reduces the energy in each beam, correspondingly, providing that each split beam is to operate simultaneously. If output pulses from the laser can be switched among several different optical channels at the laser repitition rate, then such beam splitting could be effective for some attack scenarios. Since it was seen that FEL's are currently many orders of magnitude below the required levels of energy for lethality, splitting is not very reasonable beyond factors of 2. This still leaves 70 facilities at the very least to combat the *current* Soviet force quality and level. With fast burn boosters and proliferation by 2 that becomes 210 again.[28]

The power requirement for such an extensive array of FEL's is significant. With each FEL outputting about 1 GigaJoule in 1 second, and with "wall to laser" efficiencies of 10%, each FEL must be powered by 10 GigaWatts input for at least 3 minutes. For the more modest estimate of 140 ground based FEL's the total power must be 1400 GigaWatts or several times the total power producing capacity of the U.S. With the higher estimate of 210 to 420, the total power gets up to an order of magnitude higher than the total U.S. capacity. To prepare this system for deployment therefore necessitates a large increase of national power capacity. There have been proposals to substantially increase the short term power available on demand by developing energy storage facilities. Large banks of capacitors or massive mechanical systems, like flywheels have been suggested, but the feasibility and expense of such a large scale storage network are only speculative. The cost of the lasers alone will be enormous. At $3 billion each,[27] the 210 to 420 FEL stations would cost $630 billion to $1.3 trillion.

How many space mirrors will be needed? It may be necessary to have about one-third as many geosynchronous mirrors as ground based FEL stations. Certainly the number of low orbit battle mirrors within view of a target at any time must equal the number of operational FEL's. With an absentee rate[20] of 5% that gives a lower limit of about $20 \times 8 = 160$ (for 8 FEL's). With a factor of 6 for acquisition, assessment and retargeting along with account being taken of proliferation, that number of battle mirrors could be up to 1,000. The cost of the mirrors alone will exceed the cost of effective offensive proliferation countermeasures by a considerable factor.[22] Higher orbit basing of battle mirrors will lower the numbers required, but will lower survivability as well. While such considerations continue to be debated, the estimates of numbers of battle mirrors agree within factors of 2 or 3, so that the conclusions about cost effectiveness are unchanged.

IV. BEAM DIVERGENCE AND MIRROR REQUIREMENTS

An important consideration in determining the necessary intensity of the laser output is the beam divergence. For significant damage to result, the energy over the area, or the *fluence*, at a somewhat hardened target is estimated[29] to be about 20 kiloJoules/cm^2. It should be noted that this estimate of fluence is conservative. Relatively simple countermeasures to *harden* the missile skin against laser damage have been suggested, *e.g.* increasing the surface reflectivity by polishing or using alternate methds, giving the missile a rapid spin to disperse the beam energy over a larger area, or encasing the shell in a layer of highly absorbent material. Such countermeasures could increase the requisite fluence by an order of magnitude.[19,20] The increasing area occupied by the diverging beam over the full path of the beam from source to target is crucial. A major constraint on this divergence will be the diffraction limit of the beam upon reflection at the geosynchronous or the relay mirror, transmitting and targeting or battle mirrors. By operating at optical wavelengths, the FEL has an advantage in this regard compared to other high intensity infrared lasers. The diffraction angle is roughly

$$\theta_d = 1.2\frac{\lambda}{D}$$

where D is the mirror diameter.

As a relevant example, a 10 meter mirror leads to $\theta_d = 6 \times 10^{-8}$ radians for 0.5 micron light. Over a total path length of about 40,000 Km from the geosynchronous mirror the diffraction limited beam diameter will increase by 2.4 meters. For a perfectly reflecting plane mirror, the 10 m beam becomes 12 m and the fluence required is 40% larger at the mirror. The 20 KJ/cm^2 fluence at the target thereby would require a total energy being reflected from this mirror of $20 \times \pi\ (12.4 \times 10^2)^2/4 = 2.4 \times 10^7$ KJ. For 1 second dwell time, the requisite power reflected from the mirror would be 24 GigaWatts. For a perfect 10 m diffraction limited spherical focusing mirror, whose construction will be seen to be very unlikely, the beam will be narrowed to 2.4 m. To achieve 20 KJ/cm^2 at the focus would require $20 \times \pi\ (2.4 \times 10^2)^2/4 = 9 \times 10^8$ Joules at the mirror, or a power reflected from the mirror of 0.9 GigaWatts over a 1 second dwell time. Hence, a factor of 20 improvement would be obtained if diffraction limited optics could be achieved. While the Livermore FEL facility has achieved 1 GW output power, that was for a 10 to 20 nanosecond pulse (at a repetition rate of 0.5 p.p.s.) in the microwave region of 8.7 mm.[30] The corresponding average power over 1 second is 10 to 20 Watts — eight orders of magnitude away from the required average power.

The FEL source itself introduces divergence in the beam that will be determined by the design of the optical cavity. Whether an oscillator or an amplifier is being discussed, the size of transmitting, reflecting or partially reflecting mirrors is constrained by the ability of those mirrors to withstand very high intensity light without degradation. This is a formidable requirement. The mirrors in the system must withstand at least an order of magnitude more intensity than the intensity necessary for target destruction. Such mirrors do not exist, although research is being conducted.

To minimize mirror damage a maximum mirror diameter is desirable. At this time the largest reasonable projected size[22] is about 10 m. This upper limit constrains the geometry of an FEL amplifier or oscillator optical cavity. A laser beam of minimum diameter d at the center of the oscillator cavity (or at one end of the wiggler in an amplifier) will diverge by roughly λ/d radians. Over a distance L that beam will spread to $D \simeq L \times \lambda / d$ (up to factors of 2π for specific geometries). Then with typical electron beam diameters 1 mm, the lengths required to present the mirrors with a 10 m beam are 20 Km. Such lengths are being proposed at this time.[3] The cost of building optical cavities of these macroscopic dimensions will be considerable.

Even with very large diameter mirrors it is not known how resistant to FEL harmonics the materials will be; degradation has been observed at quite low intensities.[31] Harmonics of the fundamental FEL frequency will inevitably accompany the laser beam due to the non-sinusoidal wiggler and resulting electron motion. For the primary output in the visible region, harmonics in the ultraviolet (UV) can be quite pronounced. This has two undesirable effects. A significant amount of the electron beam energy can be concentrated unavoidably into UV rather than visible electromagnetic radiation. While this may provide a means for achieving workable UV lasers[32] for research and commercial technology, the strong atmospheric absorption of UV radiation discourages the use of UV lasers as boost phase weapons. Secondly, high intensity UV radiation will be strongly absorbed by metallic surfaces when $\lambda \leq 1000$ Å, the wavelength corresponding to the typical plasma frequency for metals. When absorbed by the electrons in the metal, the UV energy is transformed into heat which will lead to distortion, melting and/or fracture of the reflecting surface.[18] Whether or not more UV damage resistant material can be fabricated in the future remains to be seen.

How much power will the laser need to emit at the desired wavelength? Previously, estimates were made of the power that must be reflected from 10 meter diameter diffraction limited geosynchronous mirrors in order to supply the battle mirrors with sufficient power to inflict damage on the targets. For perfectly reflecting mirrors, the power reflected is exactly the power incident. But for real mirrors the reflection coefficient or reflectivity is less than unity and so the incident power is the reflected power divided by the reflectivity. Then, ignoring all other losses of power, the laser must emit the estimated 0.9 to 24 GigaWatts divided by the product of the reflectivities of the transmission, geosynchronous and battle mirrors. With the ideal 10 meter diffraction limited transmission mirror there is no further loss due to diffraction since all of the intensity, in principle at least, can be focused to a width of 2.4 m within the geosynchronous mirror surface. Of course, such large diffraction limited mirrors are far from realization.[22] It is extremely optimistic to assume their existence. In any case, atmospheric losses will require further increases in output energy as will be discussed.

The geosynchronous mirror ideally would be a focusing mirror of some kind so that the 10 m beam, ignoring diffraction, could be narrowed at the low orbit region of the relay and target mirrors. However, any slight aberration would have an enormous effect over the long path length. As an example, suppose the 10 m mirror is spherical with a radius of curvature chosen so that the focal plane for parallel, nearly on-axis incident light is at 40, 000 Km. For a spherical mirror, in the small angle approximation, the radius of curvature is twice the focal distance, 80, 000 Km. Such a radius of curvature would require that the center of the mirror be 0.03 microns or 300 Angstroms deeper than the plane of the mirror's circumference — 3 parts per billion — a severe constraint to satisfy uniformly over the mirror surface. A parabolic mirror that maintains parallelism rather than focusing the beam will have the same order of tolerance — hundredths of microns. The tolerances involved here suggest that even pure plane mirrors will be problematic.

Another consideration is that the geosynchronous mirrors would have to be capable of directing the beam to relay or battle mirrors in low orbits for which the aiming requirement is extremely stringent. To get within 1 m of the low orbit mirror will require accuracy of 1 m/40, 000 Km or 25 nanoradians — half the diffraction limit. Any jitter of the mirror system on the order of half a wavelength over the full diameter of 10 m will vitiate the aim. How can such stability be maintained while being subjected to nearly destructive intensities of coherent light, following the battle mirror at about 7 Km/sec or 200, 000 nanoradians/sec, and running cooling systems? These are open questions.

Lastly, the question of orbital stability of the geosynchronous mirrors must be addressed. Associated with a beam of intensity I, normal to the surface, is a radiation pressure exerted on a perfectly reflecting mirror of $2I/c$. That pressure exerted over the large surface of the mirror accelerates the mirror away from the beam. Over the time of operation, in minutes, even these very massive mirrors can acquire considerable velocities that will change their orbits. Take, for example, the energy requirement at the mirror of 20 GigaJoules in 1 sec (over a 10 m diameter) the plane mirror case worked out above. The corresponding intensity is 27 KW/cm^2 and the radiation pressure is 1.8×10^{-4} N/cm^2, which is small compared to 1 atmosphere (10 N/cm^2). However, in the vacuum of space such a pressure is considerable. A 10 metric ton system like the Hubble Space Telescope,[22] exposed to such a pressure for 1 minute, will acquire a retreating velocity of 0.8 m/s. This is comparable to the orbital velocity of 2.4 m/s. Hence, the orbit will shift so appreciably in 1 minute that the relaying and redirecting functions of the mirror will be jeopardized. Inertial guidance combined with compensatory steering rockets must be designed. But the extreme sensitivity of the aiming task could be breached by the steering rocket operation. This problem will be quite difficult to circumvent.

V. ATMOSPHERIC PROPAGATION

Electromagnetic waves undergo many different kinds of interactions with constituents of the atmosphere. Most of the phenomena that are significant for very intense coherent beams are thoroughly studied and are well-known.[33]

Rayleigh scattering by molecules and particulates removes photons from the coherent beam with cross sections that are inversely proportional to the fourth power of the wavelength until the wavelength is the size of the scatterer. For shorter wavelengths the cross section is a constant proportional to the scatterer's area. In the visible region, about 0.5 microns, the net effect of Rayleigh scattering for a vertically propagating beam is to transmit about 87% under perfectly clear conditions. This is a comparatively small reduction.

Molecular and particulate absorption are rapidly varying functions of wavelength in the near infrared; less so in the visible. This variability is a result of molecular rotational and vibrational excitations. Absorption and scattering by water droplets is quite severe and causes the impenetrability of clouds by near IR or visible light.

At high intensities, non-linear effects begin to appear. Dielectric breakdown occurs at the 10^7 Watt/cm^2 intensity regime. The heating that accompanies light absorption can alter the index of refraction sufficiently to bend the beam outward thereby increasing the beam diameter and decreasing the intensity. This is the "thermal blooming" effect.[19] Both effects can be avoided by staying below threshold intensities on the order of Megawatts/cm^2 and using short pulses. Whether or not those limitations will be too severe for adequate beams to reach their targets depends on the particulars of design.[33] If the purely diffraction limited case of the previous section is taken as the most optimistic example, with 10 m diameter mirrors for transmission and geosynchronous relay, the power output of the laser must be in the GigaWatts. Over the 10 m mirror that gives an average intensity of KiloWatts/cm^2. But if that laser energy is concentrated in several millisecond pulses in each second, the intensity will peak in the MegaWatts/cm^2 anyway. There is an optimization problem here, which may not have a solution for the high energies required for effective weapons.

Normal atmospheric turbulence produces variations in density and hence in the index of refraction that tends to destroy the coherence of a large diameter beam. A measure of that phenomenon is the transverse coherence length which can be understood as follows.[19] Consider two parallel waves, separated by a distance r, that begin propagating upward through the atmosphere with the same phase. As they travel, the random variations in index of refraction along the two upward paths alter the phases relative to an unvarying index of refraction. If r is small enough the resulting difference in the two phases will be small, *i.e.* the two rays have traversed nearly the same path, and can constructively interfere. But for separations on the order of the transverse coherence length the relative phase tends to vary by well over 90 degrees; the two waves are no longer coherent. For visible light of 0.5 microns the transverse coherence length is about 10 cm. Then an 0.5 micron beam of diameter much greater than 10 cm will be incoherent when it emerges from the upper atmosphere. This would seem to put severe limits on the size of the beam, but "adaptive optics" techniques are being developed that could alleviate such restrictions.[19]

In this context adaptive optics refers to those varied techniques for predistorting the wavefront of the emitted beam to compensate for the distortion that will result from the turbulence along the path through the atmosphere. The predistortion could be accomplished by physically altering the surface of the transmission mirror or by using various optically active apertures and lenses to alter the phase across the wavefront.[34] Of course, the pattern of density variations is changing continually and in roughly 5 millisecond, the correlation time, the pattern is essentially randomly related or uncorrelated with the previous configuration of density variations. So whatever technique is used for predistortion, it must be rapid enough to readjust the predistortion every 5 msec.

In order to accomplish the appropriate predistortion, the transmitting system must be informed of the density variations along the path at the time the beam is to be transmitted. This can be done by having a relatively low powered "calibrating" laser at the site of each geosynchronous mirror. Each calibrating laser sends an appropriate beam to the ground based laser to which it is paired. The beam is analyzed at the ground station and the requisite predistortion determined. The entire process must occur well within the millisecond time constraint, a formidable requirement for a mechanical system.

There will be turbulence effects on the downward leg as well if fast burn boosters, that finish their burn well within the atmosphere, are deployed. Compensation for the resulting loss of coherence as the beam propagates from battle mirror to target would require a similar adaptive optics loop operating between target and battle mirror. But the light beam travels only 1500 Km in the correlation time of 5 msec, and that distance is a typical distance from battle mirror to target. No time remains for analyzing the density variations. This unavoidable loss of coherence will lower the destructive capability of the beam and require even larger fluence at the target.

Another phenomenon has become of central significance for the direction of future FEL facilities. Stimulated Raman Scattering is a non-linear process that will significantly deplete a continuous beam of its coherence for intensities in the MegaWatt/cm^2 range.

Recall that Raman scattering is the inelastic scattering of light by a molecule in which the radiated molecule undergoes a transition to a different energy level while the outgoing photon's energy is shifted correspondingly (Stokes or anti-Stokes lines). For visible light the probability for Raman scattering of a *single* photon off a *single* atmospheric molecule is quite small (cross sections[35] of order microbarns). However, for scattering in the atmosphere the process can be sizeable. The probability for such a single scattering in the atmosphere is proportional to the number of incoming photons of both incident and shifted frequency, *i.e.* stimulated scattering occurs and more energy shifted photons emerge after each scattering. For high enough intensities, the incident and outgoing photons effectively interact in the molecular medium; the situation becomes analogous to stimulated emission in laser media.

At sufficiently high intensities in the MegaWatt/cm^2 region, the stimulated inelastic scattering enters a gain regime. Then the scattering depletes the incoming collimated coherent beam significantly since Raman scattering favors large angle scatters. It has been found that atmospheric N_2 has rotational excitations that enhance this phenomenon.[36] For altitudes below 45 Km (where pressure broadening determines the N_2 resonance absorption width) and visible light, the gain per unit length per unit intensity is roughly (5 to 10) $\times 10^{-6}$ per cm per MegaWatt/cm^2. Over a full vertical path of 45 Km that becomes 23 to 45 per MegaWatt/cm^2. When the overall exponential gain reaches about 25 the stimulated Stokes line beam achieves roughly 1% of the intensity of the incoming laser beam. The intensity of the laser beam for which that 1% is reached is the *threshold* intensity. Above that intensity the non-linearity of the system of incident and scattered intensities drives the incident beam to depletion. For the case in point, the threshold[36] is at only 1 to 2 MW/cm^2, far below the required weapon intensity!

What becomes of the inelastically scattered light, which is also coherent? Because of the large angle scattering, that beam diverges from the collimated incident beam. The divergence is simply related to the width of the incident beam aperature, D, and the height of atmosphere over which the gain-regime scattering occurs, h. Consider that cylinder of diameter D and height h essentially resting with its base at the laser transmission mirror. Just at threshold intensity the 1% stimulated scattered light rays follow paths that traverse the full height of the cylinder. Hence, the scattered rays are contained in a diverging cone of apex angle roughly D/h. For the visible light of concern, with a 10 m transmission mirror the diverging angle is $10\,\text{m}/45\,\text{Km} = 2.2 \times 10^{-4}$ rad. This angle is quite large compared to $\lambda/D \simeq 5 \times 10^{-8}$ for diffraction of the laser beam, and essentially destroys the effectiveness of the laser for intensities over the threshold. It is easy to see that as the incident intensity increases above threshold the height of the cylinder decreases proportionally (to maintain constant gain of 25) and so the conic divergence angle increases directly as the intensity above threshold.

Now these considerations apply to light beams that are effectively continuous, which means that the beam pulses are of long duration compared to the molecular relaxation times. The relevant N_2 rotational level relaxation times ($1/2\pi\Delta\nu$, where $\Delta\nu$ is the pressure broadened line width of Fig. 5.22 in Ref. [35]) are in the range 30 psec to 20 nsec from the bottom of the atmosphere up to 45 Km.

What effect will this Stimulated Raman Scattering have on ground based FELs? With 1 GW *average* powers required for lethality and 10 m mirrors, the average intensities will be $\sim 1\,\text{KW/cm}^2$, well below threshold. However that power is output in pulses. As an example of this point suppose the necessary 1 GW average laser output is concentrated into 1000 pulses of 50 nanosecond each second, for a duty factor of 5×10^{-5}.

The average power in the pulse is then 2×10^7 MW. With a 10 m transmission mirror the output intensity will be 26 MW/cm^2, far above the threshold.

Now this last example is especially relevant for the design of a ground based FEL boost phase weapon. As will be discussed in the next section, the two favored systems involve an induction linac and an rf linac. The induction linac produces very high current electron bunches with pulse times of 10's of nanoseconds, long by the molecular relaxation time comparison, and at very slow repetition rates. The rf linac produces considerably lower current bunches, but with 10's of picosecond pulse times at very high repetition rates. The parameters chosen in the last example are those appropriate for a scaled up induction linac FEL amplifier[37] (repetition rates of several Hertz are available now rather than the projected KiloHertz). With the pulse times obtainable however, the example shows that there will be so disastrous a depletion as to make the scaled up *induction linac* FEL amplifier design a doubtful venture unless repitition rates can be brought to the 100 KHz range and intensity can be reduced by spreading the beam at the source. This latter strategy would require narrowing the beam at the relay mirror without losing coherence within the expanded beam in the atmospheric traversal; quite a daunting problem. For the much shorter times and greater duty cycles of an *rf linac* FEL oscillator, this effect may be circumvented. Then, however, the system is inherently less efficient. Whatever the case for the two proposed scaled up FEL's, this Stimulated Raman Scattering is a critical factor in the successful functioning of the system.

VI. PRESENT ACHIEVEMENTS AND FUTURE PROSPECTS

At this time there are essentially two basic schemes for high power and short wavelength FEL's: the FEL amplifier using an induction linac; an FEL oscillator using an rf linac. Of these, the most spectacular successes to date have been the induction linac experiments by the Livermore group.[30,37,38,39] These will be reviewed first.

Induction linear accelerators are very high current machines. At Livermore currents of Kiloamps are routine. The most noteworthy results[30] reported that extremely high efficiencies of 30% to 40% were obtained. This was achieved by using active tapered wigglers consisting of electromagnets that were adjusted on-line for optimization. While this was certainly an impressive technical achievement, the implications for space weapon development are not clear as an examination of the parameters of this FEL reveal. With the 3.5 MeV linac (ETA), and the relatively widely spaced wiggler electromagnets, the laser wavelength was 9 mm — microwave. The output peak power reported was about 1 GW, while the input peak power was about 3 GW (electron energy in eV times peak current in amps or 3×10^6eV $\times$ 1×10^3Amps). With one 10 to 20 nsec pulse every 2 seconds the average power is only 5 to 10 Watts — nine orders of magnitude away in terms of the required power for a weapon and four orders of magnitude away from the desired wavelengths.

How have these results been scaled up in the meantime and what are future prospects for the induction FEL amplifier? Recently the higher energy linac (ATA) of 45 MeV was operated as an FEL amplifier in the IR regime of 10.6 μ, but with considerably (and predictably) lower efficiency.[37] Part of the reason for the low efficiency was due to the untapered wiggler; the tapered section had not been connected yet. However, the reported peak output power of 7 MW can be compared to the input electron peak power of 140 GW (for a beam current of 3 KAmp) to give an efficiency of 5×10^{-5}. Such a low effciency indicates that beam brightness or stability was not maintained from the injector into the wiggler region. Either current was lost or emittance increased or anticipated beam instabilities for the scaled up energies were not corrected.

Such beam instabilities have been countered experimentally by channeling through low density ions within the ATA[40], but beam quality is spoiled thereby, [41] making this technique problematic for FEL operation.

The pulse times for these machines are 10 to 50 nsec. However, because of the large massive electromagnetic induction units that need to be charged up and discharged, the repetition rate for the pulses has been only on the order of 1 p.p.s. Research into magnetic switching has produced injectors that achieve 1000 p.p.s., but an FEL that routinely runs at that rate has not been achieved.[42]

The relatively long pulse of 50 nsec gives rise to electron bunches of approximately $ct_{\text{pulse}} \simeq 15\,\text{m}$ that fill the wiggler region and act like a long amplifying medium for co-propagating EM waves of the synchronous wavelength. At high gain that "medium" could refract the optical beam producing a guiding effect, as in fiber optics, that keeps the optical beam from diffracting in the wiggler region.[43] Such "optical guiding" has been demonstrated[37] but has not given rise to the increased efficiency that was expected.

With this mixture of interesting achievements and daunting problems, can the induction FEL results be improved? The repetition rate and electron energy must be increased. This requires adding more induction units that are magnetically switched. A factor of 4 or 5 in the beam energy is necessary to bring the laser into visible which means a much longer system. With the ATA at 80 m length, the linear increase of output energy with length necessitates more than one-third Km of induction units. Over such distances the beam quality and stability will be difficult to maintain. The emittance must be smaller for the electrons to safely traverse a longer beam channel; a factor of one-fourth or one-fifth decrease in emittance will be necessary. Furthermore, it can be anticipated that at the considerably higher energies required unanticipated instabilities will arise.

Being quite optimistic about scaling up the induction system by increasing the repetition rate to 10 KHz, increasing the energy (or decreasing the wiggler spacing) to 200 MeV while keeping the same beam quality (contrary to experience) and maintaining 1 KAmp of current, the average input power, assuming 50 nsec pulses, is 100 MW while the average output will be that number multiplied by the efficiency. With the low gain, untapered wiggler falloff of efficiency as the 3/2 power of energy, an 0.20% efficiency would be expected. For a high gain, tapered wiggler system the efficiency can be considerably higher; an optimistic estimate would be 10%. Then the average power will reach 10 MW at most or some six orders of magnitude beyond the best result reported.[30] To get nearer the weapon power requirement a higher energy electron beam is needed, while maintaining a high brightness or current and a low emittance. This seems far fetched given the many years of accelerator technology development.

Next to consider is the rf linac oscillator for which the Los Alamos work is the most advanced towards high power and short wavelength[18,32] The rf linac (and the storage ring) uses an oscillator design in which the electron bunch passes through the wiggler and emits the synchronous EM radiation. As the electron bunch leaves the cavity the photon bunch is reflected and makes a round trip in time to meet the next electron bunch which is stimulated to further emission. The very delicate tuning necessary for this lasing to occur makes it difficult to achieve high efficiency: 5% to 10% is very good. The factors contributing to this relatively low efficiency involve the mismatch in size of the electron beam bunch, the wiggler and the photon bunch. With typical "micropulse" lengths in an rf linac of 20 psec separated by 45 nsec the electron bunch separations are about 1 cm in length which is only the order of one period of the wiggler. Longer electron bunches result in more stimulated emission for the single pass through the wiggler but would have lower current and brightness. Secondly, the transverse dimension of the electron bunch is considerably narrower than the optical pulse which stimulates further emission. Where the electron and photon bunches do not overlap no interaction occurs and efficiency is reduced further. These effects are characteristic of the rf linac oscillator design and limit efficiency significantly.

The Los Alamos experiments have operated[18] at 20 MeV (for a 10 μ infrared laser), with peak currents exceeding 300 Amps and have produced peak power outputs of 40 MW. While the extraction efficiency was 2%, losses in the optical cavity reduced the output by one third. However, rf linacs have short pulses, micropulses of 3.5 psec separated by 46 nsec in the reported experiment, so that the average power in *one macropulse* is 30 KW. At present LANL only produces about one macropulse of 120 μsec per second, or 3.6 J of energy in each macropulse. Then the actual average power is only 3 W. At the proposed repitition rate of 50 Hz[17], the average power would be 180 W.

Again many orders of magnitude improvements are needed in various accelerator parameters. The current obtained in these experiments was high for an rf system, 300 amps, and improvements by more than a factor of two or three are not anticipated. Increasing efficiency through recirculating the electrons can help compensate for loss of efficiency at higher electron energy linacs.[18] But overall the prospects for even three orders of magnitude increase in average power to 20 MW are rather dubious. It should be noted also, that beam instabilities and significant sideband production become problematic even at the energies already studied at Los Alamos.[32] A recent experiment at Boeing[44] achieved FEL oscillations in the visible, 0.5 microns, showing that the wavelength can be scaled down, but not demonstrating high power.

VII. CONCLUSIONS

Over the past 15 years the free electron laser has developed from a simple but elegant theoretical proposal to a laboratory instrument with the promise of great versatility and capacity. The wide range of wavelengths and relatively high powers accessible will undoubtedly prove very useful for molecular, atomic and materials research as well as for medical and industrial applications. With such achievements and future promise it is perhaps inevitable that the FEL's possibilities as a directed energy weapon would be explored.

In the preceding, the considerable requirements for a boost phase defense system centered upon ground based FEL stations have been examined. In order to thwart a nearly simultaneous launch of 1400 Soviet ICBM's, a very large contingent of FEL facilities must be placed about the globe: a lower limit was 210. The corresponding expense for the FEL's alone could reach a trillion dollars. Power needed to be immediately available would be comparable to the entire current capacity of the U.S.

The transmitting and geosynchronous mirrors in the system must be larger than any built with tolerances orders of magnitude more stringent than those satisfied at this time. Heat resistance, figure stability and steerability could be unachievable. Under the sizeable radiation pressure orbital stability and guidance become quite problematic for large diameter geosynchronous mirrors. Furthermore, while this has not been discussed extensively, it is clear that there must be as many geosynchronous mirrors as there are operating FEL's on the ground. That is, at least the $\sim$ 48 (for long boost times to $\sim$ 70 for short boosts) that will be clear of cloud cover at a given time.

Low orbit relay or battle mirrors must number in the hundreds. While they can be smaller in diameter than the transmission or geosynchronous mirrors, since they will be much closer to the targets, putting that much mass into orbit will be very long-term and expensive. This part of the system will be the most vulnerable to easily built anti-satellite weaponry.

The known difficulties with the propagation of high intensity coherent EM radiation through the atmosphere put severe limits on the intensities. At the MegaWatt per square centimeter level non-linear effects of dielectric breakdown, thermal blooming and Stimulated Raman Scattering become quite restrictive. Very short pulses or beam spreading techniques are two strategies for circumventing such limitations. But short pulse FEL's based on rf linacs are inherently less efficient than FEL amplifiers using induction linacs. And beam spreading over diameters of about 10 m may prove impossible to narrow down again with adaptive optics at the geosynchronous mirror because of the smaller scale for spatial coherence in turbulence of 10's of centimeters.

The well-known problem of absorption of near infrared and visible light by water molecules in the atmosphere imposes an insoluble problem on boost phase interception. ICBM's launched under cloud cover will be unreachable until they enter the upper atmosphere. Then with fast burn solid fuel boosters there will be little or no time for laser interception. The variability of weather conditions over the entire range of Soviet silos insures that a significant number of ICBM's will escape an FEL boost phase defense.

Finally, the projected parameters for the FEL's themselves are not encouraging for weapons applications. The two most advanced space weapon designs are the Livermore induction linac FEL amplifier and the Los Alamos rf FEL oscillator. Either of these designs needs enormous scaling up to meet the requirements of GigaWatts average power and near visible wavelengths.

In the induction scheme the energy must be increased to the hundreds of MeV's, unprecedented for such linacs. Whether that can be done while maintaining the characteristically high electron currents along with the low emittance essential for high efficiency is questionable. The necessary increase of repetition rate to increase the *average* power will be much more difficult for the many more induction units that will be needed to scale up the energy. Lastly, the 20 or so nanosecond pulse time will probably be too long to avoid serious beam degradation from Stimulated Raman Scattering for the anticipated intensities. This will be a problem at both ends, especially on the downward leg through the atmosphere to the fast burn target, where intensities will have to exceed Stimulated Raman thresholds.

The rf linac design can be scaled up in electron beam energy quite readily, and significantly higher current may be possible compared to existing intermediate energy machines. But decreased electron beam brightness will decrease the gain-efficiency product as will lower relative energy spread tolerance. Given that the efficiency of the current Los Alamos facility is less than 10%, much lower figures can be expected for a scaled up version of this design. Furthermore, the appearance of disastrous beam instabilities is not unexpected. Thus, neither the induction linac nor the rf linac can be expected to provide the extremely high average power and visible wavelength laser beam that is desirable for a ground based space weapon.

Given such pessimistic projections, all of which have been stated in the now extensive literature on directed energy weapons, the SDI program and the FEL research, are there any reasons to build a very large FEL facility? Two possibilities are to develop mid-course or terminal phase FEL weapons. Will these be more possible than boost phase interception?

For mid-course interception, the same kind of overall system would be required, but the longer targeting time available, 20 to 30 minutes, would reduce the number of independent FEL's needed, all else being *equal.* However, in mid-course there will be up to 10 times more *real* targets, the MIRV's released in the post-boost busing phase. Those RV's are seven or eight times harder than ICBM skins in order to withstand heating in reentry. Then fluences of about 150 KJ/cm^2 are needed for damage and seven or eight times higher power FEL's will be required. Furthermore, there could be 100 times more decoys in the incoming swarm. This certain large increase in targets makes the defensive task even more difficult than in boost phase; more FEL sites will be necessary.

Terminal phase defense is a very different kind of problem. Reentry vehicles will take only about 1 minute to reach their targets. FEL's designed to directly intercept these RV's must be placed so that the RV's are in line of sight. Shallow reentry trajectories make this task more difficult and lead to a design in which the FEL's form a large diameter ring defense. Such defensive rings will have to be situated around all major target cities to have any chance to provide adequate defense. However, one large yield explosion makes the entire ring defense irrelevant. Both high yield warheads and salvage fusing will be effective countermeasures. The FEL's themselves would have considerably lower average power requirements than boost phase FEL components, by one or two orders of magnitude since diffraction effects would be negligible over such relatively short distances. Nevertheless, atmospheric propagation problems would be more severe with the laser beams having to traverse much longer paths in the atmosphere. Furthermore, local cloud cover or the dust raised by a single nuclear explosion could vitiate the entire defense.

For point defense of retaliatory missile silo sites, an FEL system need not be nearly 100% successful to be significant. Only a 10% effectiveness could be sufficient to insure a "credible" retaliation. But for this task, which is restricted by treaty, other means, such as interceptor missiles, may be adequate and much cheaper.

Lastly, for the much easier role as an anti-satellite weapon (ASAT),[45] the ground based FEL could prove more able to meet the requirements. To begin with, satellites with military functions are in a variety of different orbits - low, highly elliptical, high and geosynchronous. Those in low and elliptical orbits will be vulnerable to exploding or colliding ground launched interceptor rocket devices that are currently under development by both superpowers. But directed energy weapons, travelling near or at the speed of light, could threaten hardware in any of these orbits. What makes the task easier than that of the anti- ballistic missile weapon is the long duration available for tracking, targeting, aiming and assessing the damage.

Existing satellites are much softer targets than ICBM's. Less laser energy is needed to burn or fracture the skin. Such damage can be inflicted with less power delivered to the target, by dwelling on the satellite over many minutes or even hours for high orbits. Furthermore, solar panels, photocells, optical and infrared sensors are damaged by intense, but relatively low energy pulses, e.g. light telescopes are easily blinded by existing laboratory lasers.

How significant could such a threat be? A ground based near visible FEL with several orders of magnitude lower average power than necessary for SDI purposes could easily be an ASAT weapon. The limitations on average power output and propagation through the atmosphere are not as severe. However, target satellites must be in view of the laser itself, which tightly restricts engagement time and numbers of targets, unless orbiting relay mirrors are incorporated. Relying on relay mirrors makes the system vulnerable to counter ASAT attack.

Of course protective measures like hardening or shielding of satellites could be employed, although these measures necessitate larger payloads to be lofted into orbit at greater expense. Countermeasures varying from filling relevent regions of the atmosphere with smoke to sabotage or direct attack of lasers and power supplies could be envisioned. So similar questions about survivability and cost effectiveness arise for ASAT applications as for boost phase SDI weapons. Furthermore the larger political questions of escalation and destabilization must be addressed.

In summary, the free electron laser has a dubious prognosis as an effective and extensive ground based defense against nuclear attack.[46] The proposal to attempt to scale up one of the existing FEL designs for a large test facility is difficult to justify given the many orders of magnitude by which that proposed FEL will fall short of ground based laser weapon requirements.

ACKNOWLEDGEMENTS

The author thanks Kosta Tsipis for his interest and encouragement, and J.Paton, R.Bulkeley, R.Peierls, E.Eisenhandler, G.Bekefi, J.Wurtele and E.Raitan for critically reading an earlier version of the manuscript.

REFERENCES

1. G. Bekefi, B. T. Feld, J. Permentola and K. Tsipis, *Nature* **284**,219 (1980).
2. Council on Economic Priorities, *Star Wars: The Economic Fallout* (Ballinger, New York, 1987).
3. "Army to Decide FEL Site in January," *SDI Monitor* **1** (Pasha Publications, December 1986), p. 284; W. J. Broad, "New Laser Offers Great Power and Flexibility," *Science – Times* (New York Times, September 30, 1986); P. J. Klass, "Technical Survey: Free-Electron Lasers," *Aviation Week and Space Technology* **125**, 40 (August 18, 1986); W. J. Broad, "Anti-missile Laser Project Is Delayed Nearly 2 Years," *New York Times* (April 17,1988).
4. General J. Abrahamson, Director, SDI Organization, quoted in W. J. Broad, *op. cit.*
5. S. P. Worden, Special Assistant to the Director, SDI Organization, "SDI", *Physics and Society* **15**, 2 (September 1986).
6. H. Motz, *Jour. Appl. Phys.* **22**, 527 (1951); K. M. Phillips, *Trans. IRE Elect. Dev.* **7**, 231 (1960); J. M. J. Madey, *Jour. Appl.. Phys.* **42**,1606 (1971).
7. For reviews on the fundamental principles of the FEL see T. C. Marshall, *Free Electron Lasers* (Macmillan, NY, 1985); U. Bizzarri *et al.*, *Rivista de. Nuovo Cimento* **10**, 1 (1985); W. B. Colson and A. M. Sessler, *Ann. Rev. Nucl. and Part. Sci.* **35**, 25 (1985).
8. W. B. Colsen, *Phys. Letters* **64A**, 190 (1977).
9. W. M. Kroll, P. M. Morton and M. N. Rosenbluth, *Phys. of Quant. Elect.* **7**, 115 (Addison–Wesley, Reading, MA, 1979).

10. J. M. J. Madey, *Nuovo Cimento* **40B**, 64 (1979).

11. J. M. Slater, *IEEE Jour. of Quant. Elect* **QE-17**, 1479 (1981); D. C. Quimby and J. M. Slater, *IEEE Jour. of Quant. Elect.* **QE-21**, 988 (1985).

12. G.J.Caporaso, *et.al.*, *Phys. Rev. Lett.* **57**, 1591 (1986).

13. See p. S61 in "Report to the American Physical Society of the Study Group on Science and Technology of Directed Energy Weapons" (N. Bloembergen and C. K. Patel, co-chairmen), *Rev. Mod. Phys.* **59**, S1 (1987).

14. Some of these projections are based on proposed and accepted programs reported by P. J. Klass, *Aviation Week and Space Technology* **125**, 40 (August 18, 1986), *op. cit.*

15. S. Rockwood quoted in P. J. Klass, *op. cit.*, p. 61.

16. P. J. Klass, *op. cit.* p. 66.

17. P. J. Klass, *op cit*, p. 63 and 67.

18. D. W. Feldman *et al.*, *IEEE Jour. of Quant. Elec.* **QE-23**, 1476 (1987).

19. M. Callaham and K. Tsipis, "High Energy Laser Weapons," Report No. 6, Program in Sciences and Technology for International Security (MIT Press, Cambridge, MA, November 1980).

20. S. D. Drell, P. J. Farley and D. Holloway, "The Reagan Strategic Defense Initiative," Special Report, Center for International Security and Arms Control (Stanford University, Stanford, CA, July 1984).

21. R. Bulkeley and G. Spinardi, *Space Weapons: Detterence or Delusion?* (Polity Press, Cambridge, UK, 1986); R. M. Bowman, *Star Wars: Defense or Death Star?*, Institute for Space and Security Studies (Washington, D.C., 1985).

22. G. Field and D. Spergel, *Science* **231**, 1387 (1986).

23. D. Waller, J. Bruce and D. Cook, "SDI Progress and Challenges," U.S. Senate Staff Report — Senators Proxmire, Johnston and Chiles (March 1986).

24. P. Stein, "The SDI Battle Management System," Report No. 14, Program in Science and Technology for International Security (MIT Press, Cambridge, MA, 1985).

25. The one second dwell time is a typical figure that represents a compromise among competing constraints. Longer dwell times enable lower laser power to be effective, but target tracking ability and damage dissipation on the target necessitate shorter dwell times.[19]

26. See p.S99 of Ref. [13].

27. See all articles in Ref. [3] for an estimate.

28. The APS study group considered a somewhat different configuration, including ground telescopes, and concluded that 50 - 70 devices, with 3 telescopes each, would be needed for the *current* ICBM threat. See p.S117 of Ref. [13].

29. J. C. Fletcher, "Ballistic Missile Defense," *Issues in Science and Technology* (Fall 1984) p. 25; and Refs. [11, 14].

30. T. J. Orzechowski *et al.*, *Phys. Rev. Lett.* **57**, 2172 (1986).

31. M. Billardon *et al. Phys. Rev. Lett.* **51**, 1652 (1983)..

32. C. A. Brau, *IEEE Jour. of Quant. Elect.* **QE-21**, 824 (1985).

33. See Ref. [19] and recent news item "Scientists Discover Potential Problems with Ground-Based Free Electron Lasers," *Aviation Week & Space Technology*, p. 28 (May 4, 1987); see Ref. [13] for a review.

34. See pp. S101 – S108 of Ref. [13] for a comprehensive summary.

35. See pp. S108 – S111 of Ref. [13].

36. M. Rokni and A. Flusberg, *IEEE Jour. of Quant. Elect.* **QE-22**, 1102 (1986).

37. D.Prosnitz, "FEL's Driven by Induction Linacs", *American Physical Society*, Meeting (Baltimore, April 20, 1988).

38. T. J. Orzechowski *et al.*, *IEEE Jour. of Quant. Elect.* **QE-21**, 831 (1985).

39. B. M. Schwarzchild, "Free Electron Lasers Achieve High Microwave Power," *Physics Today*, p. 17 (April 1985).

40. G.F.Caporaso, *et.al.*, *Phys. Rev. Lett.* **57**, 1591 (1986).

41. See p.S.60 of Ref. [13].

42. D.L.Birx, *et.al.*, in *Free Electron Lasers, Proceedings of the 7th International Conference* (North Holland, 1986).

43. F. Hartemann *et al.*, *Phys. Rev. Lett.* **59**, 1177 (1987); E.T.Scharlemann, A.M.Sessler and J.S.Wurtele, *Phys. Rev. Lett.* **54**, 1925 (1985).

44. D.Shoffstall,"FEL's Driven by RF Linacs",*American Physical Society*, Meeting (Baltimore, April 20, 1988).

45. U.S. Congress, Office of Technology Assessment, "Anti-Satellite Weapons, Countermeasures and Arms Control", OTA-ISC-281 (Washington, D.C., 1985); and pp.S181-S185 of Ref. [13].

46. See G.R.Goldstein, Report No. 18, Program in Science and Technology for International Security (MIT Press, Cambridge, MA, 1987) for an earlier version of this report. Subsequent developments have not altered the conclusions.

CHAPTER 13

DEVELOPMENTS IN GEODESY AND THE ACCURACY OF STRATEGIC WEAPONS

T.M. Eubanks

ABSTRACT

Recent improvements in the accuracy of ballistic missiles have produced doubts in the survivability of even the hardest military targets, such as land based Inter Continental Ballistic Missiles (ICBMs), during a nuclear war. Although most of the details of the performance of current missile guidance systems are classified, reasonable estimates of current missile accuracy can be obtained from the open literature. The basic physical principles underlying the missile accuracy problem will be discussed.

BALLISTIC MISSILES AND THEIR TRAJECTORIES

Advances in technology should make it possible to reduce the navigation errors of Inter Continental Ballistic Missiles (ICBMs) to a few tens of meters or less within the next 10 to 20 years. Although most of the details of the performance of missile guidance systems are classified, the basic physical principles involved in these accuracy increases are easily derived from the open literature. In this paper I will discuss some of the technological advances required for dekameter level missile accuracies and some of thelarger consequences of these accuracy increases.

It might be thought that the existence of radio guidance systems would remove any difficulties in missile navigation. The U.S. Government currently has under development the satellite based Global Positioning System (GPS), which should, when completed, provide 10 meter level position accuracy in real time to any location on the globe [1,2]. It is hard to see, however, what role satellite positioning systems such as the GPS would have in strategic missile guidance, although they could play an important role in the guidance of tactical weapons in lesser conflicts. No deterrence system that relies upon radio communication with a system of satellites sitting unprotected in high Earth orbit could realistically be regarded as secure from countermeasures duringan all out war. In this paper I have thus assumed that the missile guidance systems have no communication with external aids such as GPS.

As the name suggests, a ballistic missile travels on a ballistic trajectory during most of its flight. A long range ballistic missile trajectory can be divided into three separate phases, the early powered flight (ending at burnout), the free-flying ballistic flight, ending at re-entry in to the atmosphere, and the re-entry itself. Figure 1 presents a typical long range ballistic trajectory and makes it clear that a ballistic missile is really a satellite whose orbit happens to intersect the surface of the Earth. The analysis of a sub-orbital trajectory is thus essentially the same as for any space vehicle trajectory, and as such is described in great detail in the open literature[3,4,5].

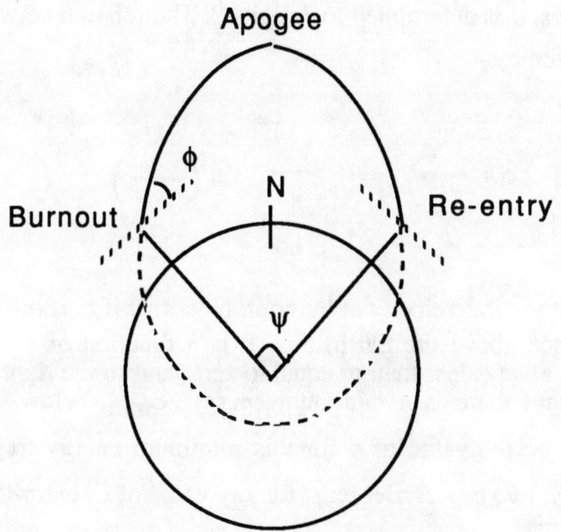

Figure 1 : Model Missile Trajectory

This paper is largely concerned with errors introduced during the flight of an intercontinental range ballistic missile. These errors will of course depend on the type of mission and the trajectory flown, and in particular they will be greatly influenced by the choice of the free-flight range angle ψ and the flight path angle ϕ, as defined in Figure 1. In this paper I have decided to analyze in detail a representative flight for which ψ is exactly 90 degrees, corresponding to a ground range of 10,053 km (assuming a spherical Earth). For a given ground range, the missile trajectory is essentially a factor of the total energy imparted during powered flight, which determines ϕ. The energy per unit mass E is given at any time by

$$E = \frac{v^2}{2} - \frac{GM_{Earth}}{r} \qquad (1)$$

where v is the spacecraft velocity, r is the distance to the center of mass of the Earth, M_{Earth} is the Earth's mass (~ 5.976×10^{24} kg) and G is the Gravitational constant (~ 6.672×10^{-11} m^3 kg^{-1} s^{-2}). GM_{Earth} appears so often in orbital mechanics that it is given the symbol μ; although G is only determined to 4 significant figures, μ is determined to 7 digits [6]. The relation between the total energy E, ψ and ϕ is given by [5]

$$\sin\left(2\phi + \frac{\psi}{2}\right) = \frac{-2E}{v_{bo}^2} \sin\left(\frac{\psi}{2}\right) \tag{2}$$

where v_{bo} is the total velocity at burnout. Note that if burnout is assumed to occur at a fixed distance above the Earth, then E is a function of v. The total energy as defined above is always less than or equal to zero, and so the right hand side is always positive. Note that there is a minimum energy, E_{min}, below which no trajectory exists; let ϕ_{min} be the value of ϕ for this minimum energy trajectory. Above the minimum energy, two trajectories exist for any value of E, one with $\phi > \phi_{min}$, and the other with $\phi < \phi_{min}$, called the high and low (or depressed) trajectories, respectively.

It is useful to have a representative trajectory to evaluate the effects of various perturbations on accuracy. Of course the actual errors for a real mission would depend upon the actual trajectory, and on the actual course flown, but use of a sample trajectory should suffice to give an idea of the magnitude of various errors.Table 1 provides details of a minimum energy trajectory for a $\psi = 90^o$ flight. It is assumed that the powered portion of the flight is under approximately "6 gee" accelerations, in which the net acceleration with respect to the ground is a constant 49 m sec^{-2}, and that the direction of this acceleration is at a constant angle ϕ_{min} with respect to the ground. It is also assumed that re-entry begins at the altitude of burnout so that the orbit is truly symmetrical. These assumptions, although crude, will suffice for our error calculations.

Table 1 : Model Missile Trajectory For a Minimum Energy Flight Path with $\psi = 90^o$	
Launch Latitude	= Re-entry Latitude = 45°
Launch Angle ϕ	= Re-entry Angle ϕ = 22.5^o
Net Acceleration	= 49 m sec^{-2}
Time of Powered Flight	= 145 seconds
Altitude at burnout	= 196 km
Downrange at burnout	= 474 km
Velocity at burnout	= 7092 m sec^{-1}
Time of free flight	= 2021 seconds
Alitude at apogee	= 1556 km
Velocity at apogee	= 5429 m sec^{-1}
Obital Eccentricity	= 0.414

MODERN MISSILE GUIDANCE SYSTEMS

Modern ballistic missiles are guided by an Inertial Navigation System (INS) which consist of a set of gyroscopes to maintain direction and accelerometers to measure net accelerations on the system[7,8,9]. If the position, velocity and orientation of the vehicle are known at the time of launch, then in principle the INS can estimate the position, velocity and orientation of that vehicle at any time in the future without any external aids. In practice no sensors are perfect and the position estimates of a real INS will slowly drift, or diverge, from the true vehicle location. A modern INS will use a Kalman filter to estimate the errors in the sensor readings, which will reduce but not remove their effects on the position estimate[10,11]. In the Kalman filter terminology, the complete set of information need to fully estimate the location of the INS unit, including its position, velocity and orientation and probably also internal variables such as accelerometer biases is known as the state or state vector for the system; this terminology will be used in this paper. It can be shown that navigation to a particular location (control of the state) requires essentially the same information as estimation of the current or future position without control (state estimation), in that the control problem introduces no further mathematical complications[10]. This paper will thus concentrate upon assessing how well a missile INS can estimate the its location, and will assume the existence of suitable rockets or other means of directing the missile to the desired target.

The strong accelerations experienced during the early powered flight of a missile, and during re-entry, make these portions of the flight quite different dynamically from the free flight between burnout and re-entry, and the errors introduced by these phases of flight must be considered separately from the free flight phase of the mission. The major difference is that strong accelerations greatly increase the errors introduced by the guidance system itself. Acceleration increases the torque on the INS gyroscopes, which then wobble, increasing orientation errors. Orientation errors during powered flight cause the thrust of the rocket system to be applied in the wrong direction, and thus will directly cause position (and velocity) errors. In addition, accelerometers themselves have errors, such as bias errors and random measurement noise, which will also integrate into INS position and velocity errors. Once the powered flight is over, these error sources essentially vanish, and the major remaining source of error is unmodeled variations in the gravitational field of the Earth. It should be noted that gravity field errors cannot be detected by a simple accelerometer system because of the principle of equivalence[12]. Gravity field errors can themselves be separated into errors associated with local mass distributions around the launch site, which largely affect the early powered flight, and by the larger scale field variations which cause errors during the unpowered part of the flight.

The simplest approximation to the growth of position errors during the unpowered phase of the flight is to simply ignore the gravitational influence of the Earth, which provides both a simple example and reasonable order of magnitude estimates of the size of errors expected during unpowered flight. In this case, the dynamics of a space vehicle can be described simply by

$$\frac{d^2 \underline{\delta r}}{dt^2} = \underline{\delta a} \tag{3}$$

where $\underline{\delta r}$ is the position error vector, and $\underline{\delta a}$ represents any unmodeled accelerations, presumably gravitational in nature. Nothing is lost in this example if only one component of $\underline{\delta r}$ is considered, which can de denoted simply by δr. Let the time derivative of this component by δv. Gravity field errors tend to be mostly high frequency in nature, and it is thus reasonable to regard δa, the unmodeled acceleration in the δr component, as being a band-limited white noise with a power spectral density of Q. Let

$$\underline{x} = \begin{bmatrix} \delta r \\ \delta v \end{bmatrix} \tag{4}$$

then

$$\frac{d^2 \underline{x}}{dt^2} = \begin{bmatrix} 0 & 1 \\ 0 & 0 \end{bmatrix} \underline{x} + \delta a = F \underline{x} + \delta a \tag{5}$$

$\underline{x}$ is the state vector of this system and F is known as the system dynamics matrix[10]. This is a simple example of a first order constant coefficient matrix stochastic differential equation and can be solved through the use of the state transition matrix Φ, given by Equation 6, where I is the identity matrix :

$$\Phi(\Delta t) = e^{F \Delta t} = I + F \Delta t + \frac{1}{2!}(F \Delta t)^2 + \frac{1}{3!}(F \Delta t)^3 + \ldots \tag{6}$$

The matrix exponential in this case is easily computed analytically, although in complicated cases numerical evaluation in generally required. The effect of the stochastic acceleration δa cannot be predicted in advance, and thus the prediction of the future position at some time t given initial conditions at time t_0 is just

$$\delta r(t) = \delta r(t_0) + (t - t_0)\, \delta v(t_0)$$

The prediction error is described by the state covariance matrix P, given by

$$P(t) = \Phi(\Delta t)\, P(t_0)\Phi(\Delta t)^T + \int_{t_0}^{t} \Phi(t - \tau)\; Q\, \Phi(t - \tau)^T\, d\tau \qquad (7)$$

The first term on the right hand side of Equation (7) represents the errors at time t due to initial condition errors at time t_0, while the second term describes the growth in error due to the random forcing δa; the superscript T indicates a matrix transpose. In terms of the error in δr, Equation 7 reduces to

$$\sigma^2_r(t) = \sigma^2_r(t_0) + \Delta t^2\, \sigma^2_v(t_0) + \frac{1}{3} Q\, \Delta t^3 \qquad (8),$$

assuming no correlation between position and velocity errors at time t_0. If the flight time is about 2000 seconds, and the desired accuracy is about 100 meters, Equation (8) indicates that a priori position errors should be kept to << 100 meters, that initial velocity errors should be kept to < 5 cm / sec, and that the power spectral density of any unmodeled accelerations should be kept to < 4×10^{-6} m^2 sec^{-3}. Although the latter constraint may appear obscure to readers unfamiliar with modern estimation and control theory, its interpretation is straight forward. Assume that conditions are those described for the model missile trajectory in Table 1 at burnout, with a horizontal missile velocity of 6.6 km sec^{-1} at an altitude of 196 km. Under these conditions the spacecraft is sensitive to gravity errors with wavelengths longer than ~ 500 km [13], which translates to perturbations with periods longer than ~ 75 seconds. Assume that the entire flight was flown under these conditions. The constraint on Q thus implies that, if the errors in the a priori gravity field are "white" (independent of frequency) at wavelengths longer than 500 km, then the root mean square (rms) error in the field should be less than 6×10^{-4} m sec^{-2}. The conventional unit in gravimetry is the Galileo, or gal (10^{-2} m sec^{-2}), and the 100 m desired error thus restricts the rms gravity field errors to be less than 60 milligals.

THE ACCURACY OF INS GUIDANCE DURING THE BOOST PHASE

An inertial guidance system maintains knowledge of the position and velocity of a vehicle using measurements of the vehicle's three dimensional acceleration vector together with estimates of the position and velocity obtained by some other means at some initial time[7]. The acceleration vector is generally measured in current systems by three orthogonally mounted single axis accelerometers, while, for a land launched missile, the required a priori information comes from a precise knowledge of the location of the missile silo, the target, and the rotation of the Earth. INS accelerometer errors integrate over time, and the INS position and velocity errors thus diverge slowly from the true position and velocity. The proper interpretation of accelerometer measurements requires knowledge of the accelerometer orientation; a priori knowledge of the INS orientation is maintained by a suite of gyroscopes, and gyroscope errors will cause a slow divergence of the INS orientation estimate, and thus the INS position estimate, from the truth. Modern accelerometers[8,14] generally measure the motion of a mass supported by a flexible beam, and appear to have accuracies of about 10^{-5} m sec^{-2}. There are a variety of different gyroscope types available for use in a modern INS [8,9], but published results indicates that typically the orientation error increase is on the order of 10^{-3} degrees per hour[7,9]. It should be noted that these error estimates are for operation on the ground, and that errors are almost certainly larger under an acceleration several times larger than surface gravity. The location of missile silos can certainly be determined to the dekameter level[1], and the orientation of the missile in the silo was measured at or below the arc second level even in the 1970's[15].

Some idea of the influence of INS acceleration and orientation errors can be gained by consideration of a simple model in which the missile is launched vertically with a constant acceleration. Suppose that, as in a real INS, the accelerometer measurements are sampled frequently. In this model a rms accelerometer error σ_a acts like a random acceleration σ_a at the time of each sample. The position and velocity errors after a time Δt are thus $1/2\ \Delta t^2\ \sigma_a$ and $\Delta t\ \sigma_a$, respectively. The accuracy goal of 0.05 m sec^{-1}, a powered flight time of ~ 150 seconds and a total acceleration of 49 m sec^{-1} together imply that σ_a should be $< 10^{-4}$ m sec^{-2}. The same accuracy goal and flight parameters require that the orientation error be less than 1-2 seconds of arc at the beginning of the flight; a gyro drift rate of 10^{-3} degrees per hour implies an orientation error growth of only ~ 0.2 seconds of arc during the entire powered flight. Errors introduced by publicly available INS system components are thus a factor of 5 to 10 times better than those required for 100 meter accuracies. A priori orientation errors could be increased by motions of the silo caused by nearby nuclear explosions[15,16], as a one second of arc tilt corresponds to only 0.1 mm for a 20 meter tall missile. It thus seems likely that orientation errors could be a problem if a missile is to be launched quite some time after an enemy attack,

A Stellar Inertial Navigation System (SINS) uses stellar observations to maintain the orientation of a standard inertial guidance system[9,17,18,19]. This system, although not completely isolated from external influences, should be very resistant to jamming. In SINS navigation, a standard inertial guidance system with gyroscopes is used to point a small telescope at selected stars, and the difference between the expected and observed positions of the stars is then used to update the navigation state vector of the spacecraft. Since, as we have seen, orientation errors are coupled with position errors in any INS, these stellar measurements can considerably improve the accuracy of the spacecraft position estimate. Further improvements can be had by measuring the angle between a star and the limb of the Earth [20]. It thus seems reasonable to assume that with SINS guidance missile velocity errors could be kept to below the 0.05 m sec^{-1} target at burnout, and position errors to well below 100 meters, and this will be assumed during the rest of this paper.

THE PROPAGATION OF ERRORS DURING UNPOWERED FLIGHT

The position and velocity errors introduced by the INS during powered flight will not remain constant during the unpowered flight of the missile, even in the absence of any additional errors introduced by the gravity field in route. Assume that the difference between the true and desired trajectory is relatively small, as it should be if the launch INS is working properly. Although either trajectory depends on the actual mass distribution within the Earth, their difference can be adequately modeled by assuming that the Earth is a point mass, and the equations of motion for the differenced trajectories can be linearized. The equations have been solved analytically for the missile guidance problem[5,21], which makes it possible to directly relate the trajectory errors at burnout and re-entry.

The behavior of trajectory errors is different in and out of the orbital plane, which makes it convenient to introduce a special coordinate system to analyze their time evolution. Since the target presumably lies at or near the surface of the Earth, one component of error at the Earth's surface is "down range", which is in the orbital plane and parallel to the Earth's surface, while the other "cross range" error is perpendicular to the orbital plane but also parallel to the Earth's surface. In flight, these errors are defined with respect to the plane tangent to the radius vector connecting the spacecraft to the center of the Earth at that time. "Along track" errors (position errors in the direction of the velocity vector) are relatively unimportant in the total error budget. For example, a 200 meter along track error translates into a 30 millisecond difference in arrival time. Even at the equator the Earth will have rotated by just 13 meters during that time, causing the longitude of impact to be in error by exactly the same amount. This effect is of course smaller at higher latitudes.

Table 2 : Relations between errors at burnout and at re-entry

Down-Range Errors at Re-entry

Initial Error at Burnout	Partial	Multiplier of Naive Estimate
Down Range Position	1.0	1.0
Burnout Altitude	$\dfrac{4\,\mu}{v_{bo}^{2} r_{bo}} \; \dfrac{\sin^2(\psi/2)}{\sin(2\phi)}$	3.3
v_{bo} magnitude	$\dfrac{8\,\mu}{v_{bo}^{3}} \; \dfrac{\sin^2(\psi/2)}{\sin(2\phi)}$	3.0
In plane $\underline{v}_{bo}$ orientation error	$\dfrac{2\sin(\psi+2\phi)}{\sin(2\phi)} - 2$	0.0

Cross-Range Errors at Re-entry

Initial Error at Burnout	Partial	Multiplier of Naive Estimate
Burnout Altitude	$\cos(\psi)$	0.0
Out of Plane Orbit Error	$\lvert \underline{r} \rvert \sin(\psi)$	1.0

Table 2, from[5,21], provides the linear relations between $\underline{\delta r}$ at burnout and along and cross range errors at re-entry (the effect of the Earth's rotation is ignored). These linear relations are typically called partials, since they are the partial derivatives of the re-entry error with respect to some initial error. The linearized re-entry error estimate for any inital error is just the partial times the initial error. Since the cross range errors at burnout do not contribute to the down range errors, while the down range errors at burnout do not contribute to the cross range error at re-entry, those zero paritals were eliminated from this table. The "naive" estimate multipliers in Table 2 are the factors required to correct the previous error estimates derived by ignoring the Earth's gravity entirely; these are derived assuming the model trajectory in Table 1. An examination of Table 2 shows that the previous naive estimates were within a factor of 3 of the correct error partials.; in particular, 100 meter accuracy at re-entry requires position and velocity errors at burnout to be on the order of 30 meters and 1-2 cm sec^{-1}, respectively, somewhat smaller than previously estimated.

PERTURBATION ANALYSIS OF FREE FLIGHT ERRORS

The effects of gravity errors on spacecraft guidance can be separated into the effects of small scale gravity field errors near the launch site on the burnout initial conditions and the effects of global scale errors on the inflight trajectory. Since a priori gravity models provide estimates of the gravity fields at long spatial wavelengths, the free flight gravity field errors depend upon the irregular variations in the gravity field at shorter wavelengths. These perturbations, being unpredictable, are generally modeled as random, or stochastic, processes [13]. The unmodeled inflight accelerations of a ballistic missile will thus perturb the true missile orbit away from the nominal trajectory. Unmodeled accelerations in free flight can generally be regarded as small, and for most purposes such bodies can be regarded as following paths close to those described by Newton's laws of gravity and of motion. I performed a linear perturbation analysis of the orbit error of a massless test particle moving about a point mass subject to random accelerations. Although accurate navigation of a space craft requires a more detailed model of the gravitational field of the parent body, a point mass model suffices for the perturbation analysis. An analytic solution was obtained for a circular orbit; considerably more complicated analytic solutions also exist for the more general case of elliptic orbits[22]. Numerical trials indicate that the error introduced by a circular orbit approximation are probably no more than a factor of 5 in this case.

The equations of motions for the test particle under Newton's laws are

$$\frac{d^2 \underline{r}}{dt^2} + \frac{\mu}{r^3} \underline{r} = \underline{\delta a} \qquad (9)$$

where $\underline{r}$ is the radius vector from the center of mass of the central body to the space craft, $r = | \underline{r} |$, and $\underline{\delta a}$ represents any disturbing acceleration, whether of gravitational or other origin. Let $\underline{r}_o$ be the radius vector of a fictitious test particle not subject to the disturbing acceleration $\underline{\delta a}$. The equation of motion for the test particle is thus

$$\frac{d^2 \underline{r}}{d t^2} + \frac{\mu}{r^3} \underline{r} = 0 \tag{10}$$

If the position and velocity of the true and fictitious test particles are equal at some time $t = t_0$ the orbits of the true and fictitious test particles are called osculating, but note that this need not be assumed in the following.

Define

$$\underline{\delta r} = \underline{r} - \underline{r_0} \tag{11}$$

then

$$\frac{d^2 \underline{\delta r}}{d t^2} + \frac{\mu}{r^3} \left[I_3 - \frac{3 \underline{r_0} \underline{r_0}^T}{r_0^2} \right] \underline{\delta r} = \underline{\delta a} \tag{12}$$

where I_3 is the 3 by 3 identity matrix and all vectors are assumed to be column vectors. $\underline{\delta r}$ can be interpreted as the motion of the spacecraft about a nominal path, and Equation 12 will hold even if the nominal path $\underline{r_0}$ includes the effects of known perturbations of the orbit. This equation makes it possible to regard perturbations about this reference orbit as a non-stationairy linear system driven by some suitable deterministic or stochastic forcing $\underline{\delta a}$.

Suppose that we pick a non-inertial reference frame aligned with the position of the spacecraft at each moment, with basis vectors in the direction of the velocity vector (along track), in the direction perpendicular to the orbital plane (cross track) and in the remaining orthogonal direction, which is roughly along the radius vector $\underline{r}$. The components of the position vector error $\underline{\delta r}'$ in this rotating coordinate system are (assuming a circular orbit)

$$\underline{\delta r} = \begin{bmatrix} \text{cross track} \\ \text{along track} \\ \text{along } \underline{r} \end{bmatrix} = \begin{bmatrix} \delta r_c \\ \delta r_a \\ \delta r_r \end{bmatrix} \tag{13}$$

After some algebra we can convert Equation 12 into

$$\frac{d^2 \underline{\delta r}'}{d t^2} + 2\Omega \begin{bmatrix} 0 & 0 & 0 \\ 0 & 0 & -1 \\ 0 & -1 & 0 \end{bmatrix} \frac{d \underline{\delta r}'}{d t} +$$

$$\Omega^2 \begin{bmatrix} 1 & 0 & 0 \\ 0 & 0 & 0 \\ 0 & 0 & -3 \end{bmatrix} \underline{\delta r}' = \underline{\delta a}' \tag{14}$$

where Ω is 2π / the orbital period = $(\mu / r_0{}^3)^{1/2}$. In this rotating coordinate system, then, the linear perturbation equations can be put into a simple form of a first order matrix differential equation. Note position and velocity errors can be viewed as giving rise to fictitious forces, analogous to the Coriolis force, even in the absence of $\underline{\delta a}$.Suppose that the random acceleration during unpowered flight can be treated as a white noise with a power spectral density Q_r (in the radial direction) and Q_h (in each horizontal direction). If the initial position and velocity errors are zero at some time , then the errors in the three components of position at some later time Δt are

$$\sigma_1^2 = \frac{1}{2\;\Omega^2}\left[\Delta t - \frac{\sin 2\Omega\Delta t}{2\Omega}\right] Q_h ,$$

$$\sigma_2^2 = Q_h \left[3\Delta t^3 + \frac{8}{\Omega^2}\,\Delta t - \frac{4}{\Omega^3}\,\sin 2\Omega\Delta t \right.$$

$$\left. + \frac{24}{\Omega}\left[\frac{\Delta t \cos \Omega\Delta t}{\Omega} - \frac{\sin \Omega\Delta t}{\Omega^2}\right]\right]$$

$$+ \frac{4\;Q_r}{\Omega^2}\left[1.5\,\Delta t + \frac{1}{\Omega}\left[\frac{\sin 2\Omega\Delta t - 2\sin \Omega\Delta t}{4}\right]\right]$$

and (15)

$$\sigma_3^2 = Q_r \left[\frac{1}{2\,\Omega^2}\,\Delta t - \frac{1}{4\Omega^3}\left[\sin 2\Omega\Delta t\right]\right]$$

$$+ \frac{4\;Q_h}{\Omega^2}\left[1.5\,\Delta t + \frac{1}{\Omega}\left[\frac{\sin 2\Omega\Delta t - 2\sin \Omega\Delta t}{4}\right]\right]$$

Comparison of these equations with the "naive" estimates derived earlier ignoring the effects of the Earth's gravity shows that the orbital dynamics has little effect in the along track direction, but inhibits the growth of errors in the cross track and radial directions, and that the growth in error variance do to gravity field errors could be one tenth or less than that predicted by the naive estimates.

THE ACCURACY OF MODERN TERRESTRIAL GRAVITY MODELS

Although high accuracy local gravity measurements can be made from the ground, and from aircraft, accurate global gravity models are a product of the space age, as only satellites can provide the required information for all parts of the globe. Each satellite responds to a particular set of spherical harmonics of the Earth's gravitational field, and tracking of an ensemble of satellites can thus be used to determine these spherical harmonics uniquely [23]. The low order field is now determined sufficiently accurately that the only errors significant for missile guidance are at small spatial wavelengths [13]. This considerably simplifies the analysis of gravity field errors, since at the small wavelengths the Earth can be regarded as flat, and the spherical harmonics can be treated as simple two dimensional Fourier coefficients. A body in free fall some distance z above the Earth is not sensitive to all gravity field variations; in the flat Earth approximation the response to a gravity field variation of amplitude A and wavelength λ is [23]

$$A e^{-2\pi z/\lambda} \qquad (16)$$

The observed gravity field is itself "red" (i.e. its power declines with increasing spatial frequency); a typical model, Kaula's rule, indicates that the power of the random accelerations caused by gravity field variations is proportional to the inverse frequency squared[24]. These facts imply that a ballistic missile in free flight is not sensitive to gravity field errors with wavelengths much shorter than a few hundred kilometers.

Since missiles fly at altitudes of a few hundred to a few thousand kilometers, calibration of gravity errors in missile guidance requires data from satellites at the same altitudes (or lower). The gravity models used for this purpose are probably developed from tracking of the Transit satellites. These satellites, some of which include drag compensation, have near circular polar orbits at altitudes of ~ 1000 km [25,26] Both the fields developed from these satellites, and their error budgets, are currently classified [26].

Figure 2 shows the approximate error in various current and planned gravity field models for various field models in the open literature[23]. The two lines marked OSU 81 and GEM L2 are for two currently existing gravity models, while the curve labeled GRM describes the field accuracy expected from a planned NASA mission, the Geopotential Research Mission[27]. The shaded curve is the level of error that would be expected in the absence of a model[24], and thus marks the high frequency limit of the usefulness of any model. Note that the error in these models is not independent of frequency but instead has a maximum near the high frequency limit of the model. The high frequency limit for modern models corresponds to times that are so short compared to the time of flight of a long range ballistic missile that it is reasonable to model the gravity field error as a bandwidth limited random walk with a high frequency limit corresponding to the limit of usefulness for the model under consideration.

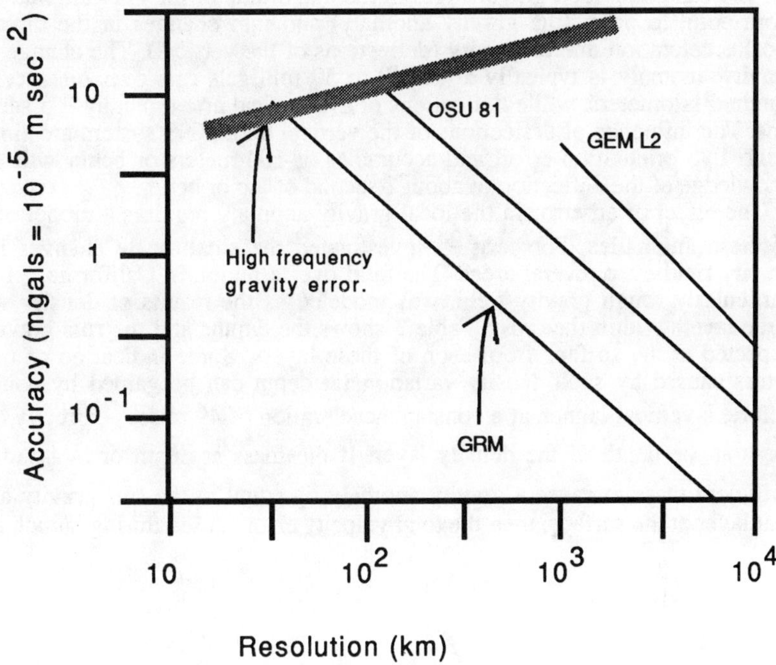

Figure 2 : Gravity Anomaly errors of existing and planned gravity models as a function of averaging interval.

Figure 2 shows that the accuracy of the OSU 81 model at a resolution of 300 km is supposed to be on the order of 1 to 10 milligals. If it is assumed that a missile is launched on a depressed trajectory, such that its altitude never reaches more than 300 km and its velocity is about 7 km sec^{-1}, then this indicates a gravity error power spectral density of $< 7x10^{-8}$ m^2sec^{-3}. The position error induced after 2000 seconds of this random acceleration is about 14 meters. It seems reasonable to conclude that gravity errors would not prevent the accurate navigation of present day ballistic missiles.

THE EFFECTS OF GRAVITY ERRORS DURING POWERED FLIGHT

A missile in powered flight is very sensitive to errors in the local gravity field due to its proximity to the surface of the Earth. Local variations in the gravity at the surface can be divided into changes in the magnitude of the gravitational acceleration from point to point (the gravity anomaly) and into changes in the direction of the local acceleration due to gravity (deflections of the vertical). The change in the local gravity anomaly is typically about 40 to 50 milligals rms over distances of several hundred kilometers, while deflections of the vertical are typically ~ 5 seconds of arc rms. The influence of deflections of the vertical on an INS system are similar to an a priori INS orientation error, and accuracies of 100 meters or better will thus require knowledge of the deflection to about 1 second of arc or better.

The effect of an error in the local gravity anomaly requires a model of the causes of these anomalies. Forsberg [24] investigated the small scale changes in the local gravity field over several areas. The field over a region in California, an area with a particularly rough gravity field, was modeled as the results of density variations at three layers within the crust; Table 3 shows the depths and the rms gravity anomaly expected at the surface from each of these layers. Some indication of the guidance errors caused by such density variations at depth can be gained by considering, as before, a vertical launch at a constant acceleration of 49 m sec^{-2} directly over a point mass at the depth of the density layer. If the mass at depth δr is assumed to have sufficient mass to cause a gravity anomaly δg equal to the rms gravity attributed to that layer at the surface, then the total velocity error caused during launch is

$$\delta v = \int_0^t \frac{\delta g \, \delta r^2}{(\delta r + 1/2\, a\, t^2)^2} dt' \tag{17}$$

The results of numerical integration of Equation 17 are also given in Table 3. (The effect of these point mass anomalies on position errors at burnout is negligibly small.) The effect of the deepest layer is large enough to cause a 17 meter position error at re-entry. This crude point mass model probably understates the induced error, since the missile would be sensitive to more of the field than that directly below the launch site.

Table 3 : Three Layer Gravity Model for California

Layer Depth km	RMS Gravity Anomaly milligals	Point Mass Velocity Error at Burnout cm sec^{-1}
2	5.5	0.04
4	6.8	0.07
26	33.9	0.86

THE ACCURACY OF SUBMARINE LAUNCHED MISSILES

Local gravity errors are probably not a problem for land launched missiles, as they can be calibrated to better than a milligal using local gravity surveys, and such surveys are done for the regions around missile silos [21]. Such surveys are difficult for the vast regions transversed by missile carrying submarines, and local gravity errors are a severe impediment to improvements in submarine launched missiles.

It is important to realize that the accuracy of submarine launched missiles depends upon knowledge of the position of the submarine at the time of launch as well as upon the accuracy of the missile guidance system itself. Modern missile carrying submarines themselves use inertial guidance systems to navigate when submerged[28], and the position error of the submarine INS will of course grow with time. Current submarine guidance systems thus rely upon occasional position determinations with satellite navigation systems[28], and, in fact, the Transit satellite system was largely developed for this purpose[29]. It is not necessary for the submarine to surface to obtain a position update, as the small antennas needed could be fitted onto the periscope or onto a small buoy.

Calibration of the local gravity field for submarine launched missiles seems to take one (or both) of two forms. The surface of the sea is nearly a geopotential, and satellite radar altimeters can be used to obtain theshape of the ocean's surface, and thus the gravity field over the oceans[30], including over the Arctic[31]. A U.S. Navy satellite called GEOSAT was launched in March, 1985, for the specific purpose of obtaining the gravity field in this manner[32,33,34]. It is also possible to put a gravity gradiometer on board the submarine itself, and the submarine can then create a model of local gravity field variations without any external aid [14,35,36]. The combination of an accurate gravity model from a satellite altimeter and a gravity gradiometer raises the possibility of directly determining the position of the submarine by matching the observed field variation with the a priori model, a process called gravity navigation[37,38]. Some combination of these techniques will be required to improve the accuracy of submarine launched missile to that obtainable from a land based missile.

THE "FIFTH" AND "SIXTH" FORCES AND MISSILE ACCURACY

Recently there has a been a great deal of discussion about the existence of possible non-Newtonian corrections to gravity at short ranges[39]. There have been a series of conflicting results indicating the existence of both attractive and repulsive forces over distances of a few meters (from laboratory experiments) to many kilometers (from geophysical experiments) which may or may not depend on the composition of the attracting bodies. If these data really do indicate the existence of a new fundamental force of nature it would be the fifth one known (after electromagnetism, the strong and weak nuclear interactions and classical gravity itself) and this research generally goes under the name of fifth force studies. It is frequently assumed that any such fifth force would obey a Yukawa potential [6], in which the total gravitation potential between two point masses m_1 and m_2 at a distance r would be

$$V = - \frac{G m_1 m_2}{r} \left[1 + \alpha e^{-r/\lambda} \right] \tag{18}$$

α being zero in classical Newtonian gravity. One recent experiment [40] is consistent with $\alpha = 0.02$ and $\lambda = 311$ meters. If, as is reasonable for such a short range force, the Earth is assumed to be a flat plane with a mean surface density of ρ (approximately 2750 kg m^{-3} over land) then such an additional potential would cause a gravity perturbation of

$$\delta g\ (r) = 2 \pi G \rho \alpha \lambda e^{-z/\lambda} \tag{19}$$

where z is the vertical distance above the ground. Evaluation of Equation 19 with the results from[40] gives a total perturbation at ground level of only 0.7 milligals decaying rapidly with r. It might be thought that such a small perturbation would not have a large effect on missile accuracy, and this is borne out by numerical integrations. As before, assume that an object is launched vertically with a total effective acceleration of 49 m sec $^{-2}$. The total induced velocity perturbation is thus given by

$$\delta v = \int_0^t \delta g(z(t'))\, dt' \tag{20}$$

where z(t) is now simply 1/2 a t^2. The total δ v from the Eckhardt et al.[40] model amounts to only ~ 4 microns / second, which is totally negligible.

Ander et al.[41] have recently tried to resolve the conflicting evidence for short range non-Newtonian forces by introducing two Yukawa potentials (the "sixth" force). In this scheme, which has some theoretical justification, the two additional forces would have different signs, and would thus nearly cancel at short ranges. The interesting aspect of this proposal for the missile accuracy question is that these two forces could conceivably have much larger ranges, up to several hundred kilometers. Ander et al. propose a potential of the form

$$V = - \frac{G m_1 m_2}{r} \left[1 - a e^{-z/v} + b e^{-z/s} \right] \tag{21}$$

The short range data imply that a - b is of the order of 0.01 and that s/v - 1 is of the order of 0.01 / a. These additional potentials would give rise to a anomalous acceleration of

$$\delta g(r) = 2 \pi G \rho \left(av\, e^{-z/v} - bs\, e^{-z/s} \right) \tag{22}$$

The total velocity perturbation assuming various choices of these parameters is given by Table 4. These two potential models, if true, would induce substantial missile targeting error, on the order of 100 to 1000 meters. The induced errors would, of course, cause a targeting bias which would depend on the distribution of mass around the launch site, and perhaps also on the composition of the local material. It is interesting to note that U.S. missile tests are generally conducted from a site near water, which could cause a very different bias from a launch from an operational silo deep in the continental heart land. This bias could of course be corrected when and if the fifth and sixth force are confirmed and their nature elucidated.

Table 4 : Velocity Errors from various two potential models

a	b	s	v	δv
		km	km	m sec^{-1}
1.0	1.01	200	202	-0.4
1.0	1.01	50	50.5	-0.06
0.1	0.11	50	55	-0.06

RE-ENTRY ERRORS AND ACTIVE RE-ENTRY VEHICLES

Re-entry errors are perhaps the fundamental limitation to the accuracy of a truly ballistic missile (one without any terminal guidance). The trajectory of a ballistic missile during re-entry depends upon the local atmospheric density and wind profile[42,43], and it has been stated that the natural variations in these quantities limit the accuracy on impact to 20-50 meters even in the absence of other sources of error [44]. Glover[42] presents linearized models of the influence of wind and pressure variations on re-entry errors which indicate that, for the trajectory in Table 1, the biggest errors are created by conditions at about 20 km altitude, and that, a 10 m sec^{-1} wind (not unusual at that altitude) would cause a 10 meter error at impact; similar errors would be caused by density variations of 2.5% at the same altitude. It seems likely that any nuclear explosions near the target could cause much larger variations in atmospheric conditions at altitude. One form of terminal defense might be simply to set off explosions high above hardened targets simply to degrade the accuracy of the incoming re-entry vehicles.

Re-entry errors could, in principle, be controlled through knowledge of the state of the atmosphere above the target. The difficulty of obtaining sufficient information remotely and in near real time, coupled with the likelihood that atmospheric conditions would be greatly modified by the attacker's own nuclear explosions in the vicinity, indicates that a reliable first strike capacity against hardened targets would probably require Maneuverable Re-entry Vehicles (MARVs).

Inertial navigation system errors for cruise missiles (or for manned aircraft) are much larger than missile INS errors for the same mission. For a typical mission aircraft flight times are a factor of 20 to 30 times longer than for a ballistic missile.

Even if the unmodeled acceleration of the aircraft INS was the same as for the space vehicle, the aircraft would have one to several hundred times the position error over the target purely due to the longer flight times. Any vehicle confined to the atmosphere, however, is subject to the large variations in gravity at high spatial frequencies over the entire mission, and gravity errors are thus much larger in general for aircraft INS. In addition, the acceleration of gravity torques the INS gyroscopes over the entire mission,. causing increased gyroscope drift errors and any accelerometer errors now will accumulate over the entire mission.

Cruise missile navigation has thus long relied upon navigation by pattern matching, or Automatic Target Recognition (ATR). One of the first automated systems was the ATRAN (Automatic Terrain Recognition and Navigation) system used on the medium range MACE missiles of the late 1950's[45]. This early analog system compared radar images with images stored on a reference film strip, had an accuracy of a few hundred meters with a maximum range of ~ 1000 km and weighed about 550 kilograms. This system was developed, in part, because Inertial Navigation Systems of the desired accuracy and weight were not available at the times. In modern cruise missiles, the ATR system is used to update an onboard INS[46] using a variety of different sensors and algorithms[47,48].

The terminal navigation of missile warheads by pattern matching is a much harder job than for cruise missiles. There is very little time available for the computer processing involved, since the warhead will be enveloped in a sheathe of incandescent gas for much of the re-entry. It is doubtful that any sensor, active or passive, could work well under such conditions. The sheathe is caused by frictional heating of the air during re-entry; such heating is roughly proportional to the density of the air times the cube of the missile velocity[43]. The performance of a body entering the atmosphere can roughly be parametrized in terms of the "ballistic factor" β , which is[43]

$$\beta = m\,g / C_D\,S \tag{22}$$

where m is the mass of the re-entry body, S is its cross-sectional area, C_D is its drag coefficient, and g is the acceleration due to gravity. Warhead impact velocities should probably be kept high to complicate any terminal defense; terminal velocities on the order of 1 to 3 km / sec imply ballistic coefficients of 10^4 to 10^5 pascals[43]. The time of maximum heating is generally before the time of maximum acceleration[43], which is at about 10 to 20 km for such high ballistic coefficients. If it is assumed that any terminal guidance system would begin operations after the time of maximum heating, then a terminal velocity of 1-3 km sec^{-1} implies that the ATR system would only have 10 to 20 seconds or less in which to work, although this time could be increased somewhat by maneuvers of the re-entry body[49]. It is clear that an ICBM ATR MARV would not be a negligible technological feat, however, there appears to be no intrinsic physical reasons preventing the development of such a system.

It should be easier to develop counter-measures against missile pattern matching navigation than against similar cruise missile navigation systems. Very little terrain is overflown after dissipation of the re-entry sheathe, and the navigation system must recognize either the target itself, or landmarks quite near to the target. Camouflage, smoke screens, chaff and other such passive defenses could be both cheap and effective if the nature of the sensor system were known, while any active sensor system (such as a terrain matching radar) could be jammed. It is also not clear if any system could reliably locate a target through the clouds of debris raised by previous nuclear explosions in the vicinity of the target. (Such explosions might again serve as a useful defensive measure for hardened targets.) Reliable missile ATR would thus be enhanced if the warhead computers had the ability to integrate the results from more than one sensor, and also to take measures to avoid the response from any terminal defense. The end result of an arms race between MARVs and terminal defense measures is thus likely to be the development of "smart warheads" with a fair capacity for autonomous action.

CONCLUSIONS

It is my conviction that the major strategic problems facing the United States today are not technical problems, but political problems, and that these cannot be solved by purely technical means. No political solution, however, can be considered sound unless it takes into account the present and future state of military technology, and,conversely, the development of military technology should not proceed in a vacuum, but should be influenced by informed political debate. The purpose of this article has been to further that informed political debate in an unbiased fashion. I feel that it is appropriate in this section to offer some thoughts about the implications of the advances in the accuracy of strategic missiles.

It seems clear to me that modern guidance systems can be be reasonably expected to meet the 100 to 200 meter accuracy expected of them. It should be possible, with some work, to reduce this still further, however, uncertainties in the state of the atmosphere will probably limit such improvements to the 20-50 meter level. While unknown sources of errors might cause the degradation of the accuracy of all missile systems, and might cause a few systems to completely miss their targets, it seems likely that the actual accuracy of ballistic missile has greatly improved over the past two decades and should continue to improve in the future. A 200 kiloton warhead exploding at the surface would cause a crater of 100 meters or more in radius [50]. As the expected missile accuracy continues to decrease below 100 meters, then at some point the actual accuracy should reach the 100 meter level; it is reasonable to expect that the increases in true accuracy will slowly but surely translate into a greatly increased ability to destroy hard targets with a given size warhead

Dekameter level or better accuracies should make it possible to attack even hardened targets with very low yields (perhaps even without nuclear explosives). The general reduction in warhead yields seen during the last two decades[51] will thus probably continue. In the absence of any form of arms control, this reduction in yields might eventually prove to be destabilizing during a crisis, as it would reduce the amount of fallout, soot and dust during a nuclear attack, and thus possibly make a nuclear first strike seem as less of a suicidal option. Coupled with arms control and a general reduction in the number of warheads, however, the results could be a great reduction in the consequences of any war that did break out. Perhaps the best use for the emerging dekameter level accuracy technology would be as a bargaining chip during the next round of arms reduction talks with the USSR.

The opinions expressed in this paper are the author's own. This paper was prepared without any assistance by the U.S. Government or by the Jet Propulsion Laboratory, financially or otherwise, and the opinions expressed are not necessarily those of either the U.S. Government or the Jet Propulsion Laboratory.

REFERENCES

1. Milliken, R.J., and Zoller, C.J., Navigation, 25, 95-106 (1978).
2. Martin, E.H., Navigation, 25, 201-210 (1978).
3. Battin, R.H., An Introduction to the Mathematics and Methods of Astrodynamics (AIAA, New York, 1987).
4. Roy, A.E., Orbital Motion, (Adam Hilger, Ltd., Bristol, 1982).
5. Bate, R.R., Mueller, D.D., and White, J.E., Fundamentals of Astrodynamics (Dover Publications, New York, 1971).
6. Stacey, F.D., Tuck, G.J., Moore, G.I., Holding, S.C., Goodwin, B.D., and Zhou, R., Rev. Modern Physics, 59, 157-174 (1987).
7. Schwarz, K.P., Rev. Geophys. Space Phys., 21, 878-890 (1983).
8. Ragan, R.R., edt., IEEE Trans. AES, AES-20, 414-440 (1984).
9. Kuritsky, M.M., and Goldstein, M.S., Proc. IEEE, 71, 1156-1176 (1983).
10. Gelb, A. (edt.), Applied Optimal Estimation, (M.I.T. Press, Cambridge, 1974).
11. Maybeck, P.S., Stochastic Models, Estimation and Control, Chap. 6 (Academic Press 1979).
12. Forward, R.L., IEEE Trans. AES, AES-17, 511-519 (1981).
13. Nash, R.A. and Jordan, S.K., Proc. IEEE, 66, 532-550 (1978).
14. Jekeli, C., EOS Trans. AGU, 69, 105-117 (1988).
15. Wuerth, J.M., Navigation, 23, 64-75 (1976).
16. McKinley, H.L., AIAA Guidance and Control Conference 1975, AIAA Paper No. 75-1064 (AIAA, New York, 1975).
17. White, R.L., Adams, M.B., Geisler, E.G., and Grant, F.D., IEEE Trans. AES, AES-11, 195-202 (1975).
18. Bar-Itzhack, I.Y., J. Guidance and Control, 1, 305-312 (1978).
19. Tillotson, R., Treder, A., Norris, R., and Scott, P.J., Proc. National Aerospace Meeting 1982 1-51 (Institute of Navigation, Washington, D.C., 1982).
20. Gounley, R., White, R. and Gai, E., J. Guidance, 7, 129-134 (1984).
21. Bunn, M. and Tsipis, K., Ballistic Missile Guidance and Technical Uncertainties of CounterSilo Attacks, Report Number 9 of the M.I.T. Program in Science and Technology for International Security (M.I.T., Cambridge, 1983).
22. Stern, R.G., Report No. 62 at the Fourteenth Congress of the International

Astronautical Federation (1963).
23. Rapp, R.H., Chapter 2.2 in Space Geodesy and Geodynamics, A.J. Anderson and A. Cazenave, edt., (Academic Press, London, 1986).
24. Forsberg, R. Scientific Report No. 6, AFGL-TR-84-0214 (Air Force Geophysics Laboratory, Hanscom AFB, 1984).
25. Anderele, R.J., Chapter 3.1 in Space Geodesy and Geodynamics, A.J. Anderson and A. Cazenave, edt., (Academic Press, London, 1986).
26. Balmino, G., Chapter 2.1 in Space Geodesy and Geodynamics, A.J. Anderson and A. Cazenave, edt., (Academic Press, London, 1986).
27. Keating, T., J. Astronautical Sci., 32, 145-158 (1984).
28. McKelvie, B., and Galt, H., Navigation, 25, 310-322 (1978)
29. Weiffenbach, G.C., IEEE Trans. AES, AES-22, 474-481 (1986).
30. Stewart, R., Fu. L.L., and Lefebvre, M., Science Opportunities from the Topex/Poseidon Mission, JPL Publication 86-18 (Jet Propusion Laboratory, Pasadena, 1986).
31. Anderson, A.J., Marquart, G., and Schernick, H.G., EOS Trans. AGU, 69, 873-881 (1988).
32. Miller, L., Cheney, R.E., and Douglas, B.C., Science, 239, 52-54 (1988).
33. Cheney, R.E., and Miller, L., EOS Trans. Agu, 69, 754-758 (1988).
34. Cheney, R.E., Douglas, B.C., Agreen, R.W., Miller, L., Milbert, D., and Porter, D.L., EOS Trans. Agu, 67, 1354 (1986).
35. Zondek, B., J. Guidance and Control, 2, 173-178 (1979).
36. Britting, K.R., Madden, S.J., and Hildebrant, R.A., AIAA Guidance and Control Conference 1972, AIAA Paper No. 72-850 (AIAA, New York, 1972).
37. Clancy, T., The Hunt For Red October, 97-98 (Naval Institute Press, 1984).
38. May. M.M., Position Location And Navigation Symposium 1978 Record, 212-218 (IEEE 1978).
39. Scharzchild, B., Physics Today, 41, 21-24 (1988).
40. Eckhardt, D.H., Jekeli, C., Lazarewicz, A.R., Romaides, A.J., and Sands, R.W., Phys. Rev. Let., 60, 2567-2570 (1988).
41. Ander, M.E., Goldman, T., Hughes, R.J., and Nieto, M.M., Phys. Rev. Let., 60, 1225-1228 (1988).
42. Glover, L.S., J. Spacecraft, 9, 483-484 (1972).
43. Regan, F.J., Re-entry Vehicle Dynamics, AIAA Education Series (AIAA, New York, 1984).
44. King, H.H., J. Spacecraft, 17, 240-247 (1980).
45. Koch, R.F., and Evans, D.C., SPIE Volume 238, Image Processing for Missile Guidance, 2-9 (SPIE 1980).
46. Carr., J.R., and Sobek, J.S., SPIE Volume 238, Image Processing for Missile Guidance, 36-41 (SPIE 1980).
47. Bhanu, B., IEEE Trans. AES, AES-22, 364-379 (1986).
48. Conrow, E.H., and Ratkovic, J.A., SPIE Volume 238, Image Processing for Missile Guidance, 426-453 (SPIE 1980).
49. Bunn, M. Technology Review, 91, 28-36 (1988).
50. Tsipis, K., Arsenal, App. F (Simon and Schuster, New York, 1983).
51. McNamara, R.S., Blundering into Disaster (Pantheon Books, New York, 1986).

CHAPTER 14

Comparison of Basing Modes for Land-Based Intercontinental Ballistic Missiles

by Peter D. Zimmerman
Senior Associate, Carnegie Endowment for International Peace

Washington, DC 20036

Intercontinental ballistic missiles based in silos have become relatively vulnerable, at least in theory, to counter-force attacks. This theoretical vulnerability may not, in fact, be a serious practical concern; it is nonetheless troubling both to policy-makers and to the public. Furthermore, the present generation of ICBMs is aging (the Minuteman II single warhead missile will exceed its operational life-span early in the next decade) and significant restructuring of the ballistic missile force may well be necessary if a Strategic Arms Reduction Treaty (START) is signed. This paper compares several proposed schemes for modernizing the ICBM force. Because the rail-garrison MIRVd mobile system is the least costly alternative to secure a large number of strategic warheads, it receives a comparatively large amount of attention.

* * * *

Determining the air blast-induced effects from a nuclear weapon is not a trivial task. The interactions between expanding shockwave, air heated by compression, air heated by radiation, the target, and the ground beneath are non-trivial and not well represented for all cases by any simple analytic formula. Nevertheless, it is precisely those difficult-to-model effects which determine the damage which a given target -- at a particular range from a weapon of specified yield exploded at a selected altitude -- will suffer. And, hence, it is difficult-to-model effects which determine the survival or failure of second-strike ballistic missiles.

We recognize immediately that, dimensionally, energy density and pressure are equal, in appropriate units. An explosion of given yield (energy release) Y (in kilotons) fills an expanding sphere of radius r with energy. The energy density (or equivalently, pressure), at a given radius should thus vary roughly as:

$$P \approx Y/r^3 \qquad (1)$$

Alternatively, the radius at which a given overpressure occurs for weapons of varying yields scales as

$$R = R_1 * Y^{1/3} \qquad (2)$$

where R_1 is the radius at which the pressure occurs for a weapon of 1 kiloton (10^{12} calories).[1] Because of the interaction between direct blast and the wave reflected from the surface, a simple $1/r^3$ law does not suffice to describe the phenomenon. Comparison with figures 3.72, 3.73a-c and 3.75 in The Effects of Nuclear Weapons[2] shows the complexities introduced when real-world physics is included. Indeed, a fit to the graph of figure 3.72 "Peak overpressure from a 1-kiloton free air burst" does not reveal a clean inverse-cube dependence of pressure on radius.[3]

For each value of overpressure and weapon yield, there is, as can be seen form the figures cited in Glasstone and Dolan, a unique height of burst (HOB) for which the radius exposed to at least that overpressure is maximized. The analyst must, therefore, first select a weapon yield and a damage criterion expressed in terms of a lethal overpressure. The next step is to choose the HOB to maximize the area exposed to the selected overpressure.

As can be inferred from figure 5.146 of Glasstone, rail equipment is particularly rugged. Diesel locomotives can absorb overpressures of 90 psi end-on and 25 psi side-on from a 400 kt weapon; loaded wooden boxcars of types in use in the middle 1950s will remain on the track and functional at overpressures up to about 30 psi end-on and 5.5 psi side-on.[a] Earth moving equipment, the closest analog given to armored vehicles, is approximately as rugged as a locomotive side-on. It seems reasonable to believe that armored vehicles could be built to accept 35 psi and continue functioning. Present generation missile silos are in the 2,000 - 3,000 psi hardness class; super-hard silos are reliably believed to be capable of withstanding overpressures up to roughly 100,000 psi and (by some) also believed to be capable of remaining in place and vertical inside a blast crater.[b] These survival capabilities, then, give some criteria by which to distinguish between the three principal generic competitors for the next-generation land-based strategic missile: fixed-base in super-hard

---

[a] The estimated accuracy of these hardness figures is no better than plus or minus 20%. This degree of caution should be factored into all calculations of survivability -- but usually isn't.

[b] This does the super-hard silo little good in a two-on-one attack since a second weapon detonating anywhere in the vicinity will knock over the exposed silo.

silos; road/off-road mobile; and rail mobile.

Table I

Damage Criteria for Basing Modes

Basing mode	Criterion
Super-hard silos	100,000 p.s.i
On/off-road mobile	35 p.s.i.
Rail mobile	5 p.s.i.*

* Component of pressure perpendicular to the rails, but 2,000 psi hard when in shelter at garrison.

Super-hard silos are rugged, but their location is well known to an attacker. On/off road mobile represents a class of weapons using a hardened mobile launcher (HML) and which can be imagined to be either restricted to the road net or able to roam cross country. In general, off-road mobile is more advantageous for survival, and further calculations will assume that the off-road option is available. Rail mobile appears to be the most vulnerable by static measures, but this is not necessarily the case. The rail-garrison system offers distinct advantages.

Superhard Silos

Missile accuracies have been improving steadily with a "halving time" for the circular-error probable (CEP) of about 7 years.[4] Soviet CEPs have historically lagged American CEPs by about one halving time, but their actual lag may, in fact, have disappeared in the last generation of missile design.[5] Since the Soviet Union produces many more varieties of ICBM than does the United States, it may even have been possible for Soviet designers to construct some classes of missiles with deliberately better -- or poorer -- accuracy than other classes in order to optimize costs and striking power. American CEPs now stand at roughly 100m or less for the MX; the accuracy of the Soviet SS-18 Mod 4 must be at least comparable. Without using maneuvering warheads (MARVs) incorporating terminal guidance, a 100m CEP is just about the best ballistic accuracy which can be achieved. The unmodellable residual winds encountered on re-entry contribute roughly 75 m to the over-all error budget.[6] Nevertheless, prudence dictates the defense-conservative assumption of near-zero CEPs for attacks against silos, particularly since MARV development and testing is continuing.

The assumption of vanishing missile errors makes untenable any kind of fixed-point missile basing if second-strike invulnerability is to be preserved. This virtually excludes super-hard silos from further consideration. This remains true in spite of the advantages of fixed silo basing: time urgent control, high data rate communications permitting rapid re-targeting, and the ability to launch within seconds under attack.

John Michener[7] has shown that superhard silos are particularly vulnerable to sub-surface bursts, and that high yield earth penetrating weapons could enable a single attacking warhead to destroy more than one superhard silo. If earth penetrator warheads can be constructed, even patterned arrays (the current name for "Dense Pack") become supremely vulnerable. One can, that is, imagine an attack in which all weapons land and penetrate before any are detonated, thus eliminating the fratricide effects which made patterned arrays attractive in the first place. Additionally, closely spaced silos reduce the attack price to keep a near-constant rain of warheads exploding at altitudes of a few kilometers in order to pin down the retaliatory launch. Finally, the construction of additional silo launchers is forbidden by the SALT II Treaty, and might be forbidden in a future START Treaty.

On/Off-Road Mobile

Mobility promotes missile survival by providing moving targets, thus making it difficult or impossible for an attacker to aim his warheads at the forces he wants to destroy. In a given amount of time, t, a mobile force able to disperse uniformly overland at an average velocity v must lie in a region of area A:

$$A = \pi * (vt)^2, \qquad (3)$$

but the exact location of the target within the region is uncertain. This is advantageous because the area which can be destroyed by a nuclear weapon of yield $Y \approx Y^{2/3}$ (that is, the area is proportional to the equivalent megatonnage or EMT) and because of a missile designer's rule of thumb that the EMT which a given rocket can carry is fixed, regardless of the "fractionation" of that yield into MIRVs.[8] In general, then, a given missile can barrage a fixed area to a given overpressure, but a mobile launcher can generate an increasing area, depending upon its dash speed, constraints imposed by the landscape (canyons, rivers, dunes, steep slopes, etc., about which see below) and, most critically, the amount of warning time it has in which to make its location uncertain. If enough time is available, a mobile system can generate an area so great that the survivability of a large fraction of the force is beyond question.

A 400 kt weapon (roughly the yield of the RV on an SS-18) exploded at the optimum altitude of about 1,100 m can produce an overpressure of 35 psi over a circle of radius about 1250 m according to the output of DNA nuclear weapons effects program used.[c] An SS-18 carrying 10 warheads can, therefore, destroy HMLs over a total area of $10*\pi*(1250)^2$, or roughly 50 km^2. Hobson[9] finds that average dash speeds over a reasonable mix of terrains for the HML prototype is 47 km/h, which we round to 50 km/h. The dash time needed to generate an area greater than can be barraged by a single SS-18 with 10 RVs (or its EMT equivalent) is, therefore, around 5 minutes. Hobson's specific numerical results differ from these because (a) HML hardness in this paper is greater by 5 psi than his assumed hardness and (b) the SS-18 warhead is assumed in this paper to have a yield of 400 kt vice 500 kt, and (c) there are minor differences in the modelling of the dependence of the overpressure as a function of radius.

Barraging a large area with many warheads is not a trivial problem. Each warhead must be aimed at a specific grid point and must fall close to its target. In addition, extra warheads should be allocated to the grid so that even if a single warhead fails to arrive, most of its primary area of responsibility is pummeled by the neighbors which do arrive. Placing target points in a hexagonal closest-packed array results in the greatest efficiency, but leaves 9% of the area unbarraged to the full level of the destruction criterion. Since CEP is at best a statistical estimate, more warheads must be allocated to ensure coverage of areas assigned to warheads which miss their targets by more than one CEP. Furthermore, the grid must be made more dense to account for areas left completely uncovered because of booster failure or other mishaps. One may plausibly argue, of course, that a target subjected to several shocks each near to its critical hardness will, in fact, be neutralized; an offense-conservative argument, however, requires exceeding the damage criterion for each point in the array, which requires, in turn, the allocation of more warheads per unit barrage area than the hex-pack minimum.

If the CEP of the attacking missile is comparable to the radius of destruction of its warhead, one may expect to find additional areas which are not barraged and some areas barraged to unnecessarily high levels. For the case of the SS-18 and an assumed 100 m CEP, this requires roughly 2% more weapons in the barrage; for a 250 m CEP, the effect increases to require approximately an additional 5%. Other Soviet missiles with less accuracy (e.g. SS-17 and SS-19 missiles and older SLBMs) will leave still larger areas statistically naked. The effect is a relatively small one for the SS-18 and a hexagonal array where 9% of the territory is already conceded to the victim outright; it is higher for target grids which guarantee meeting or exceeding

---

[c] It is worth noting that the DNA program does not find the same optimum HOB as one would arrive at using the scaling rules and graphs in Glasstone; the optimum found with the program is significantly lower.

the damage criterion for all points in the deployment area. The reliability of the delivery system and its warheads are a more significant factor in sizing the needed offense.

Soviet missiles may have a reliability between 50% and 100% (Operational missiles with reliabilities less than 50% hardly make sense; 100% is an ideal upper limit.), with 80% a plausible estimate. The potential victim must calculate a deployment area based on the assumptions that (a) the aggressor uses a hex-pack target grid, (b) that HMLs in the interstitial spaces will be destroyed, (c) the CEP of the attacking missiles is low while (d) their reliability is very high. The offense must, in turn, assume that (a) HMLs in the interstitial region will survive, so that (b) an overlapping of destructive circles is necessary, (c) that the CEP of his warheads is no higher than that achieved on the test range and probably lower, and, finally, (d) that his reliability is at the low end of the probable range. These two sets of assumptions obviously lead to significantly different figures for the number of missiles needed to barrage. It must, however, be emphasized that the potential victim cannot size his force and deployment area on an offense-conservative basis. It may be possible for experts in nuclear weapons design to estimate the yield of a warhead based on its mass and size which may be observed at some degree of confidence using ground-based radar. The only parameter of the offensive force which the defense can know with high confidence is the number of warheads.

Operational factors require a significantly longer dash time to reach a safe deployment than the simple calculations above. First, most of the dash must take place over roads if a high average speed is to be achieved, the vehicles leaving the roadway only in the last minute or two before anticipated warhead impact. Second, after reaching its alert station the HML will require some time interval to deploy whatever ground-hugging equipment it uses to prevent being overturned by the high winds from the nuclear blasts, a period which subtracts from the actual dash time. Third, significantly less than the entire alert area is actually available to HMLs because the terrain is unsuitable. The attacker must be presumed to have a good idea which areas can be excluded and so can be presumed not to waste warheads on those regions.

A reasonable time line for randomizing the locations of the HML force begins not less than 2 minutes after Soviet missiles are launched, and must allow at least 5 minutes transmit the "go" signal to the HMLs and to get missile crews into their HMLs, and an additional minute to start engines and move from the initial point. The time line probably ends 5 minutes before the anticipated landing of enemy warheads to allow time for the HML to tie itself to the ground and for the tractor to get clear. Thus, 13 minutes of the total 30 minute flight time for an ICBM are not available for the dash.

Even so, it would appear to be theoretically possible to generate twice the area which the SS-18 force can barrage given reasonable dash times of 10-12 minutes and assuming no constraints imposed by the landscape (but see below). Such an HML force could expect to have 50% of its missiles survive a first strike; if 500 Midgetman small ICBMs were so deployed, the surviving force of 250 missiles should be adequate for deterrence, even in the absence of the other two legs of the Triad.

It may not be reasonable, however, to rely upon the Soviets using SS-18s or other ICBMs to attack an HML system. Submarine launched missiles with 15 minute flight times provide far less warning, and hence less dash time. Unless the HMLs are on fully generated alert-- crews on board and awake, engines running -- and unless the time needed to transmit the emergency message to the HMLs can be reduced to near zero, very little dash time is available in the event of an SLBM attack. Indeed, the HML force has nearly the same vulnerabilities against submarine attack as do manned bombers, thus violating one of the principles embodied in the doctrine of the U.S. strategic Triad.

There are, however, operational complications not accounted for in the above calculation:

The author had the opportunity to visit the Grand Forks Air Force Base missile deployment area recently. Most silos are essentially adjacent to two-lane paved roads of reasonable quality. The roads are aligned on a nearly rectangular grid at intervals of 6 to 12 miles, with much poorer roads in between. Although one tends to think of the Great Plains as being flat, more like "planes", that is not the case. The Grand Forks missile deployment area is criss-crossed with stream beds, hills, rock outcroppings and the like. It includes large marshy areas as well. All of these make off-road mobile missiles less attractive operationally than they might seem from simply looking at a small-scale map.

Even more complicating for the notion of off-road mobile in this area, however are the long rows of trees planted during the Dust Bowl era to provide windbreaks. These have now grown to form almost impenetrable boundaries between fields. The actual area available for off-road mobile operations is almost certainly less than half of the apparent area. Worse, still, is the fact that the landscape, including the windbreak forests, will tend to channel the movement of HMLs along fairly predictable lines. Indeed, the survivability of HML mobile missiles in the Grand Forks area will more closely resemble that of rail-garrison missiles which generate track (i.e. linear uncertainty because they will be confined to the roads) rather than the hypothetical area generation used in the usual calculations of HML mobile survivability.

Rail-Garrison

The Air Force's primary option for modernization of the land-based leg of the Triad is the "Rail-Garrison MX" system. In this plan, MX missiles are based on trains which are permanently stationed on Strategic Air Command bases. Under conditions of alert, the missiles are "flushed" onto the rail net of the nation where they quickly "generate enough track" to be beyond the barrage capabilities of the Soviet missile force. At present the Air Force proposes to use 25 trains each carrying two missiles; each train would consist of two locomotives (for redundancy and speed), 2 security cars, 2 launchers, and one launch control car. The trains would be garrisoned at up to 17 different installations in the United States.

Consider the physics of attacking such a train. Trains are at least 5 psi hard, side-on and up to 90 psi end-on. We take as a damage criterion a 5 psi component of overpressure normal to the track. Average train speed is assumed to be 30 mph, and we adopt the Air Force estimate based on the number of intersections and spurs in the U.S. rail net that starting from 7 garrisons 120,000 miles (192,000 km) of track are filled in 24 hours; from 13 garrisons, 13,000 miles (20,800 km) of track are filled in 3 hours. Roughly, then 165 miles (265 km) are filled per train per hour. The actual calculation is subtle and depends on the actual rail lines available and the precise starting points of the trains.

For the given damage criterion it is clear that weapons should not be targeted to hit the track, but should be given an offset to either side. For RVs with a 400 kt yield, exploded at an HOB which optimizes the 30 psi over-pressure (and hence extends the 5 psi perpendicular component over a relatively large length of track), the optimum offset is 2.8 km, and 4.5 km of track is barraged, per warhead. A defense-conservative kill criterion of 2.5 psi normal to the tracks permits a single 400 kt warhead to destroy 6.8 km of track; if the warhead yield is increased to 500 kt the length of track barraged increases to 7.1 km (same HOB); if the kill criterion is reduced to 2 psi normal to the track, a 400 kt warhead can neutralize 7.8 km and a 500 kt warhead clears 8.2 km.

The present SS-18 force numbers 3080 warheads without START constraints. The force is, therefore, capable of barraging between 13,900 km and 25,250 km of track. Even using the most defense-conservative criteria, 25 trains cannot be barraged by the entire SS-18 force after 3.8 hours. After 8 hours, under the same assumptions, more than 50% of the rail-mobile force must survive a barrage from the entire SS-18 force. Under the optimistic set of assumptions, the 50% survivability is reached in less than 4 hours.

In all probability these estimates of rail-garrison survivability after sortie on to the tracks are unduly pessimistic. An attack on any component of this land-based system must be accompanied by an attack on all other components. It is highly probable that the United States and the Soviet Union will conclude a START agreement limiting the total number of ballistic missile warheads to 4,900 and the number of land-based ICBM warheads to some fraction of the total -- probably about 3,000-3,300. The Soviet Union, according to the communique from the Moscow Summit, has agreed that it will not deploy more than 1540 warheads on no more than 154 SS-18s. Even with a START agreement in force the United States will retain at least 450 single-warhead ICBMs (e.g. the present Minuteman II force which will remain usable at least until the mid 1990s), and possibly as many as 800 land-based missiles -- some of them MIRVd. The silo-based missiles will, therefore, tend to siphon off ballistic missile attacks which might, otherwise, have been aimed at a mobile system.

The START ceiling on ballistic missile warheads, **the sum of ICBM and SLBM weapons**, is 4,900. If Soviet missiles must also engage U.S. silo-based missiles with a two-on-one (two warheads per silo) attack to give a very high probability of destruction, that will require roughly 1,600 RVs, leaving only 3,300 to barrage the mobile system. However, no prudent strategist would fail to hold back a significant reserve, perhaps one third of his force. Under such circumstances only 3,300 warheads could be used in the initial attack; if 1,600 were used against the silos, only 1,700 would be available for use against the mobile system.

It is presently accepted that only the 1,540 SS-18 Mod 4 missiles which would be retained under START are accurate enough for counter-silo attacks. Under a START agreement any system which can only be attacked by a barrage is virtually guaranteed survivable. HMLs based at Minuteman silos, however, are far less survivable in the first few minutes after an alert than rail-garrison missiles because the HMLs would be only 35 psi hard, and would be destroyed as "collateral" damage by the missiles aimed at the silo itself.

If a U.S. response to a Soviet attack is delayed pending negotiations or evaluation, the retaliatory system must have the ability to endure for hours or perhaps days. Submarines and undamaged silos clearly can operate for such periods; it is difficult to envision a HML with a two-man crew having the resources to endure, move to avoid detection, and continue to remain operational for long periods after a nuclear attack. Radiation effects, alone, on the crew might argue against such long endurance.

Trains equipped with two locomotives for redundancy and a tank car for diesel fuel can operate for an indefinite period. The size of the missile crew and security force is large enough to permit operation in shifts as well as mutual psychological support. Because trains are softer than HMLs, the weapons used to attack them will burst, statistically, at greater ranges from the trains. This, in turn, may reduce prompt crew casualties due to radiation.

How does fractionation of the ICBM throw-weight affect the track length barraged by a single missile? Assume the throw-weight is fractionated into f warheads and that the quantity $f*Y^{2/3}$ is a constant equal to C. The length of track, L, barraged by a single warhead of yield Y is $k*Y^{1/3}$, where k is an appropriate dimensional constant.

$$L = k*f*Y^{1/3} \quad (4)$$

$$L^2/(k^2 * f) = f*Y^{2/3} = C \quad (5)$$

so,

$$L^2 = C*k^2*f$$

and, finally,

$$L = k*(C*f)^{1/2} \ \alpha \ f^{1/2} . \quad (6)$$

Increased fractionation does, therefore, slowly increase the barrage capability of an ICBM force but at some cost in operational and testing difficulties. This avenue will be foreclosed in a START agreement by limitations on missile loadings, limitations which will be verified by some kind of inspection, as well as flight test restrictions as was done under SALT II. The START Treaty will automatically discourage the deployment of large numbers of highly fractionated missiles because it will limit warheads, not launchers. Thus, in order to present as large a number of targets as possible to an attacker's forces, the two sides will almost surely choose to deploy single-RV missiles or missiles with only two or three warheads.

Some argue that rail-mobile missiles complicate the verification of arms control agreements. These points are raised: rail-mobile missiles can roam all over the country, be hidden in tunnels and buildings, or be concealed in civilian-appearing trains. But rail-mobile missiles are just that -- confined to the railroad net. Rail lines are easily recognized features when seen in satellite photography, even at the 10 meter resolution provided by the French SPOT 1 satellite. The Soviet rail net is particularly easy to inspect, since it is a skeleton whose back bone is formed by the Trans-Siberian and BAM rail lines stretching across 7 time zones, but which has few ribs running perpendicular to the spine. Trains on tracks are identifiable in high-resolution imagery, but such pictures necessarily cover only fairly narrow angles of view. Hence, they cannot be used effectively to search for trains which have been ordered to sortie from garrison.

Trains cannot be hidden for long periods in tunnels; concealing a missile there would block an entire rail line, and construction of special tunnels would probably be observed. Trains cannot circulate at random, but must follow schedules -- particularly on a single-track main line such as the Trans-Siberian -- since trains in opposite directions can only meet where sidings exist. Rails rarely enter buildings; they stop at loading docks. The relatively few points at which tracks enter buildings in a suspicious way could be inspected if an arms control agreement limited the number of mobile missiles and their deployment sites.

Finally, the facilities at which missiles are mated to rail cars will be very distinctive, easily recognized in satellite photography. Relatively straight forward analysis based on imagery of such a facility can provide good estimates of its maximum through-put, and hence an upper limit to the number of deployed rail-mobile missiles. The number of such facilities in each country can readily be made a part of any future arms limitation or reduction agreements.

Consider the strategic signal sent when, in a developing crisis, the trains are "flushed" -- ordered to leave their garrisons and sortie onto the rail net. Some, including Hobson,[10] argue that would "put the Soviets on notice that the United States was very serious about the likelihood of nuclear war, and that they would have just one or two more hours to destroy our MX force." In fact, the Soviet interpretation of the signal given by flushing the trains may well be just the opposite. By taking steps to place second-strike forces beyond the reach of an attacker, flushing the trains would communicate that the United States is preparing to ride out an attack and does not seek a first strike.

That signal is amplified by the fact that the missile's accuracy would suffer for several hours after the trains stopped moving, making it impossible to use rail-garrison MX missiles against time-urgent hard targets such as Soviet ICBM silos. Invulnerable missiles of relatively low accuracy do not invite pre-emption. Finally, dispersal from garrisons is totally reversible; until the missiles are fired, the trains can be recalled, thus making alerting a mobile missile force far more like scrambling manned bombers than like generating an alert of silo-based ICBMs or bring SSBNs to missile-firing depth.

If the US does not choose to announce that the trains have been flushed, there is no reason that the Soviets should know about the action for periods of hours[d]. Suppose one assumes that the Soviets use two imaging satellites 12 hours apart, each reporting through a Soviet version of the US Tracking and Data Relay Satellite system (TDRSS). Each satellite makes fewer than 3 passes over the US each day, half of which are in darkness, for a total of 3 passes in daylight. The time interval between the passes is thus about 4 hours, and the mean time to detect the sortie of a single train, assuming the satellite is targeted on a garrison and sees evidence of the train in motion is about 2 hours.

The time to relay the information to the Soviet Union is vanishingly small, but the processing time at far end is probably not less than 15 minutes, and more probably an hour (this includes constructing the image from the down-linked signal, looking at it, and making some kind of decision.). The minimum time one can expect to have before detection of sortie is thus about 2 or 3 hours, quite comparable to the time for the trains to reach a safe level of dispersal -- particularly when the 30 minute ICBM flight time is included. If the Soviets choose to execute a first strike against the rail-mobile system with short flight time SLBMs, the time delay <u>increases</u> because of the difficulty of communicating with the submarines.

The situation is really more favorable to the rail-mobile system than that, since a satellite cannot see terribly far off-nadir, 27° for the French SPOT remote sensing satellite, and hence cannot observe more than one quarter of the width of the US in any one pass. The Department of Defense estimates that the intelligence cycle time for the United States in a similar situation would be approximately 10 hours. The intelligence collection time can be reduced by adding satellites, but only at relatively great cost.

Raising the orbital altitude of the satellites increases the area

---

[d] It might be possible for the Soviets to plant spies near the gates of the rail garrison or to locate sensors on or near the tracks. While the use of <u>Spetznaz</u> troops cannot be excluded, it would present difficult operational problems for the Soviets, and would be extremely embarrassing if any such teams were detected prior to war breaking out. The question of sensors is more difficult to assess since they could, in principle, be quite small, relying upon low-power radio transmission to locations where clandestine satellite up-links could be placed. Nonetheless, these sensors would have to be so well concealed that they were not detected in the course of regular inspections of the rail lines for many kilometers in each direction from the point at which the dedicated rail lines (on the missile base) join the main line.

which can be surveyed at one time much more rapidly than it increases the period of the satellite.[e] Increasing the area in a single frame is not without complications, because, the resolution must decrease while the amount of data which must be transmitted each second increases. To gain back the resolution with more advanced sensor elements requires a great increase in the data-transmission capability.

Nevertheless, one former high official in the U.S. intelligence community privately suggests that the minimum practical re-visit interval for selected points is about 90 minutes. That does not mean that the revisit time for all points can be shrunk to that low value; in particular, one cannot hope to scan enormous areas at that rate while retaining the 5 meter resolution needed to locate trains on pre-surveyed rail lines nor the (roughly) 1 meter resolution needed to identify a missile train moving along with other rail traffic.

In addition, it is not obvious that a sortie will be detected unless the satellite is looking at the garrison just as the train leaves its horizontal silo, since an empty silo will closely resemble a full one. Furthermore, it is always possible to select the exact time at which the train leaves the silo and merges with the normal rail traffic for a moment when no observation satellite is over the horizon, since the ephemerides of most satellites are well known.

Because it is improbable that an attacker would know from national technical means the exact moment at which the trains were flushed onto the rail net, the attacker would always have to assume that the sortie took place right after his last inspection of the garrison. In effect, therefore, the attacker would have to barrage a length of track roughly corresponding to 90 minutes generating time even if the attack happened to land at the moment the trains left their shelters. The draft environmental impact statement for the rail garrison program indicates that the trains will be housed in massive shelters, perhaps having a 2,000 psi hardness.

These same difficulties in satellite reconnaissance make finding or tracking trains on the rails extremely difficult, and hence maintain the invulnerability of a dispersed mobile system. The same assertion is true, of course, for On/Off-road mobile systems, but the planned deployment schemes restrict their mobility to relatively small areas. In contrast, the track generated by rail mobile systems increases until the entire U.S. rail net is filled -- a total of nearly

---

[e] The area which can be seen for a given off-nadir pointing capability increases as the square of the average altitude above the surface of the earth; the period of a satellite in circular orbit varies as the 3/2 power of the radius of the orbit itself -- including the 6,400 km radius of the earth.

200,000 route kilometers -- plus sidings, and yards. Even if the trains are required to avoid populated areas, the system will remain survivable.

Communicating with the trains in the pre-strike environment is relatively straight-forward since the U.S. railroads have invested heavily in track-side communications systems, primarily fiber optics based. Additionally, the rails themselves form a transmission line. Richard Garwin, among others, has pointed out that after a strike the rails and fiber optics lines might well be severed. While it is not easy to sever rails with a nuclear weapon off-set from the tracks, the possibility cannot be excluded. In that event communication with the trains is increased in difficulty to talking to HMLs which have been widely dispersed after the same kind of barrage. Because HMLs are harder than trains, the barrage must be denser, so trailed fiber optics cables may not survive.

Cost of the System

It is generally conceded by partisans of both rail-garrison and HML deployment that the cost of the rail system will be on the order of $10 billion while that of a mobile midgetman system will be roughly $40 billion over a 10-year life cycle for similar numbers of warheads. The differences in the pricetags originate in, primarily, two areas. The HML system uses many more missiles for the same number of warheads since it is not MIRVd, and the Midgetman system uses far more expensive launch vehicle since the HMLs must be hardened to a considerable extent, must carry crew and communications equipment, and must be mobile. The differences in salaries alone is a significant increment.

It may also be unrealistic to consider a 10 year life-cycle. Most American strategic weapons systems have been retained in service for far longer -- up to 30 years are now scheduled for the Minuteman II and III. While the amortized differences in capital costs would appear to be less of a factor over a longer life-cycle, the differences in manpower will loom far more important than they would for a 10 year period.

Rail equipment, in contrast to HMLs, is cheap. Acquisition of equipment and construction of facilities for a 25 train, 50 missile, MX rail-garrison force has been estimated at $10.4-$12.1 billion in 1988 dollars.[11] Used but adequate locomotives can be bought on the open market for $100,000; new locomotives cost around $2 million. The incremental cost per train, not including the missiles, and not amortizing r&d funds over trains beyond the planned initial deployment of 25, is roughly $160 million.

The rail system needs no more than 4,300 people to operate it, year round;[12] the HML system requires not fewer than 5,000 crew members <u>in the HML cabs</u> as well as many more for support, command, and control.[f] Excluding the salaries for peripheral personnel, the rail-garrison system will cost $200 million/year for operations and support.

* * * *

In Summary

Both On/Off-road-mobile and Rail Garrison-mobile ICBMs are survivable now and for the future against ICBM barrage, given strategic warning. If attacked without warning, both systems are moderately vulnerable to SLBM attack; the HML system is less vulnerable to ICBM barrage given only tactical warning after the launch of enemy missiles than is the rail-mobile system (on the assumption that HML systems generate area rather than "track"), but nuclear "bolt out of the blue" attacks are improbable. The rail-mobile system possesses four distinct advantages over the HML system:

- Pre-attack communications with the rail system can take place either through the rails or through the fiber-optics system which parallels most main-line tracks. In this case "pre-attack" means the period until warheads land.

- The location of a car on the rail lines can be routinely ascertained with existing rail systems (not satellites) to better than one meter.

- The endurance of a dispersed rail system can be very high. A train using two locomotives and a diesel fuel car can stay on alert for weeks at a time if necessary.

- The cost of a rail mobile system, calculated either in absolute terms or per warhead, is far less than that of an On/Off Road-mobile system.

---

[f] The calculation goes as follows: 500 HMLs, each with a crew of 2, require 1,000 people on duty at all times. To man a single post continuously in peacetime requires 5 people for each post when 8 hour shifts, 40 hour weeks and 30 days annual leave plus holidays are included. Therefore, not fewer than 5,000 billets will be required just to fill the crew cabs of the HML system. The actual number could easily turn out to be double that when maintenance, communications and command personnel are included.

ENDNOTES

1. Frank von Hipple in **Physics, Technology and the Nuclear Arms Race**, D. W. Hafemeister and D. Schroeer, editors, American Institute of Physics (New York: 1983), p. 5..

2. S. Glasstone and P.J. Dolan. **The Effects of Nuclear Weapons**, third edition. Prepared and published by the United States Departments of Defense and Energy (Washington: 1977).

3. Although they are not widely advertised, the Defense Nuclear Agency does on occasion make publicly available excellent programs to calculate nuclear weapons effects on personal computers. All calculations of weapons effects as functions of yield, distance and height of burst used in this paper were made with the assistance of the DNA program "Weapon Effects".

4. Dietrich. Schroeer, **Science, Technology and the Nuclear Arms Race**, John Wiley and Sons (New York: 1984), fig. 6.8, p. 148.

5. **Soviet Military Power** and other U.S. DoD public estimates for the SS-18.

6. T.M. Eubanks (Jet Propulsion Lab), private communication.

7. In forthcoming Forum study.

8. Art Hobson has shown that for sets of warheads of similar shapes and different sizes, each set constrained to fit within the same diameter circle -- the bus ring of the missile -- this rule is plausible. This reasoning neglects certain other effects, but buttresses a conclusion justified primarily because it works up to levels of about 10-20%.

9. A. Hobson, "Survivability of Mobile Midgetman" in Forum Study.

10. A. Hobson, "How the MX Missile Would Increase Risks", **Arkansas Gazette**, March 20, 1988.

11. Lt.Col. T. Maxwell, USAF. Rail-garrison system project officer. 2 March, 1988. Private communication.

12. Table S-1, "Direct Employment...", Draft Environmental Impact Statement, Peacekeeper Rail Garrison Program, United States Air Force, June, 1988.

CHAPTER 15

INTRODUCTION TO ANTISUBMARINE WARFARE TECHNOLOGY*

Mark Sakitt**

Physics Department
Brookhaven National Laboratory, Upton, New York

and

The Center For International Security and Arms Control
Stanford University, Stanford, California

ABSTRACT

This paper is an introduction to the acoustical techniques that are used in strategic antisubmarine warfare. We discuss the signals that are emitted from submarines, how they propagated and how they are attenuated in the ocean. Background noise in both the deep oceans and in the ice-covered Arctic regions are described. The general problems of searching and detection are outlined with specific examples evaluated for the U.S. and Soviet ballistic missile submarines.

I. INTRODUCTION

The subject of antisubmarine warfare is both important and interesting.

Its importance is based upon the fact that the major threat to the United States Navy's ability to carry out its basic missions is the Soviet submarine fleet. The mission to protect the sea lines of communication in the North Atlantic during a conventional war in Europe between NATO and the Warsaw Treaty countries can be strongly contested by Soviet nuclear attack submarines (SSNs). While there are additional threats such as Soviet naval aviation, the Backfire bombers and the older Bears, at the longer ranges it is the SSNs which will be the hardest to defeat. For the U.S. naval mission to maintain a survivable strategic missile submarine force, our SSBNs, again the major threat is from the Soviet SSNs, although that threat is currently not considered a serious problem since we can patrol far from the Soviet coast. In our New Maritime Strategy, we will seek out the

*This work is based, in part, upon a more detailed monograph "Submarine Warfare in the Arctic: Option or Illusion" by Mark Sakitt, published by The Center for International Security and Arms Control, Stanford CA, May 1988

**Permanent address: Brookhaven National Laboratory

Soviet SSBNs in their bastions during a conventional war with the Soviets. In the Arctic, under the ice, we will again be opposed by Soviet SSNs. In the more open waters Soviet surface and air assets will be also in play. Lastly our navy has the nonsoviet role of power projection in other areas of the world which has become a more frequent type of warfare than U.S Soviet warfare. The major threat to that mission is starting to be the increased proliferation of modern high performance diesel electric submarines with the ability to fight quietly in coastal waters.

Submarines have become quieter and will become even quieter. The problems of ASW become not only harder but different. The ability of the navy to carry out the above mission will degrade, but differently for the different missions. Proper responses must be taken based upon careful analysis of the details of the missions, if we are going to maintain our current capability.

Submarine warfare is interesting because it combines some of the very elegant areas of science and engineering. It relies on some of the classical work of acoustics going back to Rayleigh as well as the work of modern oceanography. It combines the latest work in signal processing with the field of operations research.

What we will do here is try to give the reader the flavor of the type of work that is used in ASW so he or she will have an intuitive feel for the problems in this field. In order to do this we will make approximations of many of the physical effects that enable us to rapidly get some results. These results will be correct enough for this purpose but not sufficient for actual use. As is often the case life is much more complicated than what will be described. We will only concentrate on the acoustical detection of submarines and focus on the deep ocean and the Arctic regions with the view toward U.S Soviet conflicts. Nonacoustical techniques are used for short range detection and there is considerable research on other methods but the dominant method currently is acoustical.

We will examine the possiblity of either side finding the opposing strategic ballistic submarines. When we deploy our SSBNs in the large oceans, we will assume the the only Soviet threat that will survive the early stages of war in the open oceans is the Soviet submarine attack fleet. For the Soviet case, since they are tending to deploy their SSBNs near their shores or under the Arctic ice, we will assume the major threat to them is the U.S. SSN fleet. In all our discussions we will focus on the future when Soviet submarine fleets will consist mainly of their latest classes which are extremely quiet.

II. OCEAN ACOUSTICS

Since Soviet ballistic and attack submarines can deploy in the large oceans the navy would like to have a detection system that has a long detection range and that would work in all weather and sea states. In principle, one could detect a submarine leaving port or passing through a choke point and then trail it in the large oceans. Such efforts do take place but leakage through choke points and aggressive countermeasures to break trails still leave the requirement that detection be done at a large distance. This is particularly needed for finding ballistic missile submarines. For attack submarines, since they must approach their target to carry out their mission, one could rely, as a last resort, on effective local ASW forces to protect a convoy or screen a surface task force to defeat the enemy submarines. This type of ASW requires the complex operational integration of underwater, surface and air based platforms for sensors and weapons and is not the subject of this article.

Sonar can be divided into two classes, active and passive. In the active case one sends an acoustical signal out and detects the reflection from the target. The sender and receiver are usually the same platform but that is not necessary. When they are separated it is called a bistatic system. The passive systems simply listen for the noise that the target is making. In the active case, since the reflected wave generates a new wave front the propagation function is different than in the passive case. If the propagation function in the passive case is f(r), then for the active case, on a single platform, the propagation function is $f(r)^2$. Since the function is usually a power law fall off, the active sonar has a steeper fall off function. One could presume that passive sonar is more useful for long range detection, and that is usually the case. However one has to keep in mind that the range also depends on the relative strength of the targets noise versus the amount of power that you can put into an active sonar system and still be useful. At some point backscattering from the ocean and other objects limits the active system. Today the long range systems in use are passive but research is pursuing the possibility of active low frequency systems. We will later come back to this point when we discuss the attenuation of signals in the water.

A. TARGET SIGNALS

The acoustical signals that a submarine gives can be classified into three groups: machinery noise, cavitation and blade tonals. The noise due to machinery is both broad band and discrete. Some of the discrete lines are dependent on the speed of the submarine and others are not. Some reactors are designed so that at low speeds they will naturally convect and pumps can be shut down. The large size of modern submarine allows much room for insulation of machinery from the hull. Nuclear submarines are just now approaching the quietness of a diesel electric when it runs on its battery. For modern state of the art submarines the machinery noise should not be significant if one can afford to move slowly.

Cavitation noises are due to the shearing of the water which produces bubbles which then noisily collapse. This is a phenomenon occurs when a submarine is going at high speed and is causing turbulence in the water and has a rapid onset. Since the speed at which a particular submarine will start to cavitate is known, the solution is to keep below that speed unless it is necessary to go fast for tactical reasons. Because it is related to shear, the deeper the submarine, the higher the ambient water pressure, the harder it is to cause cavitation. One can go faster before cavitating if the submarine is deeper. Some tactical situations may not allow the submarine the luxury of only slow operation. Keeping up with a fast moving carrier task force or rushing to get into position to attack a task force might require increased noise levels by SSNs in order for them to be successful. For SSBNs slow operation are more normal and low noise levels are critical for the SSBNs main mission, their survival.

After one has reduced the machinery noise and avoided cavitation, the remaining source of noise is blade tonals. This is a low level noise caused by the propeller modulating a nonuniform flow of water. There is a coupling of the energy of the propulsion into acoustical energy. This is distinct from cavitation and occurs at a much lower level. The noises generated are at discrete frequencies, n*w*B, where w is the rotational frequency of the propeller, B is the number of blades on the propeller and n is the harmonic number. The presence of discrete lines that are harmonically related make this source an attractive candidate for searching. Note that typical blade numbers are 3 to 5 and typical rotational frequencies are a few per second. This results in emission lines in the region of a few hertz to less than 100 hertz. "Normal" propellers have efficiencies in the 10^{-10} to 10^{-8} range, so for shaft powers of about 10**7 watts one would have an acoustical radiation of less than 1/10 of a watt. For submarines which have been carefully designed to minimize this residual source of noise one should expect to do much better than this estimate. Modern submarines, when they are operating in their quiet mode, can mainly be detected, at distances, by their blade tonals.

B. SIGNAL PROPAGATION

In discussing the propagation of sound in waters where submarines can be deployed there are substantial differences expected for the shallow versus the deep regions and for the completely ice covered versus the open regions. For some of the possible warfare areas, like the entrances to the Barents Sea, currents can greatly complicate temperature and velocity characteristics. Our expanded security interest in the northern waters has resulted in a vigorous research program supported by the United States Navy, which has led to significant new data for these northern waters.

Our understanding of the deep oceans is that the upper portions of the water have a velocity variation that is dominated by the temperature variation. As the temperature decreases with depth it leads to a velocity that decreases with depth. Below some depth the temperature is relatively constant and the increasing pressure leads to an increase in the velocity. These two effects combine to produce a velocity minimum, as shown in figure 1a. In contrast to the well studied oceans, the temperature of the water in the ice covered Arctic regions is expected to be almost constant and thus the variation in sound velocity would be mainly due to the pressure. This would produce a rather flat profile looks, as shown in figure 1b.

For the idealized case in which the velocity profile is composed of straight lines the trajectories are pure semicircles shown in figure 1b, (for the deep ocean one can approximate the profile by two lines and in the ice covered regions by a single line). This is easy to understand. Referring to that figure, if at two points we have two different sound velocities we can use Snell's law and write

$$\cos(\phi(z))/c(z) = k$$

where k is some constant. Then taking derivatives we obtain

$$-\sin(\phi(z)) \cdot d\phi(z)/dz = k \cdot dc(z)/dz$$

or

$$d\phi(z)/ds = k \cdot dc(z)/dz.$$

Therefore, if the velocity gradient is linear, the curvature will be constant; that is, we will have circles for the ray paths.

For the deep oceans the circles will be tight on the upper paths and looser on the bottom paths as is shown in the figure. Propagation of sound with such a profile

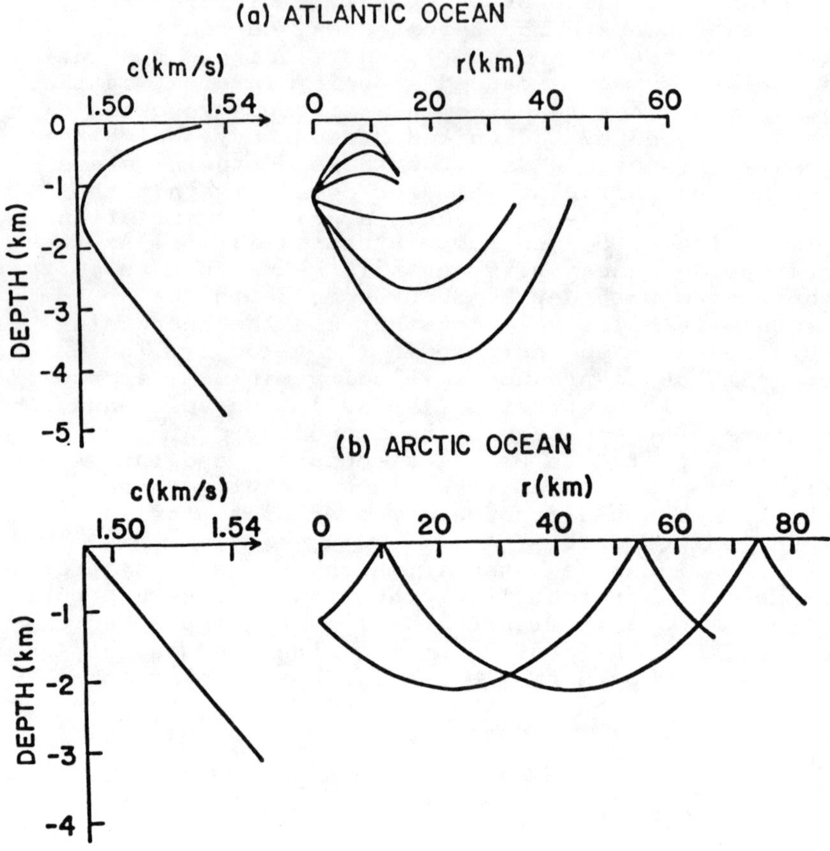

Fig. 1. Sound velocity dependence on depth and typical ray paths for (a) deep, open oceans and (b) ice-covered Arctic region

leads to a focusing and channelling of the sound that produces enhancements in the sound intensities of several orders of magnitude at long distance[1]. Significant efforts have gone into understanding methods of parameterizing the velocity profiles and to analytically solve the ray trace problem[2]. Typical propagation rays for that velocity profile are shown in figure 1b.

This simple picture illustrates the major questions. What fraction of the emitted sound avoids the losses due to a bottom bounce? What will be the losses due to reflections off the surface of the water or the underside of the ice? After detecting a submarine, you need to locate it, so you must understand how the propagation mechanisms vary in both time and space and how predictable such variations are.

The distance dependence of the sound intensity is readily calculated for the ideal case of propagation for that bundle of rays that do not hit the bottom or the top surface. For short distances it must propagate via spherical spreading since the solid angle effect dominates. At larger distance, once the bottom and top surfaces no longer affect the problem, the propagation would be via cylindrical spreading. This change from r^2 to only r dependence on the fall off allows long range detection.

In addition to the radial dependence of the spreading, the velocity minimum also causes a convergence effect. When one plots the rays for many of the angles they tend to cross roughly every 20 kilometers. The exact distances and the tightness of the convergence are obviously dependent on the details of the velocity profile. These convergence zones lead to enhancements in the intensity of the sound by between one and two orders of magnitude. ASW operations commonly use this feature of convergence zones.

For the case where the sound velocity varies linearly with the depth,

$$c(z) = c_0 * (1+a*z),$$

there is a maximum angle for which a ray will propagate just missing the bottom and therefore only undergoing ice reflections. That angle simply depends on the depth of the water via:

$$\text{phi max} = (2a(h-z))^{1/2}$$

where h is the depth of the water and z is the position of the source of the sound. Typical values for the parameter a, for the hydrostatic case, are about $1.2*10^{-5}$ per meters. For those rays that miss the bottom, the maximum distance they can go before they reflect from the ice layer is simply

Dis max = 2/a * tan(phi max),

which in our case turns out to be between 6 to 12 kilometers.

There are several negative aspects of the sound channel. Since we will hear multipath signals, ones from different emission angles, there will be a spreading of the signal over time. In the case of the linear profile one can show that this spreading is proportional to the range. With arrays that can distinguish the direction of arrival of the ray with good accuracy, that information can be used to improve one's estimate of the source position and to reduce the temporal spreading. All of these arguments rely on knowing the velocity profile in the water between the source and the receiver. Furthermore we have assumed that the velocity is only a function of the depth. In general the profile is a function of all three variables, c(x,y,z) and that results in a much more complicated problem. Even if one had mapped the velocity profiles earlier they are not constant over time. There are strong day and night effects near the surface where the temperature changes noticeably, and strong seasonal effects exist. More subtle effects like the existence of internal waves and currents in the ocean all make for an interesting situation.

In particular at the entrances and exits to the Arctic region, which are the northern Norwegian and Greenland Seas, there are definite current flows that have a major effect on the velocity profiles. Where the Arctic waters meet the Atlantic there are actual polar fronts, similar to what occurs in normal everyday weather. These fronts can seriously impair our ability for long range localization of submarines. The lack of knowledge of these velocity profiles will impair longer range ASW performance and will require close-in tactics. Our ability to measure the profiles extensively during the actual combat period will be minimal. Clearly the entrance regions will be zones of intense ASW activity.

Unlike the deep oceans many areas of ASW activity, such as in the Barents Sea, are so shallow that the method used to treat the propagation of sound is by normal modes rather than ray tracing. While a simple waveguide analogy yields a lower cutoff frequency, below which we can not propagate without large losses, the optimum frequency requires more detailed calculations[3]. Since the losses in water increase with frequency, the simplest estimate is that the optimum propagation frequency would be just above the cutoff frequency. However, penetration into the bottom surface increases as the frequency decreases. In the solid or mixed bottom layer both compression and shear waves can be induced, unlike the water phase which only has compression waves. Losses in the shear mode are the major loss mechanism and push the optimum frequency to higher values. Typical optimum frequencies, measured on the Eastern Atlantic depend on the depth and show optimum values about one to two orders of magnitude above the corresponding cutoff values[3]. These results are rather insensitive to the actual ratio of the sound velocity of the water to that of the bottom surface. The optimum frequencies peak in the region of interest, below 1000 hz. These results are, of course, for shallow water without ice cover. When the upper surface is covered with ice the losses due to the irregularities of that surface greatly attenuate the higher frequencies.

C. ATTENUATION

In addition to the decrease in signal levels due to the spreading of the acoustical wave, there is some intrinsic attenuation due to the losses in the media. Both the water and the surfaces at the boundaries contribute to the losses, unless one is only considering the captured rays in the sound channel. An important question is to determine how the attenuation in the water depends on frequency.

A simple way to answer that is to consider how a small relaxation process modifies the wave equation and induces losses in the water. This approach is described by Rayleigh in his classic work on sound[4]. Recall that the wave equation comes from two basic laws coupled with an equation of state for the medium. The first is Newton's 2nd law, F=ma, which for the one dimensional case becomes

$$-dp/dx = rho*du/dt$$

where p is the pressure, u is the velocity and rho is the density of the media. The second is the law of conservation of mass which becomes

$$d(rho)/dt = -rho*du/dx.$$

We now need the equation of state to couple these two equations and form the wave equation. For an elastic medium with no losses one uses

$$dp = k^2*d(rho).$$

One can now reduce these equations to one variable, p, and obtain

$$d^2p/dx^2 = 1/k^2*d^2p/dt^2,$$

which is the classical wave equation. One can modify the equation of state of the medium to include losses by adding a simple relaxation term of the form

$$b*d(rho)/dt.$$

Then when one substitutes to get a one variable equation there is an extra term added to the wave equation, now expressed in terms of rho, instead of pressure, of the form

$$b/k^2*d^3(rho)/dx^2dt.$$

One can remove the time dependence and then extract the spatial wave functions. In the domain of interest the frequency dependence of the exponential attenuation coefficient is approximately f^2. Since the attenuation is exponential low frequencies will travel considerably further than high frequencies. Blade tonals that are in the region of ten to a hundred hertz are many orders of magnitude easier to pick up than high frequency sonar emissions of several tens of kilohertz that are used for final homing in a torpedo. The scale for detectors for long range sonar is therefore determined by the low frequency wavelengths. For ten hertz the wavelength is 150 meters and since one wants a phased towed array of many elements we need towed arrays of about a kilometer in length. This frequency dependence of the attenuation also allows for submarine communication between two submarines that are nearby via very high frequency modulated sonar beams without the prospects of enemy detection at a long range.

In the deep open oceans one can seek the signal that is trapped in the sound channel and, for long distances, not be concerned with the two boundary layers. The Arctic is different because the velocity profile is steadily increasing. The propagation is more like a duct than a channel. Reflections on the underside of the ice play an important role in the long distance attenuation of the signal. A naive belief would be that the underside of the ice was smooth which would led to perfect reflection from the ice. This is because the loss mechanisms are the induced shear waves in the ice, which have attenuation coefficients almost an order of magnitude higher than that of the compression waves. Since the grazing angles are small, using the shear velocity we would expect perfect reflection on the basis of Snell's law.

This is far from the case. Neglecting losses from the scattering of the underside of the ice grossly underestimates the propagation losses; one observes discrepancies of 20 to 40 db for the larger ranges. The surfaces are not smooth but quite complex. Early models show that the inclusion of losses due to ice reflection tended to bring theory into rough agreement with the somewhat scattered data[5]. More detailed empirical transmission loss models have been developed[6].

D. NOISE

The problem of long range detection is not so much the limit of the sensitivity of the transducer as the ambient noise of the oceans. The medium has intrinsic noise from several sources that all vary. In the deep oceans, besides for the intermittent sounds that are largely due to biological sources, the dominant ones are seismic, shipping and wave motion.

The typical spectrum of noise in the deep oceans is measured in decibels in reference to 1 micropascal in a frequency band of one hertz. For reference, a one watt acoustical output would be about 170 db at one meter distance. Below about ten hertz the main contribution is from seismic and tidal sources and is at the 80 to 90 db level. In the region from about 10 to a few hundred hertz the major sources are from shipping and vary depending on the traffic. Variations of as much as 20 db have been observed with 80 db being an average value. Above a few hundred hertz the major source is from the surface wave motion and therefore the noise strongly depends on the weather or sea state. Variations of between 20 to 30 db have been observed for wind speeds up to 30 knots peaking near 80 db.

The detailed mechanisms that produce underwater noise in the Arctic region are not as well understood. In fact, disagreement about some of the posited models is a flourishing field of research. One of the early problems was that the research often was done in limited areas for limited periods of time. The now known variability of the Arctic could lead one to mistakenly extrapolate too far on the basis of a few measurements. Also, a rather limited research program may completely miss phenomena that do not occur frequently.

Our early view was that the Arctic noise level should be lower than that of the deep oceans due to the absence of nearby shipping noise and due to the reduction of the effects of wave induced (or wind induced) noise by the ice cover. The effect of the ice cover coupled with the sound velocity profiles of the water would lead to an attenuation of the higher frequency noise producing a low frequency pass band. The effect of ice breakup should produce local noise at the edges of the ice formation region which is called the marginal ice zone (MIZ).

The marginal ice zone, the interface between the ice covered water and the open water, is a known source of Arctic noise. At the MIZ new ice is forming in the colder seasons or melting in the warmer seasons. The details of this interface vary by region and season. Due to the wave motion interacting with the ice edge there is an increase in the acoustical levels which is dependent on the wind or sea state. The background noise peaks at the interface and extends for 10 to 30 kilometers, with the higher levels on the open side of the water. Observed increases in noise levels were as high as 10 db compared to the open water and 20 db as compared to the nearby ice covered water. The frequency spectrum peaked at the lower values with the 100 hz levels 20 db higher than the 1 khz levels. These increases all become somewhat smaller as the ice water edge becomes diffuse. It is difficult to compare the effects of different sea states because other conditions also change, but the data suggests that going from a sea state of one to a sea state of five could be responsible for an increase in noise of 15 db at 100 hz.

One of the first studies of noise under the Arctic pack ice was carried out in the Canadian Archipelago near Banks Island and near the M'Clure Strait[7]. They noticed a striking impulsive type of noise that apparently was correlated with mechanical properties of the ice cover such as stress and temperature. During the spring and winter the noise mechanism was believed dominated by surface cracks due to thermal stresses. During the summer it was believed that the relative ice floe motions produced the dominant noise. For the "quiet" periods the noise levels were at or below that of an open water at a sea state of zero, but for the noisy periods the level varied between sea states of zero to three. Daily variations of 20 to 40 db were observed during these noisy periods. The thermal stress resulted in pulses of noise as high as 40 db.

A more extensive set of measurements was made more recently in the central Arctic and concentrated on the relatively low frequency region of 10 to 20 Hz[8]. The noise in this low frequency band was shown to be correlated with various parameters related to ice stress other than the thermal effect. They were correlated to composite measures of wind, current and drift. They observed correlation times of 1.4 days and recorded variations of factors of 5 within days.

During an experiment in the high Arctic Ocean one team observed the nearfield noise from an ice pressure ridge as it was forming[9]. This first measurement of an active ridge lasted only three days because the ice camp started to break up and for safety reasons the team had to be withdrawn. By the use of multiple hydrophones they were able to measure that the acoustically active length of the ridge was approximately 1 kilometer. In one day the noise in the 10 to 200 Hz interval increased by between 15 and 25 db, with the absolute level, at a distance of 100 yards, being between 95 and 100 db and rather flat throughout that frequency interval. This single unexpected and rather dramatic observation could not directly answer questions like the length of time that the activity lasted or the frequency of such active ridges.

Efficient signal processing to suppress the noise requires some understanding of the details of the structure of the noise. An attempt to characterize that structure was done and showed that the usual pure stationery Gaussian assumptions were not quite valid[10]. The experiment was done under the pack ice and looked at the frequencies below 2.5 kHz. The noise consisted of a combination of stationery Gaussian noise intermixed with random sinusoids and occasional impulsive noise bursts. The tonals at times jumped as much as 60 db during a 1/2 second period, but more often the increases were between 30 to 40 db.

They concluded that due to the random sinusoids the noise can not be considered a stationery process in the time frame of 10 minutes. Such observations are important, since it had already been shown that for ocean noise the doubling of the integration time in the signal processing did not generally lead to the expected classical 1.5 db gain in the signal to noise ratio[11].

Arctic noise is similar to other acoustical properties of the oceans in that its most constant feature is its variability. The presence of "hot spots" and short bursts of tonals could be used for hiding submarines or for covert active sonars.

III. DETECTION AND SEARCHING

The probability of detection is a function of distance and for the simplest case one can assume a cookie cutter model where the probability is unity up to some range and then falls to zero beyond that point. To estimate that range one uses the sonar equation a form of which is below

$$\text{Detection Level} = (\text{Source Level})*(\text{Transmission Loss})*(\text{Direction Gain})/(\text{Noise Level}).$$

This is usually recast in the logarithmic form and expressed in terms of decibels as

$$TL(R) = SL-NL+DI-DT$$

where the abbreviations are obvious. The first task is to determine DL, the acceptable detection level or signal to noise ratio. This is a classical signal processing problem. Given an integration time and a bandwidth, one has to determine, based upon operation requirements, what is an acceptable compromise between the false alarm rate and the detection probability. The directional gain is the enhancement one gets from a phased array such as a towed array. For example if one took SL=115 db, a quiet submarine, DI = 20 db, NL = 70 db, a rather quiet background and DT = -15 db then one would have TL = 80 db. (Note that in sonar one defines the transmission loss relative to what would be observed at one yard from the source. Sonar units are a mix of every conceivable system.) For short ranges we can use spherical spreading and the resulting range is then 10 kilometers. If the background noise level went up to 80 db the range would go down to one kilometer.

Our major concern is the vulnerability of ballistic submarines, either ours or in the case of our new maritime strategy, theirs. For a given range and a given number of searching platforms one can model the search problem. While the models discussed here are not necessarily realistic, they can form the bases for initial evaluation of the effect of varying some basic parameters of the problem, such as deployment area and detection range.

The simplest form of searching is the linear or exhaustive search without any evasion on the part of the target. The target is assumed to be fixed somewhere in a deployment area of size A. One takes the area sweep rate as twice the product of the detection range R and the speed V of the attack submarine. The probability of detection as a function of time is then simply

$$prob(t)=2*R*V*t/A.$$

This probability grows linearly with time and is exhaustive in the sense that after some fixed time period, the entire area has been searched, at which point the probability for detection becomes one. These estimates for the required search times are generally considered optimistic lower bounds. Since they are trivial to evaluate they are usually done first to eliminate the need for a more complicated analysis if the results are hopeless.

A more realistic search model assumes that the target is moving randomly in the deployment area, but is still not taking evasive action in response to the maneuvers of the attack submarine . The problem is equivalent to a fixed target and a random searcher which is easy to formulate. We divide the total deployment area into small pieces of size 2*R*V*T, which can be searched in the short time interval T. Then the probability of not detecting the target after searching n pieces in a total time period of t is

$$prob\text{-}not(t) = (1-2*R*V*T/A)^n.$$

Since n=t/T, we can take the limit to small time

intervals and obtain the well known exponential form for the probability of detection as a function of time. That is

$$prob(t) = 1-expon(-2*R*V*t/A).$$

Clearly, estimates for the required time to reach a level of detection are greater in the exponential model than in the linear model.

To give some scale to the problem let us consider the case of the Soviets searching for the U.S. Ohio class SSBNs which carry Trident ballistic missiles. The range of the Trident is about 4000 nautical miles giving a patrol area, if one wanted to be able to target Moscow of about $18*10^6$ nm. This patrol area is in fact bigger since the submarines can patrol outside of this area and then return during times of crisis. Except for the short crisis period, in which there is not enough time to conduct a search, the Soviets are forced to look in an area 2-3 times bigger. If we deployed in an area of $60*10^6$ miles and the Soviets searched at 10 knots with 50 attack submarines, then, for a detection range of 10 nm, it would take them about 250 days to find 2/3 of our submarines. Bear in mind that submarines are not out at sea for that length of time so the estimate is not too meaningful. Also the Soviets only have about 30 of the very modern type of quiet attack submarines, the Victor IIIs, the Akulas, the Mikes and the Sierras. It is commonly believed that the detection range for finding our Ohio class is closer to 1 nm than it is to 10 nm. They are very quiet submarines! If one used the one nm range the time estimate would go up to 2500 days, not a practical operation. The consensus is that assuming that our SSBNs can get free and clear from their ports into the deep oceans and that no new exotic techniques for detection of submarines are developed, our strategic submarine forces are survivable. The first of these assumptions requires an operational commitment and the second requires an ongoing research program.

The question of the vulnerability of the Soviet ballistic submarines is much more complicated and here we will only touch the subject[12]. The Soviets are deploying their modern ballistic submarines in the Arctic regions, which include the deep Arctic Ocean and some of the shallower seas bordering the Soviet Union. This deployment avoids having the submarines go through choke points where they can be picked up by U.S. ASW forces. It also allows them to hide under the ice, requiring us to use modern SSNs to seek them out or to stay in waters in which they can have assistance from the Soviet Northern Fleet's surface ships. The disadvantage is that the area of the Arctic Ocean is only about $5.4*10^6$ square miles and the nearby seas are small. The Barents Sea, for example, has an area only about 10% of that of the Arctic Ocean. It is interesting that as both sides have developed longer range submarine ballistic missiles they have developed opposite deployment strategies. We use the longer range to increase our deployment area to decrease detection. The Soviet have withdrawn to a smaller area, where they can still target us but no longer have to face the U.S. ASW forces in full strength. Both have pushed for increased survivability but in different directions justified by the asymmetry in situations.

In figure 2, we have plotted the "half time" as a function of the detection range for three different levels of the U.S. force structure. The "half time" is the time needed to find half the Soviet forces. Three half times would still leave an 1/8th of the Soviet forces undetected. We have assumed a search speed of 10 miles per hour. Note that in the next decade, if we build everything that the navy has planned we will have only 100 attack submarines of which several dozen will be allocated to protecting our large surface task forces. Some must be allocated to the Soviet far east fleet which also has ballistic submarines. The number fifty must be considered an upper limit for the number that could be committed to the Arctic. Soviet submarines are getting quieter and in the future one should expect that the detection ranges will be below 10 miles. For these shorter ranges the half times will be in the region of 30 days, considerably shorter than for the corresponding U.S. case but not yet alarming.

There are many countermeasures that make the U.S. prospects for finding and destroying Soviet submarines considerably worse than that estimate[12]. Here we will only consider one, the possibility of evading detection.

Consider the possibility that the target might hear the searcher before being detected. For the case of an active sonar search this is quite likely. In the case of a passive search it will depend on the relative noise emissions of the two submarines and the performance characteristics of the two sonar systems. As the technology on both sides becomes highly developed, the advantage tends to go to the target. This occurs because the searcher must go at a moderate speed in order to maximize the searching rate. The speed will be determined by balancing the tradeoff between the increased noise caused by the increased speed, which will decrease the detection range, and the increase in area that can be covered because of that increase in speed. On the other hand the target tends to go as slow as possible to reduce its generated noise. The most interesting, and nontrival case is one in which the range for which the target can detect the searcher, Rt, is greater than the range for which the searcher can detect the target, Rs, while the

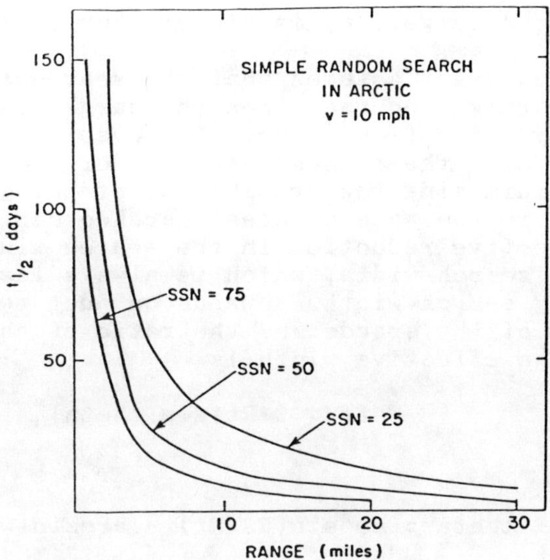

Fig. 2. Characteristic search half time as a function of detection range for three levels of U.S. forces

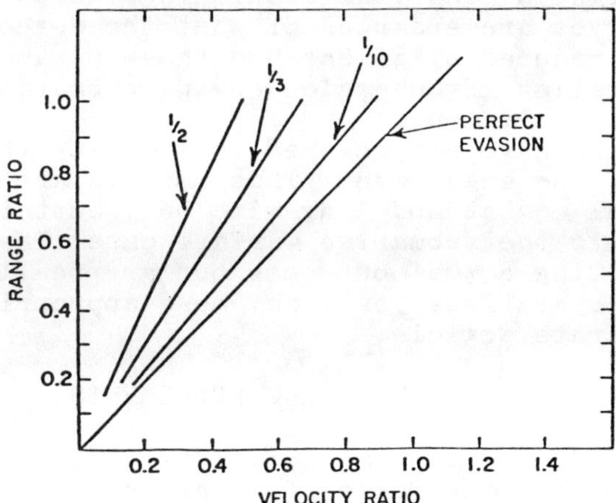

Fig. 3. Suppression factors in effective sweep rates for values of ratios of detection ranges and ratios of velocities

speed of the target, w, is slower than the searcher's speed, v.

If the target assumes that the searcher will maintain a known straight course, then the target can calculate explicitly its optimal course for evasion. That course would maximize the closest approach of the searcher, thereby minimizing the probability of being detected. The reduction in the rate of area searched can be expressed as an effective reduction in the search width. This effective search width, which is always less than the nonevading search width, depends on just two parameters, the ratio of the speeds and the ratio of the detection ranges. The effective width is

$$Weff = 2*Rt*\sin(theta)$$

where

$$theta = \arcsin(Rs/Rt) - \arcsin(w/v).$$

The effective search width can never get bigger than the nonevading width, 2*Rs, and it goes to zero when the ratio of the speeds equals the ratio of the detection ranges. Beyond that point the target always wins. The effect of evasion can thus be treated as a modification of the search widths in the nonevading models.

In figure 3 we show the reduction factors for different values of the speeds and the ratio of the detection ranges. The line at 45 degrees is the boundary for perfect evasion. All values of those ratios that lie to the right of the line result in evasion. The other three curves are examples of what the detection range would be reduced by if one had those parameters for the various ratios. Even a slow moving submarine can evade if it is a good listener!

What we have not covered is modeling the actual combat of the engagement. Those models are applications of general combat modeling with only certain specifics tailored to the submarine warfare case. They are critical to estimating actual outcomes but are really another new field of operations research, more appropriately covered in a separate article.

IV. ACKNOWLEDGEMENTS

I would like to thank Sid Drell, Ted Postol and Dave Bernstein from the Center for International Security and Arms Control at Stanford University for many insightful discussions on national security. I thank the Carnegie Corporation of New York for my support as a Science Fellow at the Center for 1986-1987 and the Brookhaven National Laboratory for granting me a leave with support during the time much of this work was done. The views expressed here are mine and do not necessarily reflect the views of my colleagues, the Center for International Security and Arms Control, Brookhaven National Laboratory, or the Carnegie Corporation.

REFERENCES

1. For general reviews see Robert J. Urick, Principles of Underwater Sound (McGraw Hill, New York 1975); Urick, Sound Propagation in the Sea (Peninsula Publishing, Los Altos, CA. 1982); and L. Brekhovskikh and Yu Lysanov, Fundamentals of Ocean Acoustics (Springer-Verlag, New York 1982).
2. Melvin A. Pedersen and DeWayne White, "Ray Theory for Sources and Receivers on an Axis of Minimum Velocity", Journal of the Acoustical Society of America 48, no. 5 (1970): 1219; and Walter H. Munk, "Sound Channel in an Exponentially Stratified Ocean with Application to SOFAR", Journal of the Acoustical Society of America 55, no. 2 (1973): 220.
3. F.B. Jensen and W.A. Kuperman, "Optimum Frequency of Propagation in Shallow Water Environments", Journal of the Acoustical Society of America 73, no. 3 (1983): 813.
4. Lord Rayleigh, The Theory of Sound, (Dover Publications, New York 1947).
5. O.I. Diachok, "Recent Advances in Arctic Hydroacoustics", Naval Research Reviews, 1976., 60.
6. Beaumont M. Buck and James Wilson, Buck/Wilson Deep Arctic Transmission Loss Model, report no. TR-58, Polar Research Laboratory, Carpinteria, CA, December 19, 1984.
7. A.R. Milne and J.H. Ganton, "Ambient Noise Under Arctic-Sea Ice', Journal of the Acoustical Society of America 36, no. 5 (1964): 855.
8. Nicholas C. Makris and Ira Dyer, "Environmental Correlates of Pack Ice Noise", Journal of the Acoustical Society of America 79, no. 5 (1986): 1434.
9. Beaumont M. Buck and James H. Wilson, "Nearfield Noise Measurements from an Arctic Pressure Ridge", Journal of the Acoustical Society of America 80 (1986): 256.
10. James G. Veitch and Allen R. Wilks, "A Characterization of Arctic Undersea Noise", Journal of the Acoustical society of America 77, no. 3 (1985): 989.
11. W.S. Hodgkiss and V.C. Anderson, "Detection of Sinusoids in Ocean Acoustic Background Noise", Journal of the Acoustical Society of America 67, no.1 (1980): 214
12. Mark Sakitt, Submarine Warfare in the Arctic: Option or Illusion?, (Center for International Security and Arms Control, Stanford, CA. May 1988).

CHAPTER 16

SEA-LAUNCHED CRUISE MISSILES: TECHNOLOGY, MISSIONS, & ARMS CONTROL VERIFICATION

Richard A. Scribner
School of Foreign Service, Georgetown University
and
Center for International Security and Arms Control, Stanford University

ABSTRACT

Sea-launched cruise missiles (SLCMs) present some particularly striking problems for both national security and arms control. These small, dual-purpose, difficult to detect weapons present some formidable challenges for verification in any scheme that attempts to limit rather than eliminate them. Conventionally armed SLCMs offer to the navies of both superpowers important offensive and defensive capailities. Nuclear armed, long-range, land-attack SLCMs, on the other hand, seem to pose destabilizing threats and otherwise have questionable value, despite strong US support for extensive deployment of them. If these weapons are not constrained, their deployment could circumvent gains which might be made in an agreements directly reducing of strategic nuclear weapons. This paper reviews the technology and planned deployments of SLCMs, the verification schemes which have been discussed and are being investigated to try to deal with the problem, and examines the proposed need for and possible uses of SLCMs. It presents an overview of the problem technically, militarily, and politically.

I. INTRODUCTION

When the issue of whether future arms control agreements can be verified is raised, control of sea-launched cruise missiles (SLCMs) is often cited as the leading example of near impossible verification requirements.[1] In fact, the general issue of how to limit cruise missiles is more than ten years old. Much has been written about the subject of SLCMs over the past few years.[2] Less critical scrutiny has been given, however, to the actual mission requirements for SLCMs, particularly long-range nuclear land-attack versions.

This paper reviews the technologies associated with sea-launched cruise missiles, the suggestions made for limiting them and the associated verification procedures that would be required, the statements made about the uses of these weapons, and conclusions that can be reached about dealing with this problem.

The matter was recently highlighted when, in the Joint US-USSR Communique issue at the Washington, 1987 Summit, the two leaders stated: "The sides will find a mutually acceptable solution to the question of limiting the deployment of long-range, nuclear-armed SLCMs," and indicated that they were "committed to establishing ceilings," and "to seek mutually acceptable and effective methods of verification of such limitations which could include: the employment of NTM, cooperative measures and on-site inspections."[3] Indeed, the Soviets have stated that further progress on START will require some understanding on limiting numbers, types, and deployment modes of SLCMs.

In general, while the technology for verification or monitoring is likely to continue to improve, weapons technologies have moved in directions that make verification more difficult. Modernization and new technologies, serving strategic and tactical objectives, have moved toward small size, dual capability,i.e., either conventional or nuclear warheads, smaller and more sophisticated warheads with insertable nuclear components, incorporation of "stealth" or low radar cross section technologies, and generally more widespread deployment including mobility and multiple possible launch platforms. A mixture of US conventional and nuclear SLCMs, for example, are slated to be deployed in the thousands on as many as 200 surface ships and submarines.[4] Apparently less ambitious Soviet plans call for numerous long-range nuclear cruise missiles to be deployed on many attack submarines.[5]

TABLE 1: Nuclear Long-Range Land Attack SLCMs: Arguments Pro & Con [6]

PRO	**CON**
• 200 nuclear strike platforms provides increased flexibilty & effectiveness	• SLCM force frees up SLBMs and enhances first strike perception and & capability
• Dispersed nuclear retaliatory capability provides improved survivability	• Nuclear SLCMs proliferated in large numbers will make ships even more important targets for Soviets
• SLCMs are a "strategic reserve force", i.e., a survivable nuclear arsenal for protracted conflict	• Such notions are ill-founded since the concept of a limited or protracted nuclear war is fallacious & dangerous
• SLCMs are cost effective; i.e., they are a low cost weapon at about $1 million apiece	• They will result in heightened nuclear tensions and reduced flexibility
• They are needed to counter Soviet SLCM threat	• US is more vulnerable to SLCMs, i.e., US centers near accessible coast
• They are needed to replace roles of Pershing IIs and GLCMs in Europe	• Further proliferation of nuclear weapons on ships will complicate relations with allies
• Arms control verification of limits on SLCMs is impossible	• Arms control limits on SLCMs are essential; verification is possible

A. Limiting SLCMs

Limiting the number of nuclear SLCMs is in the US interest, since with its large coast line it is far more vulnerable than the Soviet Union to such weapons launched from ships and submarines. There is, of course, the fundamental question of whether either navy needs such nuclear land-attack weapons. The range of disagreement regarding the need for and possible arms control limits on SLCMs is summarized in Table 1.[5]

Advances in terrain mapping guidance systems for cruise missiles make it a highly accurate weapon and, together with prospects for future versions to have ranges greater than about 3000 km and supersonic speed, give rise to speculative scenarios for strategic counterforce use of these weapons supposedly designed as a "strategic reserve force".[7] Twenty foot long cruise missiles flying low in the atmosphere are already difficult to detect and track. With refined stealth technology added, intelligence- and verification-related missile test data collection as well as early warning could be further inhibited. For practical purposes, from any distance other than close proximity to a sea launched cruise missile it does not seem possible to distinguish by sensor measurements whether such a weapon on board a ship contains a conventional or a nuclear warhead.[8] And the long-range, nuclear-armed, land-attack variant of the US Tomahawk looks exactly like the shorter range anti-ship version of the missile.

A number of clever and complicated schemes have been proposed to solve the verification problem.[9,10] Almost all of them required very intrusive inspection and monitoring of production, and assembly, as well as inspection of deployment platforms.

The situation is not hopeless by any means, but there is little doubt that verification of any limits on SLCMs will be a challenging problem. However, with the unprecedented cooperative verification provisions of the INF Treaty and those discussed for START, a new verification basis is in place which makes such future agreements more possible than was previously anticipated.[11] These new provisions coupled with an improving political climate between the two superpowers, and, most importantly, continuing opening up of the Soviet Union, means that verification need not be a barrier to further reductions and some other arms control agreements. The new flexibility on the part of the Soviets to include not only extensive intrusive inspections but also other cooperative measures, means that arms control agreements are now possible that hitherto were deemed impossible to achieve because of the presumed non negotiability of the necessary verification package, including intrusive measures.

Since verification can never be perfect and arms control measures limiting SLCMs would seem to be in the US interests, the fundamental matter is the judgement of under what conditions are the risks acceptable for the anticipated benefits.[12] For example, in the case of SLCM limits, there are recognizable technical, security and force planning, and political questions that must be addressed to make that judgement.

B. Key Questions

The technical questions relate, for example, to what kinds of detection and monitoring under what conditions may be possible to distinquish nuclear from conventional forms of the SLCMs; and how easily can conventionally armed SLCMs be converted to nuclear ones? The security questions relate to such matters as naval force posture and needs.

Whether, for example, the US Navy needs nuclear weapons at all, apart from SLBMs as an invulnerable deterrent force. And, regarding verification provisions, whether the US could allow Soviet inspection teams to board American naval ships, and what kinds of intrusive inspection is now negotatiable in view of the ground-breaking procedures adopted in the INF Treaty. There are even broader security/political questions: can the US accept looser limits on SLCMs in view of what some see as their lesser threat as preemptive first strike weapons? Is a START-type agreement including substantial reductions possible without an agreement limiting SLCMs? The issue is further clouded by the implications that inspection could have on a sensitive area with some US allies, namely, the US policy to "neither confirm not deny" the existence of nuclear weapons onboard ships porting in such countries as Australia, Denmark, and the Philippines.

II. STATUS OF THE NEGOTIATIONS

Limits on SLCMS are not presently included in the proposed 6000 strategic warhead ceiling. As indicated, the two superposers agreed at the Washington Summit to try to control SLCMs and "seek verification measures". In 1987, the Soviets proposed a ban on all SLCMs on surface ships; a limit of 400 nuclear submarine launched cruise missiles and a limit of 600 conventional submarine launched cruise missiles for a total of 1000 SLCMs on submarines only; and a limit on SLCM range of 600 km.[13] The US position has been that it will discuss limits on nuclear SLCMs, but will not discuss limits on conventional SLCMs. On April 6, 1988, Soviet negotiator Victor Karpov said the Moscow would accept an in-principle agreeement on SLCMs, and reiterated that a START agreement was not possible without limits on SLCMs.[13] As far as the most intrusive aspects of verification measures that have been put forth, the US has stated that it is unwilling to allow Soviet inspectors aboard US submarines. In an interesting and relevant aside, The Washington Post carried a story also on April 6, 1988 in which US arms control advisor Paul Nitze floated the idea of a ban on all naval nuclear weapons except SLBMs.

III. CRUISE MISSILE TECHNOLOGY

In January of 1988, the Soviet Union began deploying a long-awaited new generation of nuclear sea-launched cruise missiles, the SS-N-21s, comparable in most respects to the US Tomahawk nuclear cruise missile deployed since 1984.[14] The SS-N-21, thus far only produced as a nuclear warhead carrying missile, has a range of 3000 km and is so far only deployed on submarines. The Soviets also have a larger, longer-range, larger yield SLCM, the SS-NX-24, also for deployment on submarines "under development". Until the emergence of these weapons, the Soviets had only deployed SLCMs with ranges between 100 and 550 km.

The US has two basic types of long-range land-attack SLCMs: a nuclear version of the Tomahawk land attack missile (TLAM) designated TLAM/N and a conventional version, designated the TLAM/C. These have identical airframes (approximately 21 feet long and 21 inches in diameter; see Figure 4.). The TLAM/N has a range of 2500 km. The TLAM/C has less than half that range because conventional warheads are larger, therefore displacing fuel, and heavier. The "Bullpup" conventional warhead weighs about 1000 lbs. The US nuclear Tomahawk carries a 250 kt warhead which weighs about 270 lbs.[15] Up to 1988, the US has deployed about 1000 conventional and 100 nuclear long-range SLCMs.

The US has also deployed a shorter-range, conventionally-armed, ship attack SLCMs, designated TASM. The TASM has the same airframe as the TLAM and a range of about 450 km.

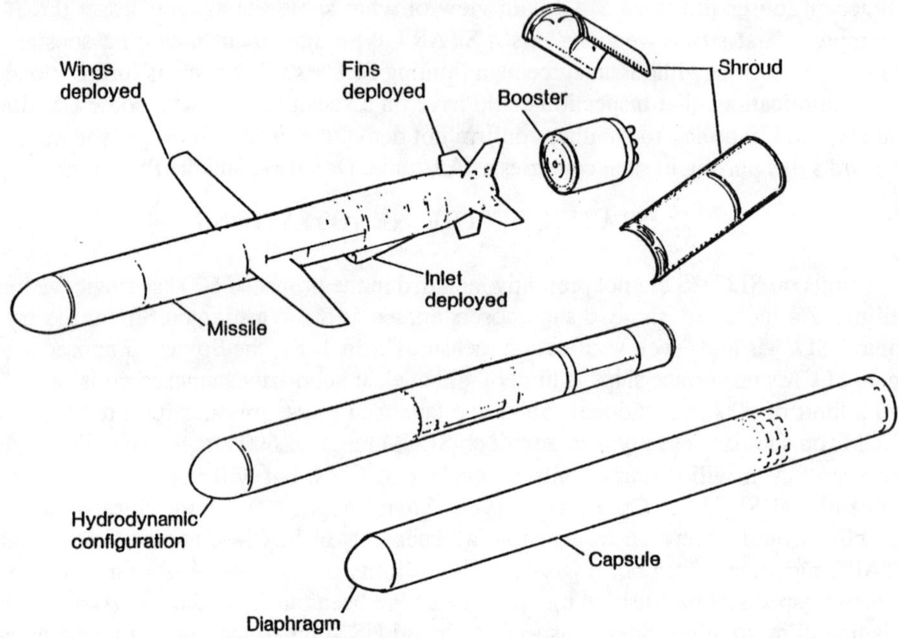

Figure 1. Tomahawk cruise missile and cannister. (Joint Cruise Missiles Project)

With a combination of inertial, radar altimeter, and terrain and target (TERCOM) guidance, the US land attack cruise missile has a reported accuracy of between 10 and 30 m CEP. The TERCOM (terrain contour matching) updates the missiles inertial guidance coordinates. The TLAM/C also employs digital scene matching in the terminal phase (DSMAC) to achieve the greater accuracy needed for conventional warhead delivery. Reportedly, Soviet SLCMs although highly accurate are not as accurate as US SLCMs.

The US Navy plans to deploy between 3000 and 4000 conventionally-armed and nuclear-armed SLCMs by 1992 on nearly 200 ships and submarines. Approximately one-quarter, or 800-1000, of the ones deployed will be nuclear-armed, long-range land-attack missiles. On surface ships the SLCMs will be launched either from deck-mounted "armored box launchers" or vertical launchers. On submarines, the SLCMs will be launched from either torpedo tubes, or vertical launchers, which will be deployed on the new Los Angeles class attack submarines.[15] According to DOD's Soviet Military Power (1987), 215 Soviet ships have nuclear cruise capability.[14]

Currently, US SLCMs fly at Mach 0.7 or about 500 mph.[15] Their speed has led to their being characterized as "slow flyers" and therefore "stabilizing". Depending on the launch location and distance to target, a land-attack missile could strike within tens of minutes of launch to as long as about 5 hours. For example, if a submarine located 200 miles east of Washington, D.C. were to launch several cruise missile against the US Capital area as part of either a disabling decapitation strike or a total first strike (however unlikely either may be), it would be problematic at best under present conditions whether such a launch would be detected (even ambiguously) and the missiles would detonate at their targets about 25 minutes later. That scenario seems neither slow nor stabilizing.

The cruise missile is carried in its canister and launched from it (see Figure 1). The booster rocket (see Figure 2) burns for about ten seconds giving the cruise missile it initial altitude and sufficient air speed for its stubby wings to provide some lift. Thereafter the air-breathing fuel-fed jet engine provides the cruising power for the missile. It may cruise at an altitude as high as 10,000 ft and then, as it approaches land, lower to perhaps 1,000 ft or less. As shown in Figure 3, the missile's TERCOM and DSMAC guidance systems bring it in very close to its target.

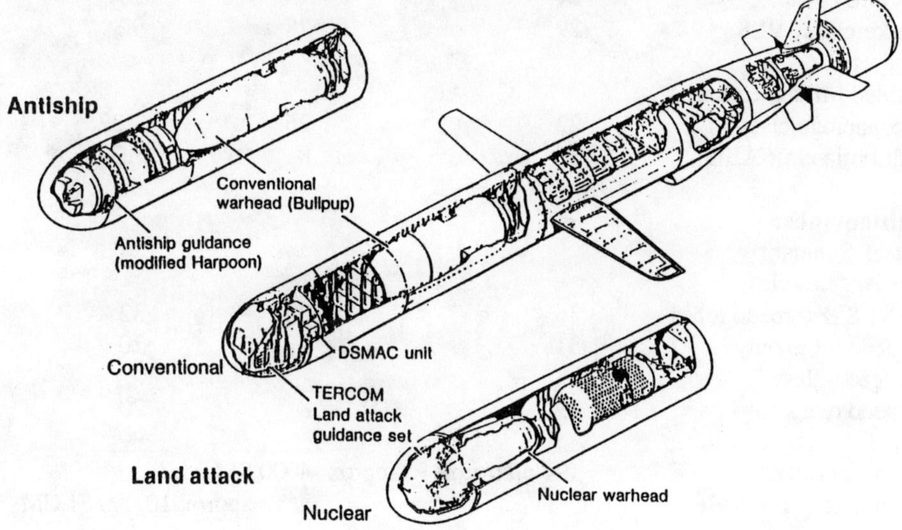

Figure 2. Antiship and Land Attack SLCMs (Joint Cruise Missiles Project)

Survivablility of a weapons system (particularly one at sea) is a critical consideration. One of the justifications given for proliferating cruise missiles on a number and variety of ships has been that it enhances their survivability. Prelaunch survivablility is of course dependent on the survivability of the launch platform. Prior to launch, SLCM carrying surface ships or attack submarines are high-value targets. The Soviet navy would go to great lengths to target them. While surface ships may thus be fairly vulnerable, submarine-based SLCMs should be highly survivable after launch. However, an attack submarine that carries SLCMs and must manuever near elements of an enemy fleet to be ready to perform its attack non-nuclear function, may not be quite so survivable as a submarine far from enemy vessels. Overall, however, it would appear the given the current Soviet capabilities, post launch survivability or penetration capability of the current generation of SLCMs is fairly high.[4]

TABLE 2: Estimated US SLCM [a] Deployments By The Mid-1990s [b]

PLATFORMS	VESSELS	SLCMS/VESSELS (est.) planned	possible
Surface Ships:			
Battleships:			
Iowa-class	4	36-60	300+
Destroyers:			
Spruance-class			
w. ABLs (2 ea.)	7	8	12
w. VLS (1 ea.)	24	45	60
Burke-class / VLS	29	28	90
Cruisers:			
Guided Missile Cruisers			
Ticonderoga-class/VLS	22	26	122
California-class/ABL	5	8	12
Submarines:			
Attack Submarines:			
Los Angeles-class			
w. VLS & torpedo tubes	36	20	32
torpedo tubes only	31	8	20
Sturgeon-class			
torpedo tubes only	39	8	20
Total planned	197 platforms [c] approx	4000 SLCMs [c]	
Estimated possible			approx 10,000 SLCMs [d]

Notes:

a. This Table refers to planned Tomahawk deployments, as contrasted with the next table which shows a deployment mixture of shorter-range and longer-range SLCMs; the US also has deployed the Harpoon ship-to-ship cruise missile, which has a range of about 100 nm and is reportedly only been deployed in a conventionally armed version.[15]

b. Drawn from 1985 estimates; for more detail see Arms Control Today, April 1986.[15]

c. Downturns in the defense budget and revised naval plans suggest that the plans could now (mid-1988) call for smaller numbers, say, 150 platforms and 3000 SLCMs.

d. The potential exists for the US Navy to expand from a planned force of a few thousand to nearly 10,000 SLCMs by modifying its deployment program.[4] A similar estimation could be done in Table 2 for the Soviet Navy (the numbers might not be so large). Absent either unilateral or bilateral control of SLCMs, the combined numbers of US and Soviet SLCMs (potentially most of which could be nuclear) could reach levels that would swamp those of reduced strategic weapons, i.e., 16,000 combined SLCMs vs. 6000 reduced strategic warheads.

TABLE 3: Estimated Soviet SLCM Deployments By The Mid-1990s [a]

PLATFORMS	VESSELS	SLCMs/VESSEL planned	potential
Surface Ships/SLCM type			
Kiev / SS-N-12	4	16	see note a
Kirov / SS-N-19	8	20	
Slava / SS-N-12	6	16	
Kresta / SS-N-12	4	4	
Kynda / SS-N-12	4	8	
subtotal	26 platforms	approx 370 SLCMs	
Submarines			
Attack / SLCM type			
Oscar / SS-N-19	12	24	
Victor III / SS-N-21	20	18	
Akula-Sierra / SS-N-21	28	18	
Mike / SS-N-21	12	18	
Dedicated			
Echo II / SS-N-12	29	8	
Yankee / SS-N-21; SS-NX-24	20	24	
Subtotal	121 platforms and	approx. 2400 SLCMs	
Total planned (est.)	147 platforms and	approx. 2800 SLCMs	
Estimated possible			approx. 6500 SLCMs

Notes:
a. Drawn from Arms Control Today, April 1986; see: this source for notes, assumptions, and original sources pertaining to the estimates in this table.[15] See note d of Table 1 for a comparison between US and Soviet SLCM deployments.

In their mission planning, both the US and the USSR rely heavily on SLCMs for anti-ship warfare. Since 1960, the Soviets have deployed a "short-range" nuclear land-attack SLCM with a range of over 500 km.[14]

Within the US and no doubt also in the Soviet Union R&D continues on 4th and 5th generations cruise missiles. Reports indicated that new features that such missiles may have include: smaller, more powerful and more efficient engines using exotic high energy fuels; intercontinental ranges; higher speeds, including supersonic; ability to evade air defenses; all-weather, day-night precision guidance systems; and new electronic counter measures (ECM) techniques, including radar masking and clutter. All of which could lead to smaller, faster, longer-range, less detectable, and more lethal nuclear-armed cruise missiles proliferating on hundreds of vessels all over the world.

Unless the two superpowers agree on SLCM restrictions, when current programs are completed, both sides will have deployed many long-range, hard-target capable, nuclear attack systems on a variety of naval vessels. Their missiles will be mixed in with other nuclear and non-nuclear systems on those vessels, often sharing the same launch system. Both the US and USSR plan to deploy "vertical launch systems" on many naval vessels. These vertical tubes are capable of holding and launching a variety of munitions, including land-attack nuclear cruise missiles and will further expand the carriers of long-range SLCMs in the two fleets.

IV. VERIFICATION CONSIDERATIONS

Now, with the establishment of the verification framework of the INF Treaty and those of the Conference on Disarmament In Europe, monitoring and verification tools such as challenge on-site inspection including ship boarding, perimeter-portal monitoring, verified data exchange, and non-destructive close-up monitoring and inspection of weapons are negotiable and viable elements of an arms control verification package. In addition, unique, tamper-proof, and relatively indestructible "tags" to mark a missile or its canister for later identification and "seals" to forestall tampering or modification of missile components may be employed. For example, a tag for USSR SLCMs would be manufactured by the US and applied during the Soviet assembly

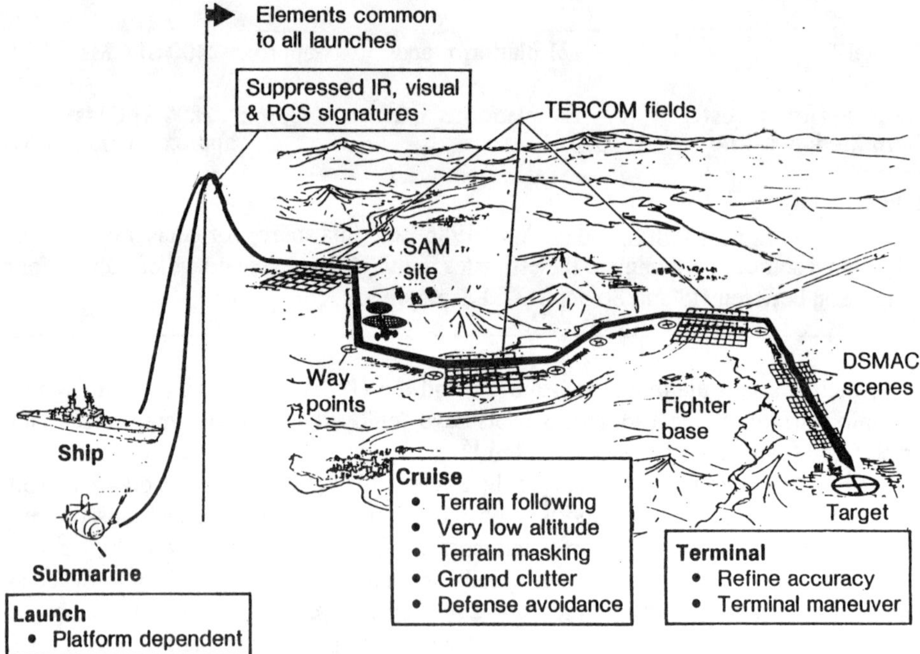

Figure 3. Land Attack Operational Concept (Joint Cruise Missiles Project)

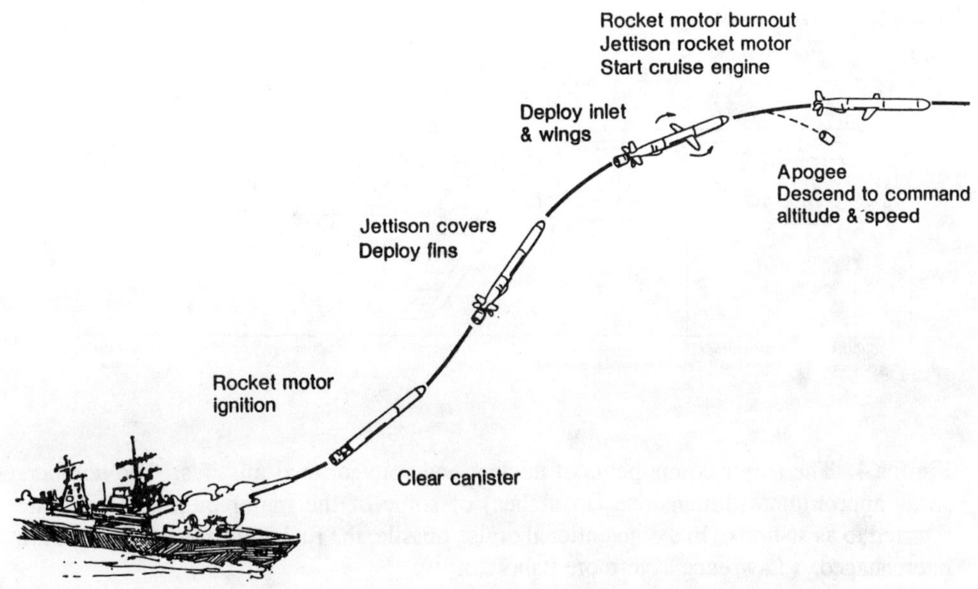

Figure 4. SLCM Launch Sequence (Joint Cruise Missiles Project)

process by US inspectors. Sandia laboratory and others are working on such tagging technologies. However, tagging is only effective (providing some of the benefits of a comprehensive ban regime) if properly used in conjunction with other monitoring provisions such as perimeter-portal monitoring. It could be an effective means of monitoring the legal SLCMs and isolating them from covert production lines.[10]
Seals could be used on conventionally-armed SLCMs to give assurance that they have not been converted to nuclear-armed ones. An example of such a seal would be an adhesive strip with an imbedded non-reproducible hologram. Any attempt to open the SLCM to replace its warhead would destroy the hologram. This would not prevent a country from converting SLCMs in a time of crisis, but would be a deterrent during normal periods since a broken seal would be clear indication of a violation.[10] Of course, if large numbers of conventional SLCMs were to exist on many launch platforms, and if such SLCMs were easily convertable to nuclear ones, since nuclear warheads are small and numbers of them could be readily concealed inside a ship, seals alone without a complex of other verification and inhibiting measures would provide little security against possible large-scale breakout.

Different groups and individuals have studied the question of what might distinguish a nuclear from a conventional SLCM. This paper won't attempt cover all of those ideas, but one obvious method is to try to detect the nuclear material in the nuclear warhead. As mentioned, remote detection is not possible under real-life situations.[8] If such detection is to be useful, it probably must be done at close range. Whether nuclear material could

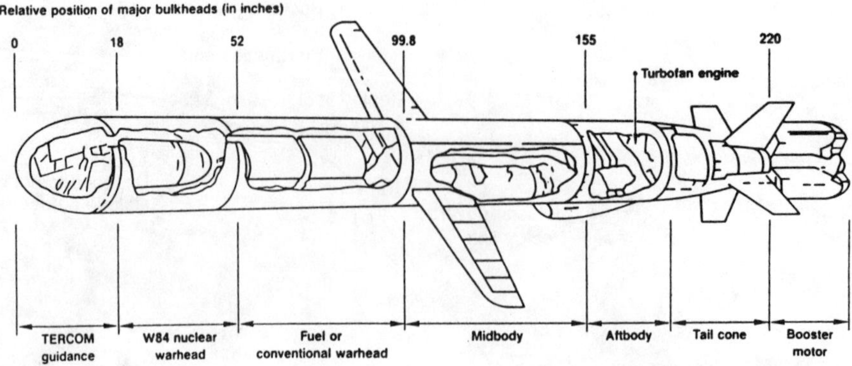

Figure 4. The major components of nuclear and conventional missiles. The numbers show approximate dimensions (in inches) of some of the major bulkheads and are referred to as stations. In a conventional cruise missile, the fuel and payload sections are interchnaged. (Lawrence Livermore Laboratory[16])

be detected by inspectors allowed to search a ship depends on the type and size of the ship (including its ability to carry shielding), the type of warhead, whether other nuclear warheads are also onboard, and the ingenuity and determination of a country which might want to hide them from inspectors.[10]

As a recent Stanford study concluded, "none of these verification techniques, by themselves, is sufficient to verify limitations on SLCMs. The methods may, however, be used in combination to increase confidence in the monitoring of an agreement limiting SLCMs."[10] Whether any synergistic subset can provide confidence that the other party is in compliance depends on the particular treaty provisions, the effectiveness of the monitoring methods, and the subjective evaluation of whomever is making the confidence assessment.

Within the US there are a several suggested solutions to the sea-launched cruise missile problem. The range of suggestions is sumarized in Table 4.

V. CRUISE MISSILE MISSION STATEMENTS

The USG views its nuclear land attack SLCMs primarily as a widely dispersed (and therefore with perhaps enhanced survivability) reserve nuclear force that makes it more difficult for the Soviets to design an attack that would compromise US retaliatory capability.[7] Despite the fact that the US has a long, open coast lines with most of itscritical assets within 200 Km of its coasts, and is therefore more vulnerable to a "de-capitation" strike by a few tens of Soviet land-attack SLCMs, which could be envisioned as the "leading edge" of a first strike, the USG has not expressed great concern about this

TABLE 4: Approaches to Possible Solutions to the SLCM Problem

1. Laissez faire	this approach is favored by those who believe arms contol only benefits the Soviets
2. Unilateral constraints	this is the "arms control without agreements" approach in which each side would restrain itself
3. Ban all SLCMs	this comprehensive arms control approach is the least difficult verification situation under a bilateral agreement, but doesn't recognize the value of conventional SLCMs
4. Limit SLCMs	this bilateral approach to limit SLCMs would bound the threat from nuclear land-attack SLCMs; verification measures are complex and intrusive; there are the following sub-approaches:

a. Limit nuclear; no limit on conventional
b. Limit numbers and platforms, e.g., no nuclear on surface ships
c. Limit both nuclear and conventional; either set specific sublimits as the Soviets have suggested or let either side be free to mix within an overall ceiling
d. Ban nuclear; either no limit on conventional or a set a ceiling on conventional

circumstance. The USG has stated that it does not want to limit conventional land-attack SLCMs, that it does not think nuclear SLCM limits are verifiable, and in any verification scheme that might result does not want Soviets on board US submarines and perhaps not on board surface ships. On the other hand, the Soviet government has expressed concern that the technically superior Tomahawk nuclear land-attack SLCM could be used in a surprise decapitation strike against the USSR and therefore in its public statements has indentified nuclear SLCMs as a threat it wishes to limit.

Former Secretary of Defense Casper Weinberger's FY 88 Annual Report to Congress states on page 27 that the US is "deploying submarine-launched (sic) cruise missiles aboard selected surface ships and submarines to make it more difficult, perhaps impossible, for the Soviets to design an attack that effectively compromises our retaliatory capability".[7] Interestingly, that report lists 337 submarines and ships expected to carry SLCMs by 1989" (p. 166). This number is 65% higher than the approximately 200 listed by others.[4,5] As indicated, the actual number may be yet smaller because of cutbacks in the plans for the US naval program. However, Weinberger's report is optimistic about the readiness for full deployment of the various SLCM systems. "Tomahawk and Harpoon cruise missiles continue to improve the fleet's antiship and land-attack capabilities, as new variants with increased range and versatility reach the fleet. Over-the-horizon targeting capability has matured and now supports employment at the weapon's full ranges" (p. 172).

The report seems to presume that the Soviets will not make compensations in both their naval attack plans and in their SLCM to deployments to adjust to US plans. It states: "The Soviet navy concentrates on protecting its SSBN forces and destroying oppostion nuclear-capable forces, such as US SLCM-equipped submarines and surface ships, and aircraft carriers. As western platforms become more numerous, capable, and dispersed, the Soviet capability to find and attack US and western SSBNs and SLCM-platforms will likely decline" (emphasis added) (p. 34). However, at the same time the report indicates that the US would move its SLCM carrying attack submarines right to Soviet naval strength. "In any major naval campaign, nuclear powered powered attack submarines would carry the brunt of the initial engagements. Early in a war, they would move into far-foward postions, including waters where the Soviet navy would be operating in strength ...." (p. 176). Presumably some and perhaps most of these attack submarines would also be carrying nuclear SLCMs as a part of the "reserve force". While operating under such conditions places a premium on stealth, quick and accurate weapons delivery, and firepower -- all of which the US submarine force is supposed to possess --, it also raises a question about the ultimate survivablility of many of the attack submarines and therefore the survivability of the SLCM strategic reserve force.

Weinberger's report suggests an underlying assumption about the need for the US to be prepared to fight a limited nuclear war, while arguing that they contribute to nuclear deterrence. SLCMs are seen as "nonstrategic nuclear forces" "linking" strategic forces to conventional ones. "The nonstrategic nuclear forces consist of systems designed to operate at less than intercontinental range. They include: ... sea-based cruise missiles. These forces enhance deterrence by proving a capability to respond to agression at the lower end of the nuclear spectrum, firmly linking the strategic forces to the conventional forces of both the United States and our allies" (p. 204). The report mixes the long-standing deterrent role of SLBMs with the less clear role of SLCMS. "Our nuclear weapons deployed at sea deter Soviet first use of similar weapons, provide a global nuclear deterrent, and contribute to the nuclear reserve force. We are requesting funds in FY 1988 to continue production of (land-attack) nuclear Tomahawk (TLAM/N) cruise missiles... (because)... the TLAM/N distributes long-range firepower throughout the fleet and confronts the Soviets with a difficult defensive task" (p. 218).

VI. EXAMINING SLCM VERIFICATION OPTIONS

Table 5 summarizes the ways in which one may know or infer SLCM inventories and deployments. Table 6 summarizes the interplay between the description of SLCM arms control measures and the articulation of naval missions for the weapons.

TABLE 5: "Knowing" SLCM Inventories and Deployments

GENERAL METHOD	**ASSESSMENT**	**IMPROVEMENT**
Intelligence Collection [a]	poor to good	cooperative measures [b]
NTM monitor production	fair to poor	
NTM monitor deployment	fair to good	
Inferring Inventories [c]	good for subs poor for ships	on-site inspection
MONITORING APPROACHES [17]	**ASSESSMENT**	**IMPROVEMENT**
Radioactive decay from outside ships	limited under practical condtns	on-board, internal inspection/detection
Functionally Related External Diff's (FRODs)	none forseen	none
Production choke points	poor with NTM alone	declared facil's & perim.-to-portal monitoring [d]
Monitor launchers	difficult at best	production monitoring
Monitor assoc. equipmnt	low reliability	production monitoring
Tagging/seals	assures ceilings on declared production kept	use in conjuction with production monitoring

Notes:

a. The liklihood of detection through NTM intelligence collection varies depending on what is being looked for and what is being used to look. For example, satelite and radar detection of a single SLCM test is problematic at best. Detection by remote monitoring of a clandestine cruise missile assembly and storage facility is far from certain. On the other hand, movement of numbers of cruise missiles to be loaded on board a large ships is likely to be detected.

b. Cooperative measures would include both external and internal arrangements: for example, agreed procedures to make monitoring easier as well as data base sharing and verification; loading and on-board ship inspection; and close inspection with agreed instrumentation, e.g., radioactive monitors. Such "regular" arrangements could be coupled with onsite challenge inspections to provide still higher confidence that all is in order.

c. Inferring inventories will probably play a role in any agreement to limit SLCMs. Estimating a carrying capacity of a platform is necessarily inaccurate, but for smaller, multiple function vessels such as submarines it could give a reasonable estimate with acceptable uncertainty (similar to counting rules for ALCMs on bombers). However, it is too imprecise for the many large surface ships.

d. Perimeter-to-portal monitoring, especially when coupled with an effective tagging system, is expected to be very effective in monitoring the production, assembly, and maintenance of SLCMs in declared facilities. However, it would be of no use in detecting covert production. Dealing with that eventuality would require such cooperative measures as agreements to enhance NTM transparency and challenge onsite inspections.

A. Verification Details for Particular Schemes

A report of a working group at Stanford University has addressed potential verification provisions for arms control limitations of long-range, nuclear-armed cruise missiles.[10] Their results for six plausible SLCM limitations approaches are summarized in Table 7.

The six options of the Stanford group span a range of possible limits and demonstrated how verification frameworks might be constructed. The options are

(1) Limit on total number of SLCMs -- this option would place an inventory limit, e.g., 1500, on long-range SLCMs, and not distinguish between those carrying nuclear warheads and those carrying conventional warheads;

(2) Limit on total number of SLCMs with a sublimit on nuclear-armed ones -- this option would place an inventory limit on the total number of long range SLCMs and place a sublimit on those which could be nuclear-armed, e.g., 3300 total including 750 nuclear which could accommodate US plans, or 1000 total including 400 nuclear which would match the Soviet proposal;

(3) Limit conventional SLCMs and ban nuclear-armed ones -- this option would ban nuclear SLCMs entirely, but allow a large number of conventional SLCMs, e.g., 2500;

(4) Ban both conventional and nuclear long-range SLCMs -- this option is patterned after the zero-zero- agreement in the INF treaty. It would ban all SLCMs with a range greater than a specified agreement, e.g., 200 km which would accommodate anti-ship SLCMs but preclude significant land-attack use;

(5) Deployment limits based on counting rules and no limits on inventories -- this option would place limits only on the deployment of SLCMs by specifying permitted classes of vessels with understood numbers of cruise missiles, for example, the number of SLCMs (either nuclear or conventional) on submarines would be set by counting rules with no inspection, and certain surface ships would be allowed an unrestricted number of only conve ntional SLCMs;

(6) Limits nuclear-armed SLCMs only -- this option would place numerical limits on nuclear SLCMs, no limits on conventional ones, and require no monitoring of deployments; for example, the production limit of long-range nuclear-armed SLCMs might be 500.

Table 6: SLCM Options: Verification & Defense Implications [4]

Option:	Verif. reqrmnt's	Monit./Ver. problems	Impl. US strat. pos.	US non-strat. pos.	Arms Cntrl implications
1. No limits	•None •NTM for tracking deployment & intell. col.	•US limited to NTM & HUMINT	•strategic reserve aug. by survivable hard target kill syst. •encourages growth in syst. likely to under-mine goal of SDI •provides hedge against Sov. BMD	•higher % US Naval forces in strategic missions •spreads nav. off. cap. agnst. land targ's to > 200 ships,vs. 12-14 carriers •tact. reach advntg US ships	•undermines goal compr. reductions strat wpns. •only unilat. control •another gray area wpn. syst. complic's fut. arms control. •may lead to instability
2. Ban Prod. & Deploy of all SLCMs	•NTM, OSI, coop. rqrd	•intrusive monit. reqrd. •disting. SLCM, GLCM, ALCM	•limits US strat. nuc. reserve •lose hedge agnst. Sov. BMD •elim. Sov. wpn. likely to under-mine US BMD	•US & Sov. give up weapon that other navies have •reinforce carrier aircraft for LA •weakens Sov. surface navy as viable mil. force	•closes maj. loophole •eliminates gray area wpn. & stems trend toward such wpns
3. Ban only nuclear SLCMs	ditto	ditto	ditto	•limits non-strat. SLCMs to conv. mun. •conv. LA cap. sust'nd •US navy tac. rch. advant.	ditto & •may enhance stability

Table 6 : SLCM Options: Verification & Defense Posture continued

Option:	Verif. require's	Monit./Ver. problems	Impl. US strat. pos.	US non-strat. pos.	Arms Cntrl implications
4. Ban long-range SLCMs	•NTM, OSI, coop. reqrd.	•close coop. rqrd. •e.g., limit testing, exch. data, ban encryp, transp. rules	•limits flex. of at sea nuc. cap. •leaves US vuln. to LA from Sov. short-range SLCMs, but protects much of USSR from US threat	•limits flex of conv. & nuc. SLCMs •reinforces US navy's depend. on carrier-based aircraft for land attack roles directed at the Soviets	•reduces threat to comp. arms control posed by long-range nuc. SLCMs •reduces gray area weapons threat
5. Ban long-range nuclear	ditto	ditto & •very intrusive OSI rqrd.	ditto	•limits non-strat. SLCMs to conv. mun. •sustains cap. for LA with non-nuc. SLCMs •restores US navy's tac. reach advan. •spreads nav. LRC offensive cap. to 200 ships	ditto
6. Limit numbers of long-range SLCMs deployed (missiles and plat-forms)	ditto	ditto & •rqrs even more extensive coop. & intrusive inspec. •e.g., prod. monitoring	•maintains US and Sov. cap., but limits scope of SLCM deployment	•limits conv. & nuc. long-range SLCMs •US navy dependent on carrier based aircraft for LA roles	•brings long range SLCMs within framework of compr. arms cntrl.

Table 6: SLCM Options: Verification & Defense Posture continued

Option:	Verif. require's	Monit./Ver. problems	Impl. US strat. pos.	US non-strat. pos.	Arms Cntrl implications
7. Limit numbers of long-range nuclear SLCMs (missiles and plat-forms)	ditto	ditto & •rqrs more intrusive measures •e.g., poss. boarding with nuclear detectors	ditto	•CLA capacity sustained •restores US navy's tac. rch. advant. •spreads off. cap. to > 200 ships	•brings long range nuc. SLCMs within framework of compr. arms control

Abbreviations:
cap.: capacity; CLA: conventional land attack; LA: land attack; LRC: long-range conventional; off.: offensive; LRN: long-range nuclear; rch.: reach; strat.: strategic; tac.: tactical

TABLE 7: Summary of Stanford SLCM Limitation and Verification Options

SLCM Limitation Options	Tags/Seals	Perimeter-Portal Monitoring	Challenge Inspections
1. Limit total number; any mix	Tag all legal SLCMs	Assembly facilities	Suspect sites & declared platforms
2. Limit total; sub-limit on nuclear	Tag & seal all legal SLCMs	Assembly & mating facilities	Suspect sites & declared platforms
3. Ban nuclear; limit conventional	Tag & seal all legal SLCMs	Assembly facilities	Suspect sites & declared platforms
4. Ban all SLCMs	none	Former assembly facilities	Suspect sites
5. Deployment limits ban nuclear on ships	none	none	Launchers only with nuc. det's
6. No limits on conventional; interim nuc. limits	Tag legal nuclear SLCMs	Nuclear mating facilities	Servicing facilities

TABLE 8: Stanford Report Options: Convertibility, Monitoring, Illegal Production

This table comments on each option, especially regarding the issue of converting conventional SLCMs to nuclear ones. The potential for conversion could translate into a breakout potential if there were a large number of conventional SLCMs deployed.

Option	Convertibility	Monitoring	Illegal Production
Option 1. SLCM Ceiling; free mix	Convertiblity not an issue	Monitoring intrusive	Illegal production difficult
Option 2. SLCM Ceiling; limit nuclear	Convertibility is an issue	Monitoring intrusive	Illegal production difficult
Option 3. Ban nuclear; limit convent'n'l	Convertibility is an issue	Monitoring intrusive	Eliminates legal nuclear SLCMs
Option 4. Ban all	Convertibility not an issue	Monitoring intrusive	Illegal production difficult; additional compliance assurance benefits from dismantled infrastructure
Option 5. SLCM ceiling; no nuclear on surface ships	Convertibility not as large an issue	Not intrusive; Verification not as difficult	Uncertainty in actual numbers; significant break-out potential
Option 6. No Conventional limit; interim limits on nuclear	Convertibility is an issue	Less intrusive verification than options 1- 4	Only interim agreement

VII. FINAL OBSERVATIONS

A. Perspectives on Arms Control Limits and Verification

The strategic nuclear arms control and verification process has developed significantly over the past twenty years. The US has created a strong foundation resting on (1) impressive technological cababilites for detecting and monitoring Soviet activities; (2) an intricate network of operational practices, agreed provisions, and procedures; and (3) the extensive experience of the US intelligence community in observing and interpreting Soviet strategic activity. The newer arrangements such as those codified in the INF Treaty promise to open up the verification monitoring system still further.

It is useful to make a distinction between arms control and verification in (1) a time of superpower cooperation and reduced tension, and (2) a time of mounting crisis and possible pending conflict. During a time of tension and conflict so great that one or both sides might decide to actively prepare for war, restraints and cooperative measures would be discarded. One of the purposes of arms control is to ensure that during the period of cooperation, no covert stockpiles or deployment modes are possible that would lead to a significant strategic advantage, i.e., a breakout or sudden change in the military balance, should a time of serious crisis develop. If the possiblility of a covert stockpile of SLCMs cannot absolutely be eliminated, then early detection capabilities should exist to provide warning in time for either political or military initiatives to halt, compensate for or nullify a possible advantage.

Refining of verification technologies and systems involved in perimeter to portal monitoring may be expected, as well as other in-country "black box" monitoring. Active and passive methods of "tagging" missiles to aid the process of counting them and deterring clandestine production will mature and probably be used in the coming years. The changes in Soviet attitudes toward verification appear to make very intrusive inspection - including even ship-boarding with measuring instruments - possible.

However, beyond the details of current arms negotiations, lay some broad and fundamental questions.

B. On Less-Than-Perfect Verification of SLCMs

Verification can never be perfect. With a weapon like SLCMs even a very stringent verification regime will not ensure that we are counting every one of them. But, is greater precision or even more certainty needed? Richard Perle is reported to have argued in the case of ICBM limits that getting at least a handle on non-deployed ballistic missiles through challenge inspections, which while intrusive and a deterrent to cheating would not conclusively prove that there is no covert stockpile, is better than having no handle at all on that problem. Surely this argument applies as well to SLCMs.

C. Conclusion

It is clear that long-range nuclear land-attack SLCMs need to be limited if not eliminated. The prospect of advanced, longer range SLCMs emphasizes the need to deal promptly with this issue. At least unilateral actions are called for to limit the number of current generation SLCMs produced and deployed. Eliminating or preventing nuclear SLCMs on surface ships would be a welcome move. Any agreement that is to be viewed as effective will require several forms of intrusive inspection probably including at least challenge inspections on board surface ships. This and other papers have laid out the possibilities. Some, particulary the work of the Stanford group, clearly indicate that while the problem is daunting and difficult it is by no means impossible. They also point up that the key technical and security issue in any scheme that would limit the number of nuclear SLCMs is to prevent ready conversion of conventional SLCMs to nuclear ones.

In the case of arms control limits on SLCMs, as is often the case in such matters, by far the more important questions, however, are political. And these relate to understanding and acceptance by the body politic in the US of the concepts of "bounded uncertainty" and "acceptable risk" in the verification of SLCMs. In general, these will likely be the key underlying issues in the SLCM and other future arms control verification debates.[12]

ACKNOWLEDGEMENTS

I am pleased to acknowledge the participants of the Stanford Workshop on SLCM Verification who are unfortunately too numerous to mention individually. From them I gained a much deeper understanding of the complexities of this issue. I would particularly like to thank Sidney Drell, John Harvey, Theodore Postol, and Sally Ride for their interest and discussion with me of these issues. I especially wish to acknowledge Sid Drell and John Lewis, my hosts at the Stanford Center for International Security and Arms Control where I was during the time most of this paper was written, and the staff of the Center who helped me immensely through my stay. I also wish to thank the Carrnegie Corporation of New York for their support during my stay at the Center.

NOTES

1. For example, a recent assertion of this point is in "The Iron Law of Arms Control," by Richard N. Perle, The New York Times,p. A-28, August 1, 1988. Perle says "...limits on sea-launched crusie missiles can't be verified as long as devices capable of launching them are mounted on ships and submarines."
2. Many authors have addressed this matter. See for example: Kosta Tsipis," Cruise Missiles", Scientific American, February 1977, p. 20.; Charles A. Sorrels, US Cruise Missile Programs -- Development, Deployment and Implications for Arms Control, McGraw-Hill Inc., 1983; "The Trident II and the Sea-Lauched Cruise Missile", The Defense Monitor, Vol. xvi, Num. 6, 1987
3. Joint US-USSR Summit Communque, December 10, 1987, Washington, D.C.
4. Stanley R. Sloan, Alva M. Bowen, Jr., and Ronald O'Rourke, "The Implications for Strategic Arms Control of Nuclear Armed Sea Launched Crusie Missiles", CRS Report No. 86-25F, December 10, 1985.
5. James Rubin, "Sea-Launched Cruise Missiles", Arms Control Today, April, 1986.
6. Much of this table is from Jeffrey S. Duncan, "Nuclear Cruise Missiles: Arguments For and Against", Oceanus, 1986.
7. Report of the Secretary of Defense, Casper W. Weinberger, to the Congress on the FY 1988-1989 Budget and FY 1988-1992 Defense Programs, January 12, 1987.

8. See for example, R. Z. Sagdeev, O. F. Prilutskii, and V. A. Frolov, "Problems of Monitoring Sea-Based Cruise Missiles Bearing Nuclear Warheads", Report of the Committe of Soviet Scientists for Protecting the World form the Nuclear Threat, presented at the Seminar on Verification Problems held jointly with the Federation of American Scientists, Key West, February 3-10, 1988; others including Robert Williams of the University of Washington have also studied this problem and reached similar conclusions.
9. See for example, Herbert Lin, "Technology for Cooperative Verification of Nuclear Weapons, Arms Control Today, April 1986; Peter D. Zimmerman and Alton Frye, "Technology to the Rescue: The SLCM Problem Solved", The Washington Post, February, 1988.
10. "Potential Verification Provisions for Long-Range, Nuclear-Armed Sea-Launched Cruise Missiles", a Workshop Report, Center for International Security and Arms Control, Stanford University, July 1988.
11. Whether their motive was to conceal strengths, weaknesses or both, the Soviets believed they had more to gain by preserving tight controls on information and access to their territory than by agreeing with the western notion that openess promotes stability and security. While the US insisted it needed such measures as data exchange and on-site inspection -- perhaps in part because of suspicions raised by Soviet intransigence --, the Soviets viewed them as instruments of espionage. In less than two years, from the time Gorbachev took office in March 1985 until the signing of the Stockholm CDE document in September of 1986, that long-held Soviet position was reversed. The INF Treaty, signed in December of 1987, codified additional intrusive verification measures and, in the words of Max M. Kampelman, head of the US delegation to the Geneva nuclear and space talks, "breaks new ground for the construction of a bold and stringent verification regime." One question to ask then when confronting the question of verification measures available to monitor limits on SLCMs and the necessarily intrusive measures that task implies is "What do we have to work with today as a result of the precedent setting agreements to date?" For example, the INF Treaty, which bans the production, testing and deployment of intermediate-range and shorter-range ballistic missiles, includes the following verification procedures (1) both sides will use national technical means to verify the Treaty's provisions, neither side will interfere, conceal, or impede such verification, and both sides will respond to requests for so-called transparency measures, e.g., sliding back roofs on silos, to make counting and inspection easier; (2) both sides will exchange detailed data (numbers of weapons, locations, etc) and provide notifications, for example, of transits; they will use the US-USSR Nuclear Risk Reduction Centers recently set up under the provisions of another agreement for the purpose of handling the exchange and for other purposes; (3) on-site inspections (a) at all missile operating bases and support facilities to verify baseline data, (b) to verify elimination of missile operating bases and support facilities, as well as (c) 10-20 other on-site inspections per year extending over 13 years; (4) Certain rights to continuous perimeter-to-portal monitoring for 13 years at a missile plant; (5) observation and inspection of the missile and support equipment elimination; and (6) creation of a Special Verification Commission to resolve questions related to compliance and agree upon measures to improve Treaty viability and effectiveness. Verification measures

of declared facilities, i.e., those identified in the data exchange, and non-declared facilities; and provisions prohibiting concealment, in particular banning all encoding of telemetry encryption so that the US has full access to all telemetric information broadcast during Soviet missile test flights. Perhaps the most radical departure from previous nuclear arms control verification practice with the Soviet Union and in its implications for operations within both countries as well as with regard to possible limts on SLCMs is the provision for challenge inspections at non-declared facilities. Clearly, both countries will have to have either veto provisions or exclusionary zones to secure their most sensitive areas against possible espionage. Such a provision would have to be managed to actually provide the intended confidence building rather than confusion and more questions some think could be its likely result. The data collected by NTM monitoring will still be a workhorse of verification, including for any SLCM agreement. And the keys to long-term confidence against covert breakout will lie (1) in provisions that would either prevent military confidence in the actually utility of a possible covert force, such as the effect of the total ban on testing in the INF Treaty, or provide for early detection and effective US response; (2) a network of information, intelligence collection, inspections, and openess about military budgeting and production; as well as (3) in improved remote monitoring and cooperative transparency measures.

12. Richard A. Scribner, " The Future of Verification", Harvard International Review, August/September 1988, p. 27. Pessimism about future verification capability has always been an element in assessing the viability of possible agreements. In the past, technological advances have managed to provide the necessary monitoring capability. Where that was not true, agreements generally were not made. In the future, technological advances will no doubt occur, but they will probably not be as important as the refining and advancing of the newly available cooperative measures, because of the nature of future agreements. The overiding problems in verification for the future will not be technical or even proceedural, they will be political, as the SLCM case illustrates. Exact counting with no uncertainties, the *sine qua non* of "effective" verification, may not be possible in some desireable and low-risk potential future agreements. This suggests that different criteria for verification may have to be accepted if future limits are to be negotiated and implemented.
13. "START Talks Moving Slowly", The New York Times, April 6, 1988.
14. Soviet Military Power, US Dept of Def., 1987.
15. Nuclear Weapons Databook, Vol 1, NRDC, Ballinger Pub. Co.,1984.
16. I am grateful to John Harvey of Livermore Laboratory for this picture. It was one of serveral presented in a seminar on SLCM verification presented at Stanford University.
17. Dean Wilkening, "Monitoring Bombers and Crusie Missiles", Verification and Arms Control, Ed. William C. Potter, Lexington Books, 1985.

CHAPTER 17

SCIENTIFIC IMPACTS ON NUCLEAR STRATEGIC POLICY: DANGERS AND OPPORTUNITIES

Spurgeon M. Keeny, Jr.
Arms Control Association, 11 Dupont Circle, NW, Washington, DC 20036

ABSTRACT

Nuclear weapons have revolutionized warfare, making a mutual capability for assured destruction a fact of life and mutual assured deterrence the underlying nuclear strategy of the superpowers. The program to find a technical solution to the threat of nuclear weapons by creating an impervious defense is fatally flawed by failure to consider responses available to a sophisticated adversary at much lower cost. Responses could involve: exploiting vulnerabilities; increased firepower; technical innovation; and circumvention. Efforts to achieve strategic defense would in fact increase risk of nuclear war by stimulating the nuclear arms race since history demonstrates neither side will allow its deterrent force to be seriously degraded. Defenses would increase instability in times of a crisis. Science has also reduced the risk of nuclear war by making possible improved control and safety of nuclear forces and predictability of US/Soviet relations, verifiability of arms control agreements, and survivable strategic systems. Science can be a tool for good or evil; mankind must be its masters not its slaves.

I.

The introduction of nuclear weapons at the end of the World War II had a truly revolutionary impact on strategic doctrine. By increasing the power of weapons by a factor of a million, science had in the course of a few years changed the nature of warfare. A single weapon could now destroy a major city. The significance of this unprecedented escalation in destructive capability was magnified by the concurrent scientific advances that made possible mass production of fissionable materials and the delivery of nuclear weapons rapidly over great distances.

Science, in its even-handed way, did not play favorites. Despite war-time devastation and relatively backward technology, the Soviet Union in 1949 also detonated an atomic bomb and a thermonuclear device in 1953, only one year after the first U.S. thermonuclear test which had a yield approximately a thousand times that of the atomic bomb that destroyed Hiroshima. We knew then that the Soviet Union was engaged in a massive program to build large-scale facilities to produce fissionable materials and was developing the means of delivering nuclear weapons.

By the mid 1950s, it was apparent that a small portion of the emerging nuclear forces of the United States and Soviet Union could inflict as many as 100 million fatalities on the other side and destroy it as a functioning society. The populations of the two nations had clearly become mutual hostages. Science had insured that all-out war between the world's major adversaries would lead to the "mutually assured destruction" of both sides unless one side could somehow successfully conduct a "disarming" first strike against the other. With the increased diversity and sophistication of delivery systems and the growth of nuclear stockpiles, such a "disarming" first strike rapidly became less and less credible.

In these circumstances, the underlying U.S. strategic policy has been to deter by threat of nuclear retaliation a Soviet nuclear attack on the United States or its allies, or other military actions, such as a massive conventional attack on NATO, that would likely escalate to general nuclear war. The rhetoric of declaratory strategic doctrine has varied over the years in efforts to increase its credibility and to reassure our allies, but the underlying U.S. strategic posture of deterrence has remained the same. During this period, there has been much esoteric discourse on the evolving nature of Soviet declaratory strategic policy, but the impact of Soviet nuclear forces on the United States has been the same--to deter a direct attack on the Soviet Union or military actions that would likely lead to that catastrophic outcome. In short, the two superpower adversaries have both operated under a common strategic policy of "mutual assured deterrence" which I have always associated with the acronym MAD when referred to as a strategic concept. "Mutual Assured Destruction," on the other hand, describes the physical reality that we will face in the event of general nuclear war.

Mutual Assured Deterrence has probably prevented a major conflict between the superpowers through more than 40 dangerous years which have seen tense confrontations involving high stakes. But living on the edge of a disaster of unprecedented magnitude in human history is not a comfortable position for humanity and hardly a prescription for the indefinite future in a world where more and more nations will be technically able to produce and deliver nuclear weapons. It is only natural, therefore, for scientists as well as politicians to look for a better basis for civilization's survival.

II.

Starting in the 50s considerable attention was given to the possibilities of passive defense against nuclear attack. At various times, blast shelters were proposed for cities, fall-out shelters for the country at large, and evacuation of urban populations to the countryside. Protection of urban populations in urban shelters is not possible if the cities are destroyed. While a well-planned evacuation might save some lives provided adequate time were available, the execution of such a plan by the attacking side would give unambiguous strategic warning and allow the other side to generate its retaliatory forces to full alert. This would result in a much more massive retaliatory strike than would have occurred with a normal day-to-day alert with the result that fatalities would probably be comparable to an unevacuated population. The evacuation might even trigger a preemptive attack that would produce even higher fatalities among the evacuating population than would occur in a retaliatory strike. In short, you can neither run nor hide from an all-out nuclear war.

From the loss of the U.S. nuclear monopoly in the early 1950s, active defenses were considered by some scientists as a possible scientific solution to the emerging Soviet nuclear threat. In the 1950s the U.S. undertook a major anti-aircraft defense deployment which was abandoned with the emergence of ballistic missiles. As a member of the Gaither Panel, which President Eisenhower convened in 1957 to assess the potential of civil defenses, I recall that the main focus of scientific interest in the group quickly changed from the dismal prospects of civil defense to the more technically intriguing question of the future potential of an active ballistic missile defense. But the prospects then, as now, were judged as poor.

The Soviet Union had sufficient interest in ballistic missile defense in the early 1960s to start the deployment of a primitive ABM system around Moscow. Largely in response to this development, the United States initiated deployment of the Sentinel ABM System under President Johnson, and then the slightly modified Safeguard System under President Nixon. I don't believe anyone involved in these programs thought they would provide a population defense or even work against the Soviet Union, since they were completely dependent on a few large radars that were totally vulnerable to nuclear attack. But we did know that the deployment of these systems would accelerate the nuclear arms race.

The U.S. reaction to the Moscow ABM system was not to change our nuclear strategy but to introduce MIRVs on our ICBMs and SLBMs to increase U.S. firepower. Apparently, the Soviet Union also concluded that nationwide ABM systems would not affect the basic strategic balance that was, for all practical purposes, approaching parity with neither side possessing a meaningful advantage. The two sides then agreed on the ABM Treaty which limited the two sides to their very limited existing ABM deployments and to only research without development and testing in areas that could conceivably lead to a rapid breakout from the treaty--in particular space-based systems.

III.

In March 1983, President Reagan sought to launch the second great scientific revolution in strategic policy when he announced his goal of eliminating the threat posed by strategic nuclear missiles and called upon the scientific community to make those nuclear weapons "impotent and obsolete." Secretary of Defense Caspar Weinberger elaborated that the administration sought a "thoroughly reliable and total" defense system that would render all attacking missiles, including low-flying cruise missiles impotent. He asserted that the envisaged system "would guarantee that there would no longer be any danger from nuclear weapons." Subsequently he offered to extend the coverage of this shield to our allies in addition to the United States. The President later offered to extend this scientific cornucopia to the Soviet Union as well.

By Presidential command, science was to eliminate the threat of Mutual Assured Destruction and with it, the strategic policy of Mutual Assured Deterrence. How this was to be done and what strategic policy would exist during the period of transition, or subsequently under the shield, were not revealed.

The misguided efforts to carry out the President's call for an impenetrable defense present a basic threat to our real security since it is an illusory solution to the fundamental military problem of our time. The pursuit of this illusion will not achieve the hoped for goal but will instead lead to much greater instability in the strategic relations of the superpowers, destroy the existing framework of arms control, and lead to an intensified qualitative and quantitative defensive and offensive arms race at a time when real progress in arms control appears possible.

IV.

Let me turn to a more detailed examination of why this scientific program designed to protect mankind presents so few opportunities and such great dangers in the coming decade. I will not surprise you with any new insights since the subject has been discussed by so many over the past two decades. In fact, one has to think carefully to identify what, if anything, is really new in a fundamental sense in the present scientific situation. There is, to be sure, a lot of exciting new technology on directed energy weapons that operate at the speed of light--but particle beams were studied extensively as potential ABM weapons in the 1960s, and lasers were an active secret program when we negotiated Article V and Agreed Statement D in the ABM Treaty with the clear intention to prevent the development, testing and deployment of these weapons in space. The current focus on attack in the boost phase is new, but by no means novel, since the Air Force in the early 1960s wanted to initiate the space-based Bambi defense system for a conventional attack on the boost phase. The fundamentals in this debate have therefore not changed except in detail over the years.

The fundamental barrier to developing a highly effective nationwide ballistic missile defense remains the incredible destructive power of thermonuclear weapons when matched against the extreme vulnerability of populations and urban society. And we live in a world where the United States and the Soviet Union each have more than 10,000 thermonuclear weapons dedicated to strategic missions. The total consequences of an exchange of even a small portion of these arsenals are impossible to comprehend. In addition to the local impact of the multiple explosions, nuclear fallout would cover most of the country with lethal levels of radioactivity.

A nationwide defense must be extremely effective at all points since leakage anywhere would be catastrophic to the defended targets. This requirement can be appreciated by comparing defense against conventional and nuclear weapons. In the past, a defense that destroyed 10 percent of attacking aircraft per sortie would be considered very successful. After ten such attacks, the bomber force would be reduced to less than one third of its initial size. The offense could normally not accept that level of attrition given the limited destructive power of conventional weapons. But in defending urban areas against nuclear weapons, destruction of 90 percent or even 99 percent of incoming warheads would be ineffective since a single penetrating weapon would destroy a target vastly more valuable than 10 or 100 warheads.

V.

The SDI would seek to achieve very high levels of attrition by a layered defense. This would involve a complex network of detectors, assessors, trackers, and exotic kill mechanisms operating from earth and space stations. The system would have to be linked and operated essentially automatically by computers dependent on unprecedentedly large and complex software programs. The system would have to work flawlessly when used for the first time in a very hostile nuclear environment. This is a staggering technical assignment even under idealized conditions.

But the concept is "fatally flawed," since in the real world the system would be operating against a sophisticated adversary using every technique at his disposal to defeat the system. There is a wide range of very effective countermeasures against such systems that the Soviet Union can develop at a fraction of the cost of the defenses that the SDI is contemplating.

(1) Vulnerability

To begin with, the offense can always attack vulnerable, critical points in the system. Extremely costly space-based battle stations, giant optically perfect mirrors, sophisticated IR sensors, and any other critical space assets would be vulnerable and lucrative targets. Such space-based systems components could be attacked at the outset of hostilities. They could also be attacked either clandestinely or openly during peacetime. It is most unlikely that space will remain a sanctuary if either side deploys major space-based weapons systems that the other side considers seriously threatening.

The operation of a multi-layered defense system will require the interaction of many vulnerable space and earth-based components. The specific failure modes of such a system will depend upon the details of its ultimate design. But one can be sure that there will be easily identifiable highly vulnerable critical nodes, destruction of which could result in the collapse of major portions of the system. A principal reason the United States rejected the early Sentinel/Safeguard ABM systems as serious candidates for nationwide defense systems against a concerted Soviet attack was the extreme vulnerability, both to direct attack and to blinding by precursor nuclear blasts, of the prime radars, on which the system depended.

(2) Firepower

In a more classic manner, the offense can overwhelm the defense with increased firepower. This can be done by the brute force approach of increasing the number of attacking missiles or increasing the number of MIRVed warheads on each missile. Many options are available in the absence of the constraints of the SALT II Treaty, which the United States has repudiated, or a START agreement, which the Soviet Union has made contingent on continued observance of the ABM Treaty. The Soviet Union, for example, could rapidly increase the total size of its missile force and the large SS-18 missile, which was formerly limited to 10 warheads by SALT II, but could easily carry as many as 30 warheads. The same objective can be accomplished much more cheaply by launching large numbers of balloons or other decoys designed to simulate real reentry vehicles. In principle, false warheads can be identified by multiple sensors; in practice this will prove very difficult to accomplish with confidence on a real time basis. Moreover, the offense has the tactical option of concentrating its firepower against the targets of its choice, while the defense must be prepared to defend all valuable targets against such a selective attack.

(3) Technical Innovation

The Soviet Union could also choose from a large range of technical opportunities to nullify or diminish the performance of the defense's sensors or kill mechanisms. An often mentioned example would be the development of a "fast burn" booster which could release its post-boost delivery vehicle--that is its warhead dispenser--in a fraction of the time required by current missile boosters. By substantially shortening the time that the highly visible and vulnerable missile boosters could be attacked outside the atmosphere, fast burn boosters could eliminate or greatly reduce the effectiveness of a defense system designed to attack the boost phase. Failure of the high leverage boost phase layer defense would be far more significant than simply the loss of one of several independent barriers since the entire weight of the attack would then have to be handled by subsequent layers. Unless they had been sized to meet this contingency, the system would be immediately overwhelmed.

Scientific ingenuity has and will produce a host of ideas that can be introduced at a fraction of the cost of the defense system and on a much shorter time scale.

(4) Circumvention

Finally, the offense has the option of circumventing the defense system by introducing an entirely different mode of attack. In the case of the SDI approach, this could be accomplished by air breathing systems that fly under the shield. For example, low-flying cruise missiles launched from submarines close to the U.S. coast would completely defeat the missile defense system and constitute a new offensive threat as difficult to defend against as ballistic missiles. The defenses problem would be further increased by incorporating stealth technology to reduce the visibility of the cruise missiles to radar. But without a very effective air defense, the defensive shield could hardly be considered to have rendered nuclear weapons "impotent and obsolete." In this connection, it is interesting to note that the United States continues to resist Soviet efforts to place limits on sea-launched cruise missiles in the START negotiations.

VI.

I believe that most independent scientists who are familiar with the program would agree with this assessment. But some do not, and one can never be certain about the future. So why not develop such a system on the chance that it might work despite the tremendous technological barriers to be overcome?

One answer is the immense expense of the undertaking both in dollars and in our utilization of our high technology resources. These dollars and brains will be diverted from other military and civilian activities where they are badly needed to maintain a balanced military force and an internationally competitive economy. The actual development of such a system, involving many hundreds of billions and ultimately even trillions of dollars, would certainly impact seriously on future military and civilian budgets. Such a reorientation of priorities is impossible to justify given the very low probability that the program could achieve its avowed goals.

The decisive argument against a program aimed at early deployment, however, is the extremely dangerous impact it will have on the stability of the U.S.-Soviet strategic military relationship on which our mutual deterrence rests. The goal of the program, simply put, is to deprive the Soviet Union of its nuclear deterrent against the United States. But neither the Soviet Union nor the United States is going to allow the other side to eliminate the deterrent value of its strategic offensive forces by precluding the possibility of retaliation. Fear of the possibility of such nuclear "impotence" will arise long before the systems are deployed.

The attempt to achieve a highly effective ballistic missile defense system will also be seen by the other side as a direct offensive threat. Today, both the United States and the Soviet Union know that neither can initiate a strategic nuclear attack against the other side without themselves being destroyed as a society by the inevitable retaliation. But if either the United States or the Soviet Union develops, or is perceived by the other side to be developing, a highly effective ballistic missile defense, the other side will fear that its adversary is positioning itself to be able to conduct a preemptive strike. Even if he recognized that his defense was far from perfect, the attacker might believe that it would work reasonably well against a reduced and disorganized "ragged" retaliatory strike. Or at least this is how a nationwide defense system would be perceived by the other side.

These fears are best reflected in statements by U.S. officials. Former Secretary of Defense Caspar Weinberger once said that the Soviet development of an effective defense "would be one of the most frightening prospects I could imagine." A White House paper on the President's Strategic Defense Initiative contains the alarming statement that if the Soviet Union deployed a nationwide ABM defensive system against ballistic missiles, "deterrence would collapse, and we would have no choice between surrender and suicide." There is no reason to think that Soviet leaders have a more relaxed view of a nationwide U.S. SDI defense system.

The deployment of a nationwide ABM system by either the Soviet Union or the United States would not in fact lead to the "surrender or suicide" of the other side. What it would do, would be to lead to a major acceleration of the nuclear arms race. The side that perceived itself threatened would move rapidly to increase the size or quality of its strategic offensive forces and would make sure that the defensive system would not result in the "collapse" of its deterrent capabilities. As I have outlined earlier, there is a broad spectrum of technical options to accomplish this.

I should add that the Soviet reaction to SDI will not be limited to improving Soviet strategic offensive forces. The Soviet Union will also certainly accelerate its own strategic defensive efforts. No matter what technical advice they might get from their scientists, Soviet politicians and military leaders will not take the chance of lagging too far behind the U.S. military technology they so admire and fear. And then in response the United States would have to build up its offensive forces.

VII.

It is not idle theoretical speculation to conclude that a build-up in strategic defenses will lead to a dangerous acceleration in the strategic arms race. In the past, when faced with Soviet strategic defense programs, the United States has always acted to insure its retaliatory capabilities. In the late 1960s the United States was concerned that the Soviet Union might be undertaking a nationwide ballistic defense system. In response, the United States introduced MIRVs on the Minuteman III and Poseidon missile to increase the firepower of our ballistic missiles to insure their ability to penetrate the projected Soviet defenses and maintain existing target coverage. There was no suggestion that we abandon or reduce our reliance on ballistic missiles.

In the same manner, even though the Soviet Union has invested vast sums on its air defense system over the past 30 years, there has never been any suggestion that the United States should abandon the air breathing leg of its triad of retaliatory forces. On the contrary, the United States has continuously upgraded the ability of its strategic bombers to penetrate Soviet air defenses. These qualitative improvements included upgrading the performance of the B-52 itself, and then electronic countermeasures and short range nuclear attack missiles (SRAMs) to insure penetration of Soviet air defenses. As Soviet air defenses improved, the U.S. introduced the air-launched cruise missiles whose low-altitude flight profile and small radar cross section greatly complicate the problem of the defense. Then came the higher performance B-1 bomber. And today we are pursuing stealth technology which will make it even harder in the future for Soviet defense radars and other sensors to detect penetrating aircraft and cruise missiles. Throughout this period the U.S. Air Force has remained confident of the ability of its aircraft and cruise missiles to penetrate Soviet defenses.

Today the actual goal of the SDI program has become very confused. Some supporters now assert that the objective is not to protect the population from nuclear weapons, but to provide a much more modest level of defense. Such a defense we are told might serve to complicate enemy war plans and contribute to the survival of our strategic deterrent. Such a goal is probably attainable, particularly since our deterrent is at present quite secure in any event. But a multi-layered ballistic defense of the type being considered by SDI would appear to be an incredibly expensive way to seek additional reinsurance for the future survival of our deterrent forces. This requirement can be achieved at much less cost and with higher confidence by such straightforward means as replacing vulnerable high-value fixed land-based MIRVed ICBMs with low-value fixed or mobile single-warhead missiles such as the Midgetman or further sea-based deployments.

However, by simply lowering the initial paper specifications for the system, we will not alter the Soviet concern that the ultimate objective of the system is a nationwide population defense. Moreover, officials at the highest level rush to reassure the world that such a limited deployment would be simply the first phase the ultimate goal of an impenetrable defense. In these circumstances, the Soviet Union would no more accept our statement of intentions than we would that of the Soviet Union.

These are not new problems. They are the same problems that we struggled with in 1972 when the ABM Treaty was negotiated with the Soviet Union. The ABM Treaty, which the Senate ratified by a vote of 88-2, committed the United States and the Soviet Union not to deploy or provide a base for a nationwide ballistic missile defense system. The treaty sought to prevent an open-ended arms race that would follow deployment of these systems and to provide a basis for future reductions in strategic offensive arms. The ABM Treaty is, of course, in direct conflict with the goals of the SDI whether they be comprehensive of limited, and the treaty will have to abrogated or openly violated if the SDI program goes beyond the present research stage.

VIII.

I have focused on the dangers rather than the opportunities presented by science because the dangers are the more significant at this time. But let me qualify this grim assessment--particularly with this audience--with the recognition that science has and will continue to contribute to maintaining the stability and safety of U.S. and hopefully Soviet strategic forces. While some developments may be two-edged swords, let me note a few critical areas:

o Command, Control, Communications, and Intelligence (C^3I)

In this area there have been a vast number of developments that have increased the control and safety of our nuclear forces and the predictability of the U.S.-Soviet relationship, both tactically through assured early warning and strategically through improved intelligence.

o Survivable Systems

Science has provided the technology that has permitted us to have the survivable systems necessary for an assured deterrence. This has made possible quiet submarines, mobile systems, cruise missiles, and stealth technology. While these have themselves become part of the problem, they have contributed to stability in the strategic relationship. They give one confidence that new capabilities will evolve as they are needed. For better or for worse, they have insured that air defense and a potential Soviet ABM system would not destroy our deterrence.

o National Technical Means of Verification

Science has made arms control an acceptable component of national security policy by providing the tools to monitor and verify compliance. In the 50s, approaches to arms control were inherently limited by our inability to verify adequately evolving Soviet strategic capabilities. The advent of space-based reconnaissance systems with a variety of sensors radically increased our knowledge of Soviet military capabilities. These capabilities have permitted us to negotiate the SALT I agreements, SALT II, INF and hopefully a START agreement reducing the level of nuclear forces. These national technical means of verification have undoubtedly contributed to the opening up of the Soviet Union. The new possibilities of on-site inspection are a further important step. But on-site inspection is not an end in itself and without powerful national technical means of intelligence collection, on-site inspection would be of only limited value. Looking to the future, science should make arms control agreements possible that would not have been considered possible in the past.

So do not despair, physics need not replace economics as the dismal science. It has the potential not only to let us understand the world, but to help mankind survive to appreciate it.

IX.

In summary, from today's perspective, I do not believe it is possible for science to provide a technical solution to eliminate the danger of nuclear war. The problem is ultimately a political problem and will only be solved in a long political journey that involves not only the United States and the Soviet Union, but the rest of the world as well. In the meantime, however, science will provide both dangers and opportunities. If abused, new scientific advances can destabilize the strategic relationship and delay the political process--or even lead to disaster. If used wisely, new scientific advances can make possible an increasingly stable strategic balance and make a framework of arms control agreements between the superpowers, and worldwide as well, that will enhance strategic stability, predictability in the evolution of military threats, and confidence in the intentions of potential adversaries. Science can be a tool for good or evil. Mankind must be its masters, not its slaves.

CHAPTER 18

THE DECLINE OF THE NUCLEAR ERA

Edward Luttwak
Center for Strategic International Studies
Washington, D.C.

From what Spurgeon Keeny had to say I understand that the building of an astrodome over America is not capable of immediate realization. I think that this news, however shocking, has been recognized for some time; this fact was understood more or less the day after the President made his speech about SDI in 1983.

Back in the primitive days, in 1940-41 on the eve of the Second World War, science had not yet attained its enormous public authority and we had not yet set up a whole machinery of scientific advice in the government. In those days the procedure by which new scientific ideas were brought to the attention of the president, or of the prime minister included such things as the famous episode of Teller serving as the chauffeur for Szilard to drive out to Einstein's place in Long Island, letters to the President which would have been disregarded except for the casual visit by a friend of his who was an economist with an interest in science, etc.

By 1983 we had a whole system for the assimilation of scientific information, governmental scientific intelligence agencies, scientific advisors and the development of procedures which tell you more or less when a modern Teller should drive a modern Szilard to Long Island. Yet on strategic defense there had in fact been policies that objectively must be described as policies of deception. What happened in the years before 1983, was that some scientists or engineers would go to the Pentagon and say "You know, you remember how we signed the ABM Treaty back in 1972. That was done on the basis of the scientific feasibilities of the late 1960's. But there has been considerable progress in computation and I've got this bright idea (whether it be mirrors or lasers or something else)." But what happened when these scientists or engineers got to the Pentagon, was that the people there felt that while this is really brilliant science, and were very impressed, they would say that there is unfortunately no military mission for this technology, that there is an ABM treaty that forcloses that mission. In other words, the military would not want that technology. Instead they wanted to spend their money on shiny new aircraft and such. So there developed a scientific feasibility to do this or that, but there was no military requirement; and if there is no requirement, there is no budget. Then the scientists would exit through one door, and through another would enter the policy maker, the policy official. The policy official would say "You know, I really detest this idea that we are violating millennia of strategic wisdom by being all offense and no defense. It would be nice to have some defense, some partial defense." But they would be told "Oh, you are right, of course. And that treaty may be renegotiable to some. But unfortunately the scientists just say that it can't be done."

But what ultimately happened was that by 1983 all sorts of technical feasibilities had accumulated. Were these feasibilities such as to allow somebody to envision realistically the construction of a population-defense system, one that would intercept virtually all of thousands of warheads? The answer was clearly no, there could be no astrodome. But could a defensive system be good enough, for example, to enable it to intercept quite a few warheads in a preferential defense system? Let us say you have a thousand silos, and you have a number of interceptors, but not a boost-phase or midcourse defense, nor a comprehensive terminal defense. And you say, that with your system, made up of a thousand interceptors, you want to defend a certain fraction of the thousand silos containing a thousand missiles. The interceptors are assigned to protect only a hundred silos. When the other fellow attacks you, he does not know which silos you are protecting. You basically play a shell game against him. The advantage, the classic advantage of the initiative held by the offense is transferred by this device from the offense to the defense. This is the benefit of preferential defense, a system that Spurgeon might recognize as something like a Sentinel Plus, or a Safeguard Plus. By 1983, such a defense became very feasible.

What is the attraction for deploying such a defense? Well, it's not an overwhelming strategic change, it's not Ronald Reagan's promise to bring a new era. It does mean that if you are going to stay in the business of land-based ballistic missiles, then you won't have to spend $30 billion in going to a mobile Midgetman missile. You won't have to go mobile at all. Maybe you won't even have to have the MX missile. Instead of deploying these missiles, it is a much better idea to build a shield of uncertain effectiveness. This effectiveness is elevated up to a much greater factor if the shield operates as a preferential defense to ward off the kind of attack that is most probable, and such a defense is much preferable to building costly ballistic missiles and have them tool all over the countryside.

The Pentagon found out about the President's SDI idea for all practical purposes only after the speech was given, since nobody there was consulted. Then the reaction was the usual one, and the bottom line has not changed since. Total perclusive defense, to intercept all warheads, simply would not work against thousands of warheads, no matter how many layers of defense you have. Such a defense could work very reliably, reliably enough to be worthwhile if the enemy has only a few warheads. An enemy might have only a few warheads because of one of two reasons. Either because he is a few-warhead enemy or because you have attacked him first. The second reason creates the preemptive aspect of SDI, which I will address shortly. But the possibility, that the enemy might have only a few warheads originally, is an interesting one. Nobody has ICBMs (missiles of intercontinental range) in what is called the Third World, but IRBMs (missiles of intermediate range) have shown up; this widespread development could have been anticipated and in fact has been anticipated. So there could be an interesting SDI that would offer a preferential defense of missile silos and a few other places, and would have a thin, a very thin, ability to intercept a small number of relatively primitive warheads. Is such a strategic defense something terribly original? No, it almost completely unoriginal. This is a reversion to the 1967-1969 Sentinel and the 1969 Safeguard systems. But it reflects a recognition of the advances in engineering possibilities since that time, advances which have been quite substantial.

Why am I not being indignant about all the terrible things that Spurgeon Keeny said about all these weapons systems that I really love. The reason is that Spurgeon took us back to 1945 with his tale and I think there is a much larger issue involved here. He took us back to the beginning of this nuclear competition. And I think that he has done so in a way that is very profoundly misleading. Not because he said anything that was inaccurate; as far as I could tell everything he said was quite accurate. But because he left out two very important phenomena that I know he knows took place and whose importance I think you recognize.

The first crucially important phenomenon was the role of nuclear weapons as a substitute for the stragetic wisdom whose absence was deplored in a famous quote, the quote that "we have gotten all these nuclear capabilities but our ways of thinking have not improved." Is it true that the ways of thinking have not improved? In 1945 you had two newly-promoted world superpowers, remote from one another, and exceptionally unprepared for international diplomacy or the management of conflict. The Soviet superpower had a regime of proven brutality on a mass scale. After all, by 1945, Stalin may still have been called Uncle Joe by some people, but it was also known by then that he had about 20-million dead under his belt. As for the American side, it was a superpower of provincialism and hysteria. While hysteria was not so much manifested in 1945, it did become manifested well before the classic years of the anti-communist witch hunts. So you have two newcomers, who are unbelievably inexperienced, and admittedly peripheral to the focus of the real conflicts. How could anyone in 1945 have failed to predict a U.S.-Soviet war, given what was then known about them? But such predictions were not even made. Why? Because of nuclear weapons. The famous quote is fundamentally wrong. Concerning the complaint that it is new technology but the old way of thinking: the new technology overcame the old way of thinking. It really did create a new way of thinking.

First of all, nuclear technology created a culture of circumspection in the political classes. Russians and Americans behaved prudently, whether their disposition was prudent or not so prudent; the presence of nuclear weapon produced a culture in which circumspection and prudence and restraint became dominant over the traditional qualities which in the past had made statesmen promotable--audacity, boldness, courage and such. Secondly, the existence of nuclear weapons killed military optimism. You couldn't have somebody come in and say, "we're going to start a new era and launch an attack by surprise, maneuver this way, maneuver that way, and I know it could go wrong but we'll cross the river and we'll make it through, don't worry, sir, just give me the go-ahead and I'll be gone." Nuclear weapons killed that approach.

Spurgeon deplored the nuclear arms race, deplored it at every stage without recognizing that the nuclear phenomenon has, I think, been the only factor that has prevented a large war, a large and cruel war that might have killed many millions of people. Spurgeon, I'm sure, instantly would recognize this point. He would then rejoin, "ah yes, but look what a toll it has taken of our nerves. It is a very dangerous way of avoiding a war." True. So what you have had is that you avoided a very great danger, a danger that would have entailed a very-high-probability, medium catastrophe like the Second World War, certainly killing millions of people. This avoidance has come by accepting a certain risk of a yet greater catastrophe. Spurgeon, of course, knows all these things. But there is one thing that perhaps he, in fact, does not know. Or which at least was not reflected in his talk. As I was listening to his mournful catalog of sucessive stages of the superpower competition, I did not hear the fact that the nuclear-arms competition itself, the very process of conceptualizing new moves, countermoves, actions, and so on, has served as a substitute for the traditional remedies of states that felt insecurity. Prior to nuclear weapons, when great powers individually felt secure to move and countermove, they would do certain things, such as preemptive attacks, such as grabbing provinces to secure themselves, such as attacking third parties to extend their basing areas. Instead of that, this time around, when we felt insecure we would launch a new weapons system.

Of course, the development of new weapons systems did create some mechanical dangers, because on every side there would be some people who would take such new weapons systems seriously and respond to them, sooner or later, to some degree. But who are these people and how did they function and when did they function. Look at 1954-55. In 1954 the United States mass-produced the B-52 bomber, with an intercontinental range, a wonderful machine, and in 1954 it enunciated a doctrine of massive retaliation. You can't think of a more dangerous situation. Here is the United States producing bombers galore, Americans are going around with crewcuts, pronouncing communism with an A, seemingly taking Congress seriously. And here is the United States Government, in the person of the Secretary of State, enunciating massive retaliation as a policy; telling the Russians "if three drunken Russians stumble across the border we're going to bomb Moscow."

What should those new weapons and that new policy have engendered according to the action-reaction phenomenon that Spurgeon has presented? It should have created really hysterical and terrible relations between the Americans and Russians. Instead, we had the 1955 Spirit of Camp David, with Khrushchev and Eisenhower meeting in the first re-establishment of contact between Russians and Americans since the Second World War. Secondly, Khrushchev unilaterally reduces troops by one-and-a-quarter million. The United States under the name of the New Look and the New-New Look carried out huge budget cuts in defense, cutting things left, right and center. The Khrushchev reductions are very substantial. Among other things, two hundred and fifty thousand officers lose their jobs. Soviet divisions are withdrawn from Finland, from their bases in Hungary, from Austria under the State Treaty, from Manchuria. Was there some massive Soviet reaction? Khrushchev did feel the need to respond. He did it by producing a handful of hand-crafted large ballistic missiles; at the time he figured that twelve would be quite a substantial number of ICBMs to have, twelve warheads. Contemporaneously, Eisenhower also resisted pressures. He figured that eighteen Atlases would be plenty.

What I am saying to you, is that the arms competition can be controlled by serious political operators, such as the very war-experienced Eisenhower and Khrushchev, who said to themselves "great, this nuclear play-acting enables us to cut back very substantially on military expenditures." Khrushchev carried out this great demobilization; Eisenhower went for the successive post-Korean reductions. And they did not take the arms race seriously. They were war-experienced leaders, and recognized that these nuclear weapons overshoot the culminating point of military effectiveness. They knew what war was about, and that the most important thing was to avoid war. They also understood that these weapons, that they themselves were happy to present as substitutes for the massive conventional forces that they were cutting, that these were not serious weapons. That is why the United States enunciated the policy of massive retaliation in 1954, which is seemingly the most aggressive, and most dangerous, possible policy. By 1972 it had developed into the policy that local wars can involve the instantaneous use of nuclear weapons. The Soviet doctrine has been that "the moment war begins we use military nuclear weapons, (not intercontinental weapons, mind you); if in a theater war we invade Germany, we will drop lots of nuclear weapons ahead of our advancing troops." And these policies were absolutely no obstacle to Khrushchev and Eisenhower sitting down together. When an obstacle did eventually arise, several years later in 1960, it was ostensibly over the crisis developing in the weapons-building program. But, actually, the obstacle developed from the fact that both Eisenhower in America and Khrushchev in the Soviet Union were under attack, and that they were under attack by the literalists, by those people who take all these things seriously, who have slide rules, who measure things, who say "My god, eighteen Atlases, what an absurd number." You know, the Wohlstetters of this world, the RAND Corporation, the metal-eaters in Russia. I am saying, that given the right political conditions, and the right political equilibrium, the nuclear arms competition is a substitute for the normal remedies of great powers in relating to one another, and this competition can coexist very easily with progress in mutual relations and with a very substantial diffusion of the heat in the relationship.

We just heard an analysis, not unconvincing, about how the American Strategic Defense Initiative should now be very extensively misinterpreted or at least interpreted pessimistically. There is a legitimate negative interpretation of SDI, once you know that while it cannot provide an astrodome defense, it can intercept some limited number of warheads, that it may be possible to use SDI to neutralize a retaliatory strike by one who has only, say, 400 warheads, a stake that could conceivably be reached by a preemptive attack. If so, the possibility exists, as Spurgeon mentioned, that SDI might in fact be one-half of an offense, combining an offensive strike with use of the defensive shield to then intercept surviving warheads. Therefore, there could then be no retaliatory Soviet attack. What should be happening now, in the light of the SDI, is the Soviet Union mass-producing ballistic missiles, saying "you are producing SDI, we've going to outrace you." In fact, Spurgeon Keeny presented that whole analysis; and as I was sitting here listening to this analysis, he made perfect sense. If we were locked in a capsule traveling in space, without access to the news, we would have to believe that Spurgeon was in fact telling us a truth without flaws. Here is SDI--it has this inherently offensive potential, and a nuclear-weapons buildup is the logical Soviet response. This is what would happen if the literalists were empowered by the slide-rule artists and McNamaras of this world, by people who take weapons seriously and do not understand war.

What is actually happening in the Soviet Union? Are they spending everything for the production of ballistic missiles to outproduce the defensive warheads? Most definitely not. There is in fact a distinctly lackadaisical attitude in the Soviet Union about this whole thing. There is the recognition that by lauching the SDI program the United States has really pulled a fast one. Here are the Russians, hardworking people, earnestly accumulating ballistic missiles, so that by about 1981-82 they have the whole world talking about their SS-18s in reverential tones. They are finally able to say that to the traditional power of the Russian army is added the great power of the Soviet ballistic-missile force. And then along come these infuriating sort-of-free-floating Americans who have launched this SDI program. There is of course a deception campaign in the U.S. about SDI, because all these scientists like Hans Bethe come out and say it's impossible. But of course that is an U.S. smokescreen, a smokescreen like the one they pulled last time with the H-bomb, when the same experts said the H-bomb was impossible. In fact, everyone penetrates that smokescreen, and the whole world is instead pervaded by images of the Americans racing ahead with a superb, wonderful, astrodome SDI system.

The Soviets recognized this political challenge of SDI. But they have definitely not responded to the SDI program, even as it evolves, with any intention of preserving the capability to respond to it in a mechanical sense. Of course there is not yet any deployed SDI defense, and the various programs and projects are in the 7-, 8-, 10- year category. But why is it that the Soviet

Union instead seems eager, very eager, to proceed to limitations through the START negotiations, which would preclude its response to a defense system that it is not, in fact, going to be able to prohibit? The reason is that the people in charge, as in various previous periods in the U.S.-Soviet relations, are in fact serious political leaders, they understand what SDI is. It is not the real thing; it a simulacrum. The SDI is a political challenge for the Soviet Union. It has flooded the world with imagery of American innovation and therefore it requires a response. But it is not something you actually respond to by mechanically doing the logical thing, which today would mean "sorry Mr. Reagan, no way we can do START; we have to reserve the right to be able to put in those 25,000 warheads which would totally exhaust your system in the proper logical way."

I am sorry to say that this rather cheerful prospect is not all that I can leave you with, because there is indeed something very serious happening. And what is happening is that this comedy is breaking down. I believe that the signs can no longer be denied that after forty years, we are leaving a period in which the phenomenon of war has been kept out of the U.S.-Soviet relationship, in which the phenomenon of war has been driven right out of Europe by nuclear weapons, in which the monster-animal of nuclear weapons has frightened away war itself. Of course war has prospered and lived very nicely outside the U.S.-Soviet relationship, and outside Europe war is doing very well. But nuclear war has not prospered. Unfortunately this phenomenon of nuclear weapons, which drove out the phenomenon of superpower war, is receding. The signs are clear that we are entering a post-nuclear era, an era in which there will, do doubt, still be many nuclear weapons around, but in which their use will be restricted and recognized only as mutual deterrence, with its pure strike-back, no longer in the way in which they have actually been engaged for forty years, namely not at all for strike-back but rather for extended deterrence. This whole competition, these frantic attempts in the last years of the current period to maintain nuclear deterrence by proliferating warheads, is like somebody trying to pay hotel rooms in Berlin in the 1920s with the depreciating currency, by bringing to the register more and more suitcases of money. This recent competition is the evident sign of the great strategic change. Nuclear weapons, like other weapons, have had their rise, their culminating point and are now in evident decline. The causes of this decline are many. There is the simple fact that nuclear weapons generally speaking overshoot the culminating point of effectiveness in striking most targets. There is the factor of inhibition acting against all things nuclear. There is the biologic fact that in the currently dominant cultures the memories of the nature of war have receeded, whereas the consciousness of the alternative risk of nuclear has become ever more clearly manifest.

The outcome, I am sorry to say, is that we are now sliding and drifting into post-nuclear conditions. In post-nuclear conditions the paraphenomenon of the nuclear competition will no longer serve to preserve stability. Instead there will be a reversion to a real phenomenon, which is the contraction of the forces themselves. The decline of nuclear weapons is manifest in many ways. First, as you know, the INF treaty symbolizes that. The Pershing and the cruise missile were like the Ferrari and the Porsche of the non-intercontinental nuclear arsenal. These were very long- range weapons, very flexible, with the Pershing being highly accurate. You know, when someone is getting rid of his Porsche and Ferrari, there's no use in trying to be reassuring by saying "oh, I've got lots of others, and I've got lots of little Izuzus and Toyotas and whatever." If a man is getting rid of his Ferraris because he claims the cars are dangerous, polluting whatever, then you worry. The INFs are gone, the short-range systems are gone, prospects for the implementation of further weapon-modernization programs are zero. There are START talks and such. Please remember, a post-nuclear condition does not mean that there are no nuclear weapons. It is not that we are entering a post-nuclear era because we're running out of nuclear weapons--we have lots of them. Instead, the willingness to get rid of them is an affirmation of their diminishing utility.

We still have a strategy of flexible response. We are supposed to defend conventionally for a while and then go nuclear. But, in fact, the point at which we will go nuclear gets endlessly postponed; more and more of the military strategies involve conventional weaponry. That is why in the Soviet Union the military preparations have become so much more realistic. The last Soviet fighter that we have observed to be finally deployed is the MiG-29, which has gill air intakes to be used in gravel airfields and such. Somebody is thinking of war over there, a real war. I mean a war in which planes land and take off, and in which airbases get broken up so that you have to go to gravel. We are sliding into a post-nuclear era, conventional preparations are getting serious. For years soldiers used to run around and play-act doing these things. Now we have sinister signs--even the Italian army is training, and the Belgians are buying ammunition. For years it was understood that nuclear weapons meant that if a war got serious it would become nuclear. Hence, whatever you guys were doing wasn't serious. Things are now definitely changing. With Gorbachev in power, should he maintain his same policies, we are going to have a Soviet Union which is very interested in leaping ahead. After some number of years we may find ourselves in a situation where the Soviet Union and the United States are still superpowers, but no longer have the substitute of the paraphenomenon of nuclear weapons to preempt the potential energies of conflict. This is what I missed in Spurgeon's catalog, his mournful catalog of action-reaction, as if that did not serve this function of absorbing the contention of two new, very powerful, very provincial, very ill-prepared superpowers that could easily have blundered into any number of confrontations. And I say, that the serious thing is that these nuclear weapons, these wonderful sweet, loving, darling nuclear weapons, which have served us so well, are fading out. They really are.

I will end by saying, that when you look back on those famous analyses of nuclear proliferation, prepared by various professors with various grants from various foundations twenty, thirty years ago, they all had little tables showing how many nuclear powers there were going to be by 1988. Or at least they would give 1980 and 85 and 1990 and project a growth to around 30 or 40 nuclear powers. Why did these people make that mistake? Was it because they were hysterical? No. But it was because they perceived the utility of nuclear weapons to be very high. These analysts viewed nuclear weapons as terribly useful weapons, so of course they thought that anybody who could get them would get them. In a sense, nuclear proliferation is a vote that says how useful nuclear weapons are. There are of course exceptions to this sense, there always are exceptions. But basically, the utility of nuclear weapons reached its apogee around, let us say 1957, a partially arbitrary year. In that year the United States spent eleven billion dollars for strategic weapons, and twelve billion dollars for what they call general-purpose forces, which are the Army, Navy and Air Force. And these too were stocked with nuclear weapons. United States forces were about 60% nuclear in 1957. Right now U.S. forces are maybe 8 to 10% nuclear. Nuclear weapons are fading out. This is very unfortunate. We may yet live to miss them.

CHAPTER 19

DISCUSSION OF TALKS BY SPURGEON KEENY AND EDWARD LUTTWAK

Spurgeon Keeny: Edward has made a number of very interesting and as always stimulating points. You observed that no one really believed in the President's vision of an impenetrable defense. And I think there is probably some truth in that. You indicated that you were instead thinking in terms of silo defense, or maybe a very small anti-nth-country defense, where such defensive capabilities would in fact be possible. Bearing in mind that this would probably be a camel's-nose-in-the-tent approach to a comprehensive defense, I would agree that technically one could achieve some amount of attrition against an attack on hard targets. However, the U.S. really does not need to enter into such an expensive system now, because, as I believe Edward himself admitted, the U.S. deterrent is really not now threatened. Furthermore, if one wanted to improve the U.S. deterrent posture, there are a host of technical improvements or basing schemes that one could implement over the next decade, schemes that would cost far less than a hard-point defense deployment, particularly if the defense were to be based in space. More importantly, on a political plane, such a deployment would appear to the Soviet Union as the first stage in an attempt to go to a major strategic defense that would take on strategic significance of the type I discussed, and therefore be part of a potential offensive threat. And this appears to be the posture that the Defense Department is adopting. Those who say that this is only the first phase, hasten to emphasize that the SDI program really is intended to proceed after deployment of this first phase towards the impenetrable defense of a true SDI system.

The question of the very minimal defense that might be useful against an accident or against an attack of some sort by a third country suffers from the same problem. Since there is really no interest in the military in such a defense, such a limited system would be looked on by the Soviet Union as a stepping stone to a more massive future deployment, with whatever impact that would have on the U.S.-Soviet relationship. In addition, such a defense would probably be of very little value against accidents, which are extremely unlikely to occur, and if they do, could easily be of a nature against which a very thin defense would not be effective. Moreover, the prospect of third countries undertaking missile attacks against the United States seems extremely remote for the foreseeable future.

I found some of Edward's initial comments on the early nuclear history very interesting, and I agree with him that a strong argument can be made, an argument that may even be true, that the existence of nuclear weapons has prevented a major conflict between the United States and the Soviet Union. There certainly was a real possibility of a dreadful, conventional war that probably would have been comparable in its destruction to the Second and First World Wars.

While I indicated in my discussion that it is legitimate to think about alternatives to nuclear weapons, I agree with Edward that I do not see the elimination of nuclear weapons in the near term. Nuclear weapons are a fact of life; they are going to be with us for a long time. Indeed, the name of the game now is to make sure that the superpower relationship is stable so that they will not be used intentionally. We must also minimize, to the extent possible, the likelihood that they might be inadvertently used. On this point, I think there is probably a great deal of agreement. And my main message to this group is that I don't think there is any magical, technical solution, including SDI, that is going to eliminate the threat of nuclear weapons in the foreseeable future.

Edward's general comment, that we are in a period where nuclear weapons are fading from the scene, and his suspicion that in some sort of nostalgic way we are going to miss the good old nuclear days, are, I think, overly pessimistic. I think that if nuclear weapons remain the ultimate deterrent, we will probably see a period of real improvements in U.S.-Soviet relations, where there will not be a major threat by either side. We should be pleased that we are moving away from the notion of nuclear weapons as a warfighting means, or as an instrument of war that one would be prepared to use in a large variety of peripheral situations. I recognize that the notion of extended deterrence may have played a role in increasing the credibility of deterrence. I think, however, the notion that nuclear weapons really are usable, that they should be introduced in the event of peripheral conflicts or of conventional conflicts between the superpowers, has always been very dangerous. I for one am pleased to see that this notion is receding, if indeed this is the case, because it is very hard to imagine any conflict between the United States and the Soviet Union and their alliances involving nuclear weapons that would not escalate into a major nuclear war. And it certainly is hard to envisage such a conflict occurring if nuclear weapons are not available for warfighting. Moreover, in the present political climate, abandoning the notion of nuclear warfighting will not significantly increase the prospects that general conventional or nuclear war will in fact occur.

Edward Luttwak: Spurgeon agrees, but nevertheless doesn't agree. My argument very simply is that weapon-system acquisition itself is the key to the maintainance and even improvement of relations. Reagan started his SDI in 1983 and did not get any major Soviet reaction as a consequence. This enables him to proceed to the idea of getting rid of those useless old ICBMs that actually do exist. Just as it has happened previously on both sides; weapon-system acquisition acts as a substitute for traditional remedies, which opens the way to the management of relations. It is a very basic disagreement: You regard each of these weapon programs in terms of the mechanical threats they can present; I argue that the political process actually works in a different way.

Keeny: I think it is a very interesting point, and you stated it in a provocative way. I agree there is a role for sandboxes for politicians, just as there is a role for sandboxes for scientists. And I don't look down on sandboxes. They serve a purpose. In my discussion, I did not say that arms control, or unilateral arms control, should be synonymous with the cessation of all innovation or all military activity. I think these should and will continue. However, these should continue in the least-provocative way possible, and on a scale and scope that does not create tensions and become a provocation in its own right.

Question from the audience: Do you truly think nuclear weapons are becoming less useful, or are we simply paying less attention to them; are we really going back to the pure second-strike deterrence posture? Do you think that, for example, the chances for a conventional war in Europe between NATO and the Soviet Union will go up?

Luttwak: Look, let us be very specific. In 1957 the United States spent a lot of money on strategic forces and only slightly more on the Army, the Navy and the Air Force. Within these forces the emphasis on nuclear weapons was such that the U.S. Army was rapidly losing the ability to fight a conventional war. The U.S. Army could not fight except with nuclear weapons; it did not have the artillery, for example. On the Soviet side, after Khrushchev's demobilization, the old armiia echelon, the front-echelon artillery, was disbanded. Khrushchev said "you do not need all of that stuff. Instead I will give you four rockets per division, each with a nuclear warhead. In a standard war you can fire them off and open the way." For reasons which did not reflect aggressiveness, but rather the opposite, both sides were equipped in such a way that they could not really fight except with nuclear weapons. Today, through the mere accumulation of conventional capabilities and the retrocession of nuclear weapons, the objective possibility of fighting a conventional war has increased from trivial border incidents toward wars in Western Europe. Have we got to the stage where our military forces are really structured and equipped to fight a big conventional war? Not quite. But what is happening is that all the good people are saying "don't spend money on SDI, spend it on conventional improvements;" even the Europeans are saying this. We are building a consensus around the delegitimization of that very nuclear deterrence which, however dangerous, was also the instrument that prevented such force buildups. These are the facts that must be recognized in budgets and force structures. He who has conventional ammunition can fight a conventional war. The precaution of not having any conventional military strength, created this alternative intermediate between having nuclear war and having none, and we had forty years of no war, and I thought it was rather nice.

Question from the audience: I thought that your analysis that weapons aquisition is a substitute for war,... [Luttwak: Not the only substitute; however a very important substitute.] ... if carried further, implies that the purchase of all these weapons is simply a charade, whose purpose is simply a competition with the Soviet Union. Did I miss something?

Luttwak: I did not use language that would suggest that I view this in a trivial light, that I was looking ex cathedra down on all these children playing foolish games. This acquisition process was a way of mediating contending interests. It was a way of buying different kinds of insurances against various eventualities. And was there an attitude of being serious? More on the American side than the Soviet. When the Soviets had to make choices between priorities--the priorities of bodies of political control and the priorities of serious weapon configurations--they always chose political control. I do not want to talk here about certain practices they have, still now and certainly in the past; details have been published, for example, about where they stored the warheads of nuclear-weapons systems, locations which were designed to do anything but... Of course, we are serious. Yet, when you reflect: How important is it that the final warhead on the last of the 1000 Minutemen ICBMs should be functioning sufficiently well to yield the maximum energy yield obtainable from that device? How important is that? Let us say that it wouldn't warrant an expenditure of five billion dollars for corrective measures. But that placement of too high an importance on weapons systems is no different from other military preparations throughout history. Just reflect on the pictures you have seen of soldiers in the sixteenth century. Was their equipment entirely functional? Their plumes were polished. And there is a reason for this; it was a mediating mechanism.

Question from the audience: There is a question I heard neither of you discuss, although it was hinted at, and that is whether the U.S. SDI program was a reaction to similar programs of the Soviets?

Luttwak: The biggest joke, of course, is the outraged, indignant propaganda out of Moscow on SDI in 1983 and 1984. Because they, of course, invented the concept of having a distinct program that would pursue unconventional, "high tech" routes toward the interception of ballistic missiles. They had the spectacular experiments and activities and facilities, including the ones that were photographed by U.S. reconnaissance satellites. These photos precipitated the famous General Keegan episode when, after his resignation from the Air Force, he talked publicly about the Soviet systems. Was there a reaction to Keegan's revelation of a Soviet SDI program? No, there was none by the U.S. program. The entire Soviet propaganda complaint against the U.S. SDI was really just theater. The U.S. program was not a reaction to the Soviet program, because the U.S. military bureaucracies regarded any such U.S. programs as a great threat to the existing traditional things that they were interested in doing. That is why, when Keegan originally went to RAND within the military system with the pictures, saying "look, the Soviets are doing this and this," he first was suppressed intra-bureaucratically. The Soviets did do such research before the United States, so their propaganda complaints about the U.S. program were outrageous propaganda. Yet, the U.S. program was not responsive to their prior SDI-style research.

Keeny: I don't really have any idea why President Reagan came to the conclusions that he did in proposing the Strategic Defense Initiative. It is my impression that it was very much his own vision, stimulated by discussions with Teller and Danny Graham and a number of other people, but I don't think that his main motivation was a reaction to Soviet programs. [Luttwak: No, Teller did not have much respect for the Soviet experiments.] Edward makes reference to General Keegan. His reputation was so low, that I think his advocacy of any particular position, or program, tended to be the kiss of death. I recall that in the early 1950s he was sort of the example par excellence of the Air Force officer who thought that a Soviet attack was coming momentarily, so that there was a constant flow of bizarre threats, scenarios and technical developments from him, from the time he was a captain and a major in the Air Force. He, of course, achieved fame as part of the input to the missile gap. And he had bizarre notions about the SDI program. But I don't think one can find the origins of the present U.S. program in reaction to the Soviet program. [Luttwak: Except for one thing. After Harold Brown made fun of the data about the laser and so on, in the same year the budget was tripled from 150 to 500 million dollars at Brown's own direction. He and the great physicists said that Keegan's physics was bad, but then he made his own assessment.] Actually Keegan was into particle beams.

Question from the audience: Some writers on arms control welcome the return to an era of assured second-strike capacity, although at a lower level.

Luttwak: How then did we get extended deterrence? Assured destruction means only strikeback, so there is then no extended deterrence at all, which means U.S. nuclear weapons are not usable strategically at all. They are then only usable when the Soviets attack us, and we strike back. That means nuclear weapons do not play a role in the balance of power. Therefore, they do not help to contain the Soviet army. Therefore, you denuclearize the alliances. Therefore, you have created a zone for conventional war. You are advocating the thing which I dread to see us drifting into, something which I think is a horrible prospect, namely the post-nuclear situation. How can you advocate that?

Keeny: I understood your question as stating that many people want to go to a minimal nuclear deterrent based solely on an assured retaliatory capability; and I would agree with you entirely. I think we are far, far from that, but I think that this is certainly a...

Luttwak: That does not solve the security problem. If you are telling me that we can only have assured destruction and are in post-nuclear conditions and nothing can be done about it, then by all means, have it go to a lower level; I am a taxpayer, too.

Keeny: There is a lot of room between a minimum-deterrent retaliatory capability, and an assured deterrent under a broad range of circumstances. I think we are in the latter state now and it is a very long time before we even approach the question of a minimum deterrent. That is not on the table at the moment. I think the first stage is to make sure we have a very stable, very safe, secure retaliatory capability under a broad range of credible circumstances. In that context one will hopefully proceed to play out a desirable political relationship between the East and the West, which will move us in the direction of a much safer relationship across the board, not just in the nuclear area but in the conventional area as well. If the political process does not evolve in that direction, clearly this arms-control process will only go so far and no further. If the political process permits, I think the constraints on further nuclear and other military buildups will follow logically. And there is a close interaction; a tension is created by excess militarization on both sides. On the other hand, you cannot solve your problems by arms control any more than by technology, since we are essentially dealing with political issues.

John Logsdon (Moderator): I think we are telling these physicists that politics is the master art, even if not the master science.

Question from the audience: I like most of the things Mr. Keeny said. So I will ask a question of Mr. Luttwak. I am surprised at his example, in which he claimed that Bethe said the H-bomb was impossible. I doubt seriously that Hans Bethe would ever say anything like that. He was essentially the first person to understand how fusion works, and that is the basis for the H-bomb. The President told us in his speech that nuclear weapons should be made impotent and obsolete. You seem to say that SDI should be working toward preferential defense of selected places. You seem to be in favor of nuclear weapons because of the fact that there has been no major war since 1945. Let me try an analogy: From 1857 to 1947 the British ruled India, the sun never set on the British Empire, and there was a lot of peaceful co-existence among people. Would that justify leaving the British Empire in place? The other point is, it is true that we have not had a major war for 43 years. But if one does come, humanity will cease to exist, according to some; and even if it should survive, it won't be the same thing. We are being held hostage to nuclear weapons.

Luttwak: First of all, in regard to Hans Bethe. I do not know what he knows or does not know. All I know is, that in the documents that describe the American decision to proceed to what was then called the superbomb, there are listed various stages of decisions. The final stage of decision was almost a judicial process, which was coordinated and supervised really by Dean Acheson. This process examined the testimony of diverse scientific voices--some people saying it should be done, some people saying it should not be done, and the people who did not like the superbomb, used to like to say that it was actually very difficult to do, and that it would not be practical. And Hans Bethe definitely was an opponent of the superbomb. In his position, he chose to express his opposition in a way that cast doubt on the feasibility of the weapon. I do not mean the scientific feasibility necessarily. I do not know what the testimony was, I have never read it. All I know is that he definitely was against the bomb. It is unlikely that he would have said "Mr. Acheson, I am pretending to be here as a scientist but actually I am a moralist or a politician." He must have said either that it could not be done, or if it could be done it would not be practical. His position on the H-bomb is on the record that is accessible to everybody, for example, in Dean Acheson's collection of documents on the subject.

As for the British Raj. The last time I was in India in Wellington, I was sitting there on the veranda of the gymkana club having a scotch, talking to the commandant of the Royal Defense College at Wellington. He was sitting there in his tweeds talking about his son who is at Oxford, doing quite well, and people would pass by wearing blazers, and then we all went for a ride, and we went pigsticking the following morning. As far as I know, the British Raj is continuing; I wanted to reassure you that this is the case.

The deeper issue is the following. Until 1945, for people like ourselves there was no choice. We tried to live in peace and sometimes they gave us war. So they bombed us and our women, our children died, and sometimes we died ourselves. Since 1945, we have had a choice: Do it the old way or have these nuclear weapons there, which kill military optimism, preclude all forms of planning that end with the word victory, and force even the most reckless and feckless and ill-taught politicians to be very careful. So we no longer have, for example, the classic Balkan war scenarios, where we are supposed to be quarreling over Middle Easterners and that will drag us into war. Instead now we tell the Middle Easterners to stop quarreling because <u>they</u> might drag us into a war. All the familiar mechanisms of war have been functioning in reverse. The price has been, that if this system broke down, then we would really be in the soup. In the old days we could cross the Atlantic or the Pacific only by means of sailboats, regularly overcome by typhoons and tornados. Now we can fly. When an airplane breaks down, you go down with no chance of survival. We made our choice. I don't pretend to be sufficiently a philosopher to give you a criteria for choosing between these two military options. But my assessment of the military balance and the evolution of budgets, of the composition of forces, and of behaviour in arms-control settings, tells me that we are loosing this choice and that we are going to go back to a situation guaranteed only by assured destruction. Assured destruction means that weapons are not in a military balance. Which means that when things start moving, there will be conventional war. So we lose the choice. We won't have a choice of flying by air with some danger that the plane might crash, or having to go in sailboats that are subjected to scurvy and sickness, regularly every voyage. We are going to lose that choice; we are going back to conventional wars. I know my friends in Wellington, and many other such friends around the world; they and my clients relish this prospect. I personally on the whole prefer the other way.

Keeny: I would like to make a comment about the Hans Bethe affairs that you brought up. I have not read the Acheson book and I don't know the quote. But just to keep it in perspective, you need to look into the background. The whole H-bomb decision was complicated by the fact that there was a specific track along which most of the thinking was going. The widespread thinking at the time was that, quite aside from ethical questions, it was a very tough question whether the H-bomb project was worth doing. It was extremely questionable whether it would work at all; and, if it did, it would have been incredibly expensive. You can always question the motives why people may have given the advice they did. But I think most scientists at the time questioned whether the then-current approach to building a superbomb made any sense, whether the superbomb was feasible--and, if it were, whether it could be weaponized. But then there was a new conceptual development by Stan Ulam and Teller, that was subsequently developed by a number of others. It changed the probability of success substantially. And I think you will find that at that point Bethe, whatever his other judgements were, did concur in its technical feasibility, and in fact played a role in the actual implementation of the ideas. But that was after they had a more technically-feasible way to design the H-bomb.

Question from the audience: Professor Luttwak, I would like to know your vision for the longer term. Do you simply see a continuation of mutually assured destruction, or deterrence, or however you call it, and of the arms race, forever? I agree there are dangers in disarmament, but aren't we going to have to face that danger sometime? We can't continue with the arms race forever.

Luttwak: Why not continue an arms race which has served as an alternative to strife, an arms race which is showing no signs of accelerating, which in fact has been decelerating?

Question from the audience: Okay, if it is decelerating, then what is going to happen if it keeps decelerating?

Luttwak: Well that is exactly what I am worried about. Look, in 1957, the United States spent $11.2 billion on strategic weapons. Do I have to tell you how many dollars that is today?

Question from the audience: Do you really want to do that for the next decade, and the next decade, and the next decade?

Luttwak: I do not know. We cannot project. Our vision of the strategic trends over the next 20 to 30 years is clouded. We may have 4, 5, 6, 7, 8 years of a Khrushchevian interval, with Mikhail Gorbachev, who is doing all the things Khrushchev was doing in terms of demobilization, cutbacks, relaunching the hopes of the Soviet system. The strategic trends seem to eventuate over much a longer span than these political changes that are now so very rapid. Nobody has given us a guarantee with any of these systems. But we have one system that we have tested and know. That system has given us a very big war every 20 years of our life. Now we have another system that has been in operation for 40 years and has not given us a war, even though it has given us a bad case of nerves. Now we are losing this second system. I do not know how you feel about the risks and benefits of the second system. Maybe it is the modern version of the human condition. In the Middle Ages, the human condition was that your child could die at any moment from any one of many disease, and you wouldn't even know it was a disease, you probably thought it was a neighbor's witchcraft. That human condition still persists, for people in certain parts of Calcutta, for example. They didn't bleed too much about it, they lived and tried and procreated and tried to do it again. Okay, we have our modern version of the human condition. I suggest that we ought not to bleed too much either. But as an analyst I have to tell you the following. You have had in recent decades two systems to choose from. My feeling is you are going to lose one of the two systems.

Question from the audience: A lot of people think there is a third system.

Luttwak: Which is what, sit down and kiss each other? [Questioner: Beginning to get along with each other.] If you could invent... [Questioner: I think that is what I am trying to do. It just takes a long time to get there. And it is dangerous where we are now.] If you could somehow resolve the political problems, the problems of political strife within even provincial committees, then you would have to ask yourself what sort of manifestations the human condition might then display? We have examples of that. And here you are not talking about political cycles of 5 or 7 years, you are not even talking about strategic trends of 20-30 years, now you are talking about secular trends. [Questioner: Conditions do change. We do not accept slavery any more; we don't accept the subjugation of women anymore.] We don't accept the subjugation of women? The Afghan rebels are about to kill one another because they insist on it. There, half the factions are going to kill the other half. Things change, but with respect to the nuclear choices not on useful time scales.

Question from the audience: Give us half the money we are spending on SDI, and things will change. [Luttwak: Two billion dollars?] On the one side you don't want to change the world by talking to each other; on the other side you don't want to change the world by spending two billion dollars. That is not fair.

Luttwak: The next time you are in Delhi, ask them how much the new Indian indigenous gun program costs. How much did a 20-mm smooth-bore cannon cost?

Question from the audience: Dr. Luttwak, you were describing the arms race as a substitute for war. As I was listening, this reminded me of Konrad Lorentz' description of the behaviour of two wolves fighting. They could inflict an enormous amount of damage on each other. Yet there comes a point where they realize the danger of mutual destruction, and the fighting becomes stylized. Is this the sort of thing your are talking about?

Luttwak: I do not believe in biological modeling of complex behavioral systems. Also, given the human record in this century, it is insulting to wolves to compare the two, right? So, I wouldn't do it. You may be so reminded, if you wish.

Question from the audience: Would it be helpful in solving nuclear-arms problems to have more cooperation with the Soviet Union on scientific/technical matters, say on environmental issues or on space exploration?

Another member of the audience: We will have to go to Mars to get there, that would solve this problem, right?

Keeny: I do not think such things will solve the underlying problem, although they can be useful. One of the real problems with the Soviets' society, particularly when they have been clearly in a very ideological adversary position, has been the closed nature of the society. This has been slowly changing; hopefully it will be changing more rapidly now. The more one penetrates a society by having different elites interact and have mutual interests in programs, the more that reduces suspicions, builds confidence, and contributes in a small way to a process that is critical to the evolution of acceptable relations between the two countries. In this way, cooperative activities are useful. I would however be careful not to exaggerate them. You might say that it is another kind of sandbox to get the U.S. and Soviet scientific communities working together on trips to Mars or environmental projects of clear mutual interest, to keep them out of other mischief. It is also inevitable, when you get to know serious people in another society, that you influence their thinking. You will also have another deterrent against activities that might not be in compliance with arms-control agreements. The more connections there are between the two societies, the less likely it is that such activities could go unnoticed, hence the greater the deterrent against such actions.

Luttwak: I fully agree. But: scientists yes, physicians no. Especially when the physicians are concerned about nuclear war. They get manipulated far too easily. And furthermore, no engineers, because we have a security concern here. The Soviets do carry out systematic intelligence exploitation of all contacts. But there is no danger in sending scientists, because they don't know anything useful [appreciative laughter]. -- Actually, they are very dangerous.

Logsdon: This last comment illustrates why these two gentlemen are in their mutual businesses.

APPENDIX A

BIBLIOGRAPHY ON PHYSICS AND THE NUCLEAR ARMS RACE: REVISITED

Dietrich Schroeer
Department of Physics and Astronomy
University of North Carolina
Chapel Hill, NC 27599-3255

ABSTRACT

This collection of article and book references is assembled to help physicists understand and teach about the nuclear arms race. Both technical and political aspects are covered. It is a continuation of a 1983 bibliography.

I. INTRODUCTION

In 1982 and 1983 D. Schroeer and J. Dowling published a "Resource Letter PNAR-1: Physics and the nuclear arms race," in the American Journal of Physics 50, 786-795 (1982), and a "Bibliography on Physics and the Nuclear Arms Race" in PHYSICS, TECHNOLOGY AND THE NUCLEAR ARMS RACE, D. W. Hafemeister and D. Schroeer (eds) (American Institute of Physics Proceedings #104, New York, 1983), pp. 271-295. This bibliography is an update of those collections of article and book references.

This bibliography includes references dealing with the physics of the arms race, the technologies generated by this physics, and how those physics-inspired technologies modify the politics of the arms race. In general, the readings included are those that one ought to have seen if one is trying to be knowledgable in this area.

This bibliography is to be an extension of the earlier collections. There is much continuity over the years in the technologies and political problems of the nuclear arms race. Therefore the format here is that of the 1983 bibliography, with the same chapter and section headings. Only a few new subheadings had to be created, e.g. on nuclear winter and SDI.

1.1 BY THE BOMB'S EARLY LIGHT: AMERICAN THOUGHT AND CULTURE AT THE DAWN OF THE ATOMIC AGE, P. Boyer, (Pantheon, New York, 1985).
1.2 "The first spin-offs from the strategic defense initiative," J. Dowling, Am. J. Phys. 56, 303-6 (1988). Some simple physics calculations related to the SDI program, useful in teaching physics.
1.3 PERSPECTIVES ON NUCLEAR WAR AND PEACE EDUCATION, R. Ehrlich (ed.), (Greenwood, New York, 1987).

II. GENERAL REFERENCES

A. Course Materials

1. Textbooks

2.1 NUCLEAR ARMS RACE: TECHNOLOGY AND SOCIETY, P. P. Craig and J. A. Jungerman (McGraw-Hill, New York, 1986). An arms-race textbook with some technical material.

2.2 WAGING NUCLEAR PEACE: THE TECHNOLOGY AND POLITICS OF NUCLEAR WEAPONS, R. Ehrlich (SUNY, Albany, NY, 1985). An arms-race textbook with a technical perspective.

2.3 TECHNOLOGY AND THE NUCLEAR ARMS RACE, with TEACHER'S MANUAL, D. Schroeer (Wiley, New York, 1984). An arms-race textbook with descriptions of technologies.

2.4 ARSENAL: UNDERSTANDING WEAPONS IN THE NUCLEAR AGE, K. Tsipis (Simon & Schuster, New York, 1983). An arms-race textbook with some technical material.

2. Strategic readers

2.5 THE ARMS RACE AND NUCLEAR WAR, D. P. Barash (Wadsworth, Belmont, CA, 1987).

2.6 THE NUCLEAR AGE READER, J. Porro (ed.), (Alfred A. Knopf, New York, 1988). Excerpts from various documents. To accompany the "War and Peace in the Nuclear Age" telecourse. See also THE NUCLEAR AGE STUDY GUIDE, B. Kirsch (Alfred A. Knopf, New York, 1988).

2.7 THE ARMS RACE AND NUCLEAR WAR, Wm. M. Evans and S. Hilgartner (eds), (Prentice Hall, Englewood Cliffs, NJ, 1987). Readings oriented toward political science, but a good selection for a supplement to a more technically-based course on the nuclear arms race.

2.8 THE NUCLEAR PREDICAMENT: A SOURCEBOOK, D. U. Gregorg (ed.), (St. Martin's, New York, 1986). Possible policy-oriented supplement for a course on the nuclear arms race.

2.9 NUCLEAR WEAPONS AND THE THREAT OF NUCLEAR WAR, J. B. Harris and E. Markusen (Harcourt Brace Jovanovich, San Diego, 1986). Readings on the nuclear arms race.

2.10 THE NUCLEAR ARMS RACE DEBATED, H. M. Levine and D. Carlton (McGraw-Hill, New York, 1986). Readings on the nuclear arms race.

III. NUCLEAR ARMS

A. Conventional Strategic War

3.1 THE NIGHT TOKYO BURNED, H. Edoin (St. Martin's, New York, 1987). Description of conventional strategic bombing of Japan during World War II.

3.2 THE RISE OF AMERICAN AIRPOWER: THE CREATION OF ARMAGEDDON, M. S. Sherry, (Yale University, New Haven, 1987). The development of American thinking about air bombardment through the end of World War II.

B. The Fission Bomb

1. Technological aspects

3.3 INTRODUCTION TO DETONATION THEORY, W. Fickett, (University of California, Berkeley, 1985). How do explosions develop?

2. Historical aspects

3.4 THE MAKING OF THE ATOMIC BOMB, R. Rhodes, (Simon & Schuster, New York, 1987). Historical account of the Manhattan Project.

C. The Fusion Bomb

1. Technical aspects

3.5 MILITARY RADIOBIOLOGY, J. J. Conklin and R. I. Walker (eds) (Academic Press, Orlando, FL, 1987).

3.6 "Atomic Bomb Doses Reassessed," L. Roberts, Science 238, 1649-51 (1987); "Genetic Effects of the Atomic Bombs: A Reappraisal," W. J. Schull, M. Otake, and J. V. Neel, Science 215, 1220-7 (1981); see also the report "Studies revise dose estimates of A-bomb survivors," Phys. Today 34 (9), 17-20 (September 1981).

2. Historical aspects

3.7 U.S. NUCLEAR WEAPONS: THE SECRET HISTORY, C. Hansen, (Orion Books, New York, 1988). Descriptions of various nuclear weapons and their operations; many interesting details.

D. Full-Scale Nuclear War

1. Technical aspects

3.8 "Worldwide radioactivity from a nuclear war," A. A. Broyles, Am. J. Phys. 54, 151-7 (1986). Back of the envelope calculations.

3.9 "Catastrophic Releases of Radioactivity," S. A. Fetter and K. Tsipis, Sci. Am. 244 (4), 41-5 (April, 1981). Which releases more radioactive fallout--the explosion of a nuclear weapon, or the worst imaginable accident at a nuclear power plant?

2. Nuclear winter

3.10 "The Climatic Effect of Nuclear War," R. P. Turco, O. B. Toon, T. P. Ackerman, J. B. Pollack, and C. Sagan, Sci. Am. 251 (2), 33-43 (August 1984). A description of nuclear winter.

3.11 "Global atmospheric effects of massive smoke injections from a nuclear war: results from general circulation model simulations," C. Covey, S. H. Schneider, and S. L. Thompson, Nature 308, 21-5 (1984).

3.12 THE COLD AND THE DARK: THE WORLD AFTER NUCLEAR WAR, P. R. Ehrlich, C. Saga, D. Kennedy, and W. O. Roberts, (W. W. Norton, New York, 1984. On the possibility of nuclear winter and its consequences.

3.13 ENVIRONMENTAL CONSEQUENCES OF NUCLEAR WAR: VOL. I: PHYSICAL AND ATMOSPHERIC EFFECTS, A. B. Pittock, et al. (Wiley, New York, 1986).

3.14 ENVIRONMENTAL CONSEQUENCES OF NUCLEAR WAR: VOL. II: ECOLOGICAL AND AGRICULTURAL EFFECTS, M. A. Harwell and T. C. Hutchinson (Wiley, New York, 1985).

3.15 "Nuclear Winter Reappraised," S. L. Thompson and S. H. Schneider, Foreign Affairs 64 (5), 981-1005 (Summer 1986). Would it be nuclear winter or nuclear fall?

3.16 "Nuclear Winter: Global Consequences of Multiple Nuclear Explosions," R. P. Turco, O. B. Toon, T. P. Ackerman, J. B. Pollack, and C. Sagan, Science 222, 1283-92 (1983). The TTAPS paper on nuclear winter.

IV. NUCLEAR BALANCE

A. Intercontinental Bombers

1. Technical aspects

4.1 INTRODUCTION TO FLIGHT (2nd Edition), J. D. Anderson (McGraw-Hill, New York, 1985). Technical description of various aspects of flight.

4.2 "The Science of Flight," P. P. Wegener, Am. Scientist 74, 268-77 (1986). Review of historical development in flight capabilities.

2. Policy aspects

4.3 WILD BLUE YONDER: MONEY, POLITICS, AND THE B-1 BOMBER, N. Kotz (Pantheon, New York, 1988). A journalistic expose.

B. Intercontinental Ballistic Missiles

1. Technical aspects

4.4 IGNITION: AN INFORMAL HISTORY OF LIQUID ROCKET PROPELLANTS, J. Clarke (Rutgers University, New Brunswick, NJ, 1972). All you ever wanted to know, and more, about liquid propellants for rockets and the history of their development.

4.5 GUIDED WEAPONS CONTROL SYSTEMS, 2nd edition, P. Garnell (Pergamon-Brassey's, New York, 1985).

C. SLBM Submarines

1. Technical aspects

4.6 STRATEGIC ANTISUBMARINE WARFARE AND NAVAL STRATEGY, T. Stefanick (Lexington Books, Lexington, MA, 1987).

4.7 "The Nonacoustic Detection of Submarines," T. Stefanick, Sci. Am. 258 (3), 41-7 (March 1988).

4.8 "Smart Weapons in Naval Warfare," P. F. Walker, Sci. Am. 248 (5), 53-61 (May 1983).

2. Political aspects

4.9 TRIDENT, D. D. Dalgleish and L. Schweikart (Southern Illinois University, Carbondale, 1984). A history of the Trident submarine system.

4.10 RUNNING CRITICAL: THE SILENT WAR, RICKOVER AND GENERAL DYNAMICS, P. Taylor (Harper and Row, New York, 1986). How the SS-N-688 U.S. attack submarines came to be fast, but at the sacrifice of some other capabilities, such as the operating depth.

4.11 NAVAL STRATEGY AND NATIONAL SECURITY: AN INTERNATIONAL SECURITY READER, S. E. Miller and S. Van Evera (Princeton University, Princeton, NJ, 1988). Has essays on stopping the sea-based counterforce threat, nuclear war at sea, etc.

D. Nuclear Deterrence and Stability

1. Quantitative aspects

4.12 "Resource Paper on the U.S. Nuclear Arsenal," Wm. M. Arkin, T. Cochran and M. M. Hoenig, Bull. At. Scientists 40 (7), 1s-15s (August/September 1984).

4.13 NUCLEAR WEAPONS DATABOOK, VOL. II. US NUCLEAR WEAPONS PRODUCTIONS COMPLEX, T. B. Cochran, Wm. M. Arkin and M. Hoenig (Ballinger, Cambridge, MA, 1984).
4.14 NUCLEAR WEAPONS DATABOOK, VOL. III. US NUCLEAR WARHEAD FACILITY PROFILES, T. B. Cochran, Wm. A. Arkin, R. S. Norris and M. Hoenig (Cambridge, MA: Ballinger Publishing Co., 1987).
4.15 "The Uncertainties of a Preemptive Nuclear Attack," M. Bunn and K. Tsipis, Sci. Am. 249 (5), 38-47 (November 1983).

2. Political aspects

4.16 NUCLEAR DETERRENCE: ETHICS AND STRATEGY, R. Hardin, J. J. Mearsheimer, G. Dworkin and R. E. Goodin (eds) (University of Chicago, Chicago, 1985). Good discussions of nuclear policies, just wars, MAD and its morality, etc.
4.17 THE SOVIET UNION AND THE ARMS RACE (2nd Edition), D. Holloway (Yale University, New Haven, CT, 1984). Good review of Soviet R&D, industry, and bureaucracy.

V. ALTERNATIVES TO NUCLEAR DETERRENCE

A. Civil (Passive) Defense

5.1 CIVIL DEFENSE: A CHOICE OF DISASTERS, J. Dowling and E. M. Harrell (eds) (American Institute of Physics, New York, 1987). Analysis of the potential and limits of civil defense.
5.2 "On Keeping them Down; or Why Do Recovery Models Recover so fast?" M. Kennedy and R. N. Lewis, in STRATEGIC NUCLEAR TARGETTING, D. Ball and J. Richelson (eds), (Cornell University Press, Ithaca, NY, 1986), pp. 194-208. Why are the recovery periods calculated for post-nuclear-war situations generally so optimistic?

B. ABM (Active) Defense

1. Technical aspects of ABM

5.3 "Phased-Array Radars," E. Brookner, Sci. Am. 252 (2), 94-103 (February 1985).
5.4 BALLISTIC MISSILE DEFENSE, A. B. Carter and D. N. Schwartz (eds) (Brookings, Washington, D.C., 1984). A review of strategic defense, largely predating the start of SDI.

3. Technology of strategic defense

5.5 "Space-based Ballistic-Missile Defense," H. A. Bethe, R. L. Garwin, K. Gottfried and H. W. Kendall, Sci. Am. 251 (4), 37-47 (October 1984). Descriptions of some SDI technologies.
5.6 "Is SDI Technically Feasible," H. Brown, For. Affairs 64, 435-545 (1986). A projection of future progress.
5.7 DIRECTED ENERGY MISSILE DEFENSE IN SPACE, A background paper prepared by A. B. Carter for the U.S. Congress, Office of Technology Assessment (U.S. Government Printing Office, Washington, D.C., April 1984). A good early evaluation of the potential of directed-energy weapons technology.
5.8 THE REAGAN STRATEGIC DEFENSE INITIATIVE: A TECHNICAL, POLITICAL AND ARMS CONTROL ASSESSMENT, S. D. Drell, P. J.

Farley and D. Holloway (Ballinger, Cambridge, MA, 1985). Essays on the technologies, politics, and arms control of strategic defense.

5.9 "Cost of Space-Based Laser Ballistic Missile Defense," G. Field and D. Spergel, Science 231, 1387-93 (1986). Evaluates the potential cost of space-based laser mirrors, and extrapolates from there to estimate the cost of laser-based strategic defense.

5.10 "How many orbiting lasers for boost-phase intercept?" R. L. Garwin, Nature 315, 286-90 (1985).

5.11 "Antisatellite Weapons," R. L. Garwin, K. Gottfried and D. L. Hafner, Sci. Am. 250 (6), 27-37 (June 1984). Descriptions of some ASAT technologies.

5.12 "Physics of a ballistic missile defense: The chemical laser boost-phase defense," C. L. Grabbe, Am. J. Phys. 56, 32-6 (1988). Technical estimates.

5.13 HOW TO MAKE NUCLEAR WEAPONS OBSOLETE, R. Jastrow (Little, Brown, Boston, 1985). R. Garwin wrote a review of this book in Phys. Today 38 (12), 75-8 (December 1985). This was followed by a spate of Letters to the Editor of Phys. Today; that debate showed that varying technical assumptions can lead to divisive technical debates in this area, for example concerning countermeasures against strategic defense, such as coating missiles--see e.g. Phys. Today 39 (1), 9-11 (January 1986), ibid (3), 9-15, 142-8 (March 1986), and ibid (11), 130-2 (November 1986).

5.14 "The Development of Software for Ballistic Missile Defense," H. Lin, Sci. Am. 253 (6), 46-53 (December 1985). A skeptic about the potential for developing computer software for SDI.

5.15 WEAPONS IN SPACE, F. A. Long, D. Hafner and J. Boutwell (eds) (W. W. Norton, New York, 1986). Some of the articles discuss the SDI technologies.

5.16 ANTI-SATELLITE WEAPONS, COUNTERMEASURES, AND ARMS CONTROL, U.S. Congress, Office of Technology Assessment (U.S. Government Printing Office, Washington, D.C., September 1985).

5.17 BALLISTIC MISSILE DEFENSE TECHNOLOGIES, U.S. Congress, Office of Technology Assessment (U.S. Government Printing Office, Washington, D.C., 1985).

5.18 SDI, TECHNOLOGY, SURVIVABILITY, AND SOFTWARE, U.S. Congress, Office of Technology Assessment (U.S. Government Printing Office, Washington, D.C., 1988).

5.19 "Software Aspects of Strategic Defense Systems," D. L. Parnas, Am. Scientist 73, 432-40 (1985). See also Letters to the Editor, ibid. 74, 12-5 (1986), and ibid 74, 222-3 (1986). The most vocal expert skeptic of SDI software.

5.20 "Strategic Defense and Directed-Energy Weapons," C. Kumar N. Patel and N. Bloembergen, Sci. Am. 257 (3), 39-45 (September 1987). An abstraction from the APS DEW study: REPORT TO THE AMERICAN PHYSICAL SOCIETY OF THE STUDY GROUP ON SCIENCE AND TECHNOLOGY OF DIRECTED ENERGY WEAPONS, DEW Study Group, Rev. Mod. Phys. 59 (3), Part II, S1-S202 (1987). A very technical report.

5.21 DIRECTED-ENERGY WEAPONS AND STRATEGIC DEFENSE: A PRIMER, D. Schroeer (International Inst. Strat. Studies, Adelphi Paper #221, London, 1987). A review and projections of future progress.

5.22 "Angle constraint for nuclear-pumped X-ray laser weapons," E. Walbridge, Nature 310, 180-2 (1984).

5.23 "Pork Bellies and SDI," P. D. Zimmerman, For. Policy (No. 63), 76-87 (Summer 1986). How uncertainties in defensive capabilities are judged assymmetrically by defender and attacker.

4. Stability calculations

5.24 "Transition to Deterrence Based on Strategic Defense," P. L. Chrzanowski, Energy and Tech. Rev. (of the Lawrence Livermore National Laboratory) (January/February 1987), pp. 31-45. Calculations involving deterrence diagrams and a crisis stability index. See also his "Strategic Defense and Crisis Stability," Lawrence Livermore Nat. Lab. Doc. UCID-20699 (December 1985).

5.25 "A Model of Defense-protected Build-down," R. Radner, in STRATEGIC DEFENSES AND ARMS CONTROL, A. M. Weinberg and J. N. Barkenbus (eds) (Paragon House, New York, 1988), pp. 111-142. Calculations of how to maintain strategic stability while building strategic defenses.

5.26 "Stabilizing Star Wars," A. M. Weinberg and J. Barkenbus, For. Policy (No. 54), 164-70 (Spring 1984). How to build defenses while maintaining strategic stability.

5.27 "Strategic defences and first-strike stability," D. Wilkening, K. Watman, M. Kennedy and R. Darilek, Survival 29 (2), 137-65 (March/April 1987). Defines a first-strike incentive parameter, then examines deterrence diagrams in terms of that parameter.

Politics of SDI

5.28 FISCAL AND ECONOMIC IMPLICATIONS OF STRATEGIC DEFENSES, B. M. Blechman and V. A. Utgoff (Westview, Boulder, CO, 1986). See also their article "The Macroeconomics of Strategic Defenses," Intl. Security 11 (3), 33-70 (Winter 1986-7).

5.29 THE SDI CHALLENGE TO EUROPE, I. H. Daalder (Ballinger, Cambridge, MA, 1987). Mostly about policy implications.

5.30 MISSILE DEFENSE IN THE 1990S, J. Gardner, E. Gerry, R. Jastrow, W. Nierenberg and F. Seitz, (George C. Marshall Inst., Washington, D.C., 2nd Edition, February 1987). Description of a missile defense for early deployment, and arguments for such an early deployment.

5.31 PERSPECTIVES ON STRATEGIC DEFENSE, S. W. Guerrier and W. C. Thompson (Westview, Boulder, CO, 1987). Nice historical introduction, but mostly about policy implications.

5.32 "Soviet Strategic Programmes and the US SDI," S. M. Meyer, Survival 27 (6), 274-92 (November/December 1985).

5.33 STAR WARS: THE ECONOMIC FALLOUT, R. Nimroody for the Council on Economic Priorities (Ballinger, Cambridge, MA, 1987). The economics of strategic defense as seen by skeptics.

5.34 TACTICAL BALLISTIC MISSILE DEFENCE IN EUROPE: FEASIBLE, AFFORDABLE, DESIRABLE? M. ter Borg and W. Smit (Amsterdam: Free University, 1987). See for example "Technical Problems of ATBM Defences," J. Altmann, pp. 48-63.

C. Chemical and Biological Warfare

5.35 "Yellow Rain," T. D. Seeley, J. W. Nowicke, M. Meselson, G. Guillemin and P. Akratanakul, Sci. Am. 253 (3), 128-37 (September 1985) Were chemical weapons used in Southeast Asia by clients of the Soviet Union?

5.36 CHEMICAL WARFARE, E. M. Spiers (University of Illinois, Champaign, 1986). History, current capabilities, negotiations.

D. Tactical Nuclear War

1. The neutron bomb

5.37 CRUISE, PERSHING, AND SS-20: THE SEARCH FOR CONSENSUS: NUCLEAR WEAPONS IN EUROPE, J. Cartwright and J. Critchley (Brassey's, London, 1985).

5.38 "Third-Generation Nuclear Weapons," T. B. Taylor, Sci. Am. 256 (4), 30-9 (April 1987).

2. Tactical nuclear war in Europe

5.39 STRATEGIC NUCLEAR TARGETTING, D. Ball and J. Richelson (eds) (Cornell University, Ithaca, NY, 1986). Includes descriptions of French and British nuclear targetting strategies.

5.40 "Ballistic Missiles in the Third World," Aaron Karp, Int. Security 9 (3), 166-95 (Winter 1984-85), and his "The frantic Third World quest for ballistic missiles," Bull. At. Scientist 44 (5), 14-20 (June 1988).

5.41 "The Strategic Nuclear Forces of Britain and France," J. Prados, J. S. Wit and M. J. Zagurek, Jr., Sci. Am. 255 (2), 33-41 (August 1986).

5.42 SURVEILLANCE AND TARGET ACQUISITION SYSTEMS, A. L. Rodgers, et al. (Brassey's, London, 1983). How good are advanced-technology weapons?

5.43 TECHNOLOGIES FOR NATO'S FOLLOW-ON FORCES ATTACK CONCEPT: A SPECIAL REPORT OF OTA'S ASSESSMENT ON IMPROVING NATO'S DEFENSE RESPONSES, U.S. Congress, Office of Technology Assessment (U.S. Government Printing Office, Washington, D.C., 1986).

5.44 "Civilian Casualties from Counterforce Attacks," F. N. von Hippel, B. G. Levi, T. A. Postol and Wm. H. Daugherty, Sci. Am. 259 (3), 36042 (Sept. 1988).

3. Command, control, communications, and intelligence

5.45 STRATEGIC COMMAND AND CONTROL, B. Blair (Brookings, Washington, D.C., 1985).

5.46 THE COMMAND AND CONTROL OF NUCLEAR FORCES, P. Bracken (Yale University, New Haven, CT, 1983).

5.47 "The Command and Control of Nuclear War," A. B. Carter, Sci. Am. 252 (1), 32-9 (January 1985).

5.48 MANAGING NUCLEAR OPERATIONS, A. B. Carter, J. D. Steinbruner and C. A. Zraket (eds) (Brookings, Washington, D.C., 1987). Includes a technical section on warning assessment sensors by J. C. Toomay, a discussion of communications technologies, etc. See also "Strategic Command, Control, Communications, and Intelligence," C. A. Zraket, Science 224, 1306-11 (1984).

5.49 THE EFFECTS OF RADIATION ON ELECTRONIC SYSTEMS, G. C. Messenger and M. S. Ash (Van Nostrand Reinhold, New York, 1986).

5.50 "Launch under Attack," J. Steinbruner, Sci. Am. 250 (1), 37-47 (January 1984).

VI. ARMS CONTROL AND DISARMAMENT

A. The Technological Imperative

6.1 "The Fundamental Physical Limits of Computing," C. H. Bennett and R. Landauer, Sci. Am. 253 (1), 48-71 (July 1985). How much further can computers improve?

6.2 INNOVATION AND THE ARMS RACE: HOW THE UNITED STATES AND THE SOVIET UNION DEVELOP NEW MILITARY TECHNOLOGIES, M. Evangelista (Cornell University, Ithaca, NY, 1988).

6.3 Special issue on "Materials for Economic Growth," Sci. Am. 255 (4) (October 1986). Describes various advanced materials and their uses.

6.4 "Quantifying Technological Imperatives," D. Schroeer, in REASSESSING ARMS CONTROL, D. Carlton and C. Schaerf (eds) (MacMillan, London, 1985), pp. 60-71. Is it possible to project the future growth of technologies?

6.5 REVIEW OF US MILITARY RESEARCH AND DEVELOPMENT, K. Tsipis and P. Janeway (eds), (Pergamon-Brassey's, McLean, VA, 1984). A survey of the status of various military technologies.

B. Nuclear Proliferation

3. Nuclear proliferation

6.6 "Pakistan's bomb-making capacity," D. Albright, Bull. At. Scientists 43 (5), 30-3 (June 1987). Numerical estimates of Pakistan's production capacities for nuclear material.

6.7 STRATEGIES FOR MANAGING NUCLEAR PROLIFERATION, D. L. Brito, M. Intriligator and A. E. Wick (eds) (Lexington Books, Lexington, MA, 1983). See for example "The Economics of Producing Nuclear Weapons in Nth Countries," T. W. Graham, pp. 9-27, and "A Statistical-Risk Model for Forecasting Nuclear Proliferation," S. M. Meyer, pp. 223-241.

6.8 "Sweden's abortive nuclear weapons project," T. B. Johansson, Bull. At. Scientists 42 (3), 31-4 (March 1986). History of a Swedish A-bomb project.

6.9 London Sunday Times, October 5, 1986. "Revelations" by an Israeli nuclear technician Mordechai Vanunu about Israel's nuclear-weapons manufacturing.

6.10 THE DYNAMICS OF NUCLEAR PROLIFERATION, S. M. Meyer (University of Chicago, Chicago, 1984).

6.11 THE INTERNATIONAL ATOMIC ENERGY AGENCY AND WORLD NUCLEAR ORDER, L. Scheinman (Resources for the Future, Washington, D.C., 1987).

6.12 THE NEW NUCLEAR NATIONS, L. S. Spector (Carnegie Endowment for Intl. Peace, Washington, D.C., 1985). An update of the potential future proliferating nations.

6.13 A. Weinberg, M. Alonso, and J. N. Barkenbus (eds), THE NUCLEAR CONNECTION: A REASSESSMENT OF NUCLEAR POWER AND NUCLEAR PROLIFERATION (New York: Paragon House, 1985). Dedicated nuclear reactors may be much better for A-bomb projects.

C. Nuclear Test Ban Treaty

1. Technical aspects

6.14 "Seismic Verification of a Comprehensive Test Ban," W. J. Hannon, Science 227, 251-7 (1985). A review of the possibilities.
6.15 "Underground nuclear weapons testing," J. Horgan, IEEE Spectrum 23 (4), 32-43 (April 1986). Description of the testing process, discussions of simulations, and a record of recent testing.
6.16 "The Yields of Soviet Strategic Weapons," L. R. Sykes and D. M. Davis, Sci. Am. 256 (1), 29-37 (January 1987). Soviet nuclear tests probably have not violated the Nuclear Threshold Test Ban treaty.
6.17 "The Verification of a Comprehensive Nuclear Test Ban," L. R. Sykes and J. F. Evernden, Sci. Am. 247 (4), 47-55 (October 1982). An optimistic view.

2. Political aspects

6.18 UNDER THE CLOUD: THE DECADES OF NUCLEAR TESTING, R. L. Miller (The Free Press, New York, 1986).

D. Arms Limitations and SALT

1. Verification

6.19 DEEP BLACK: SPACE ESPIONAGE AND NATIONAL SECURITY, Wm. E. Burrows (Random House, New York, 1986). Popular discussion of satellite surveillance, some review of techical capabilities.
6.20 "Radar Images of the Earth from Space," C. Elachi, Sci. Am. 247 (6), 54-61 (December 1982). See also his PHYSICS AND TECHNIQUES OF REMOTE SENSING (Wiley, New York, 1987); and SPACEBORNE RADAR REMOTE SENSING: APPLICATIONS AND TECHNIQUES (IEEE, New York, 1988). How useful might this technology be for verification?
6.21 "Science and society test IX: Technical means of verification," D. Hafemeister, Am. J. Phys. 54, 693-99 (1986). Back of the envelope calculations.
6.22 "The Verification of Compliance with Arms-Control Agreements," D. Hafemeister, J. J. Romm and K. Tsipis, Sci. Am. 252 (3), 38-45 (March 1985). A review of the potential for verification.
6.23 SATELLITES FOR ARMS CONTROL AND CRISIS MONITORING, B. Jasani and T. Sakata (eds) (Oxford University, New York, 1987). What might the impact of public verification, as by the French SPOT satellite system?
6.24 VERIFICATION: HOW MUCH IS ENOUGH? A. S. Krass (Lexington, Books, Lexington, MA, 1986).
6.25 REMOTE SENSING AND IMAGE INTERPRETATION (2nd Edition), T. M. Lillesand and R. W. Kiefer (Wiley, New York, 1987). A textbook.
6.26 VERIFICATION AND ARMS CONTROL, Wm. C. Potter (ed.) (Lexington Books, Lexington, MA, 1985). Includes essays on cooperative measures of verification, and verifying limits on antisatellite weapons, bombers, and cruise missiles.
6.27 "The Keyhole Satellite Programme," J. Richelson, J. Strat. Studies 7 (2), 121-53 (June 1984). An excellent review of this "spy satellite" program.
6.28 TECHNIQUES FOR IMAGE PROCESSING AND CLASSIFICATION IN REMOTE SENSING, R. A. Schowengerdt (Academic Press, Orlando, FL, 1983). A textbook.

6.29 "Space Based Radar," G. N. Tsandoulas, Science 237, 257-62 (1987). Might this be useful for verification?

6.30 ARMS CONTROL VERIFICATION: THE TECHNOLOGIES THAT MAKE IT POSSIBLE, K. Tsipis, D. W. Hafemeister and P. Janeway (eds), (Pergamon-Brassey's, McLean, VA, 1986). A review of verification technologies.

2. Politics

6.31 THE ABM TREATY AND WESTERN SECURITY, Wm. J. Durch (Ballinger, Cambridge, MA, 1987). A political review.

3. Nuclear Freeze, INF, etc.

6.32 THE NUCLEAR FREEZE DEBATE: ARMS CONTROL ISSUES FOR THE 1980S, P. M. Cole and Wm. J. Taylor, Jr. (eds) (Westview, Boulder, CO, 1983). A collection of essays on the nuclear freeze.

6.33 "A Bilateral Nuclear-Weapon Freeze," R. Forsberg, Sci. Am. 247 (3), 52-61 (November 1982). Enthusiasts' perspective.

6.34 "'No First Use' of Nuclear Weapons," K. Gottfried, H. W. Kendall and J. M. Lee, Sci. Am. 250 (3), 33-41 (March 1984). Enthusiasts' perspective.

6.35 THE NUCLEAR WEAPONS FREEZE AND ARMS CONTROL, S. E. Miller (ed.) (Ballinger, Cambridge, MA, 1984). A collection of essays on the nuclear freeze.

6.36 "Stopping the Production of Fissile Materials for Weapons," F. von Hippel, D. H. Albright and B. G. Levi, Sci. Am. 253 (3), 40-47 (September 1985). What are the possibilities of verification?

E. Disarmament

6.37 HAWKS, DOVES, AND OWLS: AN AGENDA FOR AVOIDING NUCLEAR WAR, G. T. Allison, A. Carnesale and J. S. Nye, Jr. (eds), (McGraw-Hill, New York, 1985). A set of essays on arms control.

6.38 SUPERPOWER GAMES: APPLYING GAME THEORY TO SUPERPOWER CONFLICT, S. J. Brams, (Yale University, New Haven, CT, 1985).

6.39 MILITARY APPLICATIONS OF MODELING: SELECTED CASE STUDIES, F. P. Hoeber (Gordon & Breach, New York, 1981). Some nice examples, e.g. costing exercises, penetration of bombers, etc.

6.40 THE ARMS RACE: ECONOMIC AND SOCIAL CONSEQUENCES, H. G. Mosley (Lexington Books, Lexington, MA, 1985).

6.41 THE NUCLEAR DILEMMA AND THE JUST WAR TRADITION, Wm. V. O'Brien and J. Langan (eds) (Lexington Books, Lexington, MA, 1986). A conference prompted by the debate about the U.S. Catholic bishops' pastoral letter about nuclear war.

6.42 "Chaos--a model for the outbreak of war," A. M. Saperstein, Nature 309, 303-5 (1984). Is the outbreak of war similar to thw switch in physical systems from stability to chaos?

6.43 MATHEMATICS OF CONFLICT, M. Shubik (ed.) (Elsevier Science, Amsterdam, 1983). See for example "Lanchester Attrition Processes and Theater-Level Combat Models," A. F. Karr, pp. 89-126.

APPENDIX B

A CHRONOLOGY OF THE NUCLEAR ARMS RACE

D.W. Hafemeister
Physics Department
California Polytechnic State University

1919

--June; Rutherford creates oxygen from nitrogen; "Talk softly please. I have been engaged in experiments which suggest that the atom can be artificially disintegrated. If it is true, it is of far greater importance than a war." ($^4He + ^{14}N > ^1H + ^{17}O$)

1920

--Rutherford speculates on the existence of the neutron at the Royal Society.

1931

--November; Urey discovers deuterium.

1932

--February; Chadwick discovers the neutron;
($^4He + ^9Be > 3\ ^4He + ^1n$)

1933

--January 28; Hitler becomes Chancellor of Germany.
--March 4; Roosevelt becomes President of the U.S.
--April; Born, Courant, Franck and many other scientists are compelled to leave the University of Gottingen because of their "Jewish physics;" The "Aryan physics" of Stark and Lenard is not widely accepted by physicists.
--October; Szilard recollects that "it occurred to me in October, 1933 that a chain reaction might be set up if an element could be found that would emit two neutrons when it swallowed one neutron." This idea became a classified British patent in 1935 before the fission of U was discovered.

1934

--Artificial radioactivity discovered by Curie/Joliet (bombardment with alpha particles) and by Fermi (bombardment with neutrons).

1938

--December; Fermi receives the Nobel prize for the discovery of the transuranic elements (actually fission of uranium) and departs for the "new world."
--December 22; Hahn and Strassman (later Meitner and Frisch) conclude that the identification of barium implies that the uranium nucleus has been been fissioned by neutrons.

1939

--January to May; Many experiments on uranium fission (a brief list is at the end of this appendix);
--April 29; Conference in Berlin to consider a German nuclear reactor and bomb.
--August 2; Szilard and Teller obtain a letter from Einstein on the possibility of a uranium weapon; Roosevelt receives the letter on October 11, 1939 from Sachs.

--September 1; Hitler invades Poland.

1940

--June 3; Hartek fails to observe neutron multiplication with his reactor in Hamburg (185 kg of natural uranium oxide, 15 tonnes of CO_2 ice).

1941

--January; Based on experiments with the natural uranium (300 kg) reactor in the "virus house" in Berlin, the Germans reject graphite as a moderator since neutrons did not diffuse adequately through the graphite. This mistake caused by impurities in the graphite forced the German program to rely on heavy water from Vermork, Norway.
--July; British "Maud" Committee reports that a weapon could be made with 10 kg of U-235; U.S. National Academy of Sciences endorses bomb program.
--August; Hamburg group begins construction of the ultracentrifuge to obtain U-235; some centrifuges explode in April, 1942, but they did attain 7% enrichment levels by March, 1943.
--October; Bohr and Heisenberg indirectly discuss termination of uranium research.
--December 6; Roosevelt directs substantial financial and technical resources to construct the uranium weapon.
--December 7; Japan attacks Pearl Harbor.

1942

--May; Heisenberg and Dopel observe the first multiplication of neutrons (13%) in Leipzig using 570 kg of uranium and 140 kg of heavy ice.
--December 2; First nuclear chain reaction at Chicago's Stagg Field by Fermi.

1943

--February 28; Vermork heavy water factory destroyed by the allies; a similar attack of November 19, 1942 had failed.
--March 15; Oppenheimer moves to Los Alamos.
--March; Seaborg suggests Pu weapons might be jeopardized by ^{240}Pu if it is a spontaneous neutron emitter.
--Resumption of Soviet nuclear experiments.

1944

--August 26; Bohr presents his memorandum on international control of nuclear weapons to Roosevelt (and Churchill).
--November; First batch of spent fuel obtained from Hanford reactors.
--November; Goudsmit's Alsos mission obtains documents in Strasbourg from the German atomic bomb project which imply that their rate of progress had diminished.

1945

--January; First Pu reprocessing production run at Hanford.
--January 20; First U-235 separated at the K25 gaseous diffusion plant, Oak Ridge.
--April 25; U.N. Charter signed by 50 nations in San Francisco.
--May 4; End of war in Europe.
--June 11; The Franck Report on the demonstration of the bomb and its international control was sent to the Secretary of War.
--July 16; U.S. explodes first atomic bomb, Trinity, at Alamogordo, N.M. Electronics equipment shielded by Fermi to avoid EMP pulse.

--August 6, 9; Atomic bombs dropped by U.S. on Hiroshima ("thin man," uranium, 9000 pounds, 10 feet by 28 inches in diameter) and Nagasaki ("fat man," plutonium, 11 feet by 5 feet in diameter).
--August 15; End of war in the Pacific.

1946
--June 14; Baruch presents the Acheson-Lilienthal plan to internationalize the atom to the U.N. "We are here to make a choice between the quick and the dead."
--June 30; First subsurface detonation by U.S. at Bakini.
--July; Demonstrations in Times Square, New York, against nuclear testing.
--December 31; AEC takes over nuclear weapons program from the Army.

1948
--April, May; U.S. atomic tests, Eniwetok Atoll.
--November 4, 1948; U.N. General Assembly adopts, 40 to 6, U.S. plan for international atomic control; U.S.S.R. is opposed.

1949
--April 4; NATO established.
--August 29; First Soviet atomic detonation, in the Ustyurt desert.
--October 30; General Advisory Committee of the AEC recommends that the more powerful atomic bombs should be built rather than the hydrogen bomb.

1950
--January 27; Fuchs confesses that he transmitted atomic secrets to the Soviets.
--January 31; Truman announces the decision to proceed with the H bomb.
--March; Worldwide peace offensive to "ban the bomb" (the Stockholm Appeal) signed by more than 500 million people.
--June 25, 1950 - July, 1953; North Korean army crosses the 38th parallel.

1952
--January; U.N. Security Council establishes Disarmament Committee.
--June; Lawrence-Livermore Laboratory established.
--October 3; First British atomic detonation, Monte Bello Islands, Australia.
--October 31; U.S. explodes first fusion device, Mike, of 10.4 Mt at Eniwetok (liquid deuterium, not deliverable).

1953
--March; U.S. above-ground tests start in Nevada.
--May; Representative Springfellow of Utah asks AEC to stop tests at the Nevada Test Site because of public alarm over fallout.
--August 12; First Soviet fusion device exploded on a tower in Siberia (relatively low yield, probably not deliverable, but used LiD).
--Fall; India and Australia propose a total ban on nuclear weapons to the U.N.
--December 8; "Atoms for Peace" speech by Eisenhower at U.N.

1954
--January 21; First nuclear-powered submarine, USS Nautilus, launched.
--March; The "Bravo" H-bomb test; Marshall Islanders affected by fallout.
--April; British Parliament petitions Churchill, Eisenhower, and Malenkov meet to control nuclear weapons.
--April 12 to May 6; Oppenheimer Hearings deny him access to nuclear secrets.

--August 30; Atomic Energy Act of 1954 emphasizes peaceful Uses of Atomic Energy.

1955
--January-December; Reports of increasing nuclear fallout.
--May 6; W. Germany joins NATO.
--May 14; Warsaw Pact Organization established.
--August 8-20; First International Conference on Peaceful uses of atomic energy, Geneva.
--November 23; First relatively high yield deliverable Soviet H bomb.

1956
--U.S. National National Academy of Sciences panel finds genetic effects of fallout from nuclear testing to be slight compared with those from natural radiation background; Stevenson makes fallout an issue in the U.S. Presidential campaign.

1957
--May 15; First British H bomb exploded at Christmas Island.
--July 6-11; First Pugwash Conference advocates test ban; Soviet scientists attend.
--August; AEC inaugurates Plowshare Program for peaceful uses of nuclear explosions.
--September 19; First underground test, Rainier, 1.7 kilotons.
--October 1; IAEA inaugurated in Vienna.
--October 4; First artificial Earth satellite, Sputnik I, put into orbit by the U.S.S.R.
--December 2; Shippingport electric power reactor reaches full power of 60 MW.

1958
--January 1; Euratom (European Atomic Energy Community) established.
--January 31; Explorer I, the first U.S. satellite.
--November, 1958 to September, 1961; U.S., U.K., and U.S.S.R. observe an informal moratorium on nuclear tests.

1959
--June 12; Results on Panel on Seismic Improvement made public; Geneva system assessed effective only above 20 kt; bold research program in seismology recommended.
--September 2; Vela Uniform Seismic Project established by U.S. Department of Defense.
--November 24; U.S. and U.S.S.R. sign a memorandum of cooperation for the utilization of atomic energy.

1960
--February 13; First nuclear test by France, Sahara desert.
--May 1; U.S. spy plane (U-2) shot down over the U.S.S.R.
--July 26; U.S.S.R. sugggets 3 on-site inspections/year as part of a test ban.
--November 15; First Polaris missile launched from a sub.

1961
--January 17; Eisenhower's farewell address on the military-industrial complex.
--February 1; U.S. launches Minuteman I missile.
--March 31; Pravda suggests use of H bombs to obtain fresh water from Soviet glaciers.
--April 12; Gagarin becomes the first cosmonaut in orbit.

--May 29; U.S. and U.K. agree to draft Comprehensive Test Ban Treaty (CTBT) with 12 on-site inspections/year.
--August; Installation by U.S. of first seismic stations of a network in 60 countries.
--September 1; U.S.S.R. resumes nuclear tests, including a 58 megaton explosion on Oct. 31.
--September 15; U.S. resumes nuclear tests.

1962
--February 20; Glenn becomes the first U.S. astronaut in orbit.
--July 6; Sedan excavation experiment by Plowshare; 10.4 Mt of earth displaced.
--July 8; Electromagnetic pulse (EMP) from the high altitude (400 km above Johnston Island) "Fishbowl" test (1.4 Mt) turns off 300 streetlights on Oahu, Hawaii (1200 km away).
--September 3-7; Tenth Pugwash Conference proposes "black box" for CTB verification.
--October; U.S. reduces requirement to 8-10 inspections/year.
--October 23 to November 20; Cuban missile crisis and blockade.
--December 19; Khrushchev accepts 2-3 inspections/year and 3 unmanned seismographic stations (black boxes) in the U.S.S.R.

1963
--January; Secret test-ban talks between U.S. and U.S.S.R.
--February 19; U.S. accepts 7 inspections/year provided any mysterious event can be challenged.
--April; Khrushchev withdraws offer of 3 inspections/year.
--June 20; US/USSR sign "hot line" agreement.
--August 5; Limited Test Ban Treaty signed in Moscow by U.S., U.S.S.R. and U.K.; bans nuclear explosions in atmosphere, in space, and underwater.

1964
--August 2; "Gulf of Tonkin" resolution allows U.S. troops to be sent to Vietnam.
--October 16; China (PRC) explodes first nuclear bomb.
--October 22; U.S. muffling experiment Salmon in a salt dome in Mississippi.

1965
--June; Large Seismic Array (LASA) opens in Montana to detect nuclear explosions.

1966
--January 17; U.S. B-52 bomber crashes near Palomares, Spain with 4 unarmed H bombs.
--September 24; First French H bomb, Tuamoto Islands.

1967
--January 27; Outer Space and Celestial Bodies Treaty bans nuclear weapons being placed in orbit.
--February 14; Treaty for Prohibition of Nuclear Weapons in Latin America signed in Mexico City (Tlatelolco); all the Latin nations must ratify for the treaty to enter into force.
--June 17; First Chinese H bomb exploded in the atmosphere.
--June 23-25; Johnson and Kosygin hold talks in Glassboro, N.J.
--December 10; Gasbuggy Plowshare explosion for gas production, New Mexico.

1968
--July 1; The Non-Proliferation Treaty on nuclear weapons (NPT) opened for signature.
--August; Soviet GALOSH ABM system deployed around Moscow.

1969
--July 20; U.S. Apollo 11 lands on moon.
--November 3; Committee on Disarmament reports to U.N. on comprehensive test ban; proposes worldwide exchange of seismological data.
--November to December; Preliminary SALT talks in Helsinki.

1970
--January; Nixon administration cuts Plowshare budget.
--March 2-6; Talks by 21 nations at the IAAEA on peaceful uses of nuclear explosions.
--March 5; NPT enters into force; 50 nations ratify the NPT by 1970, over 100 by 1980.
--November 30; Atlantic-Pacific Interoceanic Canal Study Commission rejects use of nuclear explosives.

1971
--March 30: U.S. deploys Poseidon SLBM.
--June 15; Izvestia mentions three Russian nuclear charges exploded in oil fields to increase productivity.

1972
--May 26; SALT I Treaties limiting ABM and offensive missiles signed by Nixon and Brezhnev in Moscow.
--June; U.N. Environmental Conference in Stockholm votes 48 to 2 to halt all testing of nuclear weapons; France and China against, U.S. and U.K. abstain, U.S.S.R absent.
--October; First U.S. detection of flight test of Soviet SS-18 ICBM.
--November; SALT II negotiations begin.

1973
--May 30; Protocol establishing regulations for the Standing Consultative Commission.
--June 22; World Court issues injunction to France not to carry out Mururoa tests.
--October 17, 1973 to March 17, 1974; OPEC embargos petroleum to U.S. and the Nethererlands.

1974
--May 18; India sets off a low-yield device (10-15 kt) under Rajasthan desert, expanding the nuclear club beyond the "big five" of World War II.
--November 24; U.S./U.S.S.R. agree to limit the number of strategic launchers (2400) and MIRV launchers (1320).

1975
--January 19; AEC reorganized into the NRC (regulatory) and ERDA (developmental, later becomes DOE on August 4, 1977).

1976
--January 29; The order for the South Korean reprocessing plant from France is cancelled. (Pakistan plant cancelled in 1978)
--October 28; Ford postpones reprocessing of spent fuel to obtain commercial Pu.

1977

--April 7; Carter postpones indefinitely reprocessing of commercial spent fuel, slows progress on the Clinch River Breeder Reactor, and calls for an International Fuel Cycle Evaluation (INFCE) to investigate ways to make the nuclear cycle more resistant to proliferation.
--October 19, 1977 to February 26, 1980; Over 40 nations prepare a report at INFCE.

1978

--January 11, The Guidelines of the Nuclear Suppliers (London meetings) are forwarded to the IAEA.
--March 10, The Nuclear Nonproliferation Act of 1978 (NNPA) was signed into law.
--April 4; Camp David Accord signed by Egypt and Israel in Wshington.
--April 27; Carter issues an Executive Order to export low-enriched uranium fuel to India; this is the first override of the NRC as allowed for in the NNPA; the Congress does not veto this action.
--July 13; Euratom ends the the embargo of nuclear fuel from the U.S. (required by NNPA) by agreeing to discuss renegotiation of the US-Euratom agreement on the issue of retransfer and reprocessing rights.

1979

--April, 6; U.S. cuts off economic and military aid to Pakistan after concluding that Pakistan is building an enrichment plant to produce weapons grade uranium.
--June 8; First Trident SLBM launched.
--June 18; The SALT II Treaty is signed in Vienna by Brezhnev and Carter.
--December 12; NATO adopts dual track with INF weapons, negotiations to restore INF balance with Soviets and to modernize with Pershing II and GLCM.
--December 26; U.S.S.R. invades Afganistan; SALT II treaty removed from consideration by the Senate.

1980

--July 15; Presidential Office of Science and Technology Policy reports that the light signals recorded by a VELA satellite over the South Atlantic on September 22, 1979 were probably not from a nuclear explosion.
--December; Tomahawk submarine-launched cruise missile flies out to sea and returns 600 miles to land; later it obtains an accuracy of about 5 meters on a flight from Pt. Mugu to Nevada.

1981

--April 12-14; Space shuttle Columbia flies 36 orbits.
--June 7, Israel destroys Iraq's Osirak reactor.
--November 13; Senate ratifies Protocol I of Tlatelolco Treaty; U.S. cannot store, deploy, or use nuclear weapons in Puerto Rico, Virgin Islands, or Guantanomo Naval Base.
--November 30; US tables Zero-Zero INF proposal in Geneva.

1982

--June 29; Strategic Arms Reduction Talks (START) begin in Geneva.
--July 16; Nitrze and Kuitsinsky take a walk in the woods, trying to cut INF weapons by 50%.
--1982; U.S.S.R. places about 300 SS-20 missiles on its western border.
--November; Reagan adopts the dense pack basing mode for the MX missile.

1983
--March 23; Reagan announces SDI program "to make nuclear weapons impotent and obsolete."
--April 6; Scowcroft Commission recommends the Midgetman (single-warhead, mobile missiles) as well as the deployment of 100 MX as bargaining chip.
--summer; The large phased-array radar under construction at Krasnoyarsk is discovered.
--November 23; Soviets leave INF negotiations as Pershing II and GLCM are deployed. START talks suspended on December 8.

1984
--January; The Strategic Defense Initiative Organization (SDIO) is created.
--January 23; Reagan's first report to Congress on "Soviet Noncompliance" is released amid some controversy.
--January 30; Soviet transmits aide memoire to State Department on U.S. noncompliance.
--June 10; The Homing Overlay Experiment (HOE) intercepts a missile in midcourse in a limited test of kinetic kill vehicles.

1985
--January 8; Shultz and Gromyko agree to begin negotiations on START, INF, and Defense and Space in Geneva. They begin on March 12.
--February 20; Nitze outlines his criteria for the deployment of SDI; reliability, survivability, and cost effectiveness at the margin.
--October 6; The broad interpretation of the ABM Treaty, which would allow tests of other physical principles in space, is encouraged by the Reagan Administration.
--October 14; Shultz states that "... that a broader interpretation of our authority is fully justified. This is, however, a moot point, our SDI research program has been structured and, as the President said, will continue to be conducted in accordance with a restrictive interpretation of the treaty's obligations."
--November 21, 1985. Reagan and Gorbachev agree in Geneva to focus on "the principle of 50% reductions" in strategic weapons and on the "idea of an interim INF agreement."

1986
--July; Natural Resources Defense Council sets up seismographs in the Soviet Union.
--September; First B-1B bomber squadron becomes operational.
--October 11-12; Reagan and Gorbachev agree to broad formula to reduce strategic weapons by 50%, as well as a tentative goal of zero amid some confusion.
--November 28; The 131st U.S. B-52 bomber with ALCMs is deployed, exceeding the 1320 limit on MIRVed launchers, thus negating the SALT II Treaty. U.S. is first country to publicly renounce an arms control agreement.
--December; First 10 MXs become operational in Minuteman III silos in Wyoming.

1987
--February; Soviet unilateral moratorium on nuclear testing ends after 18 months.
--April 23; APS Directed Energy Weapons Study (DEWS) Group "estimates that even in the best of circumstances, a decade or more of intensive research would be required ... for an informed decision ...on DEWS."

--September 15; Shultz and Shevardnadze sign the Nuclear Risk Reduction Accord. Soviets present a list of permitted criteria for testing under the ABM treaty.
--December 8; Reagan and Gorbachev sign the INF Treaty.

1988
--May 27; Senate ratifies INF Treaty (93-5), the first ratification of an arms control treaty since 1972.
--June 28; Office of Technology Assessment testifies before the House Committee on Foreign Affairs that it disagrees with the Reagan Administration's charge of "likely violation of the TTBT."
--August/Sept.; Joint Verification Experiments on nuclear tests carried out by US and Soviets.

ACKNOWLEDGEMENT AND REFERENCES

In preparing this list of chronological data, we would like to acknowledge the lists and suggestions of others; in particular we would like to thank Bruce Bolt, Prentice Dean, Warren Donnelly, Dietrich Schroeer, Richard Scribner, and Spencer Weart.
Some references that we have found to be useful are as follows:

1. "Key Dates" in Arms Control Today, Arms Control Association, Washington, DC.
2. B. Bolt, Nuclear Explosions and Earthquakes, Freeman, San Francisco, 1976.
3. P. Dean, Energy History Chronology from World War II to the Present, U.S. Dept. of Energy, DOE/ES-0002, Washington, D.C., 1982.
4. W. Donnelly and D. Kramer, Nuclear Proliferation Factbook, Congressional Research Service, Library of Congress, 1980.
5. G. Duffy, Compliance and the Future of Arms Control, Ballinger, Cambridge, MA, 1988.
6. B. Goldschmidt, The Atomic Complex, Am. Nuc. Soc., La Grange Park, IL, 1982.
7. R. Hewlett and O. Anderson, A History of the U.S. Atomic Energy Commission; The New World: 1939-46, Penn. St. Univ. Press, University Park, PA, 1962. R. Hewlett and F. Duncan, Atomic Shield: 1947-52, Penn. St. Univ. Press, University Park, PA, 1969.
8. D. Irving, The German Atom Bomb, Simon and Schuster, NY, 1967.
9. D. Kevles, The Physicists, Vintage, N.Y., 1979.
10. H. Smyth, "Atomic Energy for Military Purposes," Rev. Mod. Phys. 17, 351 (1945).
11. H. York, Making Weapons, Talking Peace, Basic Books, NY, 1987.
12. A Chronology of United States Arms Control and Reduction Initiatives (1946-1987), USACDA, 1987.
13. Strategic Defense Initiative: A Chronology (1983-1988), USDOD, 1988.

APPENDIX C

DATA TABLES ON THE NUCLEAR BALANCE

Dietrich Schroeer
Department of Physics and Astronomy
University of North Carolina
Chapel Hill, NC 27599-3255

Table 1. The long-range and medium-range strategic bombing forces of the U.S.S.R. and the U.S.A. Most of the numbers are from the MILITARY BALANCE: 1987-1988 of the London International Institute for Strategic Studies, 1987. Some of the parameters come from the tables in the 1983 AIP Proceedings #104.

Bombers	Range	Top Speed	Bomb Payload	Number	EMT[1]
			UNITED STATES		
B-52G	9,900 km	0.95 Mach[2]	20 tonnes; 4x1 Mt +4 SRAMx170 kt +12 ALCMx200 kt	98	914 EMT
B-52H	16,000 km	0.95 Mach[2]	20 tonnes; non-nuclear	(69)	
B-52H	16,000 km	0.95 Mach[2]	20 tonnes; 4x1 Mt +4 SRAMx170 kt	56	293 EMT
B-52H	12,000 km	0.95 Mach[2]	20 tonnes; 4x1 Mt +4 SRAMx170 kt +12 ALCMx200 kt	46	429 EMT
B-1B	12,000 km	1.25 Mach[3]	30 tonnes; 10x1 Mt or 20 SRAMx170 kt or 12 ALCMx200 kt[4]	54	421 EMT
FB-111A	4,700 km	2.5 Mach[5]	15 tonnes; 2x1 Mt +4 SRAMx170 kt	(56)	
Tanker		0.95 Mach		(615)	
			U.S. TOTAL	317[6]	2057 EMT[6]
			SOVIET UNION		
Tu-95ABC	12,800 km	0.9 Mach	18 tonnes; 3x1 Mt	80	240 EMT
Tu-95G	12,800 km	0.9 Mach	18 tonnes; 2 ALCMx1 Mt	20	40 EMT
Tu-95H	12,800 km	0.9 Mach	18 tonnes; 8 ALCMx250 kt	50+	159 EMT
Mya-4	11,200 km	0.94 Mach	9 tonnes; 2x1 Mt	40	80 EMT
Tu-22	6,200 km	1.4 Mach	5 tonnes; 1x1 Mt +1 ALCMx1 Mt	(165)	
Tu-26	11,000 km	1.92 Mach[3]	8 tonnes; 2x1 Mt +1 ALCMx1 Mt	(290)	
Tankers				(~150)	
			U.S.S.R. TOTAL	190[6]	519 EMT[6]

[1] EMT (equivalent megatons) is equal to $Y^{2/3}$, with Y in megatons. For the 170-kt SRAM the EMT=0.307 EMT, for the 200 kt ALCM the EMT=0.342.

[2] At high altitudes; the penetration speed at low altitude is 0.66 Mach.

[3] At high altitudes; the maximum low-altitude speed is around 0.9 Mach.

[4] Each of these payloads is about 1/2 of the maximum load it can carry of that particular weapon. See J. W. R. Taylor, JANE'S ALL THE WORLD'S AIRCRAFT 1987-88 (Jane's, New York, 1987). In calculating the EMT caried by the B-1B, a payload of 12 ALCM and 12 SRAM is assumed, for a total EMT of 7.8 EMT.

[5] At high altitudes.

[6] Counting only truly international bombers; i.e. excluding the FB-111A, the Tu-22 Blinder, and the Tu-26 Backfire.

Table 2. ICBM missiles presently deployed or being deployed. All data are from the MILITARY BALANCE: 1987-1988 of the London International Institute for Strategic Studies, 1987. Some of the weapons parameters come from the tables in the 1983 AIP Proceedings #104.

Missile	Stages	Launch	Range	Accuracy	Payload
		UNITED STATES			
Minuteman II	3	hot	11,300 km	370 m	0.7 tonnes
Minuteman III	3	hot	14,800 km	220 m	1.0 tonnes
Minuteman IIIA	3	hot	12,900 km	220 m	1.2 tonnes
Peacekeeper MX	3[1]	cold	11,000 km	100 m	3.2 tonnes
		SOVIET UNION			
SS-11 (MOD 1)	2[2]	hot	9,600 km	1400 m	1.0 tonnes
(MOD 2)	2[2]	hot	13,000 km	1100 m	1.1 tonnes
(MOD 3)	2[2]	hot	10,600 km	1100 m	1.1 tonnes
SS-13 (MOD 2)	3	hot	9,400 km	1800 m	0.6 tonnes
SS-17 (MOD 3)	2[2]	cold	10,000 km	400 m	3.0 tonnes
SS-18 (MOD 4)	2[2]	cold	11,000 km	250 m	7.6 tonnes
SS-19 (MOD 3)	2[2]	hot	10,000 km	300 m	3.4 tonnes
SS-24	3	n.a.[3]	10,000 km	200 m	3.6 tonnes
SS-25	3	n.a.[3]	10,500 km	200 m	0.7 tonnes

[1] The post-boost bus has a rocket fueled by stable liquid propellant.
[2] These rockets are liquid fueled.
[3] n.a. means that this column is not applicable to this weapon. In this case the launch of the mobile missile is not from a silo, so the terms hot and cold launch have no meaning.

Table 2 continued.

Missile	Warheads	# ICBMs	# RV	Mt	EMT[4]
	UNITED STATES				
Minuteman II	1 @ 1.5 Mt	450	450	675	590
Minuteman III	3 MIRV @ 170 kt	227	681	116	209
Minuteman IIIA	3 MIRV @ 335 kt	300	900	302	434
Peacekeeper MX	10 MIRV @ 300 kt	23	230	69	103
	U.S. TOTAL	1000	2261	1162	1336
	SOVIET UNION				
SS-11 (MOD 1)	1 @ 1 Mt	20	20	20	20
(MOD 2)	1 @ 1 Mt	360[6]	360	360	360
(MOD 3)	3 MRV @ ~300 kt[7]	60[6]	180	54	81
SS-13 (MOD 2)	1 @ 600 kt	60	60	36	43
SS-17 (MOD 2)	4 MIRV @ 500 kt	150	600	300	378
SS-18 (MOD 4)	10 MIRV @ 500 kt	308	3080	1540	1940
SS-19 (MOD 3)	6 MIRV @ 550 kt	360	2160	1188	1450
SS-24	8-10 MIRV @ 100 kt	( )[8]			
SS-25	1 @ 550 kt	~80	80	44	54
	U.S.S.R. TOTAL	1398	6540	3542	4326

[4] EMT (equivalent megatons) is equal to $Y^{2/3}$, with Y in megatons.

[5] Swedish International Peace Research Institute, WORLD ARMAMENTS AND DISARMAMENT 19S87-88 (Oxford University, New York, 1987).

[6] Reference 5. The MILITARY BALANCE says 420 for MOD 2 and MOD 3 combined.

[7] The MILITARY BALANCE says 100-300 kt; Ref. 5 says 250-350 kt.

[8] Being deployed to replace SS-11s.

Table 3. Listing of nuclear-tipped SLBMs carried on submarines. All numbers are from THE MILITARY BALANCE 1987-1988 by the London International Institute for Strategic Studies, 1987. Some of the weapons parameters come from the tables in the 1983 AIP Proceedings #104.

Missiles	Range	Payload (tonnes)	RV/SLBM	CEP	RVs	EMT[1]
			UNITED STATES			
Poseidon	4,000 km	1.5	10 MIRV @ 40 kt	450 m	256x10	299
Trident I	7,400 km	1.3	8 MIRV @ 100 kt	450 m	384x8	662
Trident II	11,000 km	?	14 MaRV @ 150 kt	220 m[2]	0	0
			U.S. TOTAL		5,632	961
			SOVIET UNION			
SS-N-6[3]	3,000 km	0.7	1 @ 1 Mt	1300 m	272x1	272
SS-N-8[3]	≥7,000 km	0.7	1 @ 1 Mt	≥900 m	292x1	292
SS-N-17	3,900 km	1.1	1 @ 500 kt	1400 m	12x1	8
SS-N-18[3]	6,500 km	2.3	5 MIRV @ 500 kt	900 m	224x5	706
SS-N-20	8,300 km	?	6 MIRV @ 100 kt	500 m	80x6	103
SS-N-23[3]	8,300 km	?	10 MIRV @ 100 kt	<900 m	48x10	103
			U.S.S.R. TOTAL		1596	1484
			UNITED KINGDOM			
Polaris	4,600 km	0.7	3 MRV @ 200 kt	900 m	64x3	66
			U.K. TOTAL		192	66
			FRANCE			
MSBS M-20	3,000 km	?	1 @ 1 Mt	?	80x1	80
MSBS M-4	6,000 km	?	6 MIRV @ 150 kt	?	16x6	27
			FRENCH TOTAL		176	107

[1] EMT (equivalent megatons) is equal to $Y^{2/3}$, with Y in megatons.
[2] The projected Trident II missiles are often said to be comparable in accuracy to the Minuteman III missiles.
[3] These missiles are liquid fueled.

Table 4. The strategic nuclear balance between the U.S. and the U.S.S.R.; a summary of Tables 1-3.

Delivery Systems	←-----Quantities Number of Launchers	Throw weight tonnes	Y(Mt)	Y(EMT)	Qualities-----→ Number of Warheads	Median CEP
			UNITED STATES			
ICBMs	1,000	692	1,162	1,336	2,261	220 m
Submarines	640	883	410	961	5,632	450 m
Bombers	317	5,620	1,521	2,057	4,624	n.a.[1]
US TOTAL	1,957	7,195	3,093	4,354	12,517	n.a.
			SOVIET UNION			
ICBMs	1,398	4,589	3,542	4,326	6,540	300 m
Submarines	928	1,122[2]	1,220	1,484	2,596	900 m
Bombers	190	3,060	460	519	760	n.a.
USSR TOTAL	2,516	8,771	5,222	6,329	9,896	n.a.

[1] n.a. means not applicable in this case.
[2] Assuming a payload of 1.1 tonnes for the solid-fueled SS-N-20 and 2.3 tonnes for the liquid-fueled SS-N-23.

Table 5. Details of the strategic balance between the U.S. and the U.S.S.R. Damage area is taken as 30 mi^2 (75 km^2) per equivalent megaton. The missile-silo blast resistances are taken as 300 psi for the SS-11 missiles, 2000 psi for hot-launched missiles and the MX, and 5000 psi for the Soviet cold-launched SS-17 and SS-18. After Tables 1-3.

Missiles	# Warheads	Deterrence Capability			
		Mt per Warhead	EMT per Warhead	Area Destroyed /RV	Area Destroyed Total
UNITED STATES					
Minuteman II	450x1	1.5 Mt	1.31 EMT	39 mi^2	17,700 mi^2
Minuteman III	227x3	170 kt	0.31 EMT	9 mi^2	6,300 mi^2
Minuteman IIIA	300x3	335 kt	0.48 EMT	14 mi^2	13,000 mi^2
Peacekeeper MX	23x10	300 kt	0.45 EMT	13 mi^2	3,100 mi^2
Poseidon C-3	256x10	40 kt	0.12 EMT	4 mi^2	9,000 mi^2
Trident I	384x8	100 kt	0.22 EMT	6 mi^2	19,800 mi^2
Bombers: bombs	200x4	1 Mt	1 EMT	30 mi^2	24,000 mi^2
: SRAM	1,448	170 kt	0.31 EMT	9 mi^2	13,300 mi^2
: ALCM	2,376	200 kt	0.34 EMT	10 mi^2	24,400 mi^2
U.S. TOTAL	12,517				130,600 mi^2
SOVIET UNION					
SS-11 (MOD 1)	20x1	1 Mt	1 EMT	30 mi^2	600 mi^2
(MOD 2)	360x1	1 Mt	1 EMT	30 mi^2	10,800 mi^2
(MOD 3)	60x3	300 kt	0.45 EMT	13 mi^2	2,400 mi^2
SS-13	60x1	600 kt	0.71 EMT	21 mi^2	1,300 mi^2
SS-17 (MOD 3)	150x4	500 kt	0.63 EMT	19 mi^2	11,300 mi^2
SS-18 (MOD 4)	308x10	500 kt	0.63 EMT	19 mi^2	58,200 mi^2
SS-19 (MOD 3)	360x6	550 kt	0.67 EMT	20 mi^2	43,500 mi^2
SS-25	80x1	550 kt	0.67 EMT	20 mi^2	1,600 mi^2
SS-N-6	272x1	1 Mt	1 EMT	30 mi^2	8,200 mi^2
SS-N-8	292x1	1 Mt	1 EMT	30 mi^2	8,800 mi^2
SS-N-17	12x1	1 Mt	1 EMT	30 mi^2	400 mi^2
SS-N-18	224x5	500 kt	0.63 EMT	19 mi^2	21,200 mi^2
SS-N-20	80x6	100 kt	0.22 EMT	6 mi^2	3,100 mi^2
SS-N-23	48x10	100 kt	0.22 EMT	6 mi^2	3,100 mi^2
Bombers: bombs	320	1 Mt	1 EMT	30 mi^2	9,600 mi^2
: ALCM	40	1 Mt	1 EMT	30 mi^2	1,200 mi^2
: ALCM	400	250 kt	0.40 EMT	12 mi^2	4,800 mi^2
U.S.S.R. TOTAL	9,896				190,100 mi^2

Table 5 continued.

Missiles	Vulnerability (K required) /RV-90%	-Total	/RV-97%	-Total	Available Capability K for Counterforce CEP	/RV	Total
UNITED STATES							
Minuteman II	80	36,000	122	54,900	370 m	33	14,800
Minuteman III	80	18,200	122	27,700	220 m	22	14,800
Minuteman IIIA	80	24,000	122	36,600	220 m	22	19,600
Peacekeeper MX	80	1,800	122	2,800	100 m	154	35,400
Poseidon C-3[1]	n.a.	n.a.	n.a.	n.a.	450 m	2	5,100
Trident II[1]	n.a.	n.a.	n.a.	n.a.	450 m	4	11,200
Bombers: bombs[2]	n.a.	n.a.	n.a.	n.a.	n.a.	n.a.	n.a.
: SRAM[2]	n.a.	n.a.	n.a.	n.a.	n.a.	n.a.	n.a.
: ALCM[2]	n.a.	n.a.	n.a.	n.a.	n.a.	n.a.	n.a.
U.S. TOTAL		80,000[3]		122,000			100,900
SOVIET UNION							
SS-11 (MOD 1)	23	500	35	700	1400 m	2	0
(MOD 2)	23	8,300	35	12,600	1100 m	3	1,000
(MOD 3)	23	1,400	35	2,100	1100 m	1	200
SS-13	80	4,800	122	7,300	1800 m	1	100
SS-17 (MOD 3)	148	22,200	226	33,900	400 m	14	8,100
SS-18 (MOD 4)	148	45,600	226	69,600	250 m	35	106,600
SS-19 (MOD 3)	80	28,800	122	43,900	300 m	26	55,300
SS-25[1]	n.a.	n.a.	n.a.	n.a.	200 m	58	4,600
SS-N-6[1]	n.a.	n.a.	n.a.	n.a.	1300 m	2	600
SS-N-8[1]	n.a.	n.a.	n.a.	n.a.	≥900 m	4	1,200
SS-N-17[1]	n.a.	n.a.	n.a.	n.a.	1400 m	1	0
SS-N-18[1]	n.a.	n.a.	n.a.	n.a.	900 m	3	3,000
SS-N-20[1]	n.a.	n.a.	n.a.	n.a.	500 m	3	1,400
SS-N-23[1]	n.a.	n.a.	n.a.	n.a.	<900 m	1	400
Bombers: bombs[2]	n.a.	n.a.	n.a.	n.a.	n.a.	n.a.	n.a.
: ALCM[2]	n.a.	n.a.	n.a.	n.a.	n.a.	n.a.	n.a.
: ALCM[2]	n.a.	n.a.	n.a.	n.a.	n.a.	n.a.	n.a.
U.S.S.R. TOTAL		111,600		170,100			182,500

[1] These mobile missiles are not vulnerable to attack by nuclear missiles, no matter how accurate, hence the vulnerability parameters are not applicable here (n.a.).

[2] Bombers are not vulnerable if launched on warning, and they are too slow to have a counterforce capability against enemy missile silos, hence neither the vulnerability nor the counterforce parameters are applicable here (n.a.).

Table 6. The U.S.S.R./U.S. strategic balance expressed in terms of the area each country's forces can destroy, the counterforce available to each country, and the counterforce each country would need for a 90% or 97% successful attack on the enemy's missile silos. A summary of Tables 1-5.

Delivery Systems	Area Destroyable	Silo Kill factor K: Available	Silo Kill factor K: Required to attack enemy silos: Reqd. for 90%	Silo Kill factor K: Required to attack enemy silos: Reqd. for 97%
		UNITED STATES		
ICBMs	40,100 mi^2	84,600 mi^2	80,000[1]	122,000
Submarines	28,800 mi^2	16,300 mi^2	n.a.	n.a.
Bombers	61,700 mi^2	n.a.	n.a.	n.a.
US TOTAL	130,600 mi^2	100,900 mi^2	80,000	122,000
		SOVIET UNION		
ICBMs	129,700 mi^2	175,900 mi^2	111,600[2]	170,100[2]
Submarines	44,800 mi^2	6,600 mi^2	n.a.	n.a.
Bombers	15,600 mi^2	n.a.	n.a.	n.a.
USSR TOTAL	190,100 mi^2	182,500 mi^2	111,600[2]	170,100[2]

[1] n.a. means not applicable in this case.

[2] The SS-25 mobile ICBMs actually might survive an attack by U.S. missiles, no matter how good the accuracy of the attacking missiles.

APPENDIX D

PRINCIPAL FINDINGS OF THE OFFICE OF TECHNOLOGY ASSESSMENT OF THE U.S. CONGRESS

OTA has produced a number of reports that are relevant to the issue of international security and arms control:

NUCLEAR PROLIFERATION AND SAFEGUARDS (1977)
THE EFFETS OF NUCLEAR WAR (1979)
MX MISSILE BASING (1981)
DIRECTED ENERGY MISSILE DEFENSE IN SPACE (1984)
TECHNOLOGY TRANSFER TO THE MIDDLE EAST (1984)
REMOTE SENSING AND THE PRIVATE SECTOR (1984)
INTERNATIONAL COOPERATION AND COMPETITION IN CIVILIAN SPACE ACTIVITIES (1985)
BALLISTIC MISSILE DEFENSE TECHNOLOGY (1985)
ANTI-SATELLITE WEAPONS, COUNTERMEASURES, AND ARMS CONTROL (1985)
U.S.-SOVIET COOPERATION IN SPACE (1985)
DEFENDING SECRETS, SHARING DATA (1987)
COMMERCIAL NEWSGATHERING FROM SPACE (1987)
TECHNOLOGY TRANSFER TO CHINA (1987)
NEW TECHNOLOGY FOR NATO (1987)
SDI: TECHNOLOGY, SURVIVABILITY AND SOFTWARE (1988)
SEISMIC VERIFICATION OF NUCLEAR TEST ING TREATIES (1988)
LAUNCH OPTIONS FOR THE FUTURE (1988)

The principal findings of the first three reports, on proliferation (1977), the effects of nuclear war (1978), and MX (1981), are contained in Appendix D, PHYSICS, TECHNOLOGY AND THE NUCLEAR ARMS RACE, AIP 104 (1983). Because of the limitations on space, we will only list the principal findings from three reports in this volume:

I. SEISMIC VERIFICATION OF NUCLEAR TESTING TREATIES (1988)

II. SDI: TECHNOLOGY, SURVIVABILITY, AND SOFTWARE (1988)

III. NEW TECHNOLOGY FOR NATO (1987)

I. SEISMIC VERIFICATION OF NUCLEAR TESTING TREATIES (1988)

HOW LOW CAN WE GO?

Given all the strengths and limitations of seismic methods in detecting and identifying Soviet nuclear explosions, combined with the credibility of the various evasion scenarios, the ultimate question of interest for monitoring any low-yield threshold test ban treaty is essentially: *How low can we go?* The answer to that question depends largely on what is negotiated in the treaty. As we have seen, the challenge for a monitoring network is to demonstrate a capability to distinguish credible evasion attempts from the background of frequent earthquakes and legitimate industrial explosions that occur at low yields.

The monitoring burden placed on the seismic network by various evasion scenarios can be greatly lessened if seismology gets some help. The sources of help are varied and numerous: they include not only seismic monitoring but also other methods such as satellite surveillance, radioactive air sampling, communication intercepts, reports from intelligence agents, information leaks, interviews with defectors and emigrés, on-site inspections, etc. **The structure of any treaty or agreement should be approached through a combination of seismic methods, treaty constraints, and inspections that will reduce the uncertainties and difficulties of applying seismic monitoring methods to every conceivable test situation.** Yet with these considerations in mind, some generalizations can still be made about monitoring at various levels.

Level 1—Above 10 kt

Nuclear tests with explosive yields above 10 kt can be readily monitored with high confidence.[7] This can be done with external seismic networks and other national technical means. The seismic signals produced by explosions of this size are discernible and no method of evading a seismic monitoring network is credible. However, for accurate monitoring of a 10 kt threshold treaty it would be desirable to have stations within the Soviet Union for improved yield estimation, plus treaty restrictions for handling the identification of large chemical explosions in areas where decoupling could take place.

Level 2—Below 10 kt but Above 1-2 kt

Below 10 kt and above 1-2 kt, the monitoring network must demonstrate a capability to defeat evasion scenarios. Constructing an underground cavity of sufficient size to fully decouple an explosion in this range is believed to be feasible in salt, with dedicated effort and resources. Consequently, the signals from explosions below 10 kt could perhaps be muffled. The seismic signals from these small muffled explosions would need not only to be detected, but also distinguished from legitimate chemical industrial explosions and small earthquakes. Demonstrating a capability to defeat credible evasion attempts would require seismic stations throughout the Soviet Union (especially in areas of salt deposits), negotiated provisions within the treaty to handle chemical explosions, and stringent testing restrictions to limit decoupling opportunities. If such restrictions could be negotiated, most experts believe that a high-quality, well run network of internal stations could monitor a threshold of around 5 kt. Expert opinion about the lowest yields that could reliably be monitored ranges from 1 kt to 10 kt; these differences of opinion stem from differing judgments about what technical provisions can be negotiated into the treaty, how much the use of high frequencies will improve our capability, and what levels of monitoring capability are necessary to give us confidence that the Soviet Union would not risk testing above the threshold.

Level 3—Below 1-2 kt

For treaty thresholds below 1 or 2 kt, the burden on the monitoring country would be much greater. It would become possible to decouple illegal explosions not only in salt domes but also in media such as granite, alluvium, and layered salt deposits. Although it may prove possible to detect such explosions with an extensive internal network, there is no convincing evidence that such events could be confidently identified with current technology. That is, additional work in identification capability will be required before it can be determined whether such small decoupled explosions could be reliably differentiated from the background of many small earthquakes and routine chemical explosions of comparable magnitude.

Level 4—Comprehensive Test Ban

There will always be some threshold below which seismic monitoring cannot be accomplished with high certainty. A comprehensive test ban treaty could, however, still be considered adequately verifiable if it were determined that the advantages of such a treaty would outweigh the significance of any undetected clandestine testing (should it occur) below the monitoring threshold.

ESTIMATING THE YIELD OF NUCLEAR EXPLOSIONS

For treaties that limit the testing of nuclear weapons below a specific threshold, the monitoring network not only must detect and identify a seismic event such as a nuclear explosion but also must measure the yield to determine whether it is below the threshold permitted by the treaty. This is presently of great interest with regard to our ability to verify compliance with the 150-kt limit of the 1974 Threshold Test Ban Treaty.

The yield of a nuclear explosion may be estimated from the seismic signal it produces. Yield estimation is accomplished by measuring from the seismogram the size of an identified seismic wave. When corrected for distance and local effects at the recording station, this measurement is referred to as the seismic magnitude. The relationship between seismic magnitude and explosive yield has been determined using explosions of known yields. This relationship is applied to estimate the size of unknown explosions. The problem is that the relationship was originally determined from U.S. and French testing and calibrated for the Nevada test site. As a result, Soviet tests are measured as if they had been conducted at the Nevada test site unless a correction is made. No correction would be needed if the U.S. and Soviet test sites were geologically identical, but they are not.

The Nevada test site, in the western United States, is in a geologically young and active area that is being deformed by the motion between the North American and Pacific tectonic plates. This recent geologic activity has created an area of anomalously hot and possibly even partially molten rock beneath the Nevada test site. As a result, when an explosion occurs at the Nevada test site, the rock deep beneath Nevada absorbs a large proportion of the seismic energy. The Soviet test site, on the other hand, is more similar to the geology found in the eastern United States. It is a geologically old and stable area, away from any recent plate tectonic activity. When an explosion occurs at the Soviet test site, the cold, solid rock transmits the seismic energy strongly. As a consequence, waves traveling from the main Soviet test site in Eastern Kazakhastan appear much larger than waves traveling from the Nevada test site. Unless that difference is taken into account, the size of Soviet explosions will be greatly overestimated.

The geological difference between the test sites can result in *systematic error*, or "*bias,*" in the way that measurements of seismic waves are converted to yield estimates. *Random error* is also introduced into the estimates by the measurement process. Thus, as with any measurement, there is an overall uncertainty associated with determining the size of an underground nuclear explosion. This is true whether the measurement is being made using seismology, hydrodynamic methods, or radiochemical methods. It is a characteristic of the measurement. To represent the uncertainty, measurements are presented by giving the most likely number (the mean value of all measurements) and a range that represents both the random scatter of the measurement and an estimate of the systematic uncertainty in the interpretation of the measurements. It is most likely that the actual value is near the central number and it is increasingly unlikely that the actual number would be found towards either end of the scatter range. When appropriate, the uncertainty range can be expressed by using what is called a "factor of uncertainty." For example, a factor of 2 uncertainty means that the best estimate of the yield (the "measured central value") when multiplied or divided by 2, defines a range within which the true yield will fall in 95 out of 100 cases. This is the high degree of certainty conventionally used in seismology and may or may not be appropriate in a verification context.[8]

The uncertainty associated with estimates of the systematic error can be greatly reduced by negotiating provisions that restrict testing to specific test sites and by calibrating the test sites. If such calibration were to be an integral part of any future treaty, the concern over systematic errors of this kind would become minimal. The majority of errors that would remain would be random. As discussed in chapter 2, a country considering cheating could not take advantage of the random error, because it would not be possible to predict how the random error would act on any given evasion attempt. In other words, if a country attempted to test above the threshold, it would have to realize that with every test the chances of appearing in compliance would decrease and at the same time the chance that at least one of the tests would appear in unambiguous violation would increase. For this reason, the range of uncertainty should not be considered as a range within which cheating could occur.

Most of the systematic error associated with estimating the yields of Soviet nuclear explosions is due to geological differences between the U.S. and Soviet test sites and in the coupling of the explosion to the Earth. Therefore, the single most important thing that can be done to reduce the uncertainty in yield estimation is to calibrate the test sites. Calibration could be accomplished through an exchange of devices of known yield, or through independent measures of the explosive yield such as can be provided by radiochemical or hydrodynamic methods (See box 1-B).

Our present capability to estimate seismically the yields of Soviet explosions is often cited as a factor of two.[9] While this may reflect present operational methods, it is not an accurate representation of our capability. Our capability could be greatly improved by incorporating new methods of yield estimation. Most seismologists feel that if new methods were applied, the resulting uncertainty for measuring explosive yields in the range of 150 kt at the Soviet test site would be closer to a factor of 1.5 than a factor of two.[10] Present methods are stated to be accurate only to a factor of two in part because they have not yet incorporated the newer methods of yield estimation that use surface waves and Lg waves.[11] The uncertainty of this comprehensive approach could be further reduced if calibration shots were performed and testing were restricted to areas of known geologic composition. **It is estimated that through such measures, the uncertainty in seismically measuring Soviet tests could be reduced to a level comparable to the uncertainty in seismically measuring U.S. tests. An uncertainty factor of 1.3 is the current capability that seismic methods are able to achieve for estimating yields at the Nevada test site.**

As with detection and identification, yield estimation becomes more difficult at low yields. Below about 50 kt, high-quality Lg-wave signals can only be reliably picked up by stations within the Soviet Union less than 2,000 km from the test site. For explosions below 10 kt, the uncertainty increases because small explosions do not always transmit their signals efficiently to the surrounding rock. For small explosions, the uncertainty could be reduced by restricting such tests to depths below the water table, so that their signal will be transmitted effectively.

Our capability to estimate explosive yields would depend to a large degree on the approach that was taken and what was negotiated in future treaties. If new methods of yield determination were incorporated into the measurements and calibration shots were performed, the capability of seismic methods could be improved to a point comparable to the accuracy of other methods, such as CORRTEX, that require a foreign presence and equipment at the test site. In any case, if the objective of reducing the uncertainty is to reduce the opportunities for cheating, small differences in random uncertainty do not matter.

[10]This reduction of uncertainly derives from using more than one statistically independent method. Consider as an example the situation where there are three independent methods of calculating the yield of an explosion, all of which (for the sake of this example) have a factor of two uncertainty in a log normal distribution:

# of Methods	Resulting Uncertainty
1	2.0
2	1.6
3	1.5

By combining methods, the uncertainty can be reduced below the uncertainty of the individual methods. This methodology, however, can only reduce random error. Systematic error, such as differences between the test site, will remain and limit the extent to which the uncertainty can be reduced unless calibration is performed.

SOVIET COMPLIANCE

The decision as to what constitutes adequate verification should represent a fair assessment of the perceived dangers of non-compliance. This necessarily involves a weighing of the advantages of the treaty against the feasibility, likelihood, and significance of noncompliance.

Such decisions are subjective and in the past have been influenced by the desirability of the treaty and the political attractiveness of particular monitoring systems.[12] Specific concern over compliance with test ban treaties has been heightened by findings by the Reagan Administration that:

> "Soviet nuclear testing activities for a number of tests constitute a likely violation of legal obligations under the Threshold Test Ban Treaty."[13]

Although the 1974 Threshold Test Ban Treaty and the 1976 Peaceful Nuclear Explosions Treaty have remained unratified for over 10 years, both nations have expressed their intent to abide by the yield limit. Because neither the United States nor the Soviet Union has indicated an intention *not* to ratify the treaties, both parties are obligated under international law (Article 18, the 1969 Vienna Convention on the Law of Treaties) to refrain from acts that would defeat their objective and purpose.

In examining compliance with the 150-kt threshold, seismic evidence is currently considered the most reliable basis for estimating the yields of Soviet underground nuclear explosions.[14] The distribution of Soviet tests indicates that about 10 (out of over 200) Soviet explosions since the signing of the Threshold Test Ban Treaty in 1974 *could* have estimated yields with central values above the 150 kt threshold limit, depending on how the estimate is made.[15] These 10 tests could actually be at or below the 150 kt limit, but have higher yield estimates due to random fluctuations in the seismic signals. In fact, when the same methods of yield estimation are applied to U.S. tests, approximately the same number of U.S. tests also appear to be above the 150 kt threshold limit. These apparent violations, however, do not mean that one, or the other, or both countries have cheated; nor does it mean per se that seismology is an inadequate method of yield estimation. It is inherent in any method of measurement that if several tests are performed at the limit, some of these tests will have estimated central values above the yield limit. Because of the nature of measurements (using any method), it is expected that about half the Soviet tests at 150 kt would be measured as slightly above 150 kt and the other half would be measured as slightly below 150 kt.

All of the estimates of Soviet and U.S. tests are within the 90 percent confidence level that one would expect if the yields were 150 kt or less. Extensive statistical studies have examined the distribution of estimated yields of explosions at Soviet test sites. **These studies have concluded that the Soviets are observing a yield limit consistent with compliance with the 150 kt limit of the Threshold Test Ban Treaty.**[16] [17]

[17]One of the first groups to carry out such statistical studies was Lawrence Livermore National Laboratory. Their conclusion was reported in open testimony before the Senate Armed Services Committee on Feb. 26, 1987 by Dr. Milo Nordyke, Leader of the Treaty Verification Program.

II. SDI: TECHNOLOGY, SURVIVABILITY, AND SOFTWARE (1988)

Table 1-1.—SDIO's Phase One Space- and Ground-Based BMD Architecture

Component	Number	Description	Function
First phase (approximately 1995-2000):			
Battle Management Computers	Variable	May be carried on sensor platforms, weapon platforms, or separate platforms; ground-based units may be mobile	Coordinate track data; control defense assets; select strategy; select targets; command firing of weapons
Boost Phase Surveillance and Tracking Satellite	Several at high altitude	Infrared sensors	Detect ballistic or ASAT missile launches by observing hot rocket plumes; pass information to tracking satellites
Space-based Interceptor Carrier Satellite	100s at several 100s of km altitudes	Each would carry about 10 small chemical rockets or "SBIs"; might carry sensors for tracking post-boost vehicles	On command, launch rockets at anti-satellite weapons (attacking BMD system), boosters, possibly PBVs.
Probe	10s	Ground-launched rocket-borne infrared sensors	Acquire RV tracks, pass on to ERIS interceptors
or			
Space Surveillance and Tracking System	10s	Satellite-borne infrared sensors	
or			
Space-based Interceptor Carrier Satellites	100s	Satellite-borne infrared sensors	
Exo-atmospheric Interceptors (ERIS)	1000s on ground-based rockets	Rocket booster, hit-to-kill warhead with infrared seeker	Cued by satellite-borne or rocket-borne infrared sensors, home in on and collide with RVs in late mid-course

Table 1-2.—OTA's Projections of Evolution of Ground- and Space-Based BMD Architecture

Component	Number	Description	Function
Second phase (approximately 2000-2010) replace first-phase components and add:			
Airborne Optical System (AOS)	10s in flight	Infrared sensors	Track RVs and decoys, pass information to ground battle management computers for launch of ground-based interceptors
Ground-based Radars	10s on mobile platforms	X-band imaging radar	Cued by AOS, track RVs as they enter atmosphere; discriminate from decoys, pass information to ground battle managers
High Endo-atmospheric Interceptors	1000s	Rocket with infrared seeker, non-nuclear warhead	Collide with RVs inside atmosphere, but before RV nuclear detonation could cause ground damage
Space Surveillance and Tracking Satellite (SSTS)	50-100 at few 1000s of km.	High-resolution sensors; laser range-finder and/or imaging radar for finer tracking of objects;	Track launched boosters, post-boost vehicles, and ground or space-launched ASATs; Track RVs and decoys, discriminate RVs from decoys;
		May carry battle management computers	Command firing of weapons
Space-based Interceptor Carrier	1000s at 100s of km altitudes	Each carries about 10 small chemical rockets or "KKVs"; at low altitude; lighter and faster than in phase one	On command, launch rockets at anti-satellite weapons (attacking BMD system), boosters, PBVs, and RVs
Space-based Neutral Particle Beam (NPB)	10s to 100s at altitude similar to SSTS	Atomic particle accelerator (perturber component of interactive discrimination; additional sensor satellites may be needed)	Fire hydrogen atoms at RVs and decoys to stimulate emission of neutrons or gamma rays as discriminator
Detector Satellites	100s around particle beam altitudes	Sensors to measure neutrons or gamma rays from objects bombarded by NPB; transmitters send data to SSTS and/or battle management computers	Measure neutrons or gamma rays emitted from RVs: heavier objects emit measurable neutrons or gamma rays, permitting discrimination from decoys
Third phase (approximately 2005-2115), replace second-phase components and add:			
Ground-based Lasers, Space-based Mirrors	10s of ground-based lasers; 10s of relay mirrors; 10s to 100s of battle mirrors	Several laser beams from each of several ground sites bounce off relay mirrors at high altitude, directed to targets by battle mirrors at lower altitudes	Attack boosters and PBVs

SOURCE: Office of Technology Assessment, 1988.

SENSOR TECHNOLOGY CONCLUSIONS

Phase 1

1. **A boost surveillance and tracking satellite (BSTS) could most probably be developed by the mid-1990s.** Short-wave and middle-wave infrared (S/MWIR) sensors, could provide early warning and coarse booster track data sufficient to direct SBI launches.[50]

2. **Space surveillance and tracking system (SSTS) satellites would not be available for tracking individual RVs and decoys before the late 1990s.** The ability to discriminate possible decoys in this time frame is in question. Smaller but similar sensors for a phase-one system might be placed on individual SBI platforms or on ground-based, pop-up probes.

3. **An airborne optical system could probably be available by the mid-1990s to detect and track RVs and decoys with IR sensors (although not to discriminate against a replica decoy above the atmosphere). However, its utility may be limited in performance and mission:**
   - Performance may be limited by the vulnerability and operating cost of its aircraft platform, and IR sensors might be confused during battle by IR-scattering ice crystals formed at 60 to 80 km altitude by debris reentering the atmosphere.

- The relatively short range of airborne IR sensors would limit the AOS mission to supplying data on approaching objects for endo-atmospheric interceptor radars, and possibly for exo-atmospheric interceptors a short while before RV reentry. Airborne IR sensors, unless very forward-based, could utilize only a small portion of the time available in mid-course for discrimination and therefore could not take full advantage of the fly-out range of ground-based exoatmospheric interceptors.

  In any case, an Airborne Optical System is not now included in SDIO phase-one deployment plans.

4. **Effective discrimination against more sophisticated decoys and disguised RVs in space is unlikely before the year 2000, if at all.**

Phase 2

5. **By the late 1990s at the earliest, a space surveillance and tracking system (SSTS) might furnish post-boost vehicle (PBV) and reentry vehicle (RV) track data with long-wave infrared (LWIR) above-the-horizon (ATH) sensors suitable for directing SBI launches in the mid-course.** New methods would be needed for the manufacture of large quantities of radiation-hardened focal plane arrays. Another issue is the operation of LWIR sensors in the presence of precursor nuclear explosions (including those heaving atmosphere into the ATH field of view) or other intentionally dispersed chemical aerosols. Effective mid-course SBI capability is unlikely before the late 1990s to early 2000s.

6. **There are too many uncertainties in projecting sensor capabilities and the level of Soviet countermeasures to specify a discrimination capability for SSTS.** It appears that Soviet countermeasures (penetration aids and decoys) could keep ahead of passive IR discrimination techniques:
   - Passive IR discrimination could be available by the mid-1990s, but probably would have marginal utility against determined Soviet countermeasures.
   - Active laser radar (ladar) imaging of PBV deployment offers some promise of decoy discrimination, provided that the Soviets did not mask dispersal of decoys. Space-borne imaging ladars probably would not be available until the late 1990s at the earliest.
   - Laser thermal tagging of RVs is unlikely to be practical given the need for complex, agile steering systems and given likely countermeasures such as thermal insulation of RVs and decoys.
   - Laser impulse tagging is even less likely to succeed in this phase because high-power pulsed lasers would be required.

7. **Ground-based radar (GBR) might be available by the late 1990s to direct interceptors to re-entering warheads.** There may be some questions about its resistance to RF jammers. Signal processors may have difficulty handling large numbers of targets in real-time.

Phase 3

8. Accurate IR sensors, UV ladar, or visible ladar would have to reside on each DEW platform.

9. **Interactive discrimination with neutral particle beams (NPB) appears the most likely candidate to reliably distinguish decoys from RVs, since the particles would penetrate targets, making shielding very difficult.** Before one could judge the efficacy of a total NPB discrimination system, major engineering developments would be required in: weight reduction, space transportation, neutral particle beam control and steering,[51] automated accelerator operation in space, and multi-megawatt space power.

   It is unlikely that a decision on the technical feasibility of NPB discrimination could be made before another decade of laboratory development and major space experiments. Given the magnitude of an NPB/detector satellite constellation, **an effective discrimination system against sophisticated decoys and disguised RVs would not likely be fully deployed and available for BMD use until the 2010 to 2015 period at the earliest.**

10. **Nuclear bomb-projected particles might also form the basis of an effective interactive discriminator,** if reliable space-based ladar systems were also developed and deployed to measure target velocity changes. There are too many uncertainties to project if or when this approach might succeed.

[51]Since the particle beams are invisible, novel approaches would be required to sense the direction of the beam so that it could be steered toward the target.

Weapon Technology Conclusions

Phase One

Kinetic Energy Weapons.—KEWs (or else the kinds of nuclear-armed missiles developed for BMD in the 1960s) would most likely be the only BMD weapons available for deployment in this century and possibly the first decade of the 21st century. Several varieties of non-nuclear, hit-to-kill KEW form the backbone of most near- and intermediate-term SDI architecture proposals. Considering the steady evolution of rockets and "smart weapon" homing sensors used in previous military systems, it seems likely that these KEWs could have a high probability of being able to destroy individual targets typical of the current Soviet ICBM force by the early to mid-1990s. The key unresolved issue is whether a robust, survivable, integrated system could be designed, built, tested, and deployed to intercept—in the face of likely countermeasures—a sizeable fraction of evolving Soviet nuclear weapons.

Space-Based Interceptors.—SBIs deployed in the mid to late 1990s could probably destroy some Soviet ICBMs in their boost phase. The key issue is whether the weight of the SBI projectiles could be reduced before Soviet booster burn times could be shortened, given that existing SS-24 and SS-25 boosters would already stress projected SBI constellations. **The probability of post-boost vehicle (PBV) kills is lower due to the smaller PBV size and IR signal, but SBIs might still achieve some success against current PBVs by the mid to late 1990s.**

Exo-atmospheric Reentry Interceptor System.—The ERIS, which has evolved from previous missiles, could probably be built by the early to mid-1990s to attack objects in late mid-course. The key unknown is the method of tracking and discriminating RVs from decoys. Existing radar sensors are highly vulnerable, the SSTS space-based IR sensor probably would not be available until the late 1990s to early 2000s, and the AOS airborne sensor would have limited endurance and range. This would leave either new radars or some type of pop-up, rocket-borne IR probe, which have apparently received little development effort until recently. Given the uncertainty in sensors suitable for the ERIS system, its role would probably be confined to very late mid-course interceptions and it might have limited BMD effectiveness until the late 1990s.

Phase Two

High Endo-atmospheric Defense Interceptor.—The HEDI could probably be brought to operational status as soon as the mid-1990s. To overcome the unique HEDI window heating problem, the HEDI on-board homing IR sensor needs more development than its ERIS cousin. But the HEDI system does not depend on long-range sensors to achieve its mission within the atmosphere. The HEDI could probably provide some local area defense of hardened targets by the mid-1990s against non-MaRVed RVs.[83] HEDI performance against MaRVed RVs appears questionable.

SBIs against Reentry Vehicles.—The probability that SBIs would kill RVs in the mid-course is low until the next century, given the difficulty in detecting and tracking many small, cool RVs in the presence of decoys, and given uncertainties in the SSTS sensor and battle management programs.

Phase Three

Directed-Energy Weapons.—It is unlikely that any DEW system could be highly effective before 2010 to 2015 at the earliest. No directed energy weapon is within a factor of 10,000 of the brightness necessary to destroy responsively designed Soviet nuclear weapons. (OTA has not had the opportunity to review recent SDIO suggestions for "entry level" DEWs of more modest capability. SDIO contends that effective space-based lasers of one to two orders of magnitude less than that needed for a responsive threat could be developed much sooner.) At least another decade of research would likely be needed to support a decision whether any DEW could form the basis for an affordable and highly effective ballistic missile defense. Further, it is likely to take at least another decade to manufacture, test, and launch the large number of satellite battle stations necessary for highly effective BMD. Thus, barring dramatic changes in weapon and space launch development and procurement practices, a highly effective DEW system is unlikely before 2010 to 2015 at the earliest.

Neutral Particle Beam.—The NPB, under development initially as an interactive discriminator, is the most promising mid-course DEW.[84] Shielding RVs against penetrating particle beams, as opposed to lasers, appears prohibitive for energies above 200-MeV. Although laboratory neutral particle beams are still about 10,000 times less bright than that needed for sure electronics kills of RVs in space, the necessary scaling in power and reduction in beam divergence appears feasible, if challenging. However, as discussed in chapter 4 under the topic of NPB interactive discrimination, it is unlikely that engineering issues could be re-

Figure 5-2a.—Space-Based Interceptor Mass v. Velocity

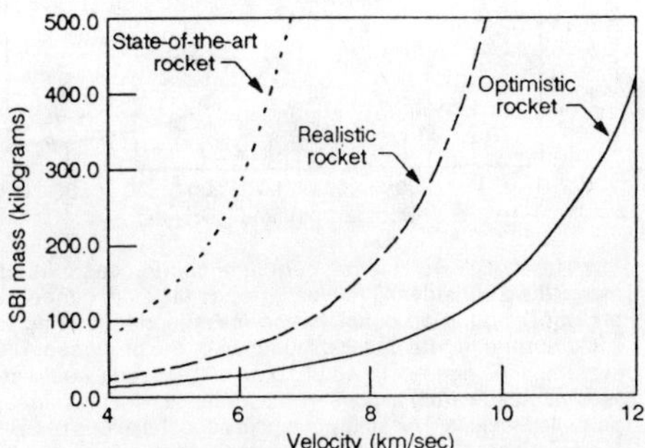

The SBI mass versus SBI velocity. These data assume 100% coverage of the current Soviet threat of 1,400 ICBMs. It should be noted that the SDIO currently proposes a substantially lower level of coverage for SBIs. Therefore, the absolute numbers in the OTA calculations are not congruent with SDIO plans. Rather, the graphs provided here are intended to show the relationships among the various factors considered. It should also be noted that numerous assumptions underlying the OTA analyses are unstated in this unclassified report, but are available in the classified version.

SOURCE: Office of Technology Assessment, 1988.

Figure 5-2b.—Number of Satellites v. SBI Velocity

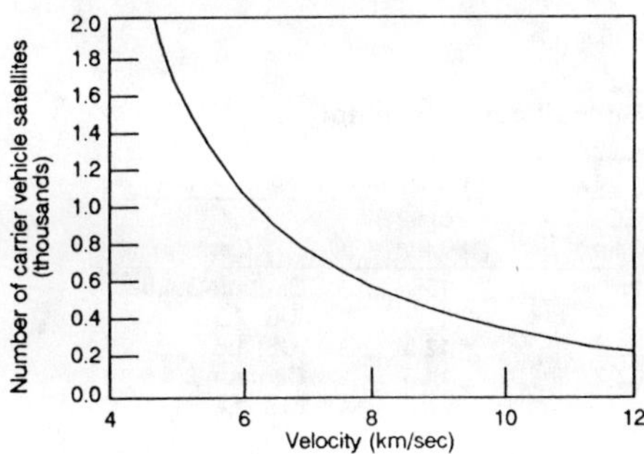

The number of SBI carrier satellites v. SBI velocity.

SOURCE: Office of Technology Assessment, 1988.

Figure 5-2c.—Number of Space-Based Interceptors v. Velocity (inclined orbits + SLBM polar orbits)

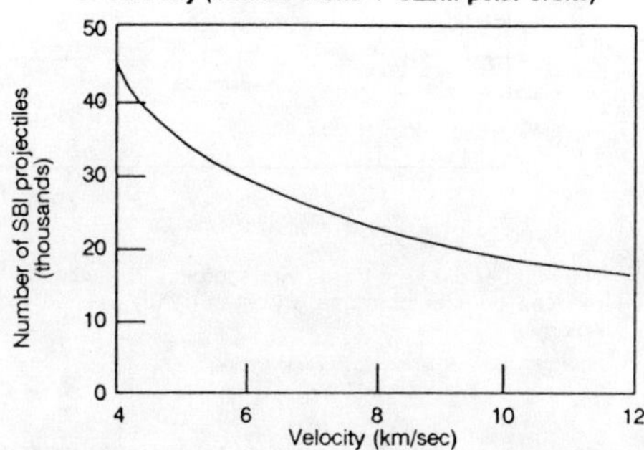

The number of space-based interceptors (SBIs) required to provide one SBI within range of each of 1,400 existing Soviet ICBMs before booster burnout.

SOURCE: Office of Technology Assessment, 1988.

Figure 5-2d.—Constellation Mass v. SBI Velocity

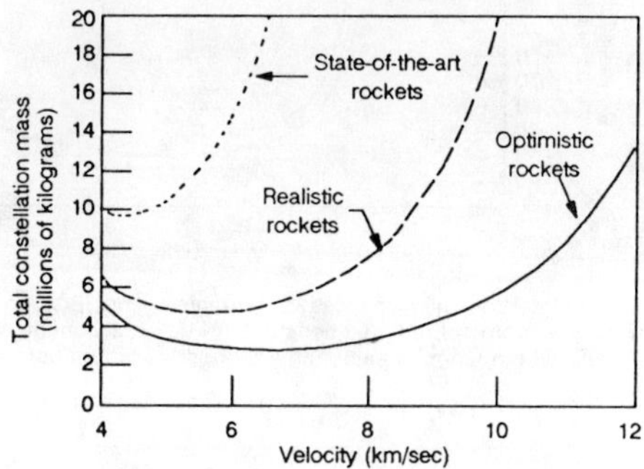

The total constellation mass in orbit (SBIs and carrier vehicles, excluding sensor satellites) v. SBI velocity. The minimum constellation mass for the "realistic" SBI to be in position to attack all Soviet boosters would be about 5.3 million kg. Faster SBIs would permit fewer carrier vehicles and fewer SBIs, but the extra propellant on faster SBIs would result in a heavier constellation. For reference, the Space Shuttle can lift about 14,000 kg into polar orbit, a 5.3 million kg constellation would require about 380 Shuttle launches, or about 130 launches of the proposed "Advanced Launch System" (ALS), assuming it could lift 40,000 kg into near-polar orbit at suitable altitudes.

SOURCE: Office of Technology Assessment, 1988.

Figure 5-3a.—Number of Projectiles v. SBI Velocity (160 second burn-time)

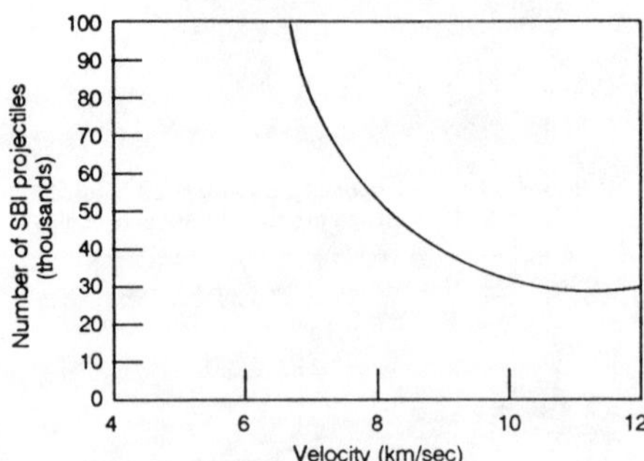

The number of space-based interceptors v. SBI velocity for reduced booster burntime (within currently applied technology).

SOURCE: Office of Technology Assessment, 1988.

Figure 5-3b.—SBI Constellation Mass v. SBI Velocity (160 second burn-time)

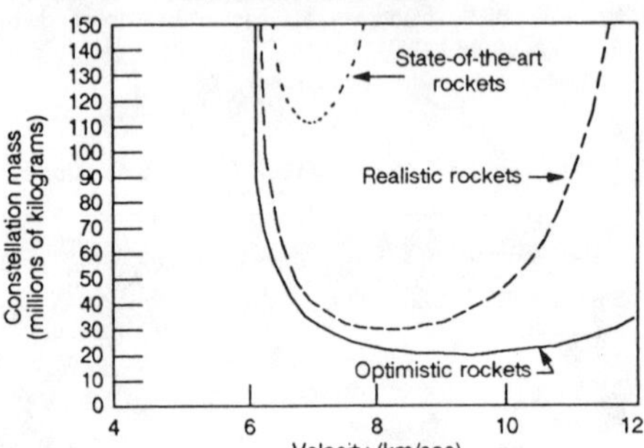

The total constellation mass (carrier vehicles and SBIs) versus SBI velocity for reduced booster burn times, assuming one SBI within range of each of 1,400 boosters before burnout.

Figure 5-4.—Total SBI Constellation Mass in Orbit v. Booster Burn Time

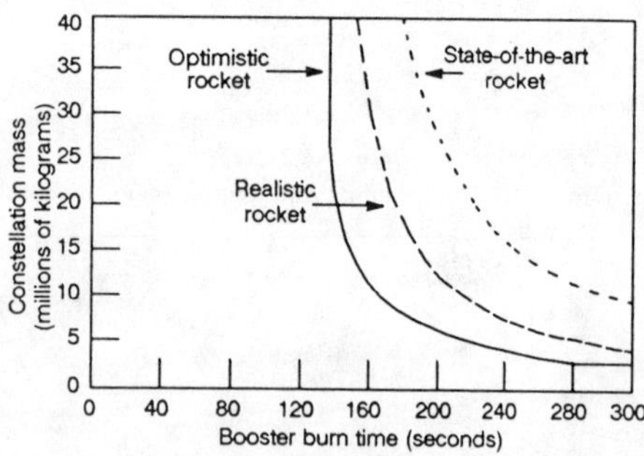

The effect of Soviet booster burn time on SBI constellation mass. If we consider 40 million kg as a maximum conceivable upper bound on constellation mass (corresponding to 2,800 Shuttle flights or 1,000 launches of the proposed ALS system), then booster times of 120 to 150 seconds would severely degrade a 100%-boost-phase defense with chemically propelled rockets. The ability of smaller constellations of SBIs to achieve lesser goals would be analogously degraded by the faster burn times.

All assumptions are the same as for the previous figures, except for the burn-out altitude, which varies with burn-time.

SOURCE: Office of Technology Assessment, 1988.

Table 5-10.—Current U.S. Space Launch Inventory[a]

	Payload per vehicle (thousands of kg)			
	Inventory quantity	LEO (180 km)	Polar (180 km)	Geo
Shuttle	3	25	15	Centaur-G:4.5 IUS: 2.3
Titan 34D	6	15.3	12.5	IUS: 1.8
Titan-4[b]	(23)	17.7	14.5	Centaur-G:4.6 IUS: 2.4
Titan II-SLV	(13)[c]	3.6	1.9	
Delta	8	2.9		
Delta (MLV)	(7)	4		1.5
Atlas	13	6		
Scout	21	.26		
(ALS)[d]	?	(50-70)	(40-55)	

[a]Parentheses indicate future systems.

[b]The Titan-4 or the Complementary Expendable Launch Vehicle (CELV) is the latest in the line of Titan missile configurations; 23 have been ordered.

[c]The Titan II-SLVs are being refurbished from the ICBM inventory. The first Titan-II may be available by 1989. An additional 56 Titan II could be refurbished from the retired ICBM fleet.

[d]The Advanced Launch System is proposed to deploy the bulk of the BMD space components.

solved before the late 1990s, which would most likely postpone deployment and effective system operation to at least 2010-2015.

Free Electron Laser.—The free electron laser (FEL) is one of the more promising BMD DEW weapon candidates. The FEL is in the research phase, with several outstanding physics issues and many engineering issues to be resolved. Even if powerful lasers could be built, the high power optics to rapidly and accurately steer laser beams from one target to the next could limit system performance. Although the basic system concept for an FEL weapon is well developed, it is too early to predict BMD performance with any certainty.

Chemical Laser.—There are too many uncertainties to project BMD performance for the chemically pumped hydrogen fluoride (HF) laser. The HF laser has been demonstrated at relatively high power levels on the ground, although still 100,000 times less bright than that needed for BMD against a responsive threat. Scaling to weapons-level brightness would require coherent combination of large laser beams, which remains a fundamental issue. This, coupled with the relatively long wavelength (2.8 micron region), make the HF laser less attractive for advanced BMD than the FEL.

Electromagnetic Launcher.—There are too many uncertainties in the EML or railgun program to project any significant BMD capabilities at this time.

Space Power Conclusions

Phase One

Power Requirements.—Nuclear power would be required for most BMD spacecraft, both to provide the necessary power levels for stationkeeping, and to avoid the vulnerability of large solar panels or solar collectors.

Dynamic Isotope Power System.—The DIPS, which has been ground-tested in the 2 to 5 kW range, should be adequate and available by the mid to late 1990s, in time for early BSTS-type sensors.

Phase Two

Nuclear Reactors.—Adequate space power may not be available for SSTS or weapon platforms with ladars before the year 2000. For BMD satellites that require much more than 10 kW of power the SP-100 nuclear reactor/thermoelectric technology would have to be developed. This is a high-risk technology, with space-qualified hardware not expected before the late 1990s to early 2000s.

Phase Three

Chemical Power.—Chemically driven energy sources (liquid oxygen and liquid hydrogen driving turbogenerators or fuel cells) could probably be available for burst powers of MW up to GW to drive weapons for hundreds of seconds by 2000-2005.

Power for Electromagnetic Launchers.—High-current pulse generators for electromagnetic launchers (EML) would require extensive development and engineering, and would most likely delay any EML deployments well into the 21st century.

High-Temperature Superconductors.—Research on high-temperature superconductors suggests exciting possibilities in terms of reducing the space power requirements and improving power generation and conditioning efficiencies. At this stage of laboratory discovery, however, it is too early to predict whether or when practical, high current superconductors could affect BMD systems.

Space Communications Conclusion

Laser communications may be needed for space-to-space and ground-to-space links to overcome the vulnerability of 60-GHz links to jamming from nearby satellites. Wide-band laser communications should be feasible by the mid-1990s, but the engineering for an agile beam steering system would be challenging.

Space Transportation Conclusions

Phase One

Mid-1990s Deployments.—Extrapolating reasonable extensions of existing space transportation facilities suggests that a limited-effectiveness, phase-one BMD system begun in the mid-1990s could not be fully deployed in fewer than 8 years.[85] Assuming that the hardware could be built to start deployment in 1994, the system would not be fully deployed until 2002. A more ambitious launcher-development program and a high degree of success in bringing payload weights down might shorten that period.

Phase Two

New Space Transportation System.—A fully new space transportation system would be required to lift the space assets of a "phase-two" BMD system. This system would have to include a vehicle with heavier lift capability (40,000 to 50,000 kg v. 5,000 kg for the Titan-4), faster launch rates (12 per year v. 3 per year per pad), and more launch pads (4 v. 1).

Optimistic Assumptions.—Even under very optimistic assumptions,[86] **the new space transportation system would be unlikely to reach the necessary annual lift requirements for a large-scale, second-phase BMD until 2000-2005, with full phase-two deployment completed in the 2008-2014 period.**

Phase Three

Ultimate DEW Systems.—It might take 20 to 35 years of continuous launches to fully deploy far-term, phase-three BMD space assets designed to counter with very high effectiveness an advanced, "responsive," Soviet missile threat. This estimate assumes deployment of the proposed ALS space transportation system and the kind of advanced space-based laser constellation suggested by SDI system architects. A set of ground-based laser installations could reduce the space launch deployment time estimate to 12-25 years.

[85]This assumes that two launch pads at Vandenberg AFB, 4-East and the SLC-6 pad intended for the Shuttle, are modified to handle the new Titan-4 complementary expendable launch vehicle (CELV), and the launch rates are increased from three Titans per year per pad up to six per year.

[86]This assumes that the SDIO bifurcated goal is met: a revolutionary space transportation system with 10 times lower cost is developed in 12 years, while a near-term component of that system yields a working vehicle of reduced capability by 1994.

SYSTEM CONCLUSIONS

Testing

1. **If the United States abandoned or achieved modification of the ABM Treaty, it could test a limited constellation of SBIs against a few ICBM's launched from Vandenberg AFB.** But this would not replicate the conditions of a massive, surprise launch of hundreds or thousands of ICBMs, ASATs, and nuclear precursors from the Soviet Union.
2. **A BMD system could not be tested against the real threat of up to thousands of ICBMs combined with defense suppression and nuclear precursors.** However, neither could such a coordinated offensive attack be fully tested.
3. **Key elements, such as IR sensors, could not be realistically tested against a background disrupted by nuclear explosions without abrogating the Limited Test Ban Treaty.**

Automation

4. **No technical barriers appear to preclude automatic operation of a space-based BMD system, but the task of operating an automatic, constantly changing constellation of sensors and weapons platforms in the face of defense suppression tactics would be a major challenge with little or no analogous experience from any other automated systems.**

Scheduling and Deployment

Phase One

5. **A near-term deployment (1995-2000) of state-of-the-art SBIs might stop up to 2,500 of an assumed constant 10,000 Soviet warhead threat in the boost and post-boost phases—if the United States devoted all of its space launch capability to lifting SBIs into orbit.** This assumes that the burn times and post-boost vehicle dispersal times of future Soviet ICBM's decrease over time in a reasonable manner. Of course, fewer SBIs could kill similar percentages of boosters if a smaller attack were assumed. The SDIO argues that defenses that are far from perfect still offer significant enhancement of deterrence (see chs. 1, 2, and 3).

Phase 2

6. **An intermediate-term or "phase-two" deployment of more advanced SBIs might kill up to 5,000 of the hypothesized fixed number of 10,000 Soviet RVs in the boost and post-boost phases, but only by orbiting from 90,000 to 160,000 SBIs.** Therefore, the United States would be unlikely to rely on SBIs for continued boost-phase interception of advanced Soviet missiles.
7. **Given the assumptions of OTA analyses, under the most optimistic conditions the Soviet Union could maintain an RV leakage into midcourse at or above the 6,000 warhead level by increasing the number of ICBMs deployed by 100 per year after the**

year 2000. Under any of the assumed conditions, the Soviet Union could increase the rate of warhead penetration against SBIs and into midcourse after 2005, reaching the pre-BMD levels of 10,000 leaking warheads by 2010. Therefore, SBIs should not be expected to achieve the strategic goal of "assured survival" against nuclear attack by a Soviet missile force unconstrained by arms reductions and limitations.

Phase 3

8. **A highly effective BMD system would require either very effective midcourse discrimination or a very effective directed-energy weapon (DEW) system, and preferably both**, since an SBI system, as limited by the most optimistic space transportation system, could never assure that fewer than 5,000 Soviet warheads and their associated decoys would leak through to the midcourse,
9. As concluded in chapter 5, it is unlikely that the United States could determine the feasibility of DEW systems by the late 1990s, and deployment probably could not begin until 2005-2010 at the earliest. **It therefore appears likely that the Soviet Union, unless constrained by offensive arms control agreements, would be able to maintain leakage rates of a few thousand nuclear warheads until at least the period 2005-2010.**

Figure 6-4a.—Number of Warheads Leaking Through Boost and Post-Boost

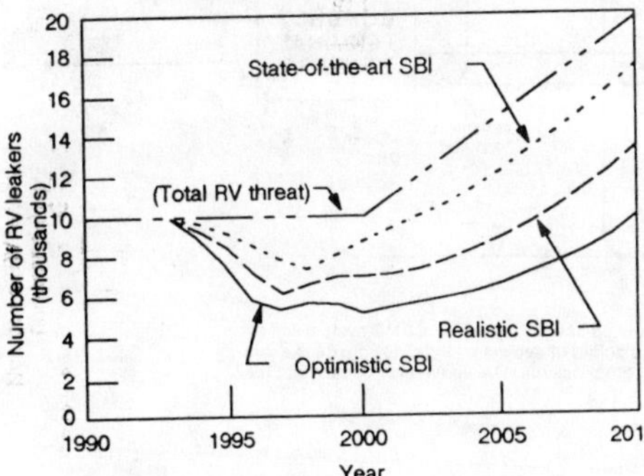

Boost and post-boost effectiveness limited only by space transportation capability, assuming that the Soviet threat increases in quality (shorter deployment times) and quantity (after 2000). Effectiveness shown for three different types of space-based interceptors against the "base case" Soviet threat.

SOURCE: Office of Technology Assessment, 1988.

Figure 6-4b.—Number of Warheads Leaking Through Boost and Post-Boost Defenses

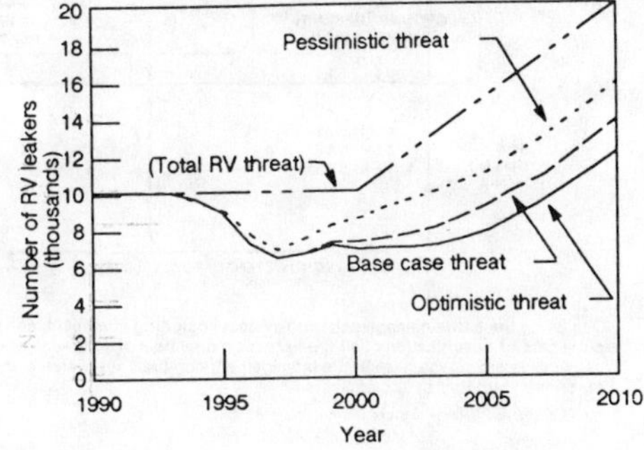

Boost and post-boost effectiveness against three different Soviet threats, all assuming "realistic" space-based interceptors.

SOURCE: Office of Technology Assessment, 1988.

System Integration and Battle Management

CONCLUSIONS

Ballistic missile defense battle management would be an extremely complex process. The number of objects, volume of space, and speed at which decisions would have to be made during a battle preclude most human participation. Aside from authorizing the system to move to alert status, prepared to fight automatically, at best the human's role would be to authorize the initial release of weapons and to change to back-up, previously-prepared, strategy or tactics. **Decisions about which weapons to use, when to use them, and against which targets to use them would all be automated.** Inclusion of human intervention points would likely add complexity to an already complex system and to compromise system performance in some situations. On the other hand, if an attacker had successfully foiled the primary defensive strategy, human intervention might allow recovery from defeat.

Battle management architectures as yet proposed are not specific enough for their claimed advantages and disadvantages to be effectively evaluated. Such evaluation must await both better architecture specifications and the development of an effective evaluation technique, perhaps based on simulation.

Figure 7-1.—Analytic Chart of Battle Management Functions in a Phase-one BMD System

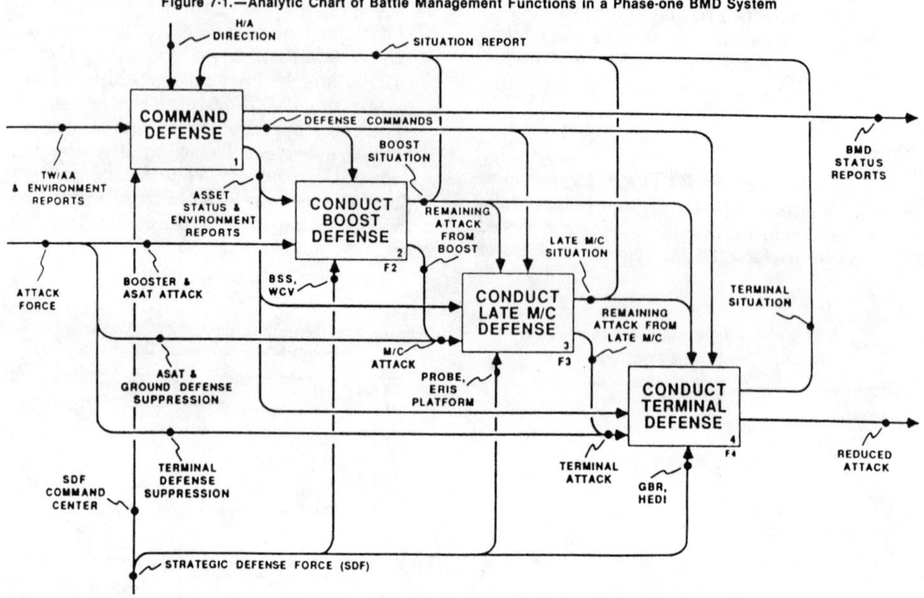

Designing the battle management sub-system (including command, control, and communications) for a BMD system will require precise specification of all the necessary channels of communication and points of decision. "Pipeline" charts like this help analysts conceptualize battle management complexities. Later, computer programs must be engineered to carry out the functions identified by the analysts.

SOURCE: Strategic Defense Initiative Organization, 1987.

Computing Technology

CONCLUSIONS

A BMD system to counter the Soviet ballistic missile threat might be the most complicated artifact ever built. It would involve the application of many different technologies; the automated interplay of thousands of different computers, sensors, and weapons; and the development of more software than has been used in any single previous project. An advanced BMD system would require computers in every fighting element of the system and in many supporting roles.

The degree of automation demanded entails not only advances in software technology (addressed in chapter 9) but also advances in secure computer networking, processing power, and radiation hardening of electronics. The extent and importance of simulations—in developing, exercising, and otherwise maintaining the system, as well as in training people in its use—would require an advance in simulation technology.

Because several difficult architectural tradeoffs have not yet been sufficiently addressed, the scope of the advances needed cannot be well predicted. Until an architectural description is available that clearly specifies battle management structure and allocates battle management functions both physically and within that structure, better predictions will not be possible.

Further discussion of the computing technology issues involved in producing an automated BMD system follows.

Table 8-1.—Computers in a Ballistic Missile Defense System

Component	Purpose of Computers
First- and Second-Phase Systems:	
Battle Management Computers[a]	Coordinate track data (e.g, maintain a track data base and correlate data from multiple sensors); maintain status of and control defense assets; select strategy; select targets; command firing of weapons; assess situation.
Boost Phase Surveillance and Tracking Satellite (BSTS)	Process signals to transform IR sensor data into digital data representing potential booster tracks; process images to recognize missile launches and to produce crudely-resolved booster tracks; communicate with battle management computers; maintain satellite platform: guidance, station keeping, defensive maneuvering; housekeeping.
Space Surveillance and Tracking Satellite (SSTS)	Process signals to transform IR, laser range-finder, and radar sensor data into digital data representing potential tracks; process images and data for fine-tracking of launched boosters, post-boost vehicles, RVs and decoys and to discriminate RVs from decoys; point sensors; communicate with other elements of the BMD system; guidance, station keeping, defensive maneuvering, housekeeping (maintain mechanical and electronic systems).
Laser Thermal Tagger	Point the laser beam; communicate with other elements of the BMD system; maintain satellite platform: guidance, station keeping, defensive maneuvering; housekeeping.
Carrier Vehicle (CV) for Space-based Interceptors (SBI)	Monitor status of SBIs; control launching of SBIs; communicate with battle manager; maintain satellite platform: guidance, station keeping, defensive maneuvering; housekeeping.
Space-based Interceptor (SBI)	Guide flight based on commands received from battle manager; Track target and guide missile home to target; communicate with battle manager; housekeeping.
Airborne Optical System (AOS)	Process signals to transform IR data into digital data representing potential tracks; process images and data for fine-tracking of post-boost vehicles, RVs and decoys and, if possible, discriminating RVs from decoys; point sensors; communicate with other elements of the BMD system; control of airborne platform; housekeeping.
Exo-atmospheric Interceptor System (ERIS)	Guide flight; process signals and images from on-board sensor for terminal guidance and target tracking; communicate with battle manager and SSTS, AOS, and probe sensors; housekeeping.
Ground-based Terminal Imaging Radar	Process signals and images to convert radar returns to target tracks; process images and data to discriminate between decoys and RVs; control radar beam; communicate with battle managers and other elements of BMD system; housekeeping.
High Endo-atmospheric Interceptors (HEDI)	Guide missile flight based on commands received from battle manager; process signals and images from on-board sensor for terminal guidance and target tracking; communicating with battle manager; housekeeping.
Third Phase, Add:	
Ground-based Laser, Space-based Mirrors	Manage laser beam generation; Control corrections to beam and mirrors for atmospheric turbulence; steer mirrors; communicate with battle manager; housekeeping.
Space-based Neutral Particle Beam (NPB)	Manage particle beam generation; steer the accelerator; track potential targets; communicate with battle manager and neutron detector; maintain satellite platform: guidance, station keeping, defensive maneuvering, housekeeping.
Radiation Detector Satellites	Discriminate between targets and decoys based on sensor inputs; communicate with battle manager and/or SSTS; maintain satellite platform: guidance, station keeping, defensive maneuvering; housekeeping.

[a]May be carried on sensor platforms, weapon platforms, or separate platforms; ground-based units may be mobile.

SOURCE: Office of Technology Assessment, 1988.

Reliable, Secure Communications

Common to all BMD systems that require human intervention at any stage is the need to provide secure, rapid communications between the human and the battle management computers. If part of the system is in space, then most likely there would be a need for space-to-ground communications. Battle management requires communications among the battle managers, the sensors, and the weapons forming a BMD system. The computers forming the communications network would digitally encode and control all the transmissions.

Achieving secure, reliable, adequate communications would call for simultaneous advances in at least two technologies. First, hardware technology, such as laser communications, needs to provide a medium that is difficult to intercept or jam and that can meet the required transmission bandwidth. Second, network technology must provide adequate, secure, survivable service for routing messages to their destinations. When damaged, the network must be able to reconfigure itself without significantly disrupting communications. Such performance would take sophisticated network control software—probably beyond the current state of the art. Proposed solutions to these problems are either untried or have only been tried in ground-based laboratory situations.

Simulations

Simulations would play a key role in all phases of a BMD system's life cycle. The SDIO is building a National Test Bed (NTB) to facilitate the development and use of BMD simulation technology. A full-scale NTB should permit experimentation with different system and battle management architectures, different battle management strategies, and different implementations of architectures. It would be the principal means of testing and predicting component, subsystem, and system reliability. Initially, the NTB would be a link among computers, each simulating a different aspect of a BMD engagement. The number of computers linked for any particular engagement would vary with the completeness and depth of detail required. Initial capabilities would not permit simulation of a full battle involving hundreds of thousands of objects. **Battle simulations on a scale needed to represent a full battle realistically have not been previously attempted. It would be crucial, but very difficult, to find a way of verifying the accuracy of such simulations, when and if they are developed.**

Technology and Architectural Trade-offs

Many difficult trade-offs have yet to be adequately addressed in the design of a BMD system to meet SDI requirements. Novel design ideas or advances in computing technology may decrease the importance of some of these trade-offs. However, no architecture has yet been specified sufficiently to permit clear trade-off studies. Issues that should be addressed include:

- simplifying software at the cost of adding computational burden to the hardware,
- simplifying battle management software by structuring it hierarchically at the expense of survivability,
- increasing survivability by decentralizing battle management at the expense of increasing communications complexity,
- customizing hardware for specific applications at the expense of increased software development cost and decreased software reliability,
- simplifying the problem of communications security at the cost of decreasing the possibilities for human intervention during battle,
- increasing the amount of human control during battle at the expense of fighting efficiency, and
- improving fighting efficiency at the cost of increasing the complexity and volume of communications (and, thereby, the risk of catastrophic communications failure).

None of these trade-offs is easy to make and few can be quantified. Compounding the difficulty is that many of the system elements—e.g., the Boost-phase Surveillance and Tracking System and Space Surveillance and Tracking System sensors, SBIs and associated CVs, high-powered lasers, and neutral particle beams—are still in the research or development stages. Moreover, no previous system has ever required the automated handling of many different devices and different kinds of devices as would an SDI missile defense. Nonetheless, tentative conclusions on some trade-offs have been reached. Most trade-offs could be properly explored by use of an appropriate simulation, such as might be provided by a full-scale National Test Bed.

Computational Requirements

Processing requirements are highly dependent on the system design, the battle management architectures, and the threat. Because detailed architectural and algorithmic specifications for an SDI BMD system are not yet available, estimates of speed and memory requirements accurate to better than a factor of 10 probably cannot be made. However, progress in processing speed has been rapid historically. If it continues at the same pace, it should yield sufficiently powerful processors to meet SDI needs within 10 years or less. Such processors might still have to be space qualified and radiation hardened.

An additional problem in providing radiation-hardened computing hardware is the lack of experience in building software tolerant of radiation-induced faults. There is little experience with complex, large-scale software systems that must operate efficiently despite the occurrence of radiation-induced transient effects in the hardware.

Software CONCLUSIONS

Based on both the preceding analysis, and the further exposition in appendix A, OTA has reached eight major conclusions.

1. **The dependability of BMD software would have to be estimated subjectively and without the benefit of data or experience from battle use. The nature of software and our experience with large, complex software systems, including weapon systems, together indicate that there would always be irresolvable questions about how dependable the BMD software was, and also about the confidence to be placed in dependability estimates.** Political decision-makers would have to keep in mind that there would be no good technical answers to questions about the dependability of the software, and no well-founded technical definition of software dependability.

   It is important to note that the Soviets would have similar problems in trying to estimate the dependability of the software, and therefore the potential performance of the system. Technical judgments of dependability would rely on peacetime tests that would be unlikely to apply to battle conditions. Political judgments about the credibility of the defense provided would therefore rest on very uncertain technical grounds.

2. **No matter how much peacetime testing were done, there would be no guarantee that the system would not fail catastrophically during battle as a result of a software error. Furthermore, experience with large, complex software systems that have unique requirements and use technology untested in battle, such as a BMD system, indicates that there is a significant probability that a catastrophic failure caused by a software error would occur in the system's first battle.**

3. It is possible that an administration and a Congress would reach the political decision to "trust" software that passed all the tests that could be devised in peacetime, despite the irresolvable doubts about whether such software might fail catastrophically the first time it was used in an actual battle. **Such a decision could be based upon the argument that the purpose of strategic forces—even defensive strategic forces—is primarily deterrence, and that a defensive system passing all its peacetime tests would be adequate for deterrence.** If deterrence succeeded, we would never know, and never need to know, whether the system would function in wartime.

4. **The extent to which BMD software would differ from complex software systems that have proven to be dependable in the past raises the possibility that software could not be created that ever passed its peacetime tests.** This a possibility exacerbated by the prospect of changing requirements caused by Soviet actions. We might arrive at a situation in which fixing problems revealed by one test created new problems that caused the software to fail the next test.

5. **No adequate models exist for the development, production, test, and maintenance of software for full-scale BMD systems.** Current DoD models of the software life-cycle and methods of software procurement appear inadequate for the job of building software as large, complex, and dependable as BMD software would have to be.

6. **The system architecture, the technologies to be used in the system, and a consistent set of performance requirements over the lifetime of the system must be established before starting software development.**[67] Otherwise, the system is unlikely even to pass realistic peacetime tests.

7. **As the strategic goals for a BMD system became more stringent, confidence in one's ability to produce software that would meet those goals would decrease as a result of the increased complication required in the software design.** Even for modest goals, such as improved deterrence, the United States could not have high confidence that the software would not fail catastrophically, whether faced with a modest threat or a severe threat. Put another way, there is no good way of knowing that BMD software would degrade gracefully rather than fail catastrophically when called on to face increasing levels of threat. Current techniques for identifying problems and detecting errors, such as simulations, would not help, although they could help to reduce the failure rate. **Furthermore, foreseeable improvements in software engineering technology would not change this situation.**

8. **The SDIO is investing relatively small amounts of money in software technology research in general, and in software engineering technology, computer security, communications networks, and fault tolerance in particular.** This investment strategy is of some concern, since particularly challenging BMD software development problems lie in these areas.

III. NEW TECHNOLOGY FOR NATO (1987)

WHAT IS THE LIKELIHOOD THAT PLAUSIBLE COMBINATIONS OF SYSTEMS WILL BE EFFECTIVE?

Plausible combinations of systems that could perform the tasks that fall under FOFA have been identified, but many of the components are still being developed. In order for any one concept to work, each piece must work (because each individual function is necessary), and they must be able to coordinate and interface.

Programs now under way are designed to overcome deficiencies in NATO's ability to attack follow-on forces, now primarily limited to: aircraft attacking fixed targets like bridges, as well as targets that, while mobile, don't spend much of their time moving;[4] and possibly aircraft flying along roads looking for columns of vehicles. Capability will improve as each deficiency is corrected, although all deficiencies need not be corrected to have a useful capability.

As each of these improvements comes on line, FOFA capability will increase incrementally.

WHAT ARE THE OUTSTANDING ISSUES BEFORE CONGRESS?

Several FOFA-related issues are likely to be matters of controversy in Congress in the next few years. These are: the Joint STARS program; the recently severely scaled-down PLSS program; Aquila and other remotely piloted vehicles (or unmanned aerial vehicles) programs; advanced anti-armor submunitions; and co-development and co-production with our European Allies.

Joint STARS

This program has been a matter of controversy between House and Senate for the past few years. By providing an ability to locate, track, and target groups of moving vehicles, Joint STARS is supposed to contribute to the commanders' awareness of the battlefield and to target engagement, which are central to many concepts for FOFA and probably very important if FOFA is to be successful. Such a capability would also be very important for identifying and analyzing the main thrusts of a Soviet offensive, and for obtaining warning of suspicious movements prior to hostilities. FOFA could be done without a system like Joint STARS, but not nearly as well.

At the heart of the controversy is the question of how survivable the E-8A (modified 707)

Deficiency	*Corrective Measures/Status*
1. Lack of ground-launched missiles	MLRS—in production ATACMS—in full scale development
2. Little ability to operate aircraft at night and in bad weather	LANTIRN[5]—in limited procurement F-15E—in procurement (MLRS, ATACMS[6])
3. Little ability to destroy masses of armored vehicles	CEM[7], DPICM[8]—in procurement (effective against all but heavily armored tanks) Smart anti-armor submunitions (sensor-fuzed weapons and terminally guided submunitions)—in development TMD[9]—in procurement NATO MSOW[10]—in development
4. Little ability to rapidly target moving combat units	Joint STARS—in full scale development Aquila RPV[11]—in full scale development other RPVs—various stages Joint Tactical Fusion Program—in full scale development
5. Little ability for Army corps to support adjacent corps	ATACMS—in full scale development
6. Enemy air defenses threaten both interdiction aircraft and surveillance aircraft	Various air defense suppression and avoidance programs in various stages Various RPV programs in various stages
7. No capability to attack very deep	B-52s carrying cruise missiles—no development yet

[5]Low Altitude Navigation and Targeting Infrared for Night. A system to aid aircraft in finding targets.
[6]These are not aircraft systems, but they can operate at night and in bad weather.
[7]Combined Effects Munition.
[8]Dual Purpose Improved Conventional Munition.
[9]Tactical Munitions Dispenser. The dispenser part of the CEM; the munition itself is the CEB (Combined Effects Bomblet).
[10]Modular Stand-off Weapon.
[11]Remotely Piloted Vehicle.

aircraft would be in a realistic combat environment. Critics contend that to be adequately survivable it would have to be operated so far from the FLOT[13] as to be virtually useless. Supporters argue that flying in protected NATO airspace with many other surveillance aircraft, benefiting from suppression of enemy air defenses, and protected by NATO fighters and SAMs,[14] it would be "survivable but not immortal."

It is likely that even with all this protection, Joint STARS would have to operate farther from the FLOT than originally envisioned in order to reduce its vulnerability. But its value would degrade slowly as it moves back, and it should be able to provide frequent coverage of broad areas out to final assembly areas, and perhaps somewhat beyond. This is the area in which frequent coverage is most needed because events will develop rapidly there. Deeper areas would be seen less frequently. This would provide a great improvement over current capabilities in area covered, frequency of coverage, timeliness, and accuracy. However, it is less than the nominal coverage usually assumed for the system. Prior to hostilities, the E-8A could operate up to the FLOT and provide much deeper coverage for indications and warning of attack.

Alternative systems that would be less detectable are possible; if operated so as to evade detection, they would also be limited in coverage, but the limitations would be different from those of the E-8A. Less area would be masked by terrain and vegetation if the platform were higher or closer to the FLOT. In combination with E-8As they might provide nearly complete coverage. *If* the alternative or complementary system were to operate in the same frequency band, it could probably use most of the radar hardware and software developed for the E-8A. **As far as OTA is aware, no detailed operations analysis that compares the FOFA capability using the E-8A Joint STARS, alternative systems, and combinations of the two has been done. If this remains an issue, such a study probably should be done, but it ought not to delay Joint STARS development.** That analysis should consider the possibility of reactive Soviet jammer development. In some cases, "customized" jammers could severely handicap either type of system, but the likelihood and practicality of such jammers needs further study.

OTA has not had access to other than general information on possible alternative systems, and cannot comment on their status. Any decision to cancel Joint STARS and begin another program should also take into account when the alternative might become available, and whether that alternative would be suitable for peacetime deployments and deployments outside Europe.

Continuation of, or Successor to, PLSS

This year Congress and the Air Force decided not to fund procurement of the PLSS and to return it to a relatively low-level developmental program. PLSS was designed to satisfy a need to quickly and accurately locate and target emitters such as the radars of modern air defense systems that would pose a threat to NATO interdiction aircraft and to surveillance systems like Joint STARS. The system was cut partly because of technical problems, and partly because the Air Force believed it was no longer worth the cost. At the time of the decision it had not achieved the specified system reliability or emitter location accuracy; however, both have now reportedly improved to near specified values. Its demonstrated target location accuracy, reporting rate, and timeliness are unsurpassed by other tactical electronic intelligence systems, but it sometimes reports one emitting target as several.

Some within the Air Force argue that other assets are adequate to do the job of locating the targets. Others argue that there are important tasks that PLSS was supposed to do that no other system can. OTA knows of no other system that can locate emitters as quickly and accurately as PLSS. Congress will have to face the question of whether a system like PLSS is needed, and, if so, whether it should be obtained by continuing PLSS or starting another program.

RPV/TADARS

The Target Acquisition/Designation Aerial Reconnaissance System (TADARS), which employs the Aquila RPV, is currently in full-scale development. Major problems that held

Table 2-1.—Illustrative Packages of Systems To Support Specific Operational Concepts
(as yet, not all the pieces exist)

	Reconnaissance, surveillance, and target acquisition		Attack	
Operational concept	Reconnaissance, surveillance, situation assessment	Target acquisition, attack control	Platform	Weapon
1. MLRS and artillery attack of regimental columns: 5 to 30 km deep	GUARDRAIL, TRS, ASARS, Joint STARS, and ASAS	AQUILA and AFATDS	8 inch Artillery and MLRS	SADARM MLRS/TGW
2. Aircraft attack of division columns: 30 to 80 km deep	GUARDRAIL, TRS, ASARS, Joint STARS, and ENSCE	Joint STARS	F-16	MSOW carrying Skeet or TGSM or CEB
3. Ballistic Missile attack of division columns: 30 to 80 km deep	GUARDRAIL, TRS, ASARS, Joint STARS, and ASAS	Joint STARS and AFATDS	MLRS launcher	ATACMS carrying DPICM or TGSM or Skeet
4. Attack with aircraft: create chokepoints and then attack the halted vehicles: 80 to 150 km deep	GUARDRAIL, TRS, ASARS, Joint STARS, and ENSCE	ASARS and GACC	F-15E F-16	AGM-130, MSOW carrying various munitions including mines
5. Air-launched cruise missile attack of rail network; 350 to 800 km deep	Various national systems	(on the weapon)	B-52	Cruise missile with various munitions

NOTES:
1. Definitions
AFATDS—Advanced Field Artillery Tactical Data System: provides target data to artillery and MLRS batteries.
AGM-130—an air-launched missile.
ASARS—Advanced Synthetic Aperture Radar System: provides images of fixed objects.
ASAS—a developmental Army center for collecting, analyzing, and disseminating surveillance data.
ATACMS—Army Tactical Missile System: a ballistic missile to be launched from MLRS launchers.
AQUILA—a remotely piloted vehicle.
CEB—Combined Effects Bomblet: Similar to DPICM; designed to be dispensed by the Combined Effects Munition (CEM).
DPICM—Dual Purpose Improved Conventional Munition: unguided submunition for use against light armor and soft targets.
ENSCE—the Air Force version of ASAS.
GACC—Ground Attack Control Center: a developmental center for controlling air attacks against ground targets.
GUARDRAIL—a tactical surveillance system.
Joint STARS—Joint Surveillance Target Attack Radar System: moving target indicator and attack control.
MLRS—Multiple Launch Rocket System.
MLRS/TGW—Terminally Guided Weapon: a smart anti-armor submunition for MLRS.
MSOW—Modular Standoff Weapon: a weapons dispenser.
SADARM—Search and Destroy Armor: a smart anti-armor submunition for artillery.
Skeet—a smart anti-armor submunition.
TGSM—Terminally Guided Submunition: a smart anti-armor submunition.
TRS—Tactical Reconnaissance System: carries various sensor suites.
2. Reconnaissance, surveillance, and situation assessment would be performed by a number of systems—particularly those shown here—feeding into the assessment center. Although all need not find the target for the attack to take place, the more there are, the greater the chances are that the target will be found, recognized, and identified with sufficient accuracy to attack it.
3. Not all the submunitions displayed in the table are necessarily being developed for deployment on the weapons shown; however, there is no fundamental reason why they could not be engineered onto those weapons.

the system up appear to have been solved. This system lacks the broad area, continuous deep coverage of Joint STARS, but could provide dedicated targeting for Army systems. TADARS can perform accurate target location as well as laser designation for artillery and laser-guided bombs. Some have proposed procuring another RPV in place of Aquila, but procuring another RPV and equipping it with Aquila's capabilities would take longer and cost more than completing TADARS development. Several types of RPVs are currently operational and under development in Belgium, Canada, France, Germany, Italy, the United Kingdom, and Israel. The U.S. Army is developing a family of advanced unmanned air vehicles, of which Aquila is the most mature, and the Navy and Air Force also have RPV programs.

Advanced Anti-Armor Submunitions

Smart anti-armor submunitions with advanced warheads—such as Skeet,[15] SADARM,[16] and MLRS/TGW[17]—may be a key to FOFA: they are the only means of killing modern tanks in significant numbers beyond the close battle.[18] But major uncertainties surround them, particularly the questions of whether technical and operational countermeasures could defeat their seekers and warheads. It will be necessary to keep a close watch on these development programs. One very valuable tool is the Chicken Little series of joint tests of munitions and munition concepts. OTA believes that this series, and others like it, ought to be supported and the results given serious consideration.

Defense Cooperation

Many of our Allies initially reacted cautiously to FOFA in part because it looked like another excuse to induce them to buy U.S. high-technology systems. They have a long-standing concern that the "two-way street" of NATO procurement favors the United States by a

[15] Smart anti-armor munition that fires a self-forging slug.

[16] Search and Destroy Armor: Smart artillery submunition that fires a self-forging slug.

[17] Terminally Guided Weapon for the MLRS: A smart submunition that carries a shaped charge warhead.

[18] Other existing and developmental munitions can destroy other armored and unarmored vehicles and have some effectiveness against tanks. Scatterable mines can delay the movement of tank units.

Figure 11-1.—Smart Anti-Armor Submunitions

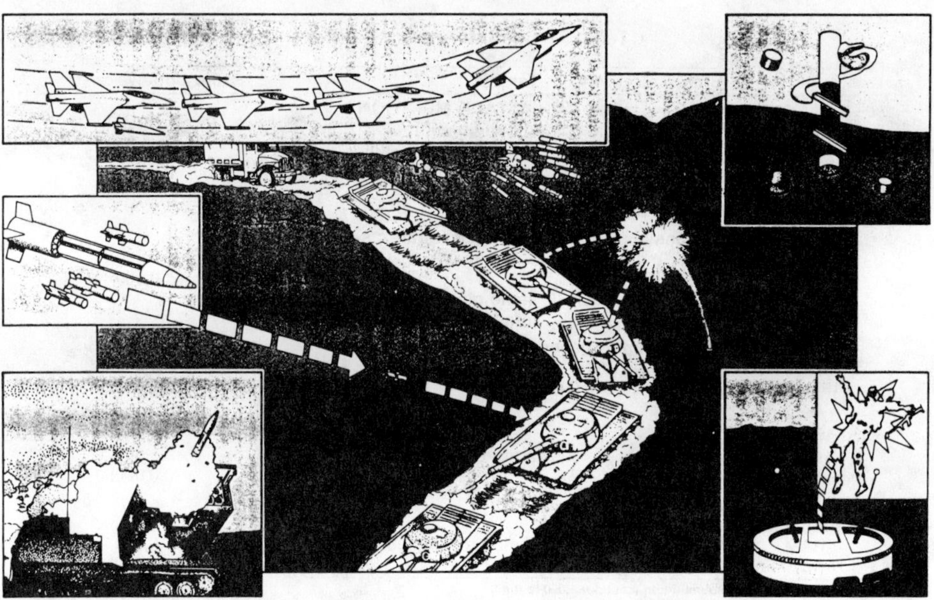

Terminally guided submunitions (TGSMs): More widely dispersed targets could be attacked by terminally guided submunitions, such as the Terminally Guided Warheads dispensed by MLRS rockets. These would search for, steer toward, and fly into tanks, detonating a shaped-charge warhead on impact. Three or six TGWs might be dispensed from each rocket (inset, left center), 12 of which could be carried and launched by each MLRS transporter-launcher (inset, below left), which could, alternatively, carry two longer range ATACMS missiles armed with TGSMs.

Sensor-fuzed weapons: As a column of tanks and trucks slows to turn and bunches up (center), it is attacked by a "four-ship" [formation] of F-16Cs (inset, above left), each of which releases a Tactical Munition Dispenser, each of which dispenses 10 Skeet dispensers (center), each of which descends by parachute and then releases four Skeet submunitions (inset, above right).

Smart mines: Two Skeet could also be lobbed toward the tanks by each of nine smart ERAM mines previously scattered along or beside the road (right center) from a TMD. As each Skeet descends, it spins and searches for tanks; if it detects one, it fires an explosively formed penetrator at the tank's engine compartment. Each mine (inset, below right) could also sense footsteps of approaching troops and lob three antipersonnel fragmentation grenades to hamper mine-clearing.

SOURCE: Office of Technology Assessment, 1987.

large margin. In recent years, Europeans have shown themselves willing to pay more for less capability to get equipment made at home. However, as the recent British decision to cancel the NIMROD and buy the AWACS demonstrates, they will not necessarily take this position to the extreme.

In the past year, the U.S. Department of Defense has been working to resolve this problem by encouraging the Europeans to identify systems they are developing and buying that could be used to support FOFA, negotiating agreements to explore co-development and co-production of U.S. systems, and encouraging the Europeans to form consortia among themselves to develop and produce FOFA-related equipment. One particular vehicle for this effort has been the 1985 Nunn Amendment authorizing funding of cooperative development projects. The European members of NATO, including France, reacted very favorably to the principle of this amendment, and to the concept of joint development of new military systems. However, it is clear that before such joint development can take place, there will have to be some major changes in existing ways of doing business. The European Allies recognize the difficulty of "harmonizing" the specific interests of the various partners in cooperative ventures. They are somewhat skeptical about the ability of the U.S. armed services to do so and about Congress committing itself to programs years in advance. However, the Europeans are increasingly unwilling to simply "buy American" systems or technologies, and indeed there are some European developments in FOFA-related technologies which the United States could profit from not having to re-invent.

This may ultimately pose a dilemma for the United States. Cooperative programs usually cost more and take longer than projects pursued solely in the United States. And, of course, sharing production or buying European systems will cost U.S. jobs. Congress will have to deal with these programs one at a time as they come up, but it might be wise to develop an overall approach to striking a balance among accommodating the desires of the Europeans, funding U.S. companies, obtaining the best systems, and obtaining the most efficient production.

Table 11-1.—Anti-Armor Munitions

	Status	Delivery	Targets	Representative cost*
TOP ATTACK MUNITIONS				
Cluster munitions (small shaped charge; unguided)				
Rockeye	inv	air	trucks, self-propelled	$100
MW-1	inv	air	arty, other light armor	(20,000)
CEM	proc	air		
ICM	proc	arty/MLRS		
Sensor-fuzed munitions (explosively formed penetrator; sensor)				
Skeet	FSD	air	same as above	5,000 (200,000)
SADARM	FSD	arty/MLRS	same plus current infantry vehicles, some tanks	
Terminally-guided munitions (shaped charge; guided)				
MLRS-TGW	R&D	MLRS	same plus	50,000
IRTGSM	R&D	ATACMS	tanks	(1,000,000)
MINES				
Scatterable mines				
GATOR	proc	air	armored vehicles	500
RAAM	proc	arty		(50,000)
AT-2	FSD	MLRS		
Smart mines				
ERAM	dead	air	armored vehicles	10,000
Army mine	R&D	arty, helo		(100,000)

*cost per individual submunition (cost per 1,000-lb load)

arty: artillery
inv: inventory
light armor: armored combat vehicles, excluding tanks.
proc: procurement
FSD: full-scale development
R&D: research & development prior to FSD

APPENDIX E

BIBLIOGRAPHICAL NOTES ON THE AUTHORS

Pierre Bescond is the President of Spot Image Corporation, the US corporation which commercializes the photos from the French Spot satellite. Bescond was formerly Director of the Launch Range in French Guiana, and Director for Satellite Operations at the Toulouse Space Center, France.

Eli Brookner is a Senior Scientist with the Raytheon Company. He was a technical consultant on the Cobra Dane, Cobra Judy and Pave Paws phased-array radars. Brookner has written the books, Radar Technology and Aspects of Modern Radar.

Paul Chrzanowski is Group Leader, Strategic Analysis Group in the Evaluation and Planning Program at the Lawrence Livermore National Laboratory. Chrzanowski has examined the issue of stability under combined offense-defense strategic systems.

Christopher Cunningham is Group Leader of the Strategic Defense Systems Study Group at Lawrence Livermore National Laboratory. Cunningham has examined the relationship between offensive and defensive strategic systems.

Seymour Deitchman is a Senior Research Associate at the Institute for Defense Analyses, where he was formerly Vice-President for Programs. Deitchman's book, Military Power and the Advance of Technology, analyzes the impact of new technology on conventional warfare, tactics and strategy.

Warren Donnelly is a Senior Specialist at the Congressional Research Service, Library of Congress. Donnelly has specialized on the issue of the proliferation of nuclear weapons, preparing periodic reports, including the Nuclear Proliferation Factbook, which are widely used by scholars in this area.

Thomas Marshall Eubanks has worked since 1987 at the U.S. Naval Observatory connected with Very Long Baseline Interferometry and on the geophysical causes of earth rotation. From 1980-87 Eubanks worked at the Jet Propulsion Laboratory on problems of a similar nature.

Anthony Fainberg is a Senior Analyst at the Congressional Office of Technology Assessment, where he participated in two studies on missile defense. Fainberg was formerly an Assistant Professor at Syracuse University and a Staff Physicist at Brookhaven National Laboratory. The views expressed are his alone and do not necessarily reflect those of OTA.

Gary Goldstein is a Professor of Physics at Tufts University specializing in intermediate and high energy physics. Goldstein has been a Visiting Scientist at the Program in Science and Technology for International Security at MIT since 1986.

David Hafemeister is a Professor of Physics at the California Polytechnic State University. He was formerly a Science Advisor in the U.S. Senate in 1975-77, a Special Assistant to an Under Secretary of State in 1977-79, and a Visiting Scientist at MIT in 1984-84, Lawrence Berkeley Lab in 1985, and the Department of State in 1987.

Spurgeon Keeny, Jr. is the President and Executive Director, Arms Control Association. Keeny served as Deputy Director (1977-80) and Assistant Director of (1969-73) of ACDA. Keeny was Technical Assistant to the President's Science Advisor (1959-69) and Senior Staff Member, National Security Council (1963-69), and he coauthored Nuclear Power: Issues and Choices, and Nuclear Arms Control: Background and Issues.

Frederick Lamb is a Professor of Physics and Astronomy at the University of Illinois and a member of University's Program in Arms Control, Disarmament, and International Security. Lamb was a Science Fellow at Stanford's Center for International Security and Arms Control, and he is a consultant for the Departments of Defense and Energy and for OTA.

Edward Luttwak holds the Arleigh Burke Chair in Strategy at the Center for Strategic and International Studies in Washington, DC. Luttwak has served as a consultant to the Intermediate Office of the Secretary of Defense, the National Security Council, and the U.S. Department of State, and he has authored 8 books, most recently, Strategy: The Logic of War and Peace.

Michael MacCracken is Division leader of the Atmospheric and Geophysical Sciences Division at the Lawrence Livermore National Laboratory. MacCracken is a coauthor of Environmental Consequences of Nuclear War, a study commissioned by the Scientific Committee of Problems of the Environment, sponsored by International Council of Scientific Unions.

Paul Richards is Mellon Professor of Natural Sciences at Columbia University, where he has taught seismology since 1971. Richards coauthored Quantitative Seismology - Theory and Methods. Richards was a MacAuthur Fellow (1981-86) and a Foster Fellow at the Arms Control and Disarmament Agency (1984-85), working on technical issues of test ban treaties.

Mark Sakitt is a Senior Physicist at the Brookhaven National Laboratory and an Adjunct Lecturer at the State University of New York at Stony Brook. In 1986-87, Sakitt was a Carnegie Science Fellow at the Center for International Security and Arms Control, Stanford University, where he wrote Submarine Warfare in the Arctic: Option or Illusion.

Dietrich Schroeer is a Professor of Physics at the University of North Carolina. Schroeer was a Research Associate at the International Institute for Strategic Studies in 1984-85. Schroeer has written Science, Technology, and the Nuclear Arms Race and Directed Energy Weapons and Strategic Defense: A Primer.

Richard Scribner is an Associate Professor, School of Foreign Service, Georgetown University. Scribner was a Visiting Scholar at Stanford's Center for International Security and Arms Control (1987-88), headed the Office of Science, Arms Control, and National Security at AAAS (1981-87), and coauthored The Verification Challenge.

Peter D. Zimmerman is a Senior Associate at the Carnegie Endowment for International Peace. Zimmerman was formerly a Professor of Physics at Louisiana State University, a William C. Foster Arms Control Fellow at ACDA (1984-86), a member of the US START delegation, and he participated in the Office of Technology Assessment study on MX basing.

AIP Conference Proceedings

		L.C. Number	ISBN
No. 1	Feedback and Dynamic Control of Plasmas – 1970	70-141596	0-88318-100-2
No. 2	Particles and Fields – 1971 (Rochester)	71-184662	0-88318-101-0
No. 3	Thermal Expansion – 1971 (Corning)	72-76970	0-88318-102-9
No. 4	Superconductivity in *d*- and *f*-Band Metals (Rochester, 1971)	74-18879	0-88318-103-7
No. 5	Magnetism and Magnetic Materials – 1971 (2 parts) (Chicago)	59-2468	0-88318-104-5
No. 6	Particle Physics (Irvine, 1971)	72-81239	0-88318-105-3
No. 7	Exploring the History of Nuclear Physics – 1972	72-81883	0-88318-106-1
No. 8	Experimental Meson Spectroscopy –1972	72-88226	0-88318-107-X
No. 9	Cyclotrons – 1972 (Vancouver)	72-92798	0-88318-108-8
No. 10	Magnetism and Magnetic Materials – 1972	72-623469	0-88318-109-6
No. 11	Transport Phenomena – 1973 (Brown University Conference)	73-80682	0-88318-110-X
No. 12	Experiments on High Energy Particle Collisions – 1973 (Vanderbilt Conference)	73-81705	0-88318-111–8
No. 13	π-π Scattering – 1973 (Tallahassee Conference)	73-81704	0-88318-112-6
No. 14	Particles and Fields – 1973 (APS/DPF Berkeley)	73-91923	0-88318-113-4
No. 15	High Energy Collisions – 1973 (Stony Brook)	73-92324	0-88318-114-2
No. 16	Causality and Physical Theories (Wayne State University, 1973)	73-93420	0-88318-115-0
No. 17	Thermal Expansion – 1973 (Lake of the Ozarks)	73-94415	0-88318-116-9
No. 18	Magnetism and Magnetic Materials – 1973 (2 parts) (Boston)	59-2468	0-88318-117-7
No. 19	Physics and the Energy Problem – 1974 (APS Chicago)	73-94416	0-88318-118-5
No. 20	Tetrahedrally Bonded Amorphous Semiconductors (Yorktown Heights, 1974)	74-80145	0-88318-119-3
No. 21	Experimental Meson Spectroscopy – 1974 (Boston)	74-82628	0-88318-120-7
No. 22	Neutrinos – 1974 (Philadelphia)	74-82413	0-88318-121-5
No. 23	Particles and Fields – 1974 (APS/DPF Williamsburg)	74-27575	0-88318-122-3
No. 24	Magnetism and Magnetic Materials – 1974 (20th Annual Conference, San Francisco)	75-2647	0-88318-123-1
No. 25	Efficient Use of Energy (The APS Studies on the Technical Aspects of the More Efficient Use of Energy)	75-18227	0-88318-124-X

No. 26	High-Energy Physics and Nuclear Structure – 1975 (Santa Fe and Los Alamos)	75-26411	0-88318-125-8
No. 27	Topics in Statistical Mechanics and Biophysics: A Memorial to Julius L. Jackson (Wayne State University, 1975)	75-36309	0-88318-126-6
No. 28	Physics and Our World: A Symposium in Honor of Victor F. Weisskopf (M.I.T., 1974)	76-7207	0-88318-127-4
No. 29	Magnetism and Magnetic Materials – 1975 (21st Annual Conference, Philadelphia)	76-10931	0-88318-128-2
No. 30	Particle Searches and Discoveries – 1976 (Vanderbilt Conference)	76-19949	0-88318-129-0
No. 31	Structure and Excitations of Amorphous Solids (Williamsburg, VA, 1976)	76-22279	0-88318-130-4
No. 32	Materials Technology – 1976 (APS New York Meeting)	76-27967	0-88318-131-2
No. 33	Meson-Nuclear Physics – 1976 (Carnegie-Mellon Conference)	76-26811	0-88318-132-0
No. 34	Magnetism and Magnetic Materials – 1976 (Joint MMM-Intermag Conference, Pittsburgh)	76-47106	0-88318-133-9
No. 35	High Energy Physics with Polarized Beams and Targets (Argonne, 1976)	76-50181	0-88318-134-7
No. 36	Momentum Wave Functions – 1976 (Indiana University)	77-82145	0-88318-135-5
No. 37	Weak Interaction Physics – 1977 (Indiana University)	77-83344	0-88318-136-3
No. 38	Workshop on New Directions in Mossbauer Spectroscopy (Argonne, 1977)	77-90635	0-88318-137-1
No. 39	Physics Careers, Employment and Education (Penn State, 1977)	77-94053	0-88318-138-X
No. 40	Electrical Transport and Optical Properties of Inhomogeneous Media (Ohio State University, 1977)	78-54319	0-88318-139-8
No. 41	Nucleon-Nucleon Interactions – 1977 (Vancouver)	78-54249	0-88318-140-1
No. 42	Higher Energy Polarized Proton Beams (Ann Arbor, 1977)	78-55682	0-88318-141-X
No. 43	Particles and Fields – 1977 (APS/DPF, Argonne)	78-55683	0-88318-142-8
No. 44	Future Trends in Superconductive Electronics (Charlottesville, 1978)	77-9240	0-88318-143-6
No. 45	New Results in High Energy Physics – 1978 (Vanderbilt Conference)	78-67196	0-88318-144-4
No. 46	Topics in Nonlinear Dynamics (La Jolla Institute)	78-57870	0-88318-145-2
No. 47	Clustering Aspects of Nuclear Structure and Nuclear Reactions (Winnepeg, 1978)	78-64942	0-88318-146-0
No. 48	Current Trends in the Theory of Fields (Tallahassee, 1978)	78-72948	0-88318-147-9

No. 49	Cosmic Rays and Particle Physics – 1978 (Bartol Conference)	79-50489	0-88318-148-7
No. 50	Laser-Solid Interactions and Laser Processing – 1978 (Boston)	79-51564	0-88318-149-5
No. 51	High Energy Physics with Polarized Beams and Polarized Targets (Argonne, 1978)	79-64565	0-88318-150-9
No. 52	Long-Distance Neutrino Detection – 1978 (C.L. Cowan Memorial Symposium)	79-52078	0-88318-151-7
No. 53	Modulated Structures – 1979 (Kailua Kona, Hawaii)	79-53846	0-88318-152-5
No. 54	Meson-Nuclear Physics – 1979 (Houston)	79-53978	0-88318-153-3
No. 55	Quantum Chromodynamics (La Jolla, 1978)	79-54969	0-88318-154-1
No. 56	Particle Acceleration Mechanisms in Astrophysics (La Jolla, 1979)	79-55844	0-88318-155-X
No. 57	Nonlinear Dynamics and the Beam-Beam Interaction (Brookhaven, 1979)	79-57341	0-88318-156-8
No. 58	Inhomogeneous Superconductors – 1979 (Berkeley Springs, W.V.)	79-57620	0-88318-157-6
No. 59	Particles and Fields – 1979 (APS/DPF Montreal)	80-66631	0-88318-158-4
No. 60	History of the ZGS (Argonne, 1979)	80-67694	0-88318-159-2
No. 61	Aspects of the Kinetics and Dynamics of Surface Reactions (La Jolla Institute, 1979)	80-68004	0-88318-160-6
No. 62	High Energy e^+e^- Interactions (Vanderbilt, 1980)	80-53377	0-88318-161-4
No. 63	Supernovae Spectra (La Jolla, 1980)	80-70019	0-88318-162-2
No. 64	Laboratory EXAFS Facilities – 1980 (Univ. of Washington)	80-70579	0-88318-163-0
No. 65	Optics in Four Dimensions – 1980 (ICO, Ensenada)	80-70771	0-88318-164-9
No. 66	Physics in the Automotive Industry – 1980 (APS/AAPT Topical Conference)	80-70987	0-88318-165-7
No. 67	Experimental Meson Spectroscopy – 1980 (Sixth International Conference, Brookhaven)	80-71123	0-88318-166-5
No. 68	High Energy Physics – 1980 (XX International Conference, Madison)	81-65032	0-88318-167-3
No. 69	Polarization Phenomena in Nuclear Physics – 1980 (Fifth International Symposium, Santa Fe)	81-65107	0-88318-168-1
No. 70	Chemistry and Physics of Coal Utilization – 1980 (APS, Morgantown)	81-65106	0-88318-169-X
No. 71	Group Theory and its Applications in Physics – 1980 (Latin American School of Physics, Mexico City)	81-66132	0-88318-170-3
No. 72	Weak Interactions as a Probe of Unification (Virginia Polytechnic Institute – 1980)	81-67184	0-88318-171-1
No. 73	Tetrahedrally Bonded Amorphous Semiconductors (Carefree, Arizona, 1981)	81-67419	0-88318-172-X

No. 74	Perturbative Quantum Chromodynamics (Tallahassee, 1981)	81-70372	0-88318-173-8
No. 75	Low Energy X-Ray Diagnostics – 1981 (Monterey)	81-69841	0-88318-174-6
No. 76	Nonlinear Properties of Internal Waves (La Jolla Institute, 1981)	81-71062	0-88318-175-4
No. 77	Gamma Ray Transients and Related Astrophysical Phenomena (La Jolla Institute, 1981)	81-71543	0-88318-176-2
No. 78	Shock Waves in Condensed Matter – 1981 (Menlo Park)	82-70014	0-88318-177-0
No. 79	Pion Production and Absorption in Nuclei – 1981 (Indiana University Cyclotron Facility)	82-70678	0-88318-178-9
No. 80	Polarized Proton Ion Sources (Ann Arbor, 1981)	82-71025	0-88318-179-7
No. 81	Particles and Fields –1981: Testing the Standard Model (APS/DPF, Santa Cruz)	82-71156	0-88318-180-0
No. 82	Interpretation of Climate and Photochemical Models, Ozone and Temperature Measurements (La Jolla Institute, 1981)	82-71345	0-88318-181-9
No. 83	The Galactic Center (Cal. Inst. of Tech., 1982)	82-71635	0-88318-182-7
No. 84	Physics in the Steel Industry (APS/AISI, Lehigh University, 1981)	82-72033	0-88318-183-5
No. 85	Proton-Antiproton Collider Physics –1981 (Madison, Wisconsin)	82-72141	0-88318-184-3
No. 86	Momentum Wave Functions – 1982 (Adelaide, Australia)	82-72375	0-88318-185-1
No. 87	Physics of High Energy Particle Accelerators (Fermilab Summer School, 1981)	82-72421	0-88318-186-X
No. 88	Mathematical Methods in Hydrodynamics and Integrability in Dynamical Systems (La Jolla Institute, 1981)	82-72462	0-88318-187-8
No. 89	Neutron Scattering – 1981 (Argonne National Laboratory)	82-73094	0-88318-188-6
No. 90	Laser Techniques for Extreme Ultraviolt Spectroscopy (Boulder, 1982)	82-73205	0-88318-189-4
No. 91	Laser Acceleration of Particles (Los Alamos, 1982)	82-73361	0-88318-190-8
No. 92	The State of Particle Accelerators and High Energy Physics (Fermilab, 1981)	82-73861	0-88318-191-6
No. 93	Novel Results in Particle Physics (Vanderbilt, 1982)	82-73954	0-88318-192-4
No. 94	X-Ray and Atomic Inner-Shell Physics – 1982 (International Conference, U. of Oregon)	82-74075	0-88318-193-2
No. 95	High Energy Spin Physics – 1982 (Brookhaven National Laboratory)	83-70154	0-88318-194-0
No. 96	Science Underground (Los Alamos, 1982)	83-70377	0-88318-195-9

No. 97	The Interaction Between Medium Energy Nucleons in Nuclei – 1982 (Indiana University)	83-70649	0-88318-196-7
No. 98	Particles and Fields – 1982 (APS/DPF University of Maryland)	83-70807	0-88318-197-5
No. 99	Neutrino Mass and Gauge Structure of Weak Interactions (Telemark, 1982)	83-71072	0-88318-198-3
No. 100	Excimer Lasers – 1983 (OSA, Lake Tahoe, Nevada)	83-71437	0-88318-199-1
No. 101	Positron-Electron Pairs in Astrophysics (Goddard Space Flight Center, 1983)	83-71926	0-88318-200-9
No. 102	Intense Medium Energy Sources of Strangeness (UC-Sant Cruz, 1983)	83-72261	0-88318-201-7
No. 103	Quantum Fluids and Solids – 1983 (Sanibel Island, Florida)	83-72440	0-88318-202-5
No. 104	Physics, Technology and the Nuclear Arms Race (APS Baltimore –1983)	83-72533	0-88318-203-3
No. 105	Physics of High Energy Particle Accelerators (SLAC Summer School, 1982)	83-72986	0-88318-304-8
No. 106	Predictability of Fluid Motions (La Jolla Institute, 1983)	83-73641	0-88318-305-6
No. 107	Physics and Chemistry of Porous Media (Schlumberger-Doll Research, 1983)	83-73640	0-88318-306-4
No. 108	The Time Projection Chamber (TRIUMF, Vancouver, 1983)	83-83445	0-88318-307-2
No. 109	Random Walks and Their Applications in the Physical and Biological Sciences (NBS/La Jolla Institute, 1982)	84-70208	0-88318-308-0
No. 110	Hadron Substructure in Nuclear Physics (Indiana University, 1983)	84-70165	0-88318-309-9
No. 111	Production and Neutralization of Negative Ions and Beams (3rd Int'l Symposium, Brookhaven, 1983)	84-70379	0-88318-310-2
No. 112	Particles and Fields – 1983 (APS/DPF, Blacksburg, VA)	84-70378	0-88318-311-0
No. 113	Experimental Meson Spectroscopy – 1983 (Seventh International Conference, Brookhaven)	84-70910	0-88318-312-9
No. 114	Low Energy Tests of Conservation Laws in Particle Physics (Blacksburg, VA, 1983)	84-71157	0-88318-313-7
No. 115	High Energy Transients in Astrophysics (Santa Cruz, CA, 1983)	84-71205	0-88318-314-5
No. 116	Problems in Unification and Supergravity (La Jolla Institute, 1983)	84-71246	0-88318-315-3
No. 117	Polarized Proton Ion Sources (TRIUMF, Vancouver, 1983)	84-71235	0-88318-316-1

No. 118	Free Electron Generation of Extreme Ultraviolet Coherent Radiation (Brookhaven/OSA, 1983)	84-71539	0-88318-317-X
No. 119	Laser Techniques in the Extreme Ultraviolet (OSA, Boulder, Colorado, 1984)	84-72128	0-88318-318-8
No. 120	Optical Effects in Amorphous Semiconductors (Snowbird, Utah, 1984)	84-72419	0-88318-319-6
No. 121	High Energy e^+e^- Interactions (Vanderbilt, 1984)	84-72632	0-88318-320-X
No. 122	The Physics of VLSI (Xerox, Palo Alto, 1984)	84-72729	0-88318-321-8
No. 123	Intersections Between Particle and Nuclear Physics (Steamboat Springs, 1984)	84-72790	0-88318-322-6
No. 124	Neutron-Nucleus Collisions – A Probe of Nuclear Structure (Burr Oak State Park - 1984)	84-73216	0-88318-323-4
No. 125	Capture Gamma-Ray Spectroscopy and Related Topics – 1984 (Internat. Symposium, Knoxville)	84-73303	0-88318-324-2
No. 126	Solar Neutrinos and Neutrino Astronomy (Homestake, 1984)	84-63143	0-88318-325-0
No. 127	Physics of High Energy Particle Accelerators (BNL/SUNY Summer School, 1983)	85-70057	0-88318-326-9
No. 128	Nuclear Physics with Stored, Cooled Beams (McCormick's Creek State Park, Indiana, 1984)	85-71167	0-88318-327-7
No. 129	Radiofrequency Plasma Heating (Sixth Topical Conference, Callaway Gardens, GA, 1985)	85-48027	0-88318-328-5
No. 130	Laser Acceleration of Particles (Malibu, California, 1985)	85-48028	0-88318-329-3
No. 131	Workshop on Polarized ^{3}He Beams and Targets (Princeton, New Jersey, 1984)	85-48026	0-88318-330-7
No. 132	Hadron Spectroscopy–1985 (International Conference, Univ. of Maryland)	85-72537	0-88318-331-5
No. 133	Hadronic Probes and Nuclear Interactions (Arizona State University, 1985)	85-72638	0-88318-332-3
No. 134	The State of High Energy Physics (BNL/SUNY Summer School, 1983)	85-73170	0-88318-333-1
No. 135	Energy Sources: Conservation and Renewables (APS, Washington, DC, 1985)	85-73019	0-88318-334-X
No. 136	Atomic Theory Workshop on Relativistic and QED Effects in Heavy Atoms	85-73790	0-88318-335-8
No. 137	Polymer-Flow Interaction (La Jolla Institute, 1985)	85-73915	0-88318-336-6
No. 138	Frontiers in Electronic Materials and Processing (Houston, TX, 1985)	86-70108	0-88318-337-4
No. 139	High-Current, High-Brightness, and High-Duty Factor Ion Injectors (La Jolla Institute, 1985)	86-70245	0-88318-338-2

No. 140	Boron-Rich Solids (Albuquerque, NM, 1985)	86-70246	0-88318-339-0
No. 141	Gamma-Ray Bursts (Stanford, CA, 1984)	86-70761	0-88318-340-4
No. 142	Nuclear Structure at High Spin, Excitation, and Momentum Transfer (Indiana University, 1985)	86-70837	0-88318-341-2
No. 143	Mexican School of Particles and Fields (Oaxtepec, México, 1984)	86-81187	0-88318-342-0
No. 144	Magnetospheric Phenomena in Astrophysics (Los Alamos, 1984)	86-71149	0-88318-343-9
No. 145	Polarized Beams at SSC & Polarized Antiprotons (Ann Arbor, MI & Bodega Bay, CA, 1985)	86-71343	0-88318-344-7
No. 146	Advances in Laser Science–I (Dallas, TX, 1985)	86-71536	0-88318-345-5
No. 147	Short Wavelength Coherent Radiation: Generation and Applications (Monterey, CA, 1986)	86-71674	0-88318-346-3
No. 148	Space Colonization: Technology and The Liberal Arts (Geneva, NY, 1985)	86-71675	0-88318-347-1
No. 149	Physics and Chemistry of Protective Coatings (Universal City, CA, 1985)	86-72019	0-88318-348-X
No. 150	Intersections Between Particle and Nuclear Physics (Lake Louise, Canada, 1986)	86-72018	0-88318-349-8
No. 151	Neural Networks for Computing (Snowbird, UT, 1986)	86-72481	0-88318-351-X
No. 152	Heavy Ion Inertial Fusion (Washington, DC, 1986)	86-73185	0-88318-352-8
No. 153	Physics of Particle Accelerators (SLAC Summer School, 1985) (Fermilab Summer School, 1984)	87-70103	0-88318-353-6
No. 154	Physics and Chemistry of Porous Media—II (Ridge Field, CT, 1986)	83-73640	0-88318-354-4
No. 155	The Galactic Center: Proceedings of the Symposium Honoring C. H. Townes (Berkeley, CA, 1986)	86-73186	0-88318-355-2
No. 156	Advanced Accelerator Concepts (Madison, WI, 1986)	87-70635	0-88318-358-0
No. 157	Stability of Amorphous Silicon Alloy Materials and Devices (Palo Alto, CA, 1987)	87-70990	0-88318-359-9
No. 158	Production and Neutralization of Negative Ions and Beams (Brookhaven, NY, 1986)	87-71695	0-88318-358-7

No. 159	Applications of Radio-Frequency Power to Plasma: Seventh Topical Conference (Kissimmee, FL, 1987)	87-71812	0-88318-359-5
No. 160	Advances in Laser Science–II (Seattle, WA, 1986)	87-71962	0-88318-360-9
No. 161	Electron Scattering in Nuclear and Particle Science: In Commemoration of the 35th Anniversary of the Lyman-Hanson-Scott Experiment (Urbana, IL, 1986)	87-72403	0-88318-361-7
No. 162	Few-Body Systems and Multiparticle Dynamics (Crystal City, VA, 1987)	87-72594	0-88318-362-5
No. 163	Pion–Nucleus Physics: Future Directions and New Facilities at LAMPF (Los Alamos, NM, 1987)	87-72961	0-88318-363-3
No. 164	Nuclei Far from Stability: Fifth International Conference (Rosseau Lake, ON, 1987)	87-73214	0-88318-364-1
No. 165	Thin Film Processing and Characterization of High-Temperature Superconductors	87-73420	0-88318-365-X
No. 166	Photovoltaic Safety (Denver, CO, 1988)	88-42854	0-88318-366-8
No. 167	Deposition and Growth: Limits for Microelectronics (Anaheim, CA, 1987)	88-71432	0-88318-367-6
No. 168	Atomic Processes in Plasmas (Santa Fe, NM, 1987)	88-71273	0-88318-368-4
No. 169	Modern Physics in America: A Michelson-Morley Centennial Symposium (Cleveland, OH, 1987)	88-71348	0-88318-369-2
No. 170	Nuclear Spectroscopy of Astrophysical Sources (Washington, D.C., 1987)	88-71625	0-88318-370-6
No. 171	Vacuum Design of Advanced and Compact Synchrotron Light Sources (Upton, NY, 1988)	88-71824	0-88318-371-4
No. 172	Advances in Laser Science–III: Proceedings of the International Laser Science Conference (Atlantic City, NJ, 1987)	88-71879	0-88318-372-2
No. 173	Cooperative Networks in Physics Education (Oaxtepec, Mexico 1987)	88-72091	0-88318-373-0
No. 174	Radio Wave Scattering in the Interstellar Medium (San Diego, CA 1988)	88-72092	0-88318-374-9
No. 175	Non-neutral Plasma Physics (Washington, DC 1988)	88-72275	0-88318-375-7

No. 176	Intersections Between Particle and Nuclear Physics (Third International Conference) (Rockport, ME 1988)	88-62535	0-88318-376-5
No. 177	Linear Accelerator and Beam Optics Codes (La Jolla, CA 1988)	88-46074	0-88318-377-3

No. 176	Intersections Between Particle and Nuclear Physics (Third International Conference) (Rockport, ME 1988)	88-62535	0-88318-376-5
No. 177	Linear Accelerator and Beam Optics Codes (La Jolla, CA 1988)	88-46074	0-88318-377-3